T0394141

Multivariable Calculus with Mathematica

Multivariable Calculus with Mathematica

Robert P. Gilbert
Michael Shoushani
Yvonne Ou

CRC Press is an imprint of the
Taylor & Francis Group, an **informa** business

First edition published 2021
by CRC Press
6000 Broken Sound Parkway NW, Suite 300, Boca Raton, FL 33487-2742

and by CRC Press
2 Park Square, Milton Park, Abingdon, Oxon, OX14 4RN

CRC Press is an imprint of Taylor & Francis Group, LLC

Library of Congress Cataloging-in-Publication Data

Names: Gilbert, Robert P., 1932- author.
Title: Multivariable calculus with mathematica / Robert P. Gilbert, Michael Shoushani, Yvonne Ou.
Description: Boca Raton : Chapman & Hall/CRC Press, 2020. | Includes bibliographical references and index.
Identifiers: LCCN 2020028017 (print) | LCCN 2020028018 (ebook) | ISBN 9781138062689 (hardback) | ISBN 9781315161471 (ebook)
Subjects: LCSH: Calculus.
Classification: LCC QA303.2 .G55 2020 (print) | LCC QA303.2 (ebook) | DDC 515--dc23
LC record available at https://lccn.loc.gov/2020028017
LC ebook record available at https://lccn.loc.gov/2020028018

ISBN: 9781138062689 (hbk)
ISBN: 9781315161471 (ebk)

Robert Gilbert dedicates this book to his daughter Jennifer, who has filled his life with joy.

Contents

Preface

This multivariable calculus book has been designed as a multi-course text. The first six chapters might be used as a third semester calculus course as it contains the usual topics, namely vectors in $\mathbb{R}^3$, quadric surfaces, curves in $\mathbb{R}^3$, continuous and differential functions defined in $\mathbb{R}^3$, extremum problems, multiple integrals and vector calculus in $\mathbb{R}^3$. The subject material has been developed using the computer algebra MATHEMATICA to augment the teaching of the course. We do not believe in using canned programs, where, for example, you just plug in the coordinates of a point and the parameters defining a plane, which then reports to you the distance of the point to the plane. This does not teach a student mathematics. We try to have the student devise MATHEMATICA functions or modules which do the calculations. The student is then required to understand the mathematical theory before setting out to devise a module or a function.

We develop some mini-programs in the MATHEMATICA sessions, which provide sufficient examples for developing the modules. These sessions are meant to teach the student mathematics, not programming; however, we feel that it is a disservice to avoid the use of a computer algebra in an undergraduate, multivariable, calculus course.

Physical examples, taken from theoretical mechanics and also from electricity and magnetism, are used to illustrate the mathematics. Moreover, there are copious examples from thermodynamics to augment the material on partial derivatives. Using multivariable Taylor series, we have added material for approximating functions of several variables. Some topics such as the Implicit Function Theorem are presented using the disclaimer that this topic is not used in the following material and can be omitted from a first reading of the text. Such material could be included in an honors course. Chapters 7 and 8 deviate from what is covered in a typical third course in the calculus. The addition of these chapters extend the use of the book to a junior or perhaps a senior level course. It is our experience from teaching third semester calculus and the introductory graduate analysis course that multivariable series and Vector Calculus have not been taught in any depth in a third semester calculus course. We suggest that Chapters 6, 7, 8 could make the basis of an upper level undergraduate course. Chapter 6 is Vector Calculus, Chapter 7 is an introduction to Tensor Analysis and Riemannian spaces; whereas, Chapter 8 is an introduction to Partial Differential Equations. There is a very nice first-order differential equations package in MATHEMATICA which we have used here. The second-order differential equations of Mathematical Physics, namely the potential, heat and wave equations are treated using separation of variables and use our modules to graph the solutions. We provided a section on the Frobenius method to show where some of the special functions used in the separation of variables method derive from.

We should mention now that all of MATHEMATICA modules and functions have the first letter capitalized; whereas we have used lower case letters to distinguish the two. Our modules will be forgotten after a session is closed and must be loaded again.

Finally, we suggest that the preferred way to teach this course is to have the MATHEMATICA session taught separately in a computer laboratory with graduate student assistance. Special assignment might be given just as for a science laboratory and the students are allowed a multi-period time interval to complete the session.

MATHEMATICA sessions are ended with a ♦ and proofs are ended with a □. As this is a book where grey scale graphics are used we added a command to some of the MATHEMATICA programs to change the color graphics to gray scale. If you run the MATHEMATICA sessions as they are given in the book you will have color graphics.

Robert Gilbert would like to thank Adi Ben Israel, his co-author on other books on symbolic algebras and calculus [1,5], for his many helpful suggestions used in the writing this book. However, any errors in the presentation belong to the current authors.

Author Bios

Robert P. Gilbert is an applied analyst, who began studying function theoretic methods applied to partial differential equations. He and Jerry Hile originated the method of generalized hyperanalytic function theory.

Professor Gilbert and James Buchanan applied this methodology to elastic materials. Further applications of the function theoretic method were made by Robert Gilbert, Steve Xu and James Buchanan to marine acoustics and inverse problems.

From this point on Dr. Gilbert's research centered on bio-medical research, in particular the study of the rigidity of human bone and the disease osteoporosis. This research was twin pronged, part being focussed on cell-biology, with James Buchanan, and the other with Alex Panchenko on the homogenization of the micro-differential equations involved with bone mechanics. Further investigations were made in collaboration with Yvonne Ou and Michael Shoushani.

Professor Gilbert became a full professor at Indiana University, Bloomington at the age of 34. At 43 he became the Unidell Foundation Chair of Mathematics at the University of Delaware. He has published over 300 articles in professional journals and conference proceedings. He is the Founding Editor of two mathematics journals *Complex Variables* and *Applicable Analysis.*

M. Yvonne Ou is an applied mathematician specializing in inverse problems, computational mathematics and rational approximations. Prior to joining the faculty of the Department of Mathematical Sciences, where she is currently a tenured associate professor, she was a research scientist member of the Computational Mathematics Group in the Oak Ridge Laboratory. This is her first book.

Michael Shoushani is an applied mathematician at Western Connecticut State University.

His research interests are in transmission and inverse problems specially related to poro-elasticity, partial differential equations, and numerical methods for partial differential equations. Professor Shoushani received his BA in mathematics from Western Connecticut State University in 2008 and his MS and PhD both in Applied Mathematics from the University of Delaware in 2014.

Chapter 1

Vectors in $\mathbb{R}^3$

1.1 Vector algebra in $\mathbb{R}^3$

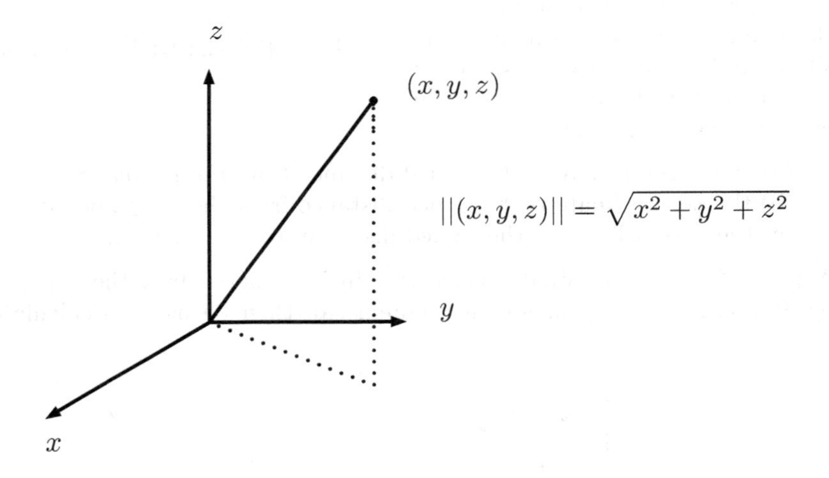

FIGURE 1.1: Cartesian coordinates in $\mathbb{R}^3$.

For all practical purposes, our world is three–dimensional[1]. It is therefore necessary to describe this three–dimensional world, and to study its properties. To this end, we borrow concepts and ideas from the two–dimensional space $\mathbb{R}^2$, the plane. Fortunately the Cartesian (x, y) coordinate system we used in the plane can be generalized to three dimensions, $\mathbb{R}^3$, as shown in Figure 1.1. We adopt the right–handed rectangular coordinate system illustrated in Figure 1.3, where the **coordinate axes** are denoted by x, y and z. The adjective "right–handed" means that by changing the orientation[2] of a single axis we obtain a left-handed system,

[1]However, relativistic physics require 4 dimensions, the three **space** dimensions and **time**.

[2]I.e. reversing the arrow, as in changing $\rightarrow$ to $\leftarrow$.

whereas changing the orientation of two axes gives back a right–handed system[3] We notice that the three coordinate axes determine three **coordinate planes**. . These are:

(a) the xy–plane, where $z = 0$,
(b) the yz–plane, where $x = 0$,
and
(c)the xz–plane, where $y = 0$.

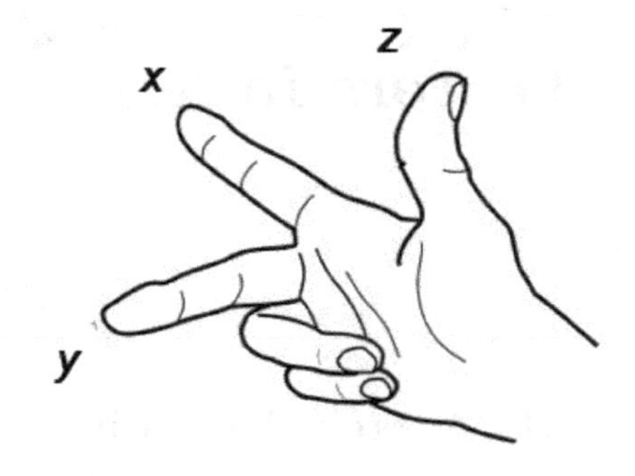

FIGURE 1.2: Right hand rule

In $\mathbb{R}^2$ we introduced rectangular coordinates for a point by taking the projections onto the x–axis as the x–coordinate, and the projection onto the y-axis as the y–coordinate. The situation is analogous in $\mathbb{R}^3$. Here the projections of the point are made onto the coordinate planes:

(a) the x–coordinate is the signed distance from the yz–plane,
(b) the y–coordinate is the signed distance from the xz–plane, and
(c) the z–coordinate is the signed distance from the xy–plane.

A point P, with coordinates (x, y, z), will be denoted by either (x, y, z), or by $P(x, y, z)$. The Pythagorean theorem can then be used to calculate the

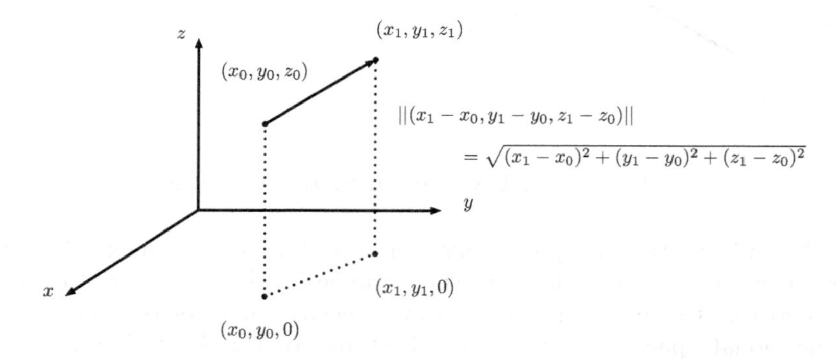

FIGURE 1.3: The distance formula in three-dimensional space.

distance of the point $P(x, y, z)$ from the origin $O(0, 0, 0)$ as

$$\|(x, y, z)\| = \sqrt{x^2 + y^2 + z^2}\ , \quad \text{see Figure 1.3.}$$

This formula translates, as in the two–dimensional case, to give the distance

[3]The reader should check this experimentally with his right and left hand.

between two points (x_1, y_1, z_1), and (x_2, y_2, z_2), as

$$\|(x_1, y_1, z_1) - (x_2, y_2, z_2)\| = \sqrt{(x_1 - x_2)^2 + (y_1 - y_2)^2 + (z_1 - z_2)^2} \,. \quad (1.1)$$

In space, as in the plane, two points determine a line. Let us see how we might find the equation of a line in space. From the diagram which we used to compute the distance between two points it is clear that the projection of a straight line segment onto any coordinate plane must be a straight line segment. Consequently the projection of a straight line onto any coordinate plane is a straight line lying in that coordinate plane. If (x, y, z) represents a variable point on the space line and (x_1, y_1, z_1), and (x_2, y_2, z_2) the two given points, then we must have, in consequence of the plane linearity of the projections,

$$\frac{x - x_1}{y - y_1} = \frac{x_2 - x_1}{y_2 - y_1} \,, \qquad \frac{z - z_1}{y - y_1} = \frac{z_2 - z_1}{y_2 - y_1} \,, \qquad \frac{x - x_1}{z - z_1} = \frac{x_2 - x_1}{z_2 - z_1} \,. \quad (1.2)$$

Alternately we might write this as

$$y - y_1 = m_1 (x - x_1) \,, \qquad z - z_1 = m_2 (x - x_1) \,.$$

If we set $x = t + x_1$, then this leads to the parametric representation

$$x = x_1 + t \,, \qquad y = y_1 + m_1 t \,, \qquad z = z_1 + m_2 t \,,$$

which reminds us of the point slope representation for plane lines. An alternate parametric representation is given by

$$x = x_1 t + x_2 (1 - t) \,, \qquad y = y_1 t + y_2 (1 - t) \,, \qquad z = z_1 t + z_2 (1 - t) \,,$$

which rearranges into

$$x = x_2 + (x_1 - x_2) t \,, \qquad y = y_2 + (y_1 - y_2) t \,, \qquad z = z_2 + (z_1 - z_2) t \,. \quad (1.3)$$

It is easy to show that these two representations of the line are equivalent. In this regard the reader should see Exercise 1.1. We notice that when $t = 0$ we are at the point (x_2, y_2, z_2), whereas when $t = 1$ we are at the point (x_1, y_1, z_1). It seems reasonable to assume that $t = \frac{1}{2}$ corresponds to the midpoint between the two given points. Use the distance formula to verify this (Exercise 1.2) In order to describe curves in space we find it convenient to use vectors. We indicate vectors by bold letters, for example $\mathbf{v}$ for a vector with components v_1, v_2, v_3, or by using

$$\text{brackets} \quad \langle v_1, v_2, v_3 \rangle, \quad \text{or parentheses} \quad \begin{pmatrix} v_1 \\ v_2 \\ v_3 \end{pmatrix} .$$

When we discuss matrices later, we shall use the brackets to represent **row**

vectors and the parentheses for **column vectors**. For the **position vector** from the origin O to the point $P(x,y,z)$ we use the notation $\mathbf{r}$

$$\mathbf{r} := < x, y, z > = \begin{pmatrix} x \\ y \\ z \end{pmatrix}^T ,$$

where the superscript T indicates that the row vector is the **transpose** of the column vector. When we work with matrices in later sections we differentiate between row and column vectors. However, in this section we need not distinguish between row and column notation for vectors. Either notation may be used here. The length of the position vector is found from the Pythagorean theorem as

$$\|\mathbf{r}\| := \sqrt{x^2 + y^2 + z^2} .$$

On the other hand, the vector from the point $Q(x_1, y_1, z_1)$ to $R(x_2, y_2, z_2)$ is

$$\mathbf{QR} = < x_2 - x_1, y_2 - y_1, z_2 - z_1 > ,$$

and its length is given by

$$|\mathbf{QR}| = \sqrt{(x_1 - x_2)^2 + (y_1 - y_2)^2 + (z_1 - z_2)^2}$$

Sometimes it is convenient to use the notation $\mathbf{x} = < x_1, x_2, x_3 >$ and $\mathbf{y} = < y_1, y_2, y_3 >$ for vectors in $\mathbb{R}^3$. From the context it should be clear when x_1 is the first component of the vector $\mathbf{x}$ and when it is the first component of the vector $\mathbf{r}_1 = < x_1, y_1, z_1 >$. In analogy to how we proceed in $\mathbb{R}^2$ we define vector addition by

$$\mathbf{x} + \mathbf{y} = < x_1, x_2, x_3 > + < y_1, y_2, y_3 > := < x_1 + y_1, x_2 + y_2, x_3 + y_3 > ,$$

and multiplication by a real scalar α by

$$\alpha \mathbf{x} = \alpha < x_1, x_2, x_3 > := < \alpha x_1, \alpha x_2, \alpha x_3 > .$$

Using these two definitions it is easy to establish that the following properties hold

$$\mathbf{x} + \mathbf{y} = \mathbf{y} + \mathbf{x} , \quad \text{commutation of vector addition}$$
$$\mathbf{x} + (\mathbf{y} + \mathbf{z}) = (\mathbf{x} + \mathbf{y}) + \mathbf{z} , \quad \text{association of vector addition}$$
$$\alpha (\mathbf{x} + \mathbf{y}) = \alpha \mathbf{x} + \alpha \mathbf{y} , \quad \text{distribution of scalar multiplication over vector addition}$$
$$(\alpha + \beta) \mathbf{x} = \alpha \mathbf{x} + \beta \mathbf{x} , \quad \text{distribution of scalar multiplication over scalar addition}$$
$$(\alpha\beta) \mathbf{x} = \alpha (\beta \mathbf{x}) , \quad \text{commutation of scalar multiplication}$$

It is useful in doing some of our calculations to express vectors in terms of their components. This can be done effectively be introducing the **unit coordinate vectors**

$$\mathbf{i} := < 1, 0, 0 > , \qquad \mathbf{j} := < 0, 1, 0 > , \qquad \mathbf{k} := < 0, 0, 1 > .$$

A space vector $\mathbf{x} := <x_1, x_2, x_3>$ can therefore be written as

$$\mathbf{x} = x_1\,\mathbf{i} + x_2\,\mathbf{j} + x_3\,\mathbf{k}$$

in terms of the unit coordinate vectors. Addition of vectors can be performed by merely adding the coefficients of the coordinate vectors

$$\mathbf{x}+\mathbf{y}=(x_1\,\mathbf{i} + x_2\,\mathbf{j} + x_3\,\mathbf{k}) + (y_1\,\mathbf{i} + y_2\,\mathbf{j} + y_3\,\mathbf{k}) = (x_1+y_1)\,\mathbf{i}+(x_2+y_2)\,\mathbf{j}+(x_3+y_3)\,\mathbf{k}\,.$$

EXERCISES

Exercise 1.1. *The formulae (1.2) and (1.3) give two representations for a line through two points. Show that these representations are equivalent.*

Exercise 1.2. *Show that the formula*

$$x = \frac{1}{2}\,(x_2 - x_1)\ , \qquad y = \frac{1}{2}\,(y_2 - y_1)\ , \qquad z = \frac{1}{2}\,(z_2 - z_1)$$

gives the midpoint on the line segment between the points (x_1, y_1, z_1) *and* (x_2, y_2, z_2) .

1.2 The inner product

The **inner product** (or **dot product**) of two vectors $\mathbf{x} = \langle x_1, x_2, x_3\rangle$ and $\mathbf{y} = \langle y_1, y_2, y_3\rangle$ is defined as

$$\mathbf{x} \cdot \mathbf{y} := x_1\,y_1 + x_2\,y_2 + x_3\,y_3$$

We list several obvious properties of the inner product which are derivable from its definition

$$\begin{aligned}
\mathbf{x}, \cdot \mathbf{x} &= \|\mathbf{x}\|^2\ , \\
\mathbf{x} \cdot \mathbf{y} &= \mathbf{y} \cdot \mathbf{x}\ , \\
\mathbf{x} \cdot (\mathbf{y} + \mathbf{z}) &= \mathbf{x} \cdot \mathbf{y} + \mathbf{x} \cdot \mathbf{z}\ , \\
(\alpha\mathbf{x}) \cdot \mathbf{y} &= \alpha\,(\mathbf{x} \cdot \mathbf{y}) = \mathbf{x} \cdot (\alpha\mathbf{y})\ .
\end{aligned}$$

Mathematica session 1.1. *Notice the syntax used for entering a vector correctly in the line below and realize that a vector is also an array. The term in the curly brackets indicates this is a vector. Bold face type indicates* MATHEMATICA *input; whereas output is presented in typewriter mode. Let note first how one inputs a vector as an* **array** *and then takes the inner product of two vectors. Notice that the second term in the* **Array** *command indicates the dimension of the vector.*

U = Array[u, 3]

$\{u[1], u[2], u[3]\}$

$\boldsymbol{V = Array[v, 3]}$

$(v[1], v[2], v[3])$

MATHEMATICA *performs the* **dot** *or inner product by placing a* **dot** *between the two vectors*

$\boldsymbol{U \cdot V}$

$u[1]v[1] + u[2]v[2] + u[3]v[3]$

We define our own inner product below. Notice we have used a small case letter to begin the name of the function. MATHEMATICA *has its own function called* **Dot**. *Can you tell the difference, if any, between these two functions?*

innerProd[A_ , B_]:=A[[1]] * B[[1]] + A[[2]] * B[[2]] + A[[3]] * B[[3]]

A = {1, 2, 3}; B = {4, 5, 6}

innerProd[A, B]

32

We may use the inner product to project a vector onto another vector. We can formulate this in MATHEMATICA

orthoProjectBontoA[{{1, 2, 3}, {4, 5, 6}}]

$\left\{\frac{12}{7}, \frac{3}{7}, -\frac{6}{7}\right\}$

direction[A_]:=A/Sqrt[Dot[A, A]]

direction[{16/7, 32/7, 48/7}]

$\left\{\frac{1}{\sqrt{14}}, \sqrt{\frac{2}{7}}, \frac{3}{\sqrt{14}}\right\}$

$\boldsymbol{Dot\left[\left\{\frac{1}{\sqrt{14}}, \sqrt{\frac{2}{7}}, \frac{3}{\sqrt{14}}\right\}, \left\{\frac{1}{\sqrt{14}}, \sqrt{\frac{2}{7}}, \frac{3}{\sqrt{14}}\right\}\right]}$

1

The length of a vector **V** is given by the Pythagorean formula

$$||\mathbf{V}|| = \sqrt{v_1^2 + v_2^2 + v_3^2}$$

Can you write a short MATHEMATICA program using **innerProd** to compute **u**? The inner product has a geometric interpretation, which is embodied in the following theorem.

Theorem 1.3. *If θ is the angle between the two space vectors* **x** *and* **y** *then*

$$\mathbf{x} \cdot \mathbf{y} = ||\mathbf{x}||\,||\mathbf{y}|| \cos\theta \,. \tag{1.4}$$

Proof. This theorem follows directly from the law of cosines as may be seen by Figure 1.4 below.

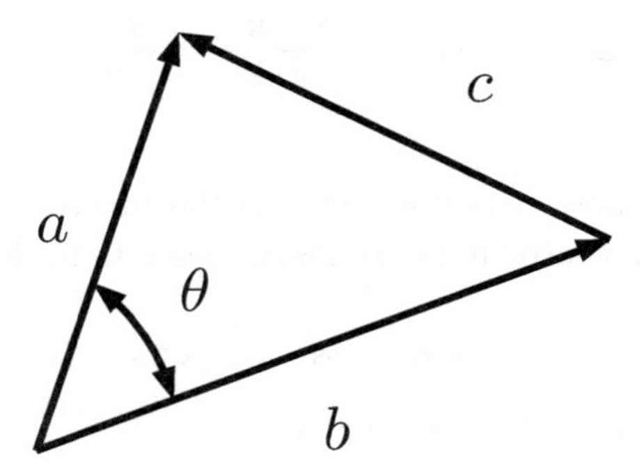

FIGURE 1.4: The law of cosines.

The proof of the above theorem may be accomplished using only formal arguments. We recall that two non colinear vectors determine a unique plane. Placing the vectors $\mathbf{x}$ and $\mathbf{y}$ (tail-to-tail) they form two sides of a triangle which, of course lie in this plane. The vector $\mathbf{x} - \mathbf{y}$ forms the third side of the triangle. The square of the length of the vector $\mathbf{x} - \mathbf{y}$ is given by

$$\|\mathbf{x}-\mathbf{y}\|^2 = (\mathbf{x}-\mathbf{y})\cdot(\mathbf{x}-\mathbf{y}) = \mathbf{x}\cdot\mathbf{x}-\mathbf{x}\cdot\mathbf{y}-\mathbf{y}\cdot\mathbf{x}+\mathbf{y}\cdot\mathbf{y} = \|\mathbf{x}\|^2\,\|\mathbf{y}\|^2-2\mathbf{x}\cdot\mathbf{y}\,.$$

However, the length of the vector $\mathbf{x} - \mathbf{y}$ may be found from the law of cosines as

$$\|\mathbf{x}-\mathbf{y}\|^2 = \|\mathbf{x}\|^2 + \|\mathbf{y}\|^2 - 2\|\mathbf{x}\|\,\|\mathbf{y}\|\,\cos\theta\,,$$

which implies after some cancellations that

$$\mathbf{x}\cdot\mathbf{y} = \|\mathbf{x}\|\,\|\mathbf{y}\|\,\cos\theta\,.$$

This is the desired result. □

From the formula (1.4) we see that if two vectors are perpendicular then their inner product is zero; moreover, rearranging this formula gives an expression for the angle between two nonzero vectors, namely

$$\theta = \operatorname{acos}\left(\frac{\mathbf{x}\cdot\mathbf{y}}{\|\mathbf{x}\|\,\|\mathbf{y}\|}\right) \tag{1.5}$$

Another way of prescribing a vector $\mathbf{x}$ is to give its length and its **direction angles.** . The direction angles are the angles α, β, γ that a vector makes with the coordinate vectors $\mathbf{i}, \mathbf{j}, \mathbf{k}$. The cosines of these angles are called the **direction cosines** and may be found from the formula

$$\cos\alpha = \frac{\mathbf{x}\cdot\mathbf{i}}{\|\mathbf{x}\|\,\|\mathbf{i}\|} = \frac{x_1}{\|\mathbf{x}\|} \tag{1.6}$$

$$\cos\beta = \frac{\mathbf{x}\cdot\mathbf{j}}{\|\mathbf{x}\|\,\|\mathbf{j}\|} = \frac{x_2}{\|\vec{x}\|} \tag{1.7}$$

$$\cos\gamma = \frac{\mathbf{x}\cdot\mathbf{k}}{\|\mathbf{x}\|\,\|\mathbf{k}\|} = \frac{x_3}{\|\mathbf{x}\|}\,. \tag{1.8}$$

From the above expressions it is clear that the direction cosines are the components of the **unit vector** $\mathbf{u} := \frac{\mathbf{x}}{|\mathbf{x}|}$, which leads to the interesting relationship

$$\cos^2\alpha + \cos^2\beta + \cos^2\gamma = 1, \tag{1.9}$$

for the components of the unit vector

$$\mathbf{u} = (\cos\alpha)\,\mathbf{i} + (\cos\beta)\,\mathbf{j} + (\cos\gamma)\,\mathbf{k}. \tag{1.10}$$

The unit vector $\mathbf{u}$ given in (1.10) is useful for computing the **component of a vector x** in the direction $\mathbf{u}$, i.e. $\mathbf{x}\cdot\mathbf{u}$.

EXERCISES

Exercise 1.4. *In* MATHEMATICA *the inner product is expressed by "." , that is by* `{x1,x2,x3}.{y1,y2,y3}`.
Write a MATHEMATICA *function for finding the angle between two vectors and apply it to the following pairs of vectors*

(a) $\langle 3,4,2\rangle$ *and* $\langle -1,0,1\rangle$

(b) $\langle -1,4,6\rangle$ *and* $\langle 3,-5,21\rangle$

Exercise 1.5. *Find the angle between the diagonal of a cube and the diagonal of one of its faces.*

Exercise 1.6. *Suppose a right pyramid has a base that is $a\times a$ square units, and has an altitude b. Find a formula for the angle between an adjacent vertical and horizontal edge.*

Exercise 1.7. *For the same right pyramid in the previous problem find a formula for the angle between two adjacent vertical edges.*

1.3 The vector product

In $\mathbb{R}^3$ there is another product that one can form with vectors known as the **vector product** or the **cross product**. This product produces a vector in contrast to the inner product which produces a scalar. The vector product of two vectors $\mathbf{x} = \langle x_1, x_2, x_3\rangle$ and $\mathbf{y} = \langle y_1, y_2, y_3\rangle$ is defined as

$$\mathbf{x}\times\mathbf{y} := (x_2\,y_3 - x_3\,y_2)\,\mathbf{i} + (x_3\,y_1 - x_1\,y_3)\,\mathbf{j} + (x_1\,y_2 - x_2\,y_1)\,\mathbf{k}\,. \tag{1.11}$$

Notice the cyclical manner in which the product is formed; namely, in the first component we find the second and third components of $\mathbf{x}$ and $\mathbf{y}$, in the second component the third and first components, and in the third component the first and second components. There is an easy way to remember this formula by writing it as a determinant. We recall the definition of a determinant of a 2×2 matrix

$$\det \begin{pmatrix} a_{11} & a_{12} \\ a_{21} & a_{22} \end{pmatrix} := a_{11}a_{22} - a_{12}a_{21} ,$$

and the determinant of a 3×3 matrix, expressed in terms of its minors,

$$\det \begin{pmatrix} a_{11} & a_{12} & a_{13} \\ a_{21} & a_{22} & a_{23} \\ a_{31} & a_{32} & a_{33} \end{pmatrix} =$$

$$a_{11} \det \begin{pmatrix} a_{22} & a_{23} \\ a_{32} & a_{33} \end{pmatrix} - a_{12} \det \begin{pmatrix} a_{21} & a_{23} \\ a_{31} & a_{33} \end{pmatrix} + a_{13} \det \begin{pmatrix} a_{21} & a_{22} \\ a_{31} & a_{32} \end{pmatrix}$$

Using the above notation we write the vector product as the determinant

$$\mathbf{x} \times \mathbf{y} := \det \begin{pmatrix} \mathbf{i} & \mathbf{j} & \mathbf{k} \\ x_1 & x_2 & x_3 \\ y_1 & y_2 & y_3 \end{pmatrix} =$$

$$(x_2y_3 - x_3y_2)\,\mathbf{i} + (x_3y_1 - x_1y_3)\,\mathbf{j} + (x_1y_2 - x_2y_1)\,\mathbf{k}$$

treating the unit vectors $\mathbf{i}$, $\mathbf{j}$, $\mathbf{k}$ (in the first row of the matrix) as numbers. Writing the cross product in this way, we can give a simple proof of the fact that the vector $\mathbf{x} \times \mathbf{y}$ is perpendicular to both $\mathbf{x}$ and $\mathbf{y}$. This is done by the formal calculation

$$\mathbf{x} \cdot (\mathbf{x} \times \mathbf{y}) = (x_1\,\mathbf{i} + x_2\,\mathbf{j} + x_3\,\mathbf{k}) \cdot \det \begin{pmatrix} \mathbf{i} & \mathbf{j} & \mathbf{k} \\ x_1 & x_2 & x_3 \\ y_1 & y_2 & y_3 \end{pmatrix}$$

$$= \det \begin{pmatrix} x_1 & x_2 & x_3 \\ x_1 & x_2 & x_3 \\ y_1 & y_2 & y_3 \end{pmatrix} = 0 ,$$

as two rows of the determinant are equal. The same argument shows that $\mathbf{y} \cdot (\mathbf{x} \times \mathbf{y}) = 0$.

The order in which the vector product is taken is important. Indeed, we have

$$\mathbf{x} \times \mathbf{y} = -\mathbf{y} \times \mathbf{x},$$

as interchanging two rows in a determinant leads to a reversal of sign.

We have yet to determine the orientation of the vector $\mathbf{x} \times \mathbf{y}$. To do this let us consider $\mathbf{i} \times \mathbf{j}$,

$$\det \begin{pmatrix} \mathbf{i} & \mathbf{j} & \mathbf{k} \\ 1 & 0 & 0 \\ 0 & 1 & 0 \end{pmatrix} = \mathbf{k} .$$

A similar calculation shows that $\mathbf{j} \times \mathbf{k} = \mathbf{i}$, and $\mathbf{k} \times \mathbf{i} = \mathbf{j}$. This shows that the cross product of two vectors produces a vector perpendicular to the two vectors but with the right–handed orientation, i.e. $\mathbf{x} \times \mathbf{y}$ is in the direction the thumb of the right hand points letting the fingers rotate from $\mathbf{x}$ to $\mathbf{y}$. The cross product also has a geometric interpretation.

Mathematica session 1.2. *In this session we do some calculations with the cross product. We introduce the vectors*

```
a = {1,2,3}
```

```
{1,2,3}
```

```
b = {4,5,6}
```

```
{4,5,6}
```

Then we take the cross product of **a** *with* **b**

```
Cross[a,b]
```

```
{−3,6,−3}
```

Let us calculate the square of the norm of the cross product of **a** *with* **b**

```
Norm[{−3,6,−3}]^2
```

```
54
```

We now begin an experiment by subtracting from this the quantity $\|\mathbf{a}\|^2\|\mathbf{b}\|^2$.

```
% − Norm[{1,2,3}]^2 * Norm[{4,5,6}]^2
```

We obtain the negative quantity

```
−1024
```

On the other hand, if we compute $\mathbf{a} \cdot \mathbf{b}$ *and square it we get the same* **positive** *quantity.*

```
Dot[a,b]^2
```

1024

Is this fortuitous, i.e is it in general true for arbitrary vectors **a** *and* **b** *that*

$$\|\mathbf{a} \times \mathbf{b}\|^2 - \|\mathbf{a}\|^2\|\mathbf{b}\|^2 + (\mathbf{a} \cdot \mathbf{b})^2 = 0? \tag{1.12}$$

Let us return to the problem at hand and compute the numerical value of $\|-3, 6, -3\|$ $\|1, 2, 3\|\|4, 5, 6\|$, *which we will verify to be* $\sin(\theta)$, *where* θ *is the angle between the vectors* **a** *and* **b**.

```
N[Norm[{−3, 6, −3}]/(Norm[{1, 2, 3}] * Norm[{4, 5, 6}])]
```

0.223814

Next we use the inner product to determine the cosine of the angle between these vectors

```
Dot[a, b]/(Norm[{1, 2, 3}] * Norm[{4, 5, 6}])
```

$\frac{16\sqrt{\frac{2}{11}}}{7}$

measured in radians. From this value we may compute the sine of θ

$\text{ArcCos}\left[N\left[\frac{16\sqrt{\frac{2}{11}}}{7}\right]\right]$

0.225726

Using this angle we compute the sine to obtain ***Sin["0.225726"]***

0.223814

Hence, we have verified, for this particular case, that (1.12) is true. Later we shall offer a mathematical proof that it is true for arbitrary vectors.

Theorem 1.8. Let $0 \leq \theta \leq \pi$ be the angle between the vectors **x** and **y** . Then

$$\|\mathbf{x} \times \mathbf{y}\| = \|\mathbf{x}\| \, \|\mathbf{y}\| \sin\theta \tag{1.13}$$

Proof. We need to use the identity

$$\|\mathbf{x} \times \mathbf{y}\|^2 = (\|\mathbf{x}\| \, \|\mathbf{y}\|)^2 - (\mathbf{x} \cdot \mathbf{y})^2 \ . \tag{1.14}$$

This identity is easy enough to verify by formally computing both sides of the equation and comparing results, which can be done e. g. with MATHEMATICA, see Exercise 1.21. For the moment we accept the result which implies

$$\begin{aligned}\|\mathbf{x} \times \mathbf{y}\|^2 &= (\|\mathbf{x}\| \, \|\mathbf{y}\|)^2 - (\|\mathbf{x}\| \, \|\mathbf{y}\| \, \cos\theta)^2 \\ &= (\|\mathbf{x}\| \, \|\mathbf{y}\| \, \sin\theta)^2 \ .\end{aligned}$$

□

From this result it is clear that two vectors are parallel if and only if

$$\mathbf{x} \times \mathbf{y} = \mathbf{0},$$

the zero vector whose three components are all zeros.

The geometric interpretation comes about by noticing that the parallelogram P with base of length $\|\mathbf{x}\|$ and altitude $\|\mathbf{y}\| \sin\theta$ has the area

$$\texttt{AREA}\,(P) \ = \ \|\mathbf{x}\| \, \|\mathbf{y}\| \, \sin\theta \ = \ \|\mathbf{x} \times \mathbf{y}\| \ , \tag{1.15}$$

whereas the triangle with this same base and altitude has one half this area. This provides us with a convenient way for calculating the area of a triangle given its three vertices.

Suppose a triangle has the vertices (a_1, a_2, a_3), (b_1, b_2, b_3), (c_1, c_2, c_3). Then the triangle has base $(b_1 - a_1, b_2 - a_2, b_3 - a_3)$ and altitude $(c_1 - a_1, c_2 - a_2, c_3 - a_3) \sin\theta$ where θ is the angle between $\mathbf{b} - \mathbf{a}$ and $\mathbf{c} - \mathbf{a}$. In order to calculate the area of the triangle we may compute the length of the cross product of these two vectors and divide by two. The cross product is given by

$$\det \begin{pmatrix} \mathbf{i} & \mathbf{j} & \mathbf{k} \\ b_1 - a_1 & b_2 - a_2 & b_3 - a_3 \\ c_1 - a_1 & c_2 - a_2 & c_3 - a_3 \end{pmatrix} \ . \tag{1.16}$$

We want to discover whether the scalar triple product of the vectors **A**, **B** and **F** depends on the order in which we place the vectors, i.e. is

$$\mathbf{A} \cdot (\mathbf{B} \times \mathbf{F}) = \mathbf{F} \cdot (\mathbf{A} \times \mathbf{B})?$$

Is there a general rule about other permutations of these vectors? By using MATHEMATICA we can check this out in the next MATHEMATICA session.

Mathematica session 1.3. *We introduce the three vectors* **A**, **B**, **F** *using the subscript notation* **control - "subscrpt number"**

$\boldsymbol{A = \{a_1, a_2, a_3\}}$

$\{a_1, a_2, a_3\}$

$\boldsymbol{B = \{b_1, b_2, b_3\}}$

$\{b_1, b_2, b_3\}$

$\boldsymbol{F = \{f_1, f_2\ , f_3\}}$

$\{f_1, f_2, f_3\}$

***Cross*[*A*, *B*]**

$\{-a_3b_2 + a_2b_3, a_3b_1 - a_1b_3, -a_2b_1 + a_1b_2\}$

The MATHEMATICA *command* **Norm**

$\boldsymbol{(Norm\,[\{a_2b_3 - a_3b_2, -a_1\,ab_3 + a_3b_1, -a_2b_1 + a_1b_2\}])\,\hat{}\,2}$

$Abs\,[-a_1\,ab_3 + a_3b_1]^{\,2} + Abs\,[-a_2b_1 + a_1b_2]^{\,2} + Abs\,[-a_3b_2 + a_2b_3]^{\,2}$

is less useful to us here than our command **myNorm** *as we have not told* MATHEMATICA *that the vector entries are real. We introduce the command* **myNorm** *in the next input line.*

***myNorm*[*V*_]:=*Sqrt*[*V*[[1]]^2 + *V*[[2]]^2 + *V*[[3]]^2]**

$\boldsymbol{Expand\,[(myNorm\,[\{a_2b_3 - a_3b_2, -a_1b_3 + a_3b_1, -a_2b_1 + a_1b_2\}])\,\hat{}\,2]}$

$a_2^2b_1^2 + a_3^2b_1^2 - 2a_1a_2b_1b_2 + a_1^2b_2^2 + a_3^2b_2^2 - 2a_1a_3b_1b_3 - 2a_2a_3b_2b_3 + a_1^2b_3^2 + a_2^2b_3^2$

We check whether $\mathbf{A}\cdot(\mathbf{B}\times\mathbf{F})$ *is equal to* $\mathbf{B}\cdot(\mathbf{F}\times\mathbf{A})$ *by subtracting one from the other to obtain zero. Notice that this was an* **even** *permutation of the vectors* **A**, (**B**, **F**). *Check this is true for the other two* **even** *permutations. Let us see what happens with an odd permutation.*

***Dot*[*A*, *Cross*[*B*, *F*]]**

$a_3\,(-b_2f_1 + b_1f_2) + a_2\,(b_3f_1 - b_1f_3) + a_1\,(-b_3f_2 + b_2f_3)$

***Dot*[*B*, *Cross*[*F*, *A*]]**

Here we have added the two triple-scalar products to see that they cancel; one is the negative of the other. Check out the other two odd permutations. What do you conclude about the scalar triple product?

$b_3(a_2f_1 - a_1f_2) + b_2(-a_3f_1 + a_1f_3) + b_1(a_3f_2 - a_2f_3)$

```
Expand[Dot[A, Cross[B, F]] - Dot[B, Cross[F, A]]]
```

0

Mathematica session 1.4. *In this session we verify that the scalar triple product can be written in determinental form. To this end we input several vectors,* **A**, **B**, **F**, *with arbitrary components.*

$\boldsymbol{A = \{a_1, a_2, a_3\}}$

$\{a_1, a_2, a_3\}$

$\boldsymbol{B = \{b_1, b_2, b_3\}}$

$\{b_1, b_2, b_3\}$

Mathematica has protected the symbols C, D and E, so we input our third vector as F.

$\boldsymbol{F = \{f_1, f_2, f_3\}}$

$\{f_1, f_2, f_3\}$

We build a matrix from these three vectors by building the array

$\boldsymbol{MatrixForm[\{\{a_1, a_2, a_3\}, \{b_1, b_2, b_3\}, \{f_1, f_2, f_3\}\}]}$

$$\begin{pmatrix} a_1 & a_2 & a_3 \\ b_1 & b_2 & b_3 \\ f_1 & f_2 & f_3 \end{pmatrix}$$

We now take the determinant of the array using the command **Det**

$$Det\left[\begin{pmatrix} a_1 & a_2 & a_3 \\ b_1 & b_2 & b_3 \\ f_1 & f_2 & f_3 \end{pmatrix}\right]$$

$-a_3b_2f_1 + a_2b_3f_1 + a_3b_1f_2 - a_1b_3f_2 - a_2b_1f_3 + a_1b_2f_3$

We next show that this is equivalent to the following scalar triple product

Dot[A, Cross[B, F]]

$a_3\,(-b_2f_1 + b_1f_2) + a_2\,(b_3f_1 - b_1f_3) + a_1\,(-b_3f_2 + b_2f_3)$

To show that they are symbolically equivalent, we subtract one from the other to obtain zero. $\boldsymbol{Expand[-a_3b_2f_1 + a_2b_3f_1 + a_3b_1f_2 - a_1b_3f_2 - a_2b_1f_3 + a_1b_2f_3 - (a_3\,(-b_2f_1 + b_1f_2) + a_2\,(b_3f_1 - b_1f_3) + a_1\,(-b_3f_2 + b_2f_3))]}$

0

By interchanging two adjacent rows in the determinant above we change the sign of the determinant, as may be seen by adding one determinant to the other obtaining zero. $\boldsymbol{MatrixForm[\{\{b_1, b_2, b_3\}, \{a_1, a_2, a_3\}, \{f_1, f_2, f_3\}\}]}$

$$\begin{pmatrix} b_1 & b_2 & b_3 \\ a_1 & a_2 & a_3 \\ f_1 & f_2 & f_3 \end{pmatrix}$$

$$Expand\left[Det\left[\begin{pmatrix} b_1 & b_2 & b_3 \\ a_1 & a_2 & a_3 \\ f_1 & f_2 & f_3 \end{pmatrix}\right] + Det\left[\begin{pmatrix} a_1 & a_2 & a_3 \\ b_1 & b_2 & b_3 \\ f_1 & f_2 & f_3 \end{pmatrix}\right]\right]$$

0

Example 1.9 (Bending of beams). Consider a horizontal beam with one end rigidly clamped, and a force **P** acting vertically on the other end, see Figure 1.5.

The **moment** caused by **P** is given by

$$\mathbf{M} = \vec{x} \times \mathbf{P}. \tag{1.17}$$

The normal load at the unsupported end of the beam will cause the beam to bend (slightly). According to Bernoulli[4] this bending occurs by stretching the fibers on the upper side of the beam and compressing the fibers on the underside of the beam in such a way that the planar cross-sections of the beam remain planar after bending as shown in Figure 1.5 Because the cross-sections remain planar it follows that the

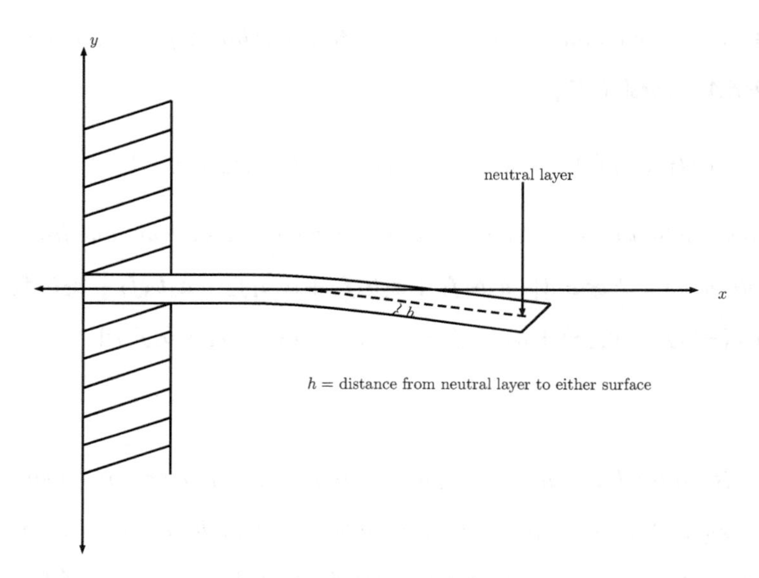

FIGURE 1.5: Illustration of bending theory.

fiber extensions decrease linearly from the top of the beam until the **neutral layer** where the extension is zero. After that layer the fiber compression increases linearly until meeting the lower surface. The distortion of fiber length per unit of length is known as the **strain**, ε. It is usually assumed to be linearly related to the stretching or compressional force, known as the **stress**, σ namely

$$\sigma = E\varepsilon. \tag{1.18}$$

As the stress varies linearly with distance y from the neutral layer, we may write

$$\sigma = \frac{\sigma_{\max}}{h}\, y, \tag{1.19}$$

where h is the distance between the neutral layer and the upper layer of the beam. We now consider the equilibrium of the moments taken about an axis normal to the plane of the beam as shown in Figure 1.5. Suppose we represent the cross-section of the beam as A and the differential cross-section by dA. The horizontal forces acting

[4] Daniel Bernoulli [1720–1782] was professor of physics at the University of Basel.

on a differential cross-section are given by σdA and hence the moment acting on this cross-section has magnitude

$$\int_A y\,\sigma\, dA = \frac{\sigma_{\max}}{h} \int_A y^2\, dA\,.$$

The integral $I := \int_A y^2\, dA$ is called the **moment of inertia** of the cross-section w.r.t. x–axis. There exist engineering tables where the moment of inertia is given for a variety of beam cross-sections. Since the moments must balance if the beam is in equilibrium we have $M = \sigma_{\max} \frac{I}{h}$; hence, we determine the stress for any point of the cross-section as

$$\sigma = \frac{M}{I} y\,. \tag{1.20}$$

If the neutral layer coincides with the y-centroid then it is easy to compute the moment of inertia for a large number of cross-sections. Indeed, for a circular yz cross-section of radius a we have

$$I = \int_A y^2\, dA = \int_0^a \int_0^{2\pi} y^2\, r\, dr\, d\theta = \int_0^a \int_0^{2\pi} r^3\, dr\, \sin^2\theta\, d\theta = \frac{\pi}{4} a^4\,.$$

In general, it can be shown that the neutral fiber does lie on the centroid. This is done by using the equilibrium condition for the force components in the horizontal direction, namely

$$\int_A \sigma\, dA = 0,$$

which implies that

$$\int_A y\, dA = 0,$$

which implies the centroid lies on the neutral fiber. We are now in a position to determine the neutral line of the deflected beam. As the neutral filament is not stretched we can form a triangle whose two adjacent sides are radii of curvature r and whose opposite side of differential length δx lies on the neutral filament. A similar triangle with adjacent sides h and opposite side $\varepsilon_{max}\delta x$ can be constructed by passing through the intersection of a side of the larger triangle with the neutral filament a line parallel to its other side. See Figure 1.6 By comparing sides of the similar triangles we get

$$\frac{1}{r} = \frac{\varepsilon_{max}}{h} = \frac{Px}{EI} \tag{1.21}$$

For small displacements the planar curvature is approximately given by $\frac{1}{r} = \frac{d^2y}{dx^2}$. Combining this with (1.21) gives us a **differential equation** for the displacement, namely

$$\frac{d^2y}{dx^2} = \frac{P}{EI}x, \tag{1.22}$$

which can be directly integrated to obtain

$$y(x) = \frac{P}{EI}\left(\frac{1}{6}x^3 + C_1 x + C_2\right), \tag{1.23}$$

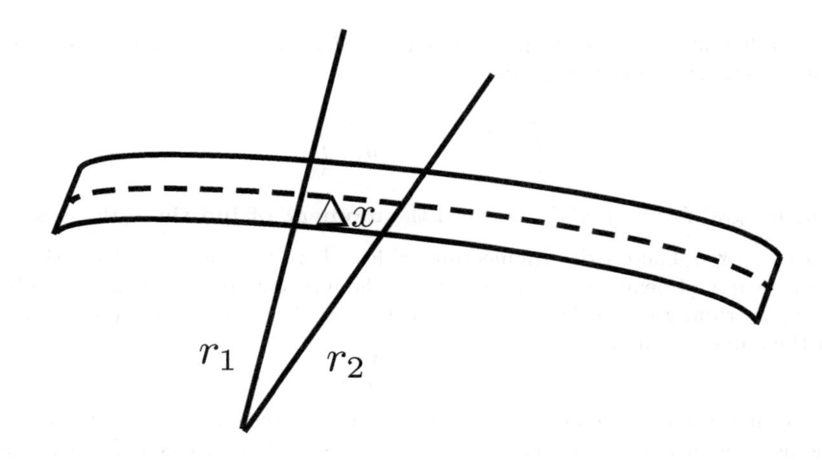

FIGURE 1.6: The deflection triangle for a beam.

where C_1 and C_2 are arbitrary constants of integration. If the beam is rigidly clamped at $x = 0$, then at this point we have the following conditions which must be met

$$y(0) = 0, \text{ and } \frac{dy}{dx}(0) = 0, \tag{1.24}$$

which imply $C_1 = 0$ and $C_2 = 0$, and hence

$$y(x) = \frac{P}{6EI}x^3.$$

The maximum deflection occurs at the free end of the beam, and is

$$y_{max} = \frac{P}{6EI}l^3. \tag{1.25}$$

We list several elementary properties of the vector product as

Theorem 1.10. *Let* $\mathbf{x}$, $\mathbf{y}$, $\mathbf{z}$ *be vectors and* α *a real number. Then*

(a) $\mathbf{x} \times \mathbf{y} = -\mathbf{y} \times \mathbf{x}$

(b) $(\alpha\,\mathbf{x}) \times \mathbf{y} = \mathbf{x} \times (\alpha\,\mathbf{y})$

(c) $\mathbf{x} \times (\mathbf{y} + \mathbf{z}) = \mathbf{x} \times \mathbf{y} + \mathbf{x} \times \mathbf{z}$

(d) $\mathbf{x} \cdot (\mathbf{y} \times \mathbf{z}) = (\mathbf{x} \times \mathbf{y}) \cdot \mathbf{z}$

(e) $\mathbf{x} \times (\mathbf{y} \times \mathbf{z}) = (\mathbf{x} \cdot \mathbf{z})\mathbf{y} - (\mathbf{y} \cdot \mathbf{x})\mathbf{z}$

Proof. Statements (a)–(c) are self evident from the definition of the vector product. The product $\mathbf{x} \cdot (\mathbf{y} \times \mathbf{z})$ listed in rule (d) is known as the **scalar triple product** of the vectors $\mathbf{x}, \mathbf{y}$, and $\mathbf{z}$. It would not make any sense if we wrote $(\mathbf{x} \cdot \mathbf{y}) \times \mathbf{z}$, as $(\mathbf{x} \cdot \mathbf{y})$ is a scalar and we can not take the cross product of a scalar and a vector. Because of this we are free to omit the parenthesis altogether as the meaning remains clear. Let us now turn to a proof of statement (d). In determinant notation the statement is equivalent to

$$\det\begin{pmatrix} x_1 & x_2 & x_3 \\ y_1 & y_2 & y_3 \\ z_1 & z_2 & z_3 \end{pmatrix} = \det\begin{pmatrix} z_1 & z_2 & z_3 \\ x_1 & x_2 & x_3 \\ y_1 & y_2 & y_3 \end{pmatrix} .$$

This is obvious as we may obtain the determinant on the right by two interchanges of rows of the determinant on the left. As each interchange provides a change of sign, two such interchanges leaves the determinant invariant.
To prove (e) we need to expand both sides of the equation and compare terms. As this is a straight forward but somewhat detailed calculation we leave it for Exercise 1.15. □

Mathematica session 1.5.

```
projectBontoA[{A_, B_}]:=Module[{mag, an}, mag[A] = Sqrt[A.A];
an = A/(mag[A])^2;
Dot[B, an] * A]
```

```
projectBontoA[{{1,2,3}, {4,5,6}}]
```

$\left\{\frac{16}{7}, \frac{32}{7}, \frac{48}{7}\right\}$

```
orthoProjectBontoA[{A_, B_}]:=Module[{mag, an}, mag[A] = Sqrt[A.A];
an = A/(mag[A])^2;
Simplify[B - Dot[B, an] * A]]
```

```
orthoProjectBontoA[{{1,2,3}, {4,5,6}}]
```

$\left\{\frac{12}{7}, \frac{3}{7}, -\frac{6}{7}\right\}$

```
direction[A_]:=A/Sqrt[Dot[A, A]]
```

```
direction[{16/7,32/7,48/7}]
```

$\left\{\frac{1}{\sqrt{14}}, \sqrt{\frac{2}{7}}, \frac{3}{\sqrt{14}}\right\}$

$\textit{Dot}\left[\left\{\frac{1}{\sqrt{14}}, \sqrt{\frac{2}{7}}, \frac{3}{\sqrt{14}}\right\}, \left\{\frac{1}{\sqrt{14}}, \sqrt{\frac{2}{7}}, \frac{3}{\sqrt{14}}\right\}\right]$

1

There is an interesting connection between the scalar triple product and the volume of a parallelepiped having the vectors $\mathbf{x}, \mathbf{y}$, and $\mathbf{z}$ as adjacent edges. Indeed, the volume is the absolute value of the scalar triple product of these vectors. We list this result as

Theorem 1.11. The volume of the parallelepiped with adjacent edges determined by the vectors $\mathbf{x}, \mathbf{y}$, and $\mathbf{z}$ is given by

$$V = |\mathbf{x} \cdot (\mathbf{y} \times \mathbf{z})| .$$

Proof. If the vectors are coplanar the parallelepiped is degenerate then $\mathbf{x}$ and

$\mathbf{y} \times \mathbf{z}$ are perpendicular; hence, the scalar triple product vanishes. So the volume of the degenerate parallelepiped is zero. We consider next the case where the three vectors are not coplanar. From Figure 1.7 we see that the area of the base of the parallelepiped is given by $A = \|\mathbf{y} \times \mathbf{z}\|$. If θ is the angle between $\mathbf{x}$ and $\mathbf{y} \times \mathbf{z}$, then either θ or $\pi - \theta$ is acute. In either case $h = \|\mathbf{x}\| \, |\cos \theta|$ is the altitude of the parallelepiped. Since $V = A \cdot h$ is the volume, we have obtained the stated result. □

In Figure 1.7 we show the geometric interpretation of the scalar triple product.

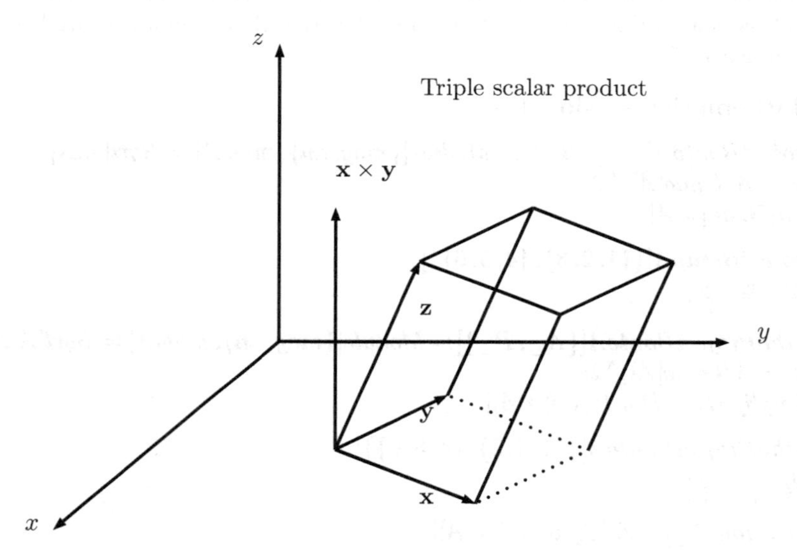

FIGURE 1.7: The scalar triple product

Mathematica session 1.6. *Recall the Mathematica function that defines a matrix, here a* 3×3 *matrix with elements* $a[i, j]$,

MatrixForm[Array[a, {3, 3}]]

$$\begin{pmatrix} a[1,1] & a[1,2] & a[1,3] \\ a[2,1] & a[2,2] & a[2,3] \\ a[3,1] & a[3,2] & a[3,3] \end{pmatrix}$$

and the determinant function `Det`,

Det[%]

$-a[1,3]a[2,2]a[3,1]+a[1,2]a[2,3]a[3,1]+a[1,3]a[2,1]a[3,2]-a[1,1]a[2,3]a[3,2]-a[1,2]a[2,1]a[3,3]+a[1,1]a[2,2]a[3,3]$

In this session we show how to work with the scalar triple product, that we compute as follows[5]

tripleScalarProduct[A_ , B_ , F_]:=Module[{p}, p = A × B; Expand[Dot[F, p]]]

which computes $(\mathbf{A} \times \mathbf{B}) \cdot \mathbf{F}$*. We apply it to three vectors*

A = {a[1], a[2], a[3]}; B = {b[1], b[2], b[3]}; F = {f[1], f[2], f[3]}

$\{f[1], f[2], f[3]\}$

to get

tripleScalarProduct[A, B, F]

$-a[3]b[2]f[1]+a[2]b[3]f[1]+a[3]b[1]f[2]-a[1]b[3]f[2]-a[2]b[1]f[3]+a[1]b[2]f[3]$

This is the determinant of the matrix that has the three vectors as its rows,

MatrixForm[{A, B, F}]

$$\begin{pmatrix} a[1] & a[2] & a[3] \\ b[1] & b[2] & b[3] \\ f[1] & f[2] & f[3] \end{pmatrix}$$

Det[%]

$-a[3]b[2]f[1]+a[2]b[3]f[1]+a[3]b[1]f[2]-a[1]b[3]f[2]-a[2]b[1]f[3]+a[1]b[2]f[3]$

Indeed,

Det [MatrixForm[{A, B, F}]] − tripleScalarProduct[A, B, F]

0

Example 1.12 (Angular Momentum)**.** The **angular momentum** or **rotational momentum** of a particle about a given point **O** is the rotational equivalent of linear momentum [6] . If **p** is the linear momentum of the particle and the vector from the point **O** to the particle is **r** then its rotational momentum around **O** is given by

$$\mathbf{L} := \mathbf{r} \times \mathbf{p} \ .$$

Let us now consider a collection of particles situated at the points $\mathbf{r}_i$ where $i = 1, 2 \ldots, n$. Let $\mathbf{p}_i$ $(i = 1, 2, \ldots, n)$ be the respective momenta. Rotational momentum is additive; hence the rotational momentum of the collection is

$$\mathbf{L}_T \ := \ \sum_{i=1}^{n} \mathbf{L}_i \ := \ \sum_{i=1}^{n} \mathbf{r}_i \times \mathbf{p}_i \ .$$

We can now show that if the total (linear) momentum of the collection is zero, then the angular momentum is independent of the point about which the moments are taken. This can be readily seen using elementary properties of the vector product. Let us denote by $\mathbf{l}_i := \mathbf{r}'_i \times \mathbf{p}_i$ the angular momentum of the ith particle with respect

[5]Remember that Mathematica function names begin with a capital letter, e.g. `Det` for determinant, so to avoid confusion we start our own function names with a low case letter.

[6]In classical physics the momentum is the product of the particle's mass and its velocity

to a point situated at the point of $\mathbf{a}$. Then the vector $\mathbf{r}'_i = \mathbf{r}_i - \mathbf{a}$ is the **moment arm** from this point to the particle, and the total angular momentum about this point is

$$\begin{aligned} \mathbf{l}_T &:= \sum_{i=1}^{n} \mathbf{l}_i := \sum_{i=1}^{n} (\mathbf{r}_i - \mathbf{a}) \times \mathbf{p}_i \\ &= \sum_{i=1}^{n} \mathbf{r}_i \times \mathbf{p}_i - \mathbf{a} \times \left(\sum_{i=1}^{n} \mathbf{p}_i \right). \end{aligned}$$

Since, by assumption the linear momentum vanishes, i.e. $\sum_{i=1}^{n} \mathbf{p}_i = 0$, we have

$$\mathbf{l}_T = \sum_{i=1}^{n} \mathbf{r}_i \times \mathbf{p}_i = \mathbf{L}_T .$$

We can also show that in an isolated system the total angular momentum is constant. We shall suppose that the individual particles in the system act on each other through forces such as gravitational, electrostatic etc., which act along lines joining the individual particles, and that the total force on the system of particles is the sum of the forces of each particle acting on another particle one at a time. The force on the ith particle by the jth particle acts along a line connecting the point $\mathbf{r}_i$ with the point $\mathbf{r}_j$. The force on the ith particle exerted by the jth particle is given by a vector function

$$\mathbf{F}_{ij} = a_{ij} \left(\|\mathbf{r}_j - \mathbf{r}_i\| \right) \frac{\mathbf{r}_j - \mathbf{r}_i}{\|\mathbf{r}_j - \mathbf{r}_i\|} ,$$

where $a_{ij} \left(\|(\mathbf{r}_j - \mathbf{r}_i)\| \right)$ is a scalar function of the distance between the two points. As the force on the jth particle by the ith particle is the opposite of the force exerted on the ith particle by the jth particle we have

$$\mathbf{F}_{ji} = -\mathbf{F}_{ij}.$$

The time derivative of the total angular momentum is

$$\begin{aligned} \frac{d\mathbf{L}}{dt} = \frac{d}{dt} \sum_{i=1}^{n} \mathbf{r}_i \times \mathbf{p}_i &= \frac{d}{dt} \sum_{i=1}^{n} m_i \, \mathbf{r}_i \times \frac{d\mathbf{r}_i}{dt} \\ &= \sum_{i=1}^{n} m_i \frac{d\mathbf{r}_i}{dt} \times \frac{d\mathbf{r}_i}{dt} + \sum_{i=1}^{n} m_i \, \mathbf{r}_i \times \frac{d^2\mathbf{r}_i}{dt^2} . \end{aligned}$$

Now the first sum vanishes because $\frac{d\mathbf{r}_i}{dt} \times \frac{d\mathbf{r}_i}{dt} = \mathbf{0}$, so

$$\begin{aligned} \frac{d\mathbf{L}}{dt} = \sum_{i=1}^{n} m_i \mathbf{r}_i \times \frac{d^2\mathbf{r}_i}{dt^2} &= \sum_{i=1}^{n} \sum_{j=1}^{n} a_{ij} \, \mathbf{r}_i \times \frac{\mathbf{r}_j - \mathbf{r}_i}{\|\mathbf{r}_j - \mathbf{r}_i\|} \\ &= \sum_{i=1}^{n} \sum_{j=1}^{n} a_{ij} \, (\mathbf{r}_i \times \mathbf{r}_j) = \mathbf{0} , \end{aligned}$$

because the $\frac{a_{ij}}{\|\mathbf{r}_j - \mathbf{r}_i\|}$ are symmetric and $\mathbf{r}_i \times \mathbf{r}_j$ is antisymmetric, i.e.

$$\mathbf{r}_i \times \mathbf{r}_j = -\mathbf{r}_j \times \mathbf{r}_i .$$

EXERCISES

Exercise 1.13. *Use* MATHEMATICA *to compute the cross products* $\mathbf{a} \times \mathbf{b}$ *for the following:*

(a) $\mathbf{a} = \langle 4, 3, 1 \rangle, \ \mathbf{b} = \langle 1, -6, 2 \rangle$

(b) $\mathbf{a} = \langle 0, -1, 1 \rangle, \ \mathbf{b} = \langle 1, -1, 0 \rangle$

(c) $\mathbf{a} = \langle -100, 0, 100 \rangle, \ \mathbf{b} = \langle \frac{1}{1000}, 0, 0 \rangle$

(d) $\mathbf{a} = \mathbf{i}, \ \mathbf{b} = \mathbf{j}$

(e) $\mathbf{a} = \mathbf{i} + \mathbf{j} + \mathbf{k}, \ \mathbf{b} = \mathbf{j}$

(f) $\mathbf{a} = 4\,\mathbf{i} + 3\,\mathbf{j} + \mathbf{k}, \ \mathbf{b} = -6\,\mathbf{j}$

(g) $\mathbf{a} = 4\,\mathbf{i} + 3\,\mathbf{j} + \mathbf{k}, \ \mathbf{b} = 2\,\mathbf{k}$

Exercise 1.14. *Use* MATHEMATICA *to show that the vector product is not* **associative**, *i.e. in general,*

$$\mathbf{a} \times (\mathbf{b} \times \mathbf{c}) \neq (\mathbf{a} \times \mathbf{b}) \times \mathbf{c}$$

Hint: Calculate the difference of the two sides in the above inequality, and use the command `vect_express`.

Exercise 1.15. *Use* MATHEMATICA *to illustrate*

$$\mathbf{x} \times (\mathbf{y} \times \mathbf{z}) \ = \ (\mathbf{x} \cdot \mathbf{z})\,\mathbf{y} - (\mathbf{y} \cdot \mathbf{x})\,\mathbf{z}.$$

Exercise 1.16. *If the vectors* **a**, **b**, **c** *are nonzero, what can you say about the vector* $\mathbf{b} - \mathbf{c}$ *if* $\mathbf{a} \times \mathbf{b} = \mathbf{a} \times \mathbf{c}$*?*

Exercise 1.17. *In the previous problem suppose* $\mathbf{a} = \mathbf{i}$ *and* $\mathbf{b} = \mathbf{i} + \mathbf{j} + \mathbf{k}$*. Find a suitable vector* **c**.

Exercise 1.18. *Find the area of the triangle with vertices at:*

(a) (1,0,0) , (0,0,1) , (0,0,1)

(b) (2,3,1) , (6,7,1) , (0,0,-1)

Exercise 1.19. *What is the volume of the parallelepiped with adjacent sides* **i**, **j**, **k***? Hence, devise a scheme for computing the volume of a parallelepiped with sides* $a\,\mathbf{i} + b\,\mathbf{j} + c\,\mathbf{k}$, **j**, **k**.

Exercise 1.20. *The volume of a parallelepiped is the area of its base multiplied by its altitude. If the base of a parallelepiped is the parallelogram with adjacent sides* $\overline{AB}$, *and* $\overline{AC}$ *then the area of the base is given by* $\|\mathbf{AB} \times \mathbf{AC}\|$*. Recall that the vector* $\mathbf{AB} \times \mathbf{AC}$ *is perpendicular to the base. If the fourth point determining the third adjacent side of the parallelepiped is* D *show that the altitude is given by*

$$\left| \mathbf{AD} \cdot \frac{\mathbf{AB} \times \mathbf{AC}}{\|\mathbf{AB} \times \mathbf{AC}\|} \right| .$$

Exercise 1.21. *Prove the identity*

$$\|\mathbf{x} \times \mathbf{y}\|^2 \ = \ (\|\mathbf{x}\| \, \|\mathbf{y}\|)^2 \ - \ (\mathbf{x} \cdot \mathbf{y})^2 \,. \qquad (1.14)$$

Exercise 1.22. *Calculate the volume of:*

(a) *the tetrahedron with adjacent sides* $\mathbf{i}$, $\mathbf{j}$, $\mathbf{k}$

(b) *the parallelepiped with sides* $a\,\mathbf{i} + b\,\mathbf{j} + c\,\mathbf{k}$, $\mathbf{j}$, $\mathbf{k}$

Exercise 1.23. *Write a* MATHEMATICA *function for computing the volume of a parallelepiped with sides*

$$a_1\,\mathbf{i} + a_2\,\mathbf{j} + a_3\,\mathbf{k}\ ,\quad b_1\,\mathbf{i} + b_2\,\mathbf{j} + b_3\,\mathbf{k}\ ,\quad c_1\,\mathbf{i} + c_2\,\mathbf{j} + c_3\,\mathbf{k}\ .$$

Exercise 1.24. *Using the results of the previous problem, what is the volume of a tetrahedron which has the points* (a_1, a_2, a_3), (b_1, b_2, b_3), (c_1, c_2, c_3).

Exercise 1.25. *Find the direction of a normal vector to the plane* $\mathcal{P}$ *through the points* $(1, 0, 0)$, $(0, 0, 1)$, $(0, 0, 1)$. *A line is drawn parallel to this normal and passing through the point* $(1, 1, 1)$. *Find the equation of the line.*

Exercise 1.26. *The altitude of a triangle with base* $\overline{AB}$ *and vertex* C *is given by* $\|\mathbf{AC} \times \mathbf{BC}\| / \|\mathbf{AB}\|$. *Use this to write a* MATHEMATICA *function which computes the distance from a fixed point to a line. Check your function on the following points given by* P *and lines given in normal form:*

(a) $P = (1, -1, 1)$ *and* $x + y + 1 = 0$

(b) $P = (0, -1, 6)$ *and* $x + 3y - 8 = 0$

(c) $P = (0, 0, 0)$ *and* $2x + 4y + 10 = 0$

(d) $P = (10, 0, -10)$ *and* $10x + 6y + 7 = 0$

(e) e $P = (1, 0, 0)$ *and* $x + y + 1 = 0$

Exercise 1.27. *Show that the determinant made up of the components of three coplanar vectors vanishes.*

1.4 Lines and planes in space

We recall that the equation for a straight line in space could be written in the parametric form

$$\begin{pmatrix} x \\ y \\ z \end{pmatrix} = \begin{pmatrix} x_0 \\ y_0 \\ z_0 \end{pmatrix} + t \begin{pmatrix} \alpha \\ \beta \\ \gamma \end{pmatrix}$$

It is easy, using this form, to write down the equation of a line L through two points P and Q. We illustrate this with an example. Let the two points be given by $P(3, -2, 1)$ and $Q(1, 4, -2)$. The vector $\mathbf{PQ}$ is given by $(-2, 6, -3)$. The line must be parallel to this vector, so we may take $\alpha = -2$, $\beta = 6$, and $\gamma = -3$. Our equation for the line may be taken in the parametric form

$$\begin{aligned} x &= 3 - 2t \\ y &= -2 + 6t \\ z &= 1 - 3t. \end{aligned}$$

On the other hand, we might have taken $\alpha = 1$, $\beta = -3$, and $\gamma = \frac{3}{2}$. This merely changes the **scale** of the parameter. The vector $\mathbf{l} := (\alpha, \beta, \gamma)$ indicates the direction of the line but not its orientation. This is true since we may replace the direction $\mathbf{l}$ by $-\mathbf{l}$ and still represent the same line. If the components of $\mathbf{l}$ are all nonzero we may replace the parameter t from our representation and obtain the **symmetric form** for the straight line, namely

$$\frac{x - x_0}{\alpha} = \frac{y - y_0}{\beta} = \frac{z - z_0}{\gamma} . \tag{1.26}$$

We turn next to the equation of the **plane**. A plane L is determined by any point $\mathbf{x}_0$ on it, and by a vector $\mathbf{n}$ orthogonal to the plane, called a **normal** to the plane. Let $\mathbf{x}$ be any point on the plane L. Then the vector $\mathbf{x} - \mathbf{x}_0$ is orthogonal to the normal, see F 1.8. This gives the equation of the plane

$$(\mathbf{x} - \mathbf{x}_0) \cdot \mathbf{n} = 0 . \tag{1.27}$$

If the normal vector is given by its coordinates as $\mathbf{n} = (a, b, c)$, then the above equation becomes

$$a\,x + b\,y + c\,z = d := a\,x_0 + b\,y_0 + c\,z_0 , \tag{1.28}$$

where d is the RHS. Every **linear equation** of this form represents a plane, except for the uninteresting case $a = b = c = 0$. A plane is also determined

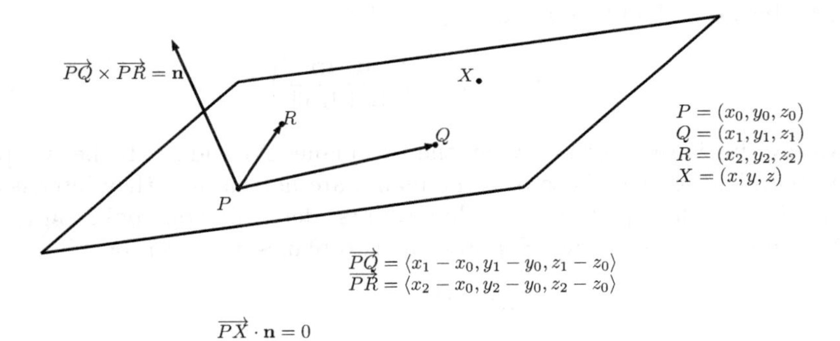

FIGURE 1.8: A plane in $\mathbb{R}^3$.

by three points on it, say $\mathbf{P} := (x_0, y_0, z_0)$, $\mathbf{Q} := (x_1, y_1, z_1)$ and $\mathbf{R} := (x_2, y_2, z_2)$. The two differences,

$$\begin{aligned} \mathbf{a} &:= \mathbf{PQ} = (x_1 - x_0, y_1 - y_0, z_1 - z_0) \\ \text{and} \quad \mathbf{b} &:= \mathbf{PR} = (x_2 - x_0, y_2 - y_0, z_2 - z_0) \end{aligned} \tag{1.29}$$

also lie in the plane, see Figure 1.8. Any vector $\mathbf{x} = (x, y, z)$ of the form

$$\mathbf{x} := P + s\,\mathbf{a} + t\,\mathbf{b} , \quad \text{where } s \text{ and } t \text{ are arbitrary} , \tag{1.30}$$

also lies in the plane. We can show, conversely, that any point on the plane is of this form. This gives a representation of the plane

$$\begin{aligned} x &= x_0 + s\,(x_1 - x_0) + t\,(x_2 - x_0) \\ y &= y_0 + s\,(y_1 - y_0) + t\,(y_2 - y_0) \\ z &= z_0 + s\,(z_1 - z_0) + t\,(z_2 - z_0) \end{aligned} \tag{1.31}$$

in terms of two parameters, s and t. Given the vectors $\mathbf{a}$ and $\mathbf{b}$ of (1.29) it is possible to determine a normal $\mathbf{n}$ to the plane as the cross product

$$\mathbf{n} = \mathbf{a} \times \mathbf{b} = \det \begin{pmatrix} \mathbf{i} & \mathbf{j} & \mathbf{k} \\ x_1 - x_0 & y_1 - y_0 & z_1 - z_0 \\ x_2 - x_0 & y_2 - y_0 & z_2 - z_0 \end{pmatrix} .$$

The equation (1.27) of the plane is then given in determinantal form as

$$\det \begin{pmatrix} x - x_0 & y - y_0 & z - z_0 \\ x_1 - x_0 & y_1 - y_0 & z_1 - z_0 \\ x_2 - x_0 & y_2 - y_0 & z_2 - z_0 \end{pmatrix} = 0 . \tag{1.32}$$

We say two planes are **parallel** if they never intersect. This is clearly the case if the normals to these planes themselves are parallel. In general, we say that the **angle** between two planes is the angle between their normals. Suppose the plane $\mathcal{M}$ has the normal $\mathbf{m}$ and the plane $\mathcal{N}$ has the normal $\mathbf{n}$. Then the angle between the two normals is given by

$$\theta = \operatorname{acos}\left(\frac{\mathbf{n} \cdot \mathbf{m}}{\|\mathbf{n}\|\,\|\mathbf{m}\|}\right) .$$

We say this is the angle between the two planes $\mathcal{M}$ and $\mathcal{N}$. If the two planes are parallel the angle is zero, if the planes are not parallel they intersect in a line. To find the equation of the line we first determine one point on the line. We can do this as follows: Suppose the two planes are given by

$$A_1 x + B_1 y + C_1 z = D_1,$$

$$A_2 x + B_2 y + C_2 z = D_2.$$

We consider an arbitrary x value x_0 and seek the y and z values which correspond to a point on the intersection. In Figure 1.9 we see the line of intersection of two nonparallel planes. Such an y and z must satisfy

$$B_1 y + C_1 z = D_1 - A_1 x_0$$

$$B_2 y + C_2 z = D_2 - A_2 x_0,$$

which has a solution providing $B_1 C_2 - B_2 C_1 \neq 0$. Let us try this on a specific case. Consider the two planes

$$x + 2y + z = -2 \text{ and } 2x + 4y + 3z = 1.$$

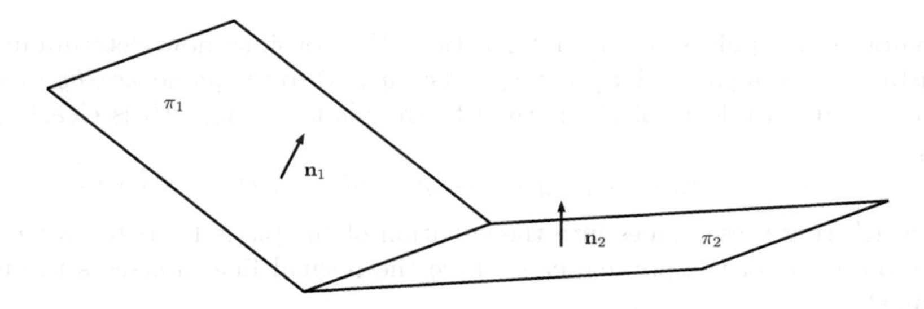

FIGURE 1.9: Intersecting planes.

Take $x_0 = 1$, then these become

$$1 + 2y + z = -2 \text{ and } 2 + 4y + 3z = 1,$$

or

$$2y + z = -3 \text{ and } 4y + 3z = -1,$$

which have as a solution $y_0 = -4$ and $z_0 = 5$. Let now find the line of intersection. The direction of the intersection line is parallel with the vector

$$\mathbf{l} = \mathbf{n} \times \mathbf{m};.$$

which in our specific case is given by $2\,\mathbf{i} - \mathbf{j}$; hence, the equation for the line of intersection may be given parametrically by

$$x = 1 + 2t\,, \quad y = -4 - t\,, \quad z = 5\,.$$

As another example we consider the planes defined by

$$x + 2\,y + z = -2 \quad \text{and} \quad 3\,x + 4\,y + 2\,z = 1\,.$$

In this case, proceeding as before with $x_0 = 1$, we obtain

$$2\,y + z = -3 \quad \text{and} \quad 4\,y + 2\,z = -2\,,$$

which are not consistent and haven't a solution. However, if instead of x we set $y_0 = 1$ we obtain

$$x + z = -4 \quad \text{and} \quad 3\,x + 2\,z = -3\,,$$

which has the solution $x_0 = 5$ and $z_0 = -9$. The line of intersection is parallel to the vector $\mathbf{j} - 2\mathbf{k}$, and the line therefore has the parametric representation

$$x = 5\,, \quad y = 1 + t\,, \quad z = -9 - 2\,t\,,$$

which never meets the plane $x = 1$.
Another way to determine the line of intersection between two planes is to

compute two points on the intersection. We consider now determining the distance from a particular point $P_0 = (x_0, y_0, z_0)$ to the plane $ax+by+cz = d$. The parametric form of the normal to the plane through P_0 is clearly given by

$$x(t) := x_0 + a\,t\;, \quad y(t) \;:=\; y_0 + b\,t\;, \quad z(t) \;:=\; z_0 + c\,t.$$

Substituting these values into the equation of the plane leads to an equation for the value of the parameter t where the normal line intersects the plane, namely

$$a\,(x_0 + a\,t) \;+\; b\,(y_0 + b\,t) \;+\; c\,(z_0 + c\,t) \;=\; d.$$

As this equation is linear in t we solve it easily and obtain $t = \frac{d-(ax_0+by_0+cz_0)}{a^2+b^2+c^2}$. Let us call this point of intersection $P_1 =< x_1, y_1, z_1 >$. The distance from P to the plane is the length $|\overline{P_0P_1}|$, which is

$$\sqrt{(x_1 - x_0)^2 + (y_1 - y_0)^2 + (z_1 - z_0)^2} = \frac{|d - (ax_0 + by_0 + cz_0)\,|}{\sqrt{a^2 + b^2 + c^2}}.$$

Mathematica session 1.7. *This session illustrates how to write a module (procedure) statement. We illustrate this by introducing a simple module for calculating the distance of a point to a plane. In future exercises you will be asked to develop more complicated modules. We use a mathematical formulae to design the module.In the first module our variables are a point* $\mathbf{P} \in \mathbb{R}^3$ *, not on the plane, a point* **Q** *on the plane and* **N** *a normal to the plane. The following module calculates the distance from* **P** *to the plane.*

```
distancePointToPlane[P_, Q_, N_]:=Module[{n, r}, n = Sqrt[Dot[N, N]]; r =
Dot[P − Q, N]; r/n]
```

Notice that we have local variables (n, r). MATHEMATICA *will not "remember" these for future calculations, one of the benefits of using the module structure.*

```
distancePointToPlane[{4, 3, 5}, {1, 0, 1}, {1, 1, 1}]
```

$\frac{10}{\sqrt{3}}$

In the next module we calculate the distance of a point to a line. The variables are a point not on the line **P** *, and a point on the line* **Q** *and the line direction* **V** *.*

$distancePointToLine[P\_, Q\_, V\_]{:=}Module[\{v,r\}, v = Sqrt[Dot[V,V]]; r = ((P-Q)\times V)/v; Sqrt[Dot[r,r]]]$

$distancePointToLine[\{4,3,5\},\{1,0,1\},\{1,1,1\}]$

$\sqrt{\frac{2}{3}}$

We give another module for calculating the distance of a point to the plane. Here **P** *is a point not on the plane. The plane is given in terms of its normal form*

$$Ax + By + Cz = D.$$

$distancePointToPlane2[P\_, A\_, B\_, C\_, D\_]{:=}Module[\{n,l\},$

$n = Sqrt[A\wedge 2 + B\wedge 2 + C\wedge 2];$

$l = Abs[P[[1]] * A + P[[2]] * B + P[[3]] * C - D];$

$l/n]$

We test this module below.

$distancePointToPlane2[\{4,5,3\},1,1,1,2]$

$\frac{10}{\sqrt{3}}$

$distancePointToPlane[\{x,y,z\},\{0,0,D/C\},\{A,B,C\}]$

$\frac{Ax+By+C\left(-\frac{D}{C}+z\right)}{\sqrt{A^2+B^2+C^2}}$

ClearAll clears all values, definitions, attributes, messages, with assigned symbols.
ClearAll["Global*"]

Mathematica session 1.8. *In this session we learn how to input arrays, determinants and matrices. These will be useful to us in the description of planes and lines. First we present he syntax for an* **array**. *We shall restrict our discussion to three variable. Let us input this system into* MATHEMATICA. *First we need to input the coefficient matrix of the system, which we designate by A. Notice the syntax used for inputting a matrix.*

$A = Array[a, \{3,3\}]$

$\{\{a[1,1], a[1,2], a[1,3]\}, \{a[2,1], a[2,2], a[2,3]\}, \{a[3,1], a[3,2], a[3,3]\}\}$

The MATHEMATICA *command* **Det** *evaluates the determinant.*

Det[A]

$-a[1,3]a[2,2]a[3,1]+a[1,2]a[2,3]a[3,1]+a[1,3]a[2,1]a[3,2]-a[1,1]a[2,3]a[3,2]-a[1,2]a[2,1]a[3,3]+a[1,1]a[2,2]a[3,3]$

We may write a linear system by utilizing the array format

Array[x, 3] *then the unknowns*

$\{x[1], x[2], x[3]\}$

Y = Array[y, 3]

$\{b[1], b[2], b[3]\}$

The operation **LinearSolve** *allows us to solve arrays of linear equations,*

$$\mathbb{A} \cdot \mathbf{X} = \mathbf{b}. \tag{1.33}$$

To see how this works we present a simple example in the MATHEMATICA *notation. The first three arrays in* **LinearSolve** *form the coefficient matrix* $\mathbb{A}$ *and the fourth array represents the vector* **b**, *namely*

LinearSolve[{{1, 3, 1}, {0, 2, 1}, {−1, 0, 4}}, {1, 2, 3}]

$\{-\frac{13}{7}, \frac{6}{7}, \frac{2}{7}\}$

We may also use **LinearSolve** *to find the solution of the system with arbitrary coefficients (1.33).*

LinearSolve[{{a[1, 1], a[1, 2], a[1, 3]}, {a[2, 1], a[2, 2], a[2, 3]}, {a[3, 1], a[3, 2], a[3, 3]}}, {y[1], y[2], y[3]}];

By putting an ; after the input line, the outputs saved but not printed. We will want to use this output later and MATHEMATICA *saves it for us. There is a formula given in terms of determinants for solving linear systems. It is known as* **Cramer's Rule**. *We will use* MATHEMATICA *to show these two procedures give exactly the same solution. The solution of (1.33) may be written as*

$$x_i = det(\mathbb{A}_i)/det(\mathbb{A}) \tag{1.34}$$

where $\mathbb{A}_i$ *is the matrix occurring by replacing the* i^{th} *column of* $\mathbb{A}$ *with the vector* **b**. *We implement this in* MATHEMATICA *once more suppressing the output. We compute the* $\mathbb{A}_i$, *for* $i = 1, 2, 3$. *next.*

Det[{{b[1], a[1, 2], a[1, 3]}, {b[2], a[2, 2], a[2, 3]}, {yb3], a[3, 2], a[3, 3]}}]/ Det[A];

Det[{{a[1, 1], b[1], a[1, 3]}, {a[2, 1], b[2], a[2, 3]}, {a[3, 1], b[3], a[3, 3]}}]/ Det[A];

Det[{{a[1, 1], a[1, 2], b[1]}, {a[2, 1], a[2, 2], b[2]}, {a[3, 1], a[3, 2], b[3]}}]/

Det*[*A*]*;

We may check to see if these two methods agree. This is accomplished by subtracting the solution given by **LinearSolve** *from the solution given by Cramer. If the solutions are equal to one another, the difference between them will be zero. The reader should do this as an exercise.*

Exercises

Exercise 1.28. *The point-slope formula for a line tells us it is sufficient to have a point* P *on a line plus the* **direction d** *of the line in order to write down its representation. Write a* MATHEMATICA *function which makes use the parametric form to describe a line passing through a prescribed point* P *having the directions* $\mathbf{d}$. *Try your* MATHEMATICA *function on the point-slope specifications below:*

(a) $P = (1, -2, 3)$ *and* $\mathbf{d} = \mathbf{i} + \mathbf{j} - \mathbf{k}$

(b) $P = (4, -2, -3)$ *and* $\mathbf{d} = 2\,\mathbf{i} + 3\,\mathbf{j} + \mathbf{k}$

(c) $P = (-12, 2, 24)$ *and* $\mathbf{d} = -4\,\mathbf{i} - \mathbf{j} + 5\,\mathbf{k}$

Exercise 1.29. *The two-point formula for a line indicates only two points are necessary to write down a representation formula for a particular line. Use this idea to write a* MATHEMATICA *function which uses this two point information to find the lines representation.*

(a) $P = (4, 2, 3)$ *and* $Q = (4, 3, 2)$

(b) $P = (1, -2, -1)$ *and* $Q = (-4, 5, 3)$

(c) $P = (12, 0, 24)$ *and* $Q = (1, 20, 32)$

Exercise 1.30. *Write a* MATHEMATICA *function for the equation of a plane which passes through a prescribed point, and which is normal to a given vector. Use this function to obtain the equations for the planes corresponding to the information given below:*

(a) $P = (1, -2, 3)$ *and* $\mathbf{n} = \mathbf{i} + \mathbf{j} - \mathbf{k}$

(b) $P = (4, -2, -3)$ *and* $\mathbf{n} = 2\,\mathbf{i} + 3\,\mathbf{j} + \mathbf{k}$

(c) $P = (-12, 2, 24)$ *and* $\mathbf{n} = -4\,\mathbf{i} - \mathbf{j} + 5\,\mathbf{k}$

Exercise 1.31. *The* **symmetric** *form of the equation of a line in space is given by*

$$\frac{x - x_0}{a} = \frac{y - y_0}{b} = \frac{z - z_0}{c}.$$

Notice if one of the constants a, b, c are zero, say a, then the corresponding numerator of this fraction $x - x_0$ *vanishes too. In this instance the line is restricted to the plane* $x = x_0$. *Redo the problems in Exercises 1.28–1.29 obtaining the equations of the lines in symmetric form.*

Exercise 1.32. *Write a* MATHEMATICA *function which gives the symmetric form of a line from two points lying on the line.*

Exercise 1.33. *Find the equation of the plane through the point* $(1, 1, 1)$ *perpendicular to the two planes*

$$x + 2y - z = 5 \quad \textit{and} \quad 2x + y + 3z = 6.$$

Exercise 1.34. *Find the equation of the line which is the intersection of the two planes indicated in the problems below:*

(a) $2x + y - z = 4$ *and.* $-x + y - z = 6$

(b) $x + 3y + 2z = 10$ *and* $4x + 7y - 5z = 1.$

(c) $-3x + 5y - 7z = 8$ *and* $x + 5y + 11z = 3$

(d) $x + 4y + 5z = 24$ *and* $6x - 2y + 3z = 16$

Exercise 1.35. *Find the equation of the line perpendicular to the given plane and passing through the given point below:*

(a) $2x + y - z = 4$ *and* $(1, 3, 4)$

(b) $x + 3y + 2z = 10$ *and* $(-1, 0, 1)$

(c) $-3x + 5y - 7z = 8$ *and* $(2, 1, 6)$

(d) $x + 4y + 5z = 24$ *and* $(6, 1, 0)$

Exercise 1.36. *Find the equation of the line lying in a prescribed plane, passing through a point in this plane, and perpendicular to another line lying in the plane.*

(a) $2x + y - z = 4$, $(1, 3, 4)$ *and* $x = 2 + t, y = 1 + t, z = 3t + 1$

(b) $x + 3y + 2z = 10$, $(-1, 0, 1)$ *and* $x = 1 + t.y = 2 + 2t, z + 7/2t - 7/2$

(c) $-3x + 5y - 7z = 8$,$(2, 1, 6)$ *and* $x = 2t - 2, y = t + 1, z = 3/7 - t/7$

(d) $x + 4y + 5z = 24$, $(6, 1, 0)$ *and* $x = 4t - 1, y = -t + 6, z = -1/5$

Exercise 1.37. *Write a* MATHEMATICA *function which computes the angle between two planes*

$$a_1 x + b_1 y + c_1 z = d_1 \quad \textit{and} \quad a_2 x + b_2 y + c_2 z = d_2 .$$

Use this function to compute the angles between the planes

(a) $2x + y - z = 4$ *and* $x + 4y + z = 2$

(b) $x + 3y + 2z = 10$ *and* $x + 4y - 2z = 4$

(c) $-3x + 5y - 7z = 8$ *and* $-x + 4y - 65z = 24$

(d) $x + 4y + 5z = 24$ *and* $x + y + z = 1$

Exercise 1.38. *Develop a* MATHEMATICA *function which produces the equation of the plane perpendicular two given planes and also passing through a specified point. Use your function to determine the planes described below:*

(a) $2x + y - z = 4$, $x + 4y + z = 2$ *and* $(1, 0, 0)$

(b) $x + 3y + 2z = 10$, $x + 4y - 2z = 4$ *and* $(1, -1, 4)$

(c) $-3x + 5y - 7z = 8$, $-x + 4y - 65z = 24$ *and* $(0, 1, -2)$

(d) $x + 4y + 5z = 24$, $x + y + z = 1$ *and* $(4, 2, 1)$

Exercise 1.39. *Develop a* MATHEMATICA *function which produces the equation of the plane passing through two specified points but parallel to the intersection of two planes. Use your function to determine the planes described below:*

(a) $(1,0,0)$, $(0,1,0)$ *with planes* $2x+y-z=4$ *and* $x+4y+z=2$

(b) $(2,-1,2)$, $(1,4,1)$ *with planes* $x+3y+2z=10$ *and* $x+4y-2z=4$

(c) $(1,0,1)$, $(3,1,-2)$ *with planes* $-3x+5y-7z=8$ *and* $-x+4y-65z=24$

(d) $(2,-4,6)$, $(1,2,-3)$ *with planes* $x+4y+5z=24$ *and* $x+y+z=1$

Exercise 1.40. *Write a* MATHEMATICA *function which computes the intersection point of a line normal to a plane and passing through a prescribed point. Calculate the intersection point in each of the following corresponding to the following prescribed plane and point:*

(a) $2x+y-z=4$ *and (1,0,0)*

(b) $x+3y+2z=10$ *and* $(1,-1,4)$.

(c) $-3x+5y-7z=8$ *and* $(0,1,-2)$.

(d) $x+4y+5z=24$ *and* $(4,2,1)$.

Exercise 1.41. *Write a* MATHEMATICA *function which determines the distance between a given point* P *and a given plane* $\mathcal{L}$. *Try this function on the points and planes below:*

(a) $2x+y-z=4$ *and* $(1,0,0)$.

(b) $x+3y+2z=10$ *and* $(1,-1,4)$.

(c) $-3x+5y-7z=8$ *and* $(0,1,-2)$.

(d) $x+4y+5z=24$ *and* $(4,2,1)$.

Chapter 2

Some Elementary Curves and Surfaces in $\mathbb{R}^3$

2.1 Curves in space and curvature

Mathematica session 2.1. *In this* MATHEMATICA *session we show how to plot curves in* $\mathbb{R}^3$ *such as are seen in the figures 2.1, 2.2, and 2.3.*

```
ParametricPlot3D[{t, t^2, t^3}, {t, 0, 1}]
```

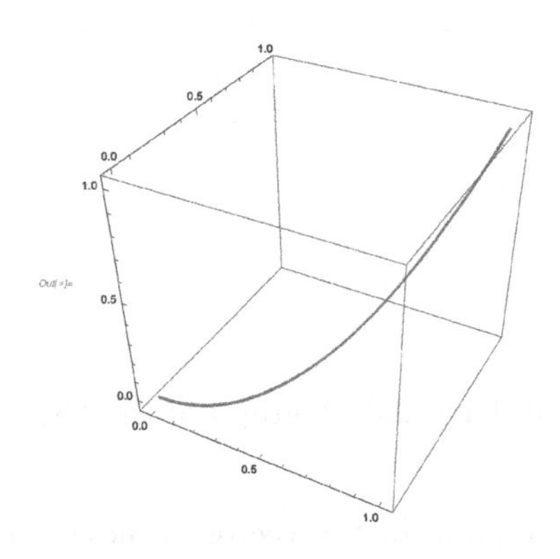

FIGURE 2.1: Plot of the curve (t, t^2, t^3) for $0 \leq t \leq 1$

```
ParametricPlot3D[{Cos[t], Sin[t], t}, {t, 0, 2 * Pi}]
```

```
ParametricPlot3D[{(Sin[20*t]+3) * Cos[t], (Sin[20 * t]+3)*Sin[t], Cos[20*t]},
{t, 0, 2 * Pi}]
```

If t is the time, then $\mathbf{r} = (x(t), y(t), z(t))$ represents a moving vector or,

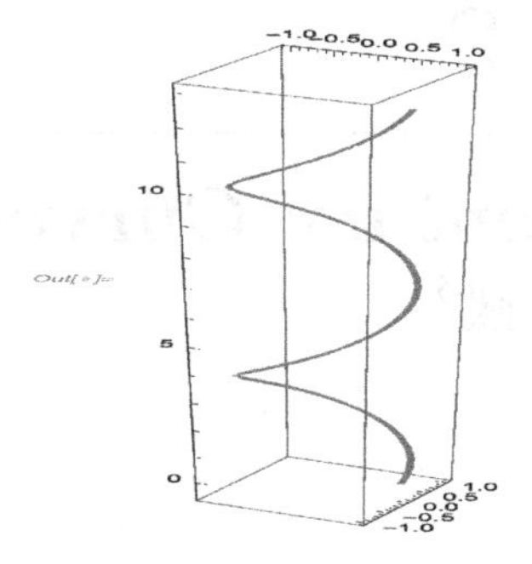

FIGURE 2.2: Helix

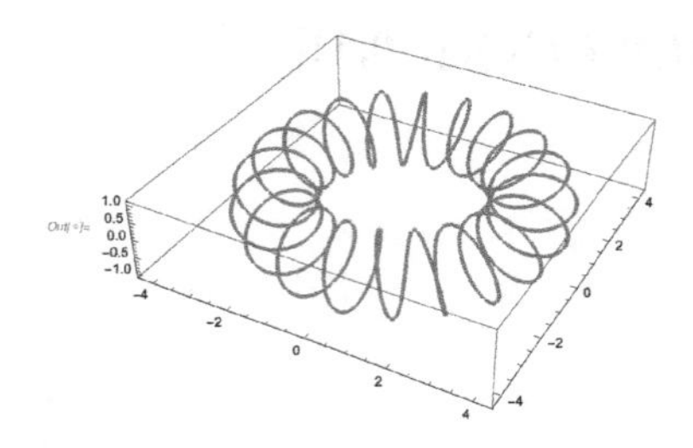

FIGURE 2.3: A graph of the "Slinky"

if we consider just the point of the vector, a moving point in space. We may speak of the **velocity** of the point also. It is defined to be the time derivative of the position, i.e.

$$\mathbf{v}(t) \ := \ \frac{d}{dt}\mathbf{r}(t) = \frac{dx(t)}{dt}\mathbf{i} + \frac{dy(t)}{dt}\mathbf{j} + \frac{dz(t)}{dt}\mathbf{k}\,. \tag{2.1}$$

Likewise, the **acceleration** may be defined as the time derivative of the velocity

$$\mathbf{a}(t) \ := \ \frac{d}{dt}\mathbf{v}(t) = \frac{d^2x(t)}{dt^2}\mathbf{i} + \frac{d^2y(t)}{dt^2}\mathbf{j} + \frac{d^2z(t)}{dt^2}\mathbf{k}\,. \tag{2.2}$$

We notice that the velocity and the acceleration are vectors. There are also scalar quantities associated with these vectors, namely the **speed** and the **scalar acceleration**. The scalar quantities are the magnitudes of the respective vectors, i.e.

$$v(t) \ := \ \|\mathbf{v}(t)\| \ = \ \sqrt{\left(\frac{dx}{dt}\right)^2 + \left(\frac{dy}{dt}\right)^2 + \left(\frac{dz}{dt}\right)^2}\,, \tag{2.3}$$

and

$$a(t) \ := \ \|\mathbf{a}(t)\| \ = \ \sqrt{\left(\frac{d^2x}{dt^2}\right)^2 + \left(\frac{d^2y}{dt^2}\right)^2 + \left(\frac{d^2z}{dt^2}\right)^2}\,. \tag{2.4}$$

The **differential arc-length** ds along the curve described by $\mathbf{r}(t)$ may be easily seen from Figure 2.4 to be given by

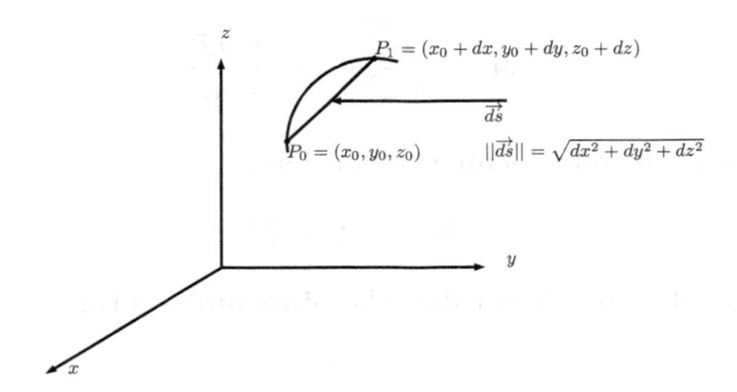

FIGURE 2.4: Arc-length differential.

$$ds \ = \ \sqrt{\left(\frac{dx}{dt}\right)^2 + \left(\frac{dy}{dt}\right)^2 + \left(\frac{dz}{dt}\right)^2}\, dt\,,$$

from which we conclude that the **arc-length** from $\mathbf{r}(a)$ to $\mathbf{r}(b)$ is given by

$$s(b) - s(a) \ = \ \int_a^b \sqrt{\left(\frac{dx}{dt}\right)^2 + \left(\frac{dy}{dt}\right)^2 + \left(\frac{dz}{dt}\right)^2}\, dt \ = \ \int_a^b v(t)\, dt\,, \tag{2.5}$$

where $v(t)$ is the speed of the moving point. If we define $S(t) \ := \ s(t) - s(a)$ as the distance along the curve from $\mathbf{r}(a)$ to the variable point $\mathbf{r}(t)$ we obtain, using the fundamental theorem of calculus, a formula for the instantaneous speed, namely

$$\frac{d}{dt} S(t) \ = \ v(t)\,.$$

Following the example of plane curves we define the unit tangent vector by using a **normalized** velocity vector, that is we divide the velocity vector by its length

$$\mathbf{T}(t) \ := \ \frac{\mathbf{v}(t)}{\|\mathbf{v}(t)\|}. \ = \ \frac{\mathbf{v}(t)}{v(t)}\,, \tag{2.6}$$

Using $\mathbf{v} = \frac{d\mathbf{r}}{dt}$ and $v = \frac{ds}{dt}$, T can be expressed as a function of the arc length s

$$\mathbf{T}(s) = \frac{d\mathbf{r}}{ds} = \frac{dx}{ds}\mathbf{i} + \frac{dy}{ds}\mathbf{j} + \frac{dz}{ds}\mathbf{k}\,. \tag{2.7}$$

From the identity $\mathbf{T} \cdot \mathbf{T} = 1$, we get upon differentiating with respect to arc length that $\mathbf{T} \cdot \frac{d\mathbf{T}}{ds} = 0$, so that $\mathbf{T}$ and $\frac{d\mathbf{T}}{ds}$ are always perpendicular. The **curvature** κ of the curve is defined to be

$$\kappa = \left\| \frac{d\mathbf{T}}{ds} \right\| = \left\| \frac{d\mathbf{T}}{dt}\frac{dt}{ds} \right\| = \frac{1}{v}\left\| \frac{d\mathbf{T}}{dt} \right\|\,. \tag{2.8}$$

The **principal normal** to $\mathbf{T}(t)$ is the unit normal vector

$$\mathbf{N} = \frac{\frac{d\mathbf{T}}{ds}}{\|\frac{d\mathbf{T}}{ds}\|} = \frac{1}{\kappa}\frac{d\mathbf{T}}{ds}\,, \tag{2.9}$$

see Figure 2.5. The unit vector $\mathbf{B}$ defined as

$$\mathbf{B} := \mathbf{T} \times \mathbf{N}\,, \tag{2.10}$$

, illustrated in Figure 2.5, is called the **binormal vector** .

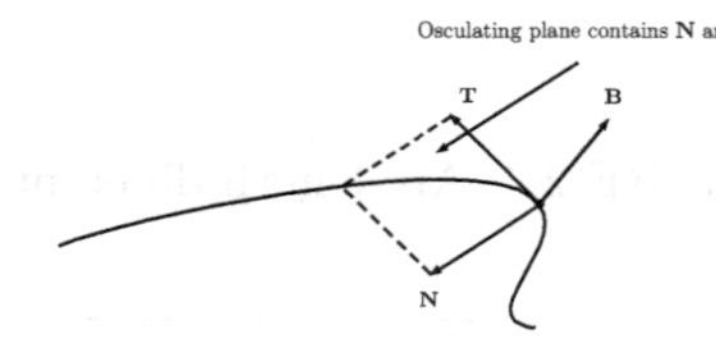

FIGURE 2.5: The tangent, normal, binormal and osculating plane for a curve.

Mathematica session 2.2. *In this session we show how we can use* MATHEMATICA *to work with curves in* $\mathbb{R}^3$*. We enter a vector* $\mathbf{r}(t)$ *as a function of the parameter* t*. Note that we use square bracket for a function* $f[t]$ *in* MATHEMATICA *notation and that vectors use the notation* $\{x, y, z\}$ *etc.*

```
r:={x[t], y[t], z[t]}
SetCoordinates[Cartesian]
SetCoordinates[Cartesian]
D[{Cos[t], Sin[t], 3 * t}, t]
{-Sin[t], Cos[t], 3}
v[{{x_, y_, z_}, t_}]:=D[{x, y, z}, t]
```

```
v[{{Cos[t], Sin[t], 3 * t}, t}]
```

$\{-Sin[t], Cos[t], 3\}$

```
tangent[{{x_, y_, z_}, t_}]:=
Simplify[D[{x, y, z}, t]/Sqrt[D[{x, y, z}, t].D[{x, y, z}, t]]]
tangent[{{Cos[t], Sin[t], 3t}, t}]
```

$\left\{-\frac{Sin[t]}{\sqrt{10}}, \frac{Cos[t]}{\sqrt{10}}, \frac{3}{\sqrt{10}}\right\}$

```
myNorm[{x_, y_, z_}]:=Sqrt[{x, y, z}.{x, y, z}]
```

```
myNorm[{1, 2, 3}]
```

$\sqrt{14}$

```
D[{x[t], y[t], z[t]}, {t, 2}]
```

$\{x''[t], y''[t], z''[t]\}$

```
myCurvatureVector[{{x_, y_, z_}, t_}]:=
Norm[CrossProduct[D[{x, y, z}, t], D[{x, y, z}, {t, 2}], CoordinateSystem
[x, y, z]]/(Norm[D[{x, y, z}, t]])^3]
myCurvatureVector[{Cos[t], Sin[t], t}, t]
```

$myCurvatureVector[\{Cos[t], Sin[t], t\}, t]$

```
D[{Cos[t], Sin[t], 3 * t}, {t, 2}]
```

$\{-Cos[t], -Sin[t], 0\}$

```
Simplify[
CrossProduct[{-Sin[t], Cos[t], 3}, {-Cos[t], -Sin[t], 0}, CoordinateSystem
([x, y, z]]/Norm[{-Sin[t], Cos[t], 3}])^3]
```

$\left\{\frac{3Sin[t]}{(9+Abs[Cos[t]]^2+Abs[Sin[t]]^2)^{3/2}}, -\frac{3Cos[t]}{(9+Abs[Cos[t]]^2+Abs[Sin[t]]^2)^{3/2}}, \frac{1}{(9+Abs[Cos[t]]^2+Abs[Sin[t]]^2)^{3/2}}\right\}$

```
(*Define function to find magnitude of a vector*)mag[vec_]:=Sqrt[vec.vec]

(*Define curvature function*)
curvature[x_, y_, z_]:=Module[{s, v, T}, s = {x, y, z};
v = D[s, t];
T = v/mag[v];
Simplify[mag[D[T, t]/mag[v]]]]
```

```
(*ApplyfunctiontoCos[t], Sin[t], t*)
curvature[Cos[t], Sin[t], t]
```

$\frac{1}{2}$

```
Simplify[{-Sin[t], Cos[t], 3} × {-Cos[t], -Sin[t], 0}/(myNorm[{-Sin[t],
Cos[t], 3}])^3]
```

$\left\{\frac{3Sin[t]}{10\sqrt{10}}, -\frac{3Cos[t]}{10\sqrt{10}}, \frac{1}{10\sqrt{10}}\right\}$

```
biNormal[R_]:=Module[{v, w, T, N},
v = D[R, t];
w = D[R, {t, 2}];
T = Simplify[v/Sqrt[v.v]];
N:=Simplify[w/Sqrt[w.w]];
Simplify[T × N]]

biNormal[{Cos[t], Sin[t], t}]
```

$\left\{\frac{Sin[t]}{\sqrt{2}}, -\frac{Cos[t]}{\sqrt{2}}, \frac{1}{\sqrt{2}}\right\}$

♦

Let us consider a point P_{t_0} on the curve $\mathcal{C}$ and the tangent vector $\mathbf{T}_{t_0}$ to this curve at P_{t_0}. If P_{t_1} is a neighboring point on $\mathcal{C}$ then secant $\overline{P_{t_0}P_{t_1}}$ and $\mathbf{T}_{t_0}$ determine a plane. If as the point $P_{t_1} \to P_{t_0}$ the plane approaches a limiting position, then the limit plane is called the **osculating plane** . If the equation of the curve is given parametrically by

$$x = \alpha(t) , \quad y = \beta(t) , \quad z = \gamma(t) ,$$

then the equation of the osculating plane, see Figure 2.5, must be given by

$$A\,(x-\alpha(t)) + B\,(y-\beta(t)) + C\,(z-\gamma(t)) = 0 . \tag{2.11}$$

Since this equation is to hold for an arbitrary point on $\mathcal{C}$, i.e. an arbitrary t, the following two conditions must hold

$$A\,\alpha'(t) + B\,\beta'(t) + C\,\gamma'(t) = 0 , \tag{2.12}$$

and

$$A\,(\alpha(t+\Delta t)-\alpha(t)) + B\,(\beta(t+\Delta t)-\beta(t)) + C\,(\gamma(t+\Delta t)-\gamma(t)) = 0 . \tag{2.13}$$

Using the **finite Taylor series**, which says that for a function $\alpha(\tau)$ with three continuous derivatives we may expand the function about the point $\tau = t$ as

$$\alpha(\tau) = \alpha(t) + \alpha'(t)(\tau-t) + \alpha''(t)\frac{(\tau-t)^2}{2} + (\tau-t)^2\varepsilon_1(t),$$

where $\varepsilon_1 \to 0$ as $t \to \tau$. A similar remark holds for the functions β and γ. Equation (2.13) may then be rewritten as

$$A\left(\alpha'(t)\Delta t + (\alpha''(t)+\varepsilon_1)\frac{(\Delta t)^2}{2}\right) + B\left(\beta'(t)\Delta t + (\beta''(t)+\varepsilon_2)\frac{(\Delta t)^2}{2}\right)$$
$$+ \quad C\left(\gamma'(t)\Delta t + (\gamma''(t)+\varepsilon_3)\frac{(\Delta t)^2}{2}\right) = 0\,. \tag{2.14}$$

Subtracting $(\Delta t) \times$ equation (2.12) from (2.14) we get

$$A\,(\alpha''(t)+\varepsilon_1) \;+\; B\,(\beta''(t)+\varepsilon_2) \;+\; C\,(\gamma''(t)+\varepsilon_3) \;=\; 0\,,$$

which in the limit as $\Delta t \to 0$ becomes

$$A\,\alpha''(t) \;+\; B\,\beta''(t) \;+\; C\,\gamma''(t) \;=\; 0\,. \tag{2.15}$$

For the system of equations (2.11), (2.12) and (2.15) to have a nontrivial solution for A, B and C, the following equation must be satisfied for any point (x,y,z) on the osculating plane.

$$\det\begin{pmatrix} x-\alpha(t), & y-\beta(t), & z-\gamma(t) \\ \alpha'(t), & \beta'(t), & \gamma'(t) \\ \alpha''(t), & \beta''(t), & \gamma''(t) \end{pmatrix} = 0\,. \tag{2.16}$$

In general, there is only one osculating plane; however, if at some point $\tau = t$ we have $\alpha''(t) = \beta''(t) = \gamma''(t) = 0$ then **all tangent planes are osculating planes**.

Remark 2.1. The binormal vector is perpendicular to the osculating plane, as it is perpendicular to both **t** and **n** as can be seen from (2.16) by interpreting the determinant as the volume of the parallelpiped spanned by any vector on the plane, the tangent vector and the normal vector. The vectors **b**, **t**, and **n**, taken in that order, form a right-handed-sytem as may be seen from Figure 2.5.

Example 2.2 (Osculating Plane)**.** Suppose we wish to find the osculating plane for the curve which is the intersections of the two cylinders 2.6

$$x^2 + y^2 = 1\,, \quad \text{and} \quad x^2 + z^2 = 1\,.$$

The intersection may be parameterized as

$$x = \cos t\,,\; y = \sin t\,, \quad \text{and } z = \sin t\,.$$

Hence, we may represent the osculating plane as

$$\det\begin{pmatrix} x-\cos t, & y-\sin t, & z-\sin t \\ \sin t, & -\cos t, & -\cos t \\ \cos t, & \sin t, & \sin t \end{pmatrix} = 0\,,$$

which simplifies to $y = z$.

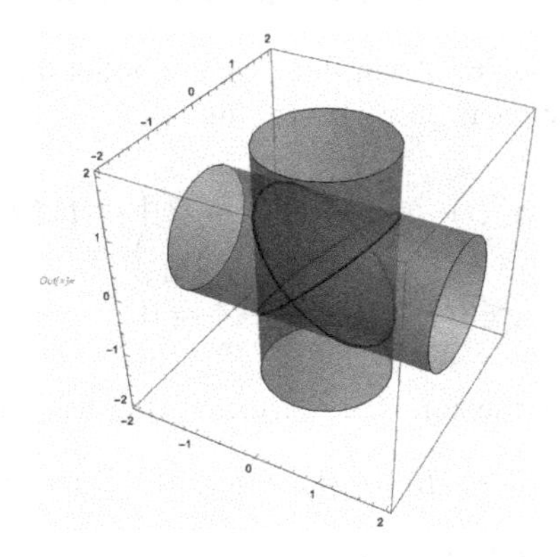

FIGURE 2.6: The intersecting cylinders $x^2 + y^2 = 1$, and $x^2 + z^2 = 1$.

The acceleration vector $\mathbf{a}(t)$ may be decomposed into a component which lies in the direction of the tangent $\mathbf{t}(t)$ and a component which lies in the direction of the principal normal $\mathbf{n}(t)$. To see this we differentiate

$$\mathbf{a} = \frac{d\mathbf{v}}{dt} = \frac{d}{dt}(v\mathbf{T}) = \mathbf{T}\frac{dv}{dt} + v\frac{d\mathbf{T}}{dt} = \mathbf{T}\frac{dv}{dt} + v\frac{d\mathbf{T}}{ds}\frac{ds}{dt},$$

which using (2.8) yields

$$\mathbf{a} = \mathbf{T}\frac{dv}{dt} + v^2\kappa\mathbf{N} := a_T\,\mathbf{T} + a_N\,\mathbf{N}.$$

So the component of $\mathbf{a}$ in the direction $\mathbf{T}$ is $\frac{dv}{dt}$; whereas the component in the $\mathbf{N}$ direction is given by $v^2\,\kappa$. We may use this decomposition to obtain useful formulae for the components. For example taking the inner product

$$\mathbf{v}\cdot\mathbf{a} = v\mathbf{T}\cdot\left(\mathbf{T}\frac{dv}{dt} + v^2\,\kappa\mathbf{N}\right) = v\,\frac{dv}{dt}.$$

On the other hand

$$\mathbf{v}\times\mathbf{a} = v\mathbf{T}\times\left(\mathbf{T}\frac{dv}{dt} + v^2\,\kappa\mathbf{N}\right) = \kappa v^3\,\mathbf{T}\times\mathbf{N}.$$

So we obtain

$$\|\mathbf{v}\times\mathbf{a}\| = \kappa v^3\,\|\mathbf{T}\times\mathbf{N}\| = \kappa v^3. \tag{2.17}$$

This gives equations for the components

$$a_T = \frac{\mathbf{v}\cdot\mathbf{a}}{v} = \frac{\dot{\mathbf{r}}\cdot\ddot{\mathbf{r}}}{\|\,\dot{\mathbf{r}}\|}, \tag{2.18}$$

and

$$a_N = \frac{\mathbf{v} \times \mathbf{a}}{v} = \frac{\dot{\mathbf{r}} \times \ddot{\mathbf{r}}}{\| \dot{\mathbf{r}} \|} . \tag{2.19}$$

It is easy to show that $\frac{d}{dt}\mathbf{b}$ is colinear with $\mathbf{n}$. Clearly since $\mathbf{b}$, $\mathbf{n}$, and $\mathbf{t}$ are mutually orthogonal there exist functions $c_1(t)$, $c_2(t)$, and $c_2(t)$ such that

$$\frac{d\mathbf{B}}{dt} = c_1(t)\,\mathbf{B} + c_2(t)\,\mathbf{N} + c_3(t)\,\mathbf{T} ,$$

where

$$c_1(t) = \frac{d\mathbf{B}}{dt} \cdot \mathbf{B} , \quad c_2(t) = \frac{d\mathbf{B}}{dt} \cdot \mathbf{N} \quad \text{and} \quad c_3(t) = \frac{d\mathbf{B}}{dt} \cdot \mathbf{T} .$$

Since $|\mathbf{B}| = 1$ we have that $\mathbf{B} \cdot \dot{\mathbf{B}} = 0$, and so $c_1(t) \equiv 0$. To prove that $c_3(t) \equiv 0$ we differentiate $\mathbf{B} := \mathbf{T} \times \mathbf{N}$ to obtain

$$\frac{d\mathbf{B}}{dt} = \frac{d\mathbf{T}}{dt} \times \mathbf{N} + \mathbf{T} \times \frac{d\mathbf{N}}{dt} .$$

Now as $\frac{d\mathbf{T}}{dt} = \kappa(t)\,\frac{ds}{dt}\,\mathbf{N}$ it follows that $\frac{d\mathbf{T}}{dt} \times \mathbf{N} = \mathbf{0}$, and therefore

$$\frac{d\mathbf{B}}{dt} = \mathbf{T} \times \frac{d\mathbf{N}}{dt} .$$

Hence we may now compute

$$c_3(t) = \frac{d\mathbf{B}}{dt} \cdot \mathbf{T} = \left(\mathbf{T} \times \frac{d\mathbf{N}}{dt} \right) \cdot \mathbf{T} = 0 .$$

So we can conclude that

$$\frac{d\mathbf{B}}{dt} = c_2\,\mathbf{N} .$$

Definition 2.3. The function $\tau(t)$ defined as

$$\frac{d\mathbf{B}}{dt} = -\tau(t)\,\frac{ds(t)}{dt}\,\mathbf{N} \tag{2.20}$$

is called the **torsion** of the curve.

Remark 2.4. The torsion plays a similar role to the curvature. As the curvature measures the tendency of a curve to vary from its tangent, the torsion measures the tendency of a line to twist away from the osculating plane. The torsion is frequently called the **second curvature**.

In the special instance where the parameter is the arc length, the formula (2.20) takes the simpler form

$$\frac{d\mathbf{B}}{ds} = -\tau(s)\,\mathbf{N} . \tag{2.21}$$

Mathematica session 2.3.

```
Clear[a]

R = {a * Cos[t], a * Sin[t], b * t}
```

$\{a\,Cos[t], a\,Sin[t], bt\}$

```
D[R, t]
```

$\{-a\,Sin[t], a\,Cos[t], b\}$

```
D[R, {t, 2}]
```

$\{-a\,Cos[t], -a\,Sin[t], 0\}$

```
D[R, t] × D[R, {t, 2}]
```

$\{ab\,Sin[t], -ab\,Cos[t], a^2\,Cos[t]^2 + a^2\,Sin[t]^2\}$

```
Dot[D[R, t] × D[R, {t, 2}], D[R, t] × D[R, {t, 2}]]
```

$a^2b^2\,Cos[t]^2 + a^2b^2\,Sin[t]^2 + \left(a^2\,Cos[t]^2 + a^2\,Sin[t]^2\right)^2$

```
Simplify[
Sqrt[Dot[D[R, t] × D[R, {t, 2}], D[R, t] × D[R, {t, 2}]]]]
```

$\sqrt{a^2\,(a^2 + b^2)}$

We now introduce a MATHEMATICA *function which computes the curvature* κ *for a arbitrary space curve, given in the form* $R(t) = \langle x(t), y(t), z(t)\rangle$.

```
kappa[R_]:=Module[{v, acel, d}, v = D[R, t];
acel = D[R, {t, 2}];
d = Sqrt[Dot[v, v]];
Simplify[Sqrt[Dot[v × acel, v × acel]]/(d)^3]]

kappa[{aCos[t], aSin[t], bt}]
```

$\frac{\sqrt{a^2(a^2+b^2)}}{(a^2+b^2)^{3/2}}$

MATHEMATICA *has a procedure for dealing with roots. We illustrate this below.*

```
Unprotect[Power];
Format[(a_^2 + b_^2)^(3/2)]:=
Defer[(a^2 + b^2) * Sqrt[a^2 + b^2]]
Protect[Power];
```

$\frac{\sqrt{a^2(a^2+b^2)}}{(a^2+b^2)^{3/2}}$

$\frac{\sqrt{a^2(a^2+b^2)}}{(a^2+b^2)\sqrt{a^2+b^2}}$

$\sqrt{a^2\,(a^2+b^2)}$/.Sqrt[a^2] → a

$\sqrt{a^2\,(a^2+b^2)}$

$\sqrt{a^2\,(a^2+b^2)}$/.$\sqrt{a^2\,(a^2+b^2)}$ → Sqrt[a^2] * Sqrt[a^2 + b^2]

$\sqrt{a^2+b^2}a$

$\frac{\sqrt{a^2(a^2+b^2)}}{(a^2+b^2)\sqrt{a^2+b^2}}$/.$\sqrt{a^2\,(a^2+b^2)}$ → Sqrt[a^2] * Sqrt[a^2 + b^2]

$\frac{a}{a^2+b^2}$

Example 2.5 (Particle in a uniform magnetic field). Let us consider the motion of a charged particle in a constant, uniform magnetic field. Suppose that the particle has the charge e and the uniform magnetic field $\mathbf{H}$ is aligned with the z axis. In Newtonian-Galilean mechanics[1] the equations of motion are

$$m\,\frac{d\mathbf{v}}{dt} = \frac{e}{c}\,\mathbf{v}\times\mathbf{H}, \tag{2.22}$$

where $\mathbf{v}$ is the particle velocity, c is the speed of light, and $\mathbf{H} := H\mathbf{k}$ is the magnetic field. The equations of motion may be expressed component-wise as

$$\frac{d\,v_x}{dt} = \omega\,v_y,\ \frac{d\,v_y}{dt} = -\omega\,v_x,\ \frac{d\,v_z}{dt} = 0, \tag{2.23}$$

where

$$\omega := \frac{e\,H}{m\,c}.$$

We are able to solve the system of equations (2.23) by using a nice trick. We multiply the second of these equations by $i := \sqrt{-1}$ and add to the first to obtain

$$\frac{d\,(v_x + i\,v_y)}{dt} = -i\,\omega(v_x + i\,v_y). \tag{2.24}$$

It is easily checked that

$$v_x + i\,v_y = a\,e^{-i\,\omega\,t}, \tag{2.25}$$

where a is a complex constant, satisfies (2.24). Indeed, just substitute it into this equation. [2] [3] We still have to determine the complex constant a. To do this we note that the initial velocity at time $t = 0$, $\mathbf{v}(0) =< v_x(0), v_y(0), v_z(0)$ and the 2-vector

[1]Galileo Galilei (1564-1642) wrote the great book *Dialogue on the Ptolemaic and Copernican Systems*, which pointed out Aristotle's error concerning the comparative speeds at which heavy and light objects fall, and his claim that the sun was the center of the world and immovable. For this he was hauled before the Pope and threatened with imprisonment and torture. Galileo was at this time sixty-seven and in poor health; hence, he recanted. The name galilean transformation refers to the the coordinate transformations between moving systems that do not take into account relativistic effects. If the velocities involved are much smaller than the speed of light, then Newtonian-Galilean mechanics are a good approximation.

[2]there is a one-to-one correspondence between the complex numbers and the vectors in a plane. In $\mathbb{R}^2$ it is designated by $\langle x, y\rangle \Longleftrightarrow x + iy$.

[3]The function given in (2.25) is referred to as a solution of the equation (2.24).

$< v_x(0), v_y(0) >$, identified with $v_x(0) + iv_y(0)$, may be written in complex form as $v_0\, e^{-i\alpha}$. Now at $t = 0$ the function (2.25) must take the form

$$(v_x + i\, v_y)(0) \;=\; a \;=\; v_0\, e^{-i\,\alpha};$$

so $a = v_0 e^{-i\alpha}$, and (2.25) becomes

$$(v_x + i\, v_y) \;=\; v_0\, e^{-i\,(\omega t+\alpha)}.$$

The real and imaginary parts of this equation are

$$\frac{d\,x}{dt} \;:=\; v_x \;=\; v_0\, \cos(\omega t + \alpha), \;\; \frac{d\,y}{dt} \;:=\; v_y \;=\; v_0\, \sin(\omega t + \alpha). \qquad (2.26)$$

Note that v_0 is the initial speed of the particle, and that furthermore,

$$v_0 = \sqrt{\left(\frac{d\,x}{dt}\right)^2 + \left(\frac{d\,y}{dt}\right)^2},$$

hence, the speed remains constant. We may integrate (2.26) once more to obtain

$$x = x_0\, + \beta\, \sin(\omega\, t + \alpha),\; y = y_0\, + \beta\, \cos(\omega\, t + \alpha), \qquad (2.27)$$

where $\beta = \dfrac{v_0}{\omega} = \dfrac{v_0\, m\, c}{e\, H}$. The third equation of (2.23) integrates to

$$z \;=\; z_0\; +\; v_z(0)\, t. \qquad (2.28)$$

We have seen from figure 2.2 that a curve represented by (2.27)–(2.28) is a **helix** .

Example 2.6 (Uniform electric and magnetic fields)**.** As before we limit our discussion to the case where $v << c$, so that we may take the momentum of the particle to be $\mathbf{p} = m\,\mathbf{v}$. The magnetic field is directed along the z axis, i.e. $\mathbf{H} = H\,\mathbf{k}$, and the electric field lies in the y–z plane i.e. $\mathbf{E} = E_y\,\mathbf{j} + E_z\,\mathbf{k}$. The equations of motion of a particle with charge e is then given by

$$m\,\frac{d}{dt}\,\mathbf{v} = e\,\mathbf{E} + \frac{e}{c}\,\mathbf{v}\, \times\, \mathbf{H}. \qquad (2.29)$$

Writing this equation in component form yields the following equations for the position vector $\mathbf{r}(t) = x(t)\,\mathbf{i} + y(t)\,\mathbf{j} + z(t)\,\mathbf{k}$

$$m\,\frac{d^2x}{dt^2} \;=\; \frac{e}{c}\,\frac{d\,y}{dt}H, \quad m\,\frac{d^2y}{dt^2} \;=\; e\,E_y - \frac{e}{c}\,\frac{d\,x}{dt}H, \quad m\,\frac{d^2z}{dt^2} \;=\; e\,E_z. \qquad (2.30)$$

The third equation of (2.30) may be integrated right away to give

$$z \;=\; \frac{e\,E_z}{2\,m}\,t^2\; +\; v_{z0}\,t\; +\; z_0, \qquad (2.31)$$

where v_{z0} is the z component of the initial velocity. Now with the first two equations we can use the same trick as we did in the previous example, namely we multiply the second equation of (2.30) by i and add it to the first equation to obtain

$$\frac{d(\dot{x} + i\,\dot{y})}{dt}\; + i\omega(\dot{x} + i\,\dot{y}) \;=\; i\frac{e}{m}\,E_y, \qquad (2.32)$$

where $\omega = \dfrac{e\,H}{m\,c}$, and where $\dot{x} := \dfrac{dx}{dt}$, and $\dot{y} := \dfrac{dy}{dt}$. It is easily seen by substitution that

$$\dot{x} + i\dot{y} \;=\; a\,e^{-i\omega\,t} \;+\; \frac{c\,E_y}{H}, \tag{2.33}$$

satisfies equation (2.32). The real and imaginary parts of (2.33) provides expressions for $\dot{x}$, and $\dot{y}$, namely

$$\dot{x} \;=\; a\,\cos\omega t \;+\; \frac{c\,E_y}{H}, \qquad \dot{y} \;=\; -a\,\sin\omega t. \tag{2.34}$$

The use of the Newtonian-Galilian formulation of these equations assumes that the velocity of the particle is much smaller than that of light. From the first of equations (2.34) it is clear for this to be the case we must have the electric field much smaller than the magnetic field, in particular we must have

$$\frac{E_y}{H} << 1.$$

We may now integrate the equations (2.34) to obtain the trajectory the charged particle moves on

$$x \;=\; \frac{a}{\omega}\,\sin\omega\,t \;+\; \frac{c\,E_y}{H\,t\;+\;x_0}, \;\; y \;=\; \frac{a}{\omega}\,(\cos(\omega t)) \;+\; y_0. \tag{2.35}$$

The graph of trajectory (2.35) is rather interesting. See Exercise 2.14 where you are asked to use MATHEMATICA to plot these curves for various values of the parameters appearing in the solution.

EXERCISES

Exercise 2.7. *Write a* MATHEMATICA *function which calculates the* **arc length** *between two points* P_0 *and* P_1 *on a parametrically described curve*

$$x \;:=\; x(t)\;, \qquad y \;:=\; y(t)\;, \qquad z \;:=\; z(t)\;.$$

Assume that the functions used in the parametric representation have continuous derivatives. Use this function to compute the arc lengths of the curves described below:

(a) $x = \sin t,\; y = \cos t,\; z = t\;,$ *from* $t = 0$ *to* $t = \pi$

(b) $x = a\,\sin t,\; y = b\,\cos t,\; z = c\,t\;,$ *from* $t = 0$ *to* $t = \dfrac{\pi}{2}$

(c) $x = e^{-t}\,\cos t,\; y = e^{-t}\,\sin t,\; z = t\;,$ *from* $t = -\dfrac{\pi}{2}$ *to* $t = \pi$

(d) $x = e^{t}\,\sin t,\; y = e^{t}\,\cos t,\; z = t\;,$ *from* $t = -\pi$ *to* $t = \pi$

(e) $x = \dfrac{1}{3}\,t^3,\; y = -\dfrac{1}{t}, z = t\,\sqrt{2}\;,$ *from* $t = 0$ *to* $t = 4$

Exercise 2.8. *Derive the following formula for the curvature of a space curve*

$$\kappa \;=\; \frac{\mathbf{v}\times\mathbf{a}}{v^3} \;=\; \frac{\dot{\mathbf{r}}\times\ddot{\mathbf{r}}}{\|\,\dot{\mathbf{r}}\|^3}\;.$$

Exercise 2.9. *Write a* MATHEMATICA *function* `OSCUL_PLANE`(α,β,γ) *which produces the osculating plane using the parametric representation of a curve in terms of* $x = \alpha(t)$, $y = \beta(t)$, *and* $z = \gamma(t)$.

Exercise 2.10. *Use the* MATHEMATICA *function from the previous exercise to compute the osculating plane of the curves*

(a) $y = x^2$ *with* $z^2 = 1 - y$

(b) $x = \ln t\ ,\ y = \sin t\ ,\ z = \dfrac{t}{(1+t^2)^{\frac{1}{3}}}$

(c) $x = e^{-t}\cos t\ ,\ y = e^{-t}\sin t\ ,\ z = t$

(d) $x = e^{t}\sin t\ ,\ y = e^{t}\cos t\ ,\ z = t$

(e) $x = \dfrac{1}{3}t^3\ ,\ y = -\dfrac{1}{t}\ ,\ z = t\sqrt{2}$

Exercise 2.11. *Prove the formula*

$$\frac{d\mathbf{N}}{dt} = -\kappa(t)\,\frac{d\,s(t)}{dt}\,\mathbf{T} + \tau(t)\,\frac{d\,s(t)}{dt}\,\mathbf{B}\,. \tag{2.36}$$

Simplify this equation for the case where the parameter is the arclength. The resulting equation is <u>*one*</u> *of the* <u>*three*</u> **Frenet equations**. *The other two are (2.9) and (2.21).* <u>*Hint*</u>*: Differentiate* $\mathbf{N} = \mathbf{B} \times \mathbf{T}$.

Exercise 2.12. *Compute* **B**, **N**, *and* **T** *for the helix*

$$\mathbf{r}(t) = (a\,\cos t)\,\mathbf{i} + (a\,\sin t)\,\mathbf{j} + (b\,t)\,\mathbf{k}\,.$$

Exercise 2.13. *For the case of a charged particle moving in a constant electric and magnetic field prove that if* $E := \|\mathbf{E}\|$ *that for the Galilean treatment to be correct, we must have the condition*

$$\frac{e\,E}{m\,c} << 1$$

holding.

Exercise 2.14. *Use* MATHEMATICA *to plot the trajectories of a particle moving in a constant electric and a constant magnetic field which were given in (2.35) by*

$$\begin{aligned} x &= \frac{a}{\omega}\,\sin\omega t + \frac{c\,E_y}{H}\,, \\ y &= \frac{a}{\omega}\,(\cos\omega t - 1)\,, \\ \text{and}\quad z &= \frac{e\,E_z}{2m}t^2 + v_{z0}t\,. \end{aligned}$$

Show that the shape of the trajectory varies significantly depending on whether the parameter a *is smaller or larger than* $|\frac{cE_y}{H}|$.

2.2 Quadric surfaces

After the plane the **quadric surfaces** are the most simple surfaces. As an example of a quadric surface we consider the **sphere** which geometrically speaking is the locus of all points equidistant from a fixed point. If we take the fixed point to be $Q := (x_0, y_0, z_0)$ and the distance to be a, then this geometric property may be analytically expressed as

$$(x - x_0)^2 + (y - y_0)^2 + (z - z_0)^2 = a^2 .$$

The surface which is the locus of all points equidistant from the z-axis is a **cylinder** $\mathcal{Z}$ whose axis coincides with the z-axis. Again if the distance is a we may describe this geometric property by $x^2 + y^2 = a^2$. Notice in this description that z does not enter at all! This means that the equation is true for **all** z-values. We may say that the circle $x^2 + y^2 = a^2$ **generates** the cylinder $\mathcal{Z}$, and that the lines parallel to the z-axis passing through points (x_0, y_0) lying on the circle are the **generators** of $\mathcal{Z}$.
Instead of using the equation of the circle in the xy-plane to generate surfaces we could use any closed curve in this plane. We refer to such surfaces as **generalized cylinders**. For example the ellipse $\frac{x^2}{a^2} + \frac{y^2}{b^2} = 1$ generates an elliptical cylinder.

The most general quadric surfaces in space, however, are the graphs of the the second degree equations

$$A\,x^2 + B\,y^2 + C\,z^2 + D\,x\,y + E\,x\,z + F\,y\,z + G\,x + H\,y + I\,z + J = 0. \quad (2.37)$$

In general, the best way to visualize the graph of such a quadratic is to consider the intersection of the surface with planes parallel to the coordinate planes. An alternate possibility is to use the plotting capabilities of MATHEMATICA. Before doing this we turn our attention first to certain special quadric surfaces. These are the

ellipsoid	$\frac{x^2}{a^2} + \frac{y^2}{b^2} + \frac{z^2}{c^2} = 1$	(2.38a)
hyperboloid of one sheet	$\frac{x^2}{a^2} + \frac{y^2}{b^2} - \frac{z^2}{c^2} = 1$	(2.38b)
hyperboloid of two sheets	$\frac{x^2}{a^2} + \frac{y^2}{b^2} + \frac{z^2}{c^2} = -1$	(2.38c)
elliptic cone	$z^2 = \frac{x^2}{a^2} + \frac{y^2}{b^2}$	(2.38d)
elliptic paraboloid	$z = \frac{x^2}{a^2} + \frac{y^2}{b^2}$	(2.38e)
hyperbolic paraboloid	$z = \frac{x^2}{a^2} - \frac{y^2}{b^2}$	(2.38f)

[4] In the case of the ellipsoid, intersection with the planes $x = x_0$ are ellipses, as are those intersections with the planes $y = y_0$, and the planes $z = z_0$. Indeed, the **trace** of the plane $z = z_0$, $(-c < z_0 < c)$, with the ellipsoid is the ellipse

$$\frac{x^2}{a^2} + \frac{y^2}{b^2} = 1 - \frac{z_0^2}{c^2}.$$

Knowing these intersections allows us to get a fairly good idea of what the ellipsoid looks like.

[4]Examples of quadic surfaces may be seen in the graphics 2.8, 2.9, 2.10,2.11,2.12,2.7.

Mathematica session 2.4. *In this session we see how to use* MATHEMATICA *to generate three-dimensional plots of surfaces. In particular we consider the quadric surfaces. The first object is an* **ellipsoid**

$$x^2 + \frac{y^2}{4} + \frac{z^2}{9} = 1$$

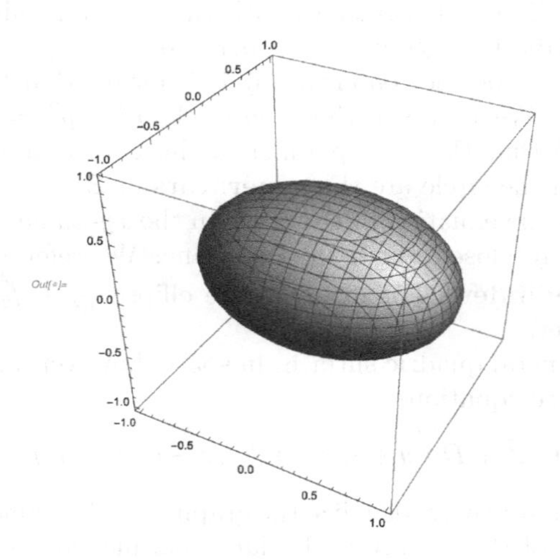

FIGURE 2.7: Oblate Spheroidal

```
Plot3D[x^2 + y^2, {x, -2, 2}, {y, -2, 2}]

Plot3D[x^2 - y^2, {x, -3, 3}, {y, -2, 2}, PlotStyle -> FaceForm[Yellow, Blue],
AxesLabel -> {x, y, z}]

ContourPlot3D[z^2 == x^2 + y^2, {x, -4, 4}, {y, -4, 4}, {z, -4, 4}]
ContourPlot3D[x^2 + y^2 - z^2 == 1, {x, -2, 2}, {y, -2, 2}, {z, -3, 3}]

ContourPlot3D[x^2 - y^2 - z^2 == 1, {x, -2, 2}, {y, -2, 2}, {z, -3, 3}]

ContourPlot3D[x^2 + 2 * y^2 + 3 * z^2==1, {x, -1, 1}, {y, -1, 1}, {z, -1, 1}]
```

For the hyperboloid of one sheet the intersection with the plane $z = z_0$ is an ellipse; see Figure 2.11. However, the intersection with either $x = x_0$ or $y = y_0$ is a hyperbola.

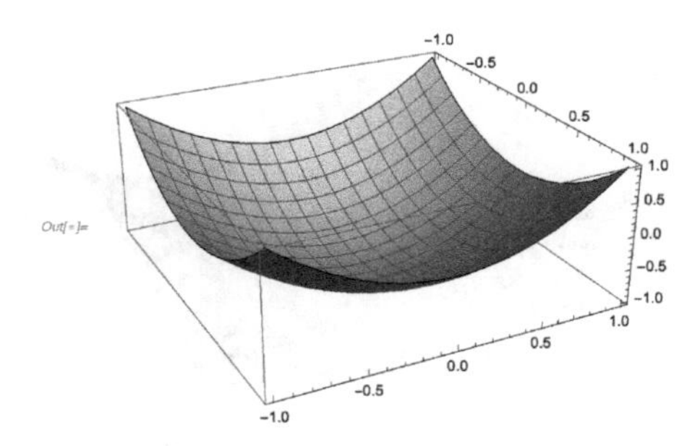

FIGURE 2.8: Elliptic-paraboloid $z = x^2 + y^2$

The two-sheeted hyperboloid consists of two separate surfaces, see Figure 2.12. Its intersection with a plane $z = z_0$ or a plane $x = x_0$ are hyperbolas, and its intersection with $y = y_0$ are ellipses or the empty set.

The elliptic cone (2.38d) has the property that its intersections with the planes $Ax + By = 0$ are cones as can be seen from the plot. To prove this analytically we eliminate one of the variables x or y from the equation $A\,x + B\,y = 0$, and substitute in (2.38d). Since at least one of the constants A or B must be nonzero, let us assume it is B. Substituting $y = -\frac{Ax}{B}$ in (2.38d) we get

$$z^2 = \frac{x^2}{a^2} + \frac{(Ax)^2}{(Bb)^2} = x^2 \left(\frac{1}{a^2} + \frac{A^2}{(bB)^2} \right) , \quad \text{which is a cone in the } xz\text{–plane} .$$

If we consider intersections of the elliptic cone 2.10 with the planes $Ax + By + C = 0$, where $C \neq 0$, then the intersections are hyperbolae. See Exercise 2.18(c).

The elliptic paraboloid has ellipses as the intersections with the planes $z = z_0 > 0$ as shown in Figure 2.8.

The intersection with $y = y_0$ is the parabola

$$z = \frac{x^2}{a^2} + \frac{y_0^2}{b^2} .$$

It is not difficult to show that the intersection with the plane $Ax+By+C = 0$

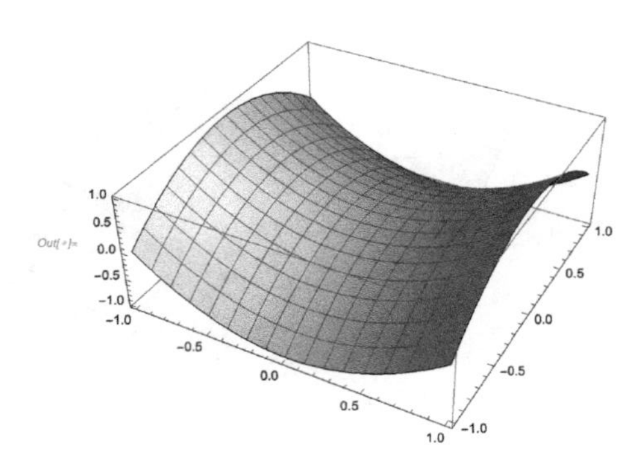

FIGURE 2.9: Hyperbolic paraboloid $z = x^2 - y^2$

is also a parabola. See Exercise 2.18(d).

Exercises

Exercise 2.15. *Plot the following quadric surfaces using* MATHEMATICA

(a) $\frac{x^2}{4}+\frac{y^2}{1}+\frac{z^2}{9}=1$ (b) $\frac{x^2}{16}-\frac{y^2}{1}+\frac{z^2}{9}=1$ (c) $\frac{x^2}{1}+\frac{y^2}{1}-\frac{z^2}{9}=1$

(d) $\frac{x^2}{9}-\frac{y^2}{1}-\frac{z^2}{4}=1$ (e) $-\frac{x^2}{4}-\frac{y^2}{9}+z^2=0$ (d) $\frac{x^2}{9}+\frac{y^2}{4}=z$

Exercise 2.16. *Plot the following quadric surfaces using* MATHEMATICA

(a) $x^2+y^2-4z=0$ (b) $16x+y^2-4z^2=0$ (c) $9x^2+4y^2+z^2=36$

(d) $2x^2+y+4z^2+1=0$ (e) $x=\frac{y^2}{4}+\frac{z^2}{3}$

Exercise 2.17. *In the following problems plot the curves with* MATHEMATICA.

(a) $(x-2)^2+(y+4)^2-4z=4$ (b) $16(y+1)+x^2-4(z+5)^2=0$

(c) $3x^2+4(y+\pi)^2+(z-\pi)^2=36$ (d) $2x^2+y+4z^2+1=0$

(e) $x=\frac{(y+7)^2}{2}+\frac{z^2}{\pi}$

Exercise 2.18. *Consider the elliptic cone*

$$z^2 = \frac{x^2}{a^2}+\frac{y^2}{b^2}.$$

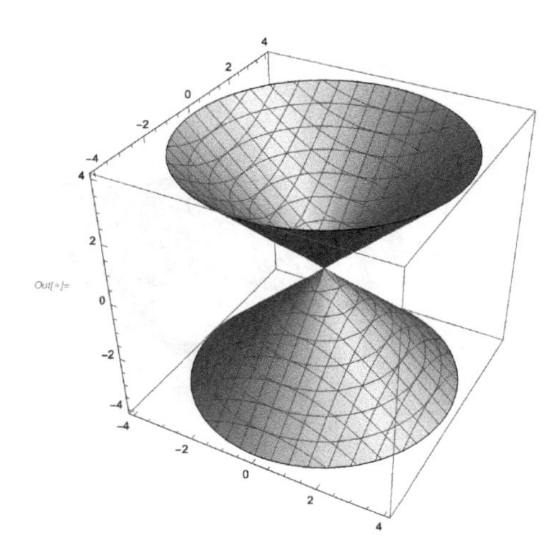

FIGURE 2.10: Cone $z^2 = x^2 + y^2$

Prove:

(a) *the intersection with a plane, intersecting the z–axis in a single point, is an ellipse.*

(b) *the intersection with a plane, containing the z–axis, is a cone.*

(c) *the intersection with a plane parallel to either the xz-plane or the yz-plane is a hyperbola.*

(d) *the intersection with a plane parallel to the z–axis is, in general, a hyperbola.*

Exercise 2.19. *Show that the intersection of the elliptic paraboloid*

$$z = \frac{x^2}{a^2} + \frac{y^2}{b^2},$$

with the plane $Ax + By + C = 0$ is, in general a parabola.

Exercise 2.20. *Find for what values of the parameters α, β, and γ the spheres $x^2 + y^2 + z^2 = 1$ and $(x-\alpha)^2 + (y-\beta)^2 + (z-\gamma)^2 = 1$ meet orthogonally.*

Exercise 2.21. *Determine the points where the curve given parametrically by $x = t^2$, $y = t^3$, and $z = t^4$ meets the surface $z = 2x^2 + y$.*

2.3 Cylindrical and spherical coordinates

We recall that, in the plane, the **polar coordinates** are sometimes more convenient than the Cartesian coordinates. In the space $\mathbb{R}^3$ it is also true

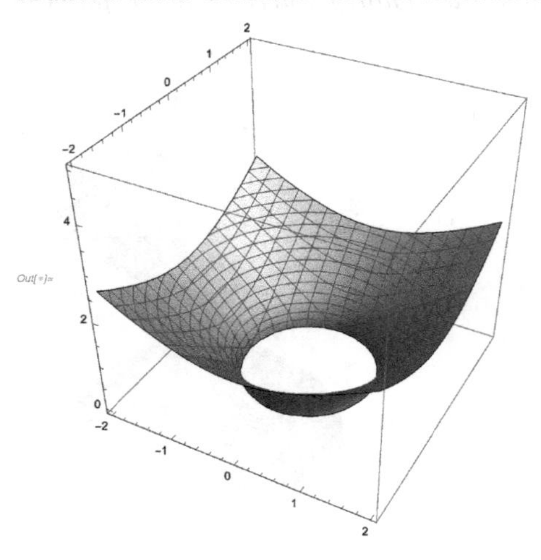

FIGURE 2.11: Hyperboloid of One Sheet

that, for some problems, an alternative coordinate system is preferred to the Cartesian system.

The analog in $\mathbb{R}^3$ to polar coordinates is the **cylindrical coordinate system**. Here we replace the x and y coordinates by the polar coordinates r and θ, and keep the z–coordinate. This gives the coordinate system

$$r := \sqrt{x^2 + y^2} \, , \quad \theta := \text{atan} \, \frac{y}{x} \, , \quad z := z \, , \tag{2.39}$$

see Figure 2.13. Conversely, the Cartesian coordinates are obtained from (r, θ, z) by

$$x := r \cos\theta \, , \quad y := r \sin\theta \, , \quad z := z \, . \tag{2.40}$$

The name **cylindrical coordinates** is suggested by the fact that the graph of $r := a$ is the (circular) **cylinder** of radius a, whose axis lies along the z–axis. Indeed, any object whose equation is $F(r, \theta) = 0$, is a (perhaps non–circular) cylinder, parallel to the z–axis.

There are other geometric forms which have a simple representation in cylindrical coordinates, in particular, the **surfaces of revolution**. For example, consider the parabola $z = x^2$ in the xz–plane. Revolving this curve about the z–axis, generates a **paraboloid of revolution**, see the left part of Figure **??**. To obtain its equation we replace, in the equation of the parabola, x by $r = \sqrt{x^2 + y^2}$, to get the paraboloid $z = x^2 + y^2$.

As another example, consider the ellipse $\frac{x^2}{a^2} + \frac{z^2}{b^2} = 1$. Revolving it about the z–axis gives an **ellipsoid of revolution**, see e.g. Figure 2.15, or a **sphere**

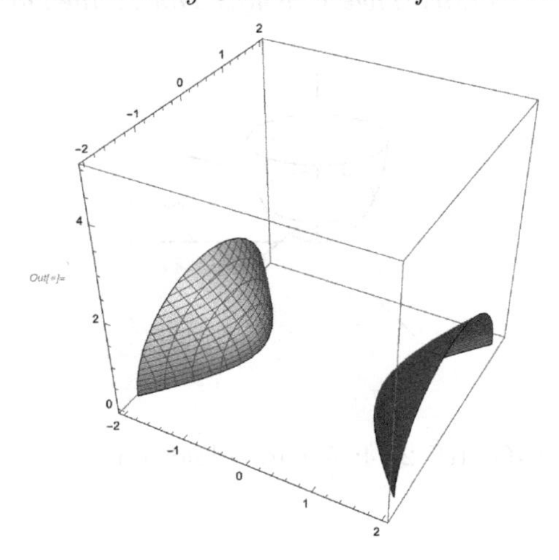

FIGURE 2.12: Hyperboloid of Two Sheets

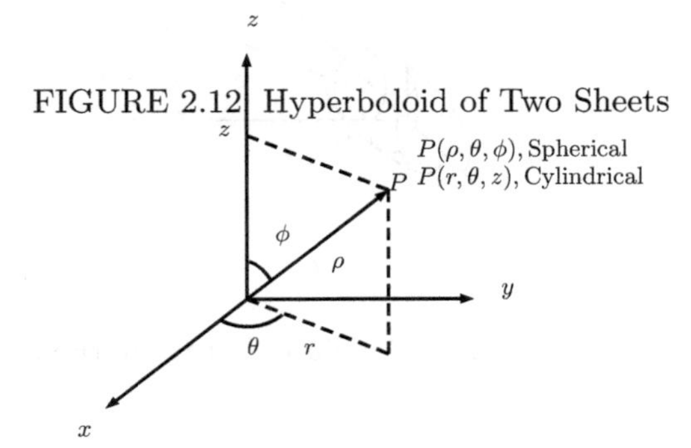

FIGURE 2.13: Cylindrical and spherical coordinates

if $a = b$,

$$\frac{r^2}{a^2} + \frac{z^2}{b^2} = 1 .$$

When $b = a$ this object becomes the **sphere** of radius a.

Another generalization of polar coordinates system is the **spherical coordinates system** (ρ, φ, θ), where ρ is the distance of the point from the origin, φ is the angle measured down from the z–axis, and θ is the same as in the cylindrical coordinates,

$$\rho := \sqrt{x^2 + y^2 + z^2}\,, \quad \varphi := \operatorname{atan}\left(\frac{\sqrt{x^2 + y^2}}{z}\right), \quad \theta := \operatorname{atan}\frac{y}{x}\,, \tag{2.41}$$

see e.g. Figure **??**(b). Conversely, we can obtain the Cartesian coordinates

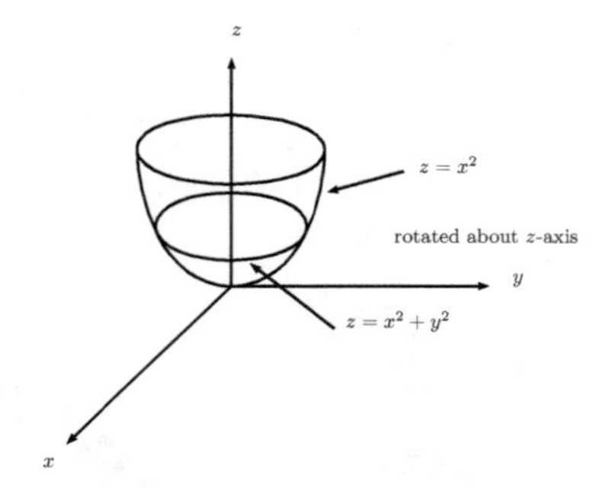

FIGURE 2.14: A paraboloid of revolution.

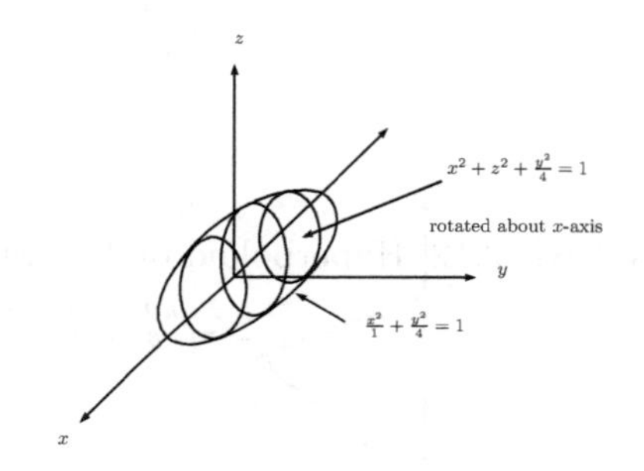

FIGURE 2.15: A ellipsoid of revolution.

from (ρ, φ, θ) by

$$x := \rho \sin\varphi \cos\theta\ , \quad y := \rho \sin\varphi \sin\theta\ , \quad z := \rho \cos\varphi\ . \tag{2.42}$$

Exercises

Exercise 2.22. *Write* MATHEMATICA *functions which convert points in rectangular coordinates to both cylindrical and spherical coordinates. Apply these functions to the following rectangular points.*

(a) $(4, 5, 7)$ (b) $(3, 4, 5)$ (c) $(1, 1, 1)$
(d) $(0, 0, 1)$ (e) $(-4, 4, 9)$ (f) $(0, -1, 0)$

Exercise 2.23. *Write* MATHEMATICA *functions which convert the following points in cylindrical coordinates to both rectangular and spherical coordinates. Apply these functions to*

(a) $(1, \pi/4, 1)$ (b) $(3, \pi, 4)$ (c) $(5, \pi/3, 12)$
(d) $(100, 5\pi/3, 100)$ (e) $(6, -\pi/2, 8)$ (f) $(3, 2\pi/3, 4)$

Exercise 2.24. *Write* MATHEMATICA *functions which convert the following points in spherical coordinates to both rectangular and cylindrical coordinates. Apply these*

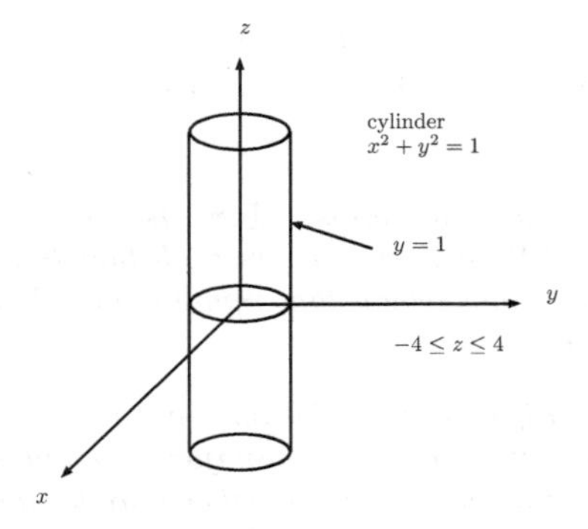

FIGURE 2.16: A cyllinder created by rotating a line about the z-axes

functions to

(a) $(1, \pi/4, 0)$ (b) $(3, \pi, \pi)$ (c) $(5, \pi/3, \pi/3)$
(d) $(100, 5\pi/3, \pi/4)$ (e) $(6, -\pi/2, \pi/2)$ (f) $(3, 2\pi/3, \pi/6)$

Exercise 2.25. *Use* MATHEMATICA *to obtain plots of the following surfaces, given in cylindrical coordinates.*

(a) $r^2 + z^2 - 4\,z = 0$ (b) $z = r^2$
(c) $\frac{r^2}{4} + \frac{z^2}{9} - 1 = 0$ (d) $r = \frac{1}{z}$

Exercise 2.26. *Convert the following descriptions of geometric objects from cartesian to cylindrical and spherical coordinates*

(a) $x^2 + y^2 + z^2 = 1$ (b) $x + y + z = 1$ (c) $x + y = 1$
(d) $x + z = 1$ (e) $x^2 + y^2 = 2x$ (f) $z = x^2 - y^2$
(g) $x^2 + y^2 + z^2 = 2\,x + 2\,y + 2\,z$ (h) $z = x^2 + y^2$

Exercise 2.27. *In the following a geometric object is described in cylindrical coordinates. Express the object in rectangular coordinates using hand calculations. Check your results with* MATHEMATICA *by plotting these functions both in rectangular and cylindrical coordinates.*

(a) $r = 5$ (b) $\theta = \pi/2$ (c) $\theta = \pi/6$
(d) $\varphi = 0$ (e) $\varphi = \pi/2$ (f) $z = r^2$
(g) $r^2 + z^2 = 100$ (h) $r = 2\sin\theta$ (i) $r^2/4 + z^2 = 1$
(j) $r^2 \sin 2\theta = z$

Exercise 2.28. *In the following a geometric object is described in spherical coordinates. Express the object in rectangular coordinates using hand calculations. Check your results with* MATHEMATICA *by plotting these functions both in rectangular and spherical coordinates.*

(a) $\rho = 1$ (b) $\theta = \pi/3$ (c) $\varphi = \pi/6$

(d) $\varphi = 0$ (e) $\rho = \sin\varphi$ (f) $\rho = 4\sin\varphi\cos\theta$

(g) $\rho = 4\sin\varphi\sin\theta$ (h) $\rho = 2\sin\theta$ (i) $\rho = \cos\varphi$

(j) $\rho^2 \sin 2\theta = z$

Exercise 2.29. *A sphere of radius* 1 [cm] *is centered at the origin. A hole of radius* 1/2[cm] *is drilled through the sphere such that the axis of the hole lies along the zraxis. Describe the solid region that remains in* **cylindrical coordinates** *and* **spherical coordinates**.

Exercise 2.30. *An ellipsoid of revolution about the z-axis has a hole drilled along the z-axis. Suppose that the z axis is the major axis and its half diameter is* 2 [cm] *the minor half axis is of length* 1 [cm]. *Furthermore let the hole diameter be* 1 [cm]. *Describe the solid that remains in cylindrical coordinates and spherical coordinates*

Exercise 2.31. *A right circular cone of radius* 5 [cm] *and height* 10 [cm] *is located with its vertex at the origin and its axis coincident with the negative x-axis. Suppose a hole is drilled along the z-axis that is* 2 [cm] *in diameter. Describe the solid that remains in cylindrical coordinates.*

Exercise 2.32. *Suppose that the ellipse*

$$\frac{x^2}{3} + \frac{y^2}{1} = 1$$

is first rotated about the x-axis to obtain an ellipsoid of revolution and then about the y-axis to obtain another another ellipsoid of revolution. Describe the region common to both objects in a suitable coordinate system

Chapter 3

Functions of Several Variables

3.1 Surfaces in space, functions of two variables

In Chapter 1 we studied planes, which are the simplest surfaces in $\mathbb{R}^3$. A plane is represented in terms of two parameters, see e.g. (1.31). Other surfaces in $\mathbb{R}^3$ can also be represented using two parameters, i.e. these surfaces are graphs $z = f(x, y)$,i.e. functions of two variables (x, y). Such functions, and surfaces, are studied in this section.

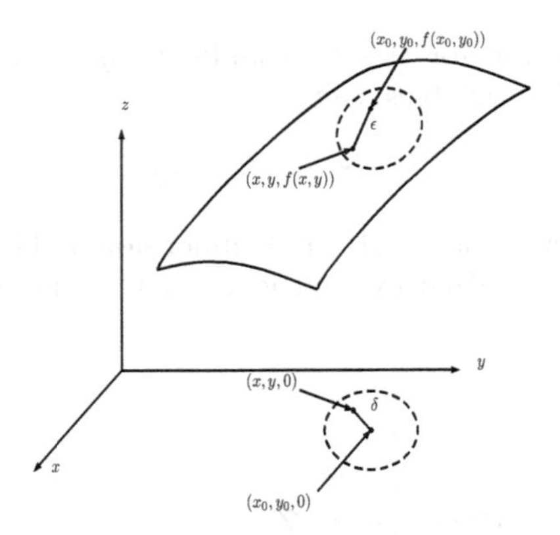

FIGURE 3.1: Illustration of a function $f(x, y)$, and its limit as $(x, y) \to (x_0, y_0)$.

Consider a function $f(x, y)$ of two variables, with domain $\mathbf{D}$ (in the xy–plane). The **graph** of f is the set of points

$$\text{GRAPH}\, f \ := \ \{(x, y, z) \, : \, z \ = \ f(x, y) \ , \ (x, y) \in \mathbf{D}\} \ ,$$

a surface in $\mathbb{R}^3$. A point $P(x_0, y_0)$ in the domain $\mathbf{D}$ thus corresponds to a point $Q(x_0, y_0, f(x_0, y_0))$ in the surface $\mathbf{S}$, see Figure 3.1.

Conversely, let $\mathbf{S}$ be a surface in $\mathbb{R}^3$ with the property that it is intersected once by any vertical line emanating from $\mathbf{D}$ (we say that $\mathbf{S}$ satisfies the **vertical line test**). Then it is clear that $\mathbf{S}$ is the graph of some function f.

Remark 3.1. The domain of $f(x, y)$ is a subset of the xy–plane, possibly the whole plane. Sometimes the domain is clear from the definition of f, for example,

$$f(x, y) \ := \ \sqrt{1 - x^2 - y^2} \quad \text{has} \quad \text{DOMAIN}\, f \ = \ \{(x, y) \ : \ x^2 + y^2 \ \leq \ 1\} \ ,$$

and

$$f(x, y) := \frac{1}{x^2 + y^2} \quad \text{has DOMAIN } f = \text{the whole } xy\text{–plane except for the origin } (0, 0) \ .$$

We can similarly describe the graph of a function $g(x, y, z)$ (of three variables)

$$\text{GRAPH}\, g \ := \ \{(x, y, z, w) \ : \ w \ = \ g(x, y, z)\} \ , \tag{3.1}$$

as a surface in 4–dimensional space. This description is of little value, because it is difficult to visualize a 4–dimensional picture. To make this easier, suppose the variable w is **time**, and consider snapshots of GRAPH g taken at times w_k , $k = 1, 2, \ldots$. Each such snapshot

$$w_k \ = \ g(x, y, z)$$

is an implicit equation for the 3 variables (x, y, z). If this equation can be solved for one of the variables, say

$$z \ = \ f_k(x, y)$$

then we get a manageable graph (in 3–dimensions). The 4–dimensional graph (3.1) can then be visualized by sufficiently many "snapshots". See figure 3.2.

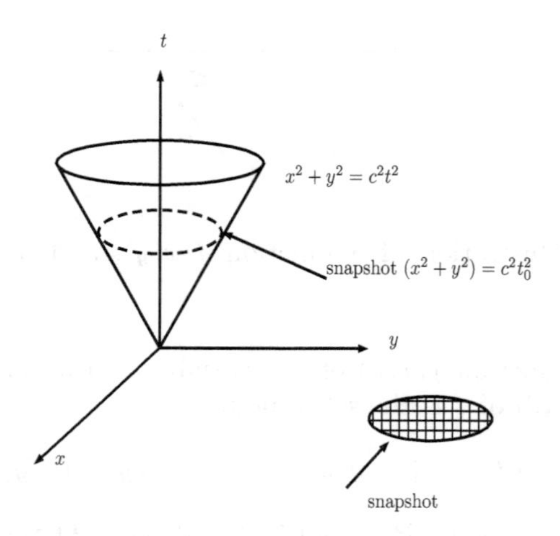

FIGURE 3.2: The light-cone and a snapshot.

Example 3.2. Relativity theory uses 4 variables, the three space variables (x, y, z) and t, the time variable. Any point (x, y, z, t) in this 4–dimensional space represents an **event**, such as receiving a light signal. The **light cone** in the $xyzt$–space is the surface given by

$$\mathbf{C} := \{(x, y, z, t) : x^2 + y^2 + z^2 - c^2 t^2 = 0\} \tag{3.2}$$

where c is the **speed of light**. Four dimensional pictures are difficult to draw, but we can still draw a reasonable 3–dimensional picture of $\mathbf{C}$ by compressing the 3 space variables (x, y, z) into a plane, see Figure 3.2.

Events (x, y, z, t) on $\mathbf{C}$ represent points (x, y, z) in space reached in time t by a light signal emanating from the origin at time 0 (event $(0, 0, 0, 0)$.) Events (x, y, z, t) inside the light cone, i.e. points (x, y, z, t) satisfying

$$x^2 + y^2 + z^2 < c^2 t^2 ,$$

are called **timelike**, because they are separated in time from $(0, 0, 0, 0)$. All observers[1] will agree on such a point that it occured before $(0, 0, 0, 0)$ (and is a **past** event), or after $(0, 0, 0, 0)$ (a **future** event). Events outside the light cone are called **spacelike**, as they are separated in space from $(0, 0, 0, 0)$.

An alternative way to draw $\mathbf{C}$ is to draw its snapshots at different times $t = t_1 , t_2 , t_3 , \ldots$, which are the surfaces

$$\{(x, y, z) : x^2 + y^2 + z^2 = c^2 t_i^2\} , \quad i = 1, 2, \ldots$$

see Figure 3.2. Some information is lost here, since $t^2 = (-t)^2$. △

From our course on single variable calculus, given a function $f(x)$ of a single variable, its behavior near a point c is given by the limit

$$\lim_{x \to c} f(x) , \quad \text{if it exists.}$$

Can we study similarly the behavior of a function $f(x, y)$, of two variables, near a point (a, b)? In particular, we need to define the limit

$$\lim_{(x,y) \to (a,b)} f(x, y) .$$

What is a reasonable definition of this limit?
A natural suggestion is to fix one of the variables, say x, and take the limit

$$\lim_{y \to b} f(x, y) , \quad \text{which if it exists, is a function of } x, \text{ say } \varphi(x) .$$

Then take the limit

$$\lim_{x \to a} \varphi(x) = \lim_{x \to a} \lim_{y \to b} f(x, y) ,$$

to get an idea of the values of f near (a, b). Alternatively, we can fix y first, compute the limit

$$\lim_{x \to a} f(x, y) , \quad \text{which if it exists, is a function of } y, \text{ say } \psi(y) ,$$

[1] Moving at a constant speed relative to the reference coordinate system.

and then the limit

$$\lim_{y\to a} \psi(x) = \lim_{y\to b} \lim_{x\to a} f(x,y) ,$$

to get another view of the behavior of f near (a,b). Two questions arise:
(a) Is it true in general that

$$\lim_{x\to a} \lim_{y\to b} f(x,y) = \lim_{y\to b} \lim_{x\to a} f(x,y) ? \tag{3.3}$$

(b) Even if (3.3) holds, is its common value a reasonable definition of the limit of f as $(x,y) \to (a,b)$?
We show now that the answer to both questions is "NO".

Mathematica session 3.1. *First we take the iterated limits in several cases where the answer seems to be it does not matter in which order we take the limits. Please note that in this case the function is a polynomial in both variables x and y.*

```
Limit[Limit[3 * x^2 + y^2 + x * y, x → 1], y → 3]
```

15

```
Limit[Limit[3 * x^2 + y^2 + x * y, y → 3], x → 1]
```

15

Next we try a rational function where we again use the iterated limits and get the same answer.

```
Limit[Limit[2 * x * y/(x^2 + y^2), x → 0], y → 0]
```

0

```
Limit[Limit[2 * x * y/(x^2 + y^2), y → 0], x->0]
```

0

However, when we change the approach to the origin by following the line $y = m * x$, it appears the limit depends on the value of the parameter m. Hence, we conclude that the limit can not exist as we do not become arbitrarily *close to the supposed limit value of* **zero**!

```
2 * x * y/(x^2 + y^2)/.y → m * x
```

$\frac{2mx^2}{x^2+m^2x^2}$

Cancelling x^2 *from numerator and denominator gives the result, which depends on* m.

```
Limit[%, x → 0]
```

$\frac{2m}{1+m^2}$

However, it may be seen that approaches using other curves give still different answers, for example suppose we approach along the parabola $y = mx^2$. *We obtain in this case*

```
f[x_, y_]:=(x^2 * y)/(x^4 + y^2)
f[x, mx^2]
```

$\frac{mx^2x^2}{mx^4+x^4}$

```
(x^2 * y)/(x^4 + y^2)/.y → m * x^2
```

$\frac{mx^4}{x^4+m^2x^4}$

```
Limit[%, x → 0]
```

$\frac{m}{1+m^2}$

Another way to determine whether a limit exists is to try **polar substitution**; *namely, we substitute*

$$x = r\cos(\theta), \quad y = \cos(\theta)$$

into the function we wish to take a limit about the origin and let $r \to 0$. *Note it is better to make the trigonometric substitution with* MATHEMATICA *before taking the limit.*

```
Limit[Limit[2 * x * y/(x^2 + y^2), x → r * Cos[Theta]], y → r * Sin[Theta]]
```

Sin[2 Theta]

```
Limit[Limit[(x^3 − y^2*x)/(x^2+y^2), x → r*Cos[Theta]], y → r*Sin[Theta]]
```

r Cos[Theta] Cos[2 Theta]

This last time the result was independent of the approach. Can you tell why?

```
r Cos[Theta] Cos[2 Theta]/.r → 0
```

We now introduce a MATHEMATICA *function that does the polar substitution and takes the limit as* $r \to 0$

```
myLinearLimit[F_, p_]:=Limit[Limit[F, y → p[[1]] + m*(x − p[[2]])], x->p[[1]]]
```

```
myLinearLimit[(x^2 − y^2)/(x^2 + y^2), {0, 0}]
```

$\frac{1-m^2}{1+m^2}$

We now give a MATHEMATICA *function which uses a parabolic approach to an arbitrary point* (p_1, p_2).

```
myQuadraticLimit[F_, p_]:=Limit[Limit[F, y → p[[1]] + m * (x − p[[2]])^2]
, x->p[[1]]]
```

We emphasize that even if this parabolic approach provides a limit, this does not ensure a limit actually exists. Another approach

```
myQuadraticLimit[(x^2 * y)/(x^4 + y^2), {0, 0}]
```

$\frac{m}{1+m^2}$

♦

Example 3.3. Consider the function

$$f(x,y) := \begin{cases} \frac{x^2-y^2}{x^2+y^2} & \text{if } (x,y) \neq (0,0) \\ 0 & \text{if } (x,y) = 0 \end{cases} \tag{3.4}$$

Then

$$\begin{aligned} \lim_{x\to 0} \lim_{y\to 0} f(x,y) &= \lim_{x\to 0} \lim_{y\to 0} \frac{x^2-y^2}{x^2+y^2} = \lim_{x\to 0} \left(\frac{x^2}{x^2}\right) = 1 \\ \lim_{y\to 0} \lim_{x\to 0} f(x,y) &= \lim_{y\to 0} \lim_{x\to 0} \frac{x^2-y^2}{x^2+y^2} = \lim_{y\to 0} \left(\frac{-y^2}{y^2}\right) = -1 \end{aligned}$$

showing that (3.3) need not hold. Indeed, if we approach the origin along the line $y := m\,x$, the limit depends on the slope m,

$$\lim_{x\to 0} \frac{x^2 - m^2\,x^2}{x^2 + m^2\,x^2} = \frac{1-m^2}{1+m^2}\,. \tag{3.5}$$

Example 3.4. Consider the function

$$f(x,y) := \begin{cases} \frac{xy}{x^2+y^2} & \text{if } (x,y) \neq (0,0) \\ 0 & \text{if } (x,y) = 0 \end{cases} \tag{3.6}$$

Then

$$\begin{aligned} \lim_{x\to 0} \lim_{y\to 0} f(x,y) &= \lim_{x\to 0} \lim_{y\to 0} \frac{xy}{x^2+y^2} = \lim_{x\to 0} 0 = 0 \\ \lim_{y\to 0} \lim_{x\to 0} f(x,y) &= \lim_{y\to 0} \lim_{x\to 0} \frac{xy}{x^2+y^2} = \lim_{y\to 0} 0 = 0 \end{aligned}$$

so (3.3) holds. However, the common value 0 cannot be the limit of f as $(x,y) \to (0,0)$, since for any α,

$$f(\alpha,\alpha) = \frac{\alpha^2}{\alpha^2+\alpha^2} = \frac{1}{2}$$

showing that $f(x,y) = 1/2$ for points (x,y) arbitrarily close to $(0,0)$.

We begin by inputting the function $f(x,y) := \frac{2xy}{x^2+y^2}$, for which we wish to determine whether a limit exists at $(0,0)$. We first compute the iterated limit

$$\lim_{x\to 0} \lim_{y\to 0} \frac{2xy}{x^2+y^2}$$

Nevertheless, it is not difficult to show that a limit does not exist, because if we approach the origin along any straight lines other than $x = 0$ or $y = 0$ we obtain a different answer. To show this we substitute in for y with $y = mx$ and take the limit along this line of slope m.

The limit is seen to depend on the slope of the line. However, this approach only shows that a limit does not exist. There are functions where the approach along a straight line is always the same, but the approach along some other curves gives different answers. For example, consider the function

$$f(x,y) := \frac{x^2y}{x^4+y^2}$$

along the family of parabolas $y = mx^2$. There is another way to determine whether a limit exists and that is to try a **polar substitution**; namely we substitute

$$x = r\cos(\theta), \text{ and } y = r\sin(\theta)$$

into the expression and take the limit as $r \to 0$. This last time the result was independent of the approach. Can you tell why?
We try another example. Once more the limit is dependent on the approach.

This first procedure does the substitution of a line through the point at which we take the limit, and then takes the limit. Examples 3.3–3.4 show that limits of functions of two variables cannot be defined as "one–variable limits". The following definition does the job.

Definition 3.5. Let the function $f(x, y)$ be defined in a neighborhood of the point (a, b) (with the possible exception of the point (a, b) itself). Then L is the **limit** of $f(x, y)$ as $(x, y) \to (a, b)$, written

$$\lim_{(x,y)\to(a,b)} f(x,y) \;=\; L\,, \tag{3.7}$$

if for every $\varepsilon > 0$ there is a $\delta > 0$ such that

$$|f(x,y) - L| \;<\; \varepsilon\,, \tag{3.8a}$$

$$\text{for all } (x,y) \text{ such that } \quad 0 < |x-a| < \delta \quad , \quad 0 < |y-b| < \delta\,. \tag{3.8b}$$

Note: the value of f at (a, b) does not appear in this definition; in fact, f needs not be defined at (a, b).

Remark 3.6. As in the one variable limit, the inequalities (3.8a)–(3.8b) state that the values of f are arbitrarily close to L for points (x, y) sufficiently close to the point (a, b). This is illustrated in Figure 3.1(b).

Examples 3.3–3.4 show functions $f(x, y)$ which do not have a limit at $(0, 0)$. Indeed

$$f(x,y) \;:=\; \frac{x^2-y^2}{x^2+y^2} \text{ has values } \pm 1 \text{ arbitrarily close to } (0,0)\text{ , and}$$

$$f(x,y) \;:=\; \frac{xy}{x^2+y^2} \text{ has values } 0, \frac{1}{2} \text{ arbitrarily close to } (0,0)\;,$$

violating (3.8a) (for all $\varepsilon < 1/2$).

The following limit–laws are analogous to the one variable laws.

Lemma 3.7 (Limit laws)**.** If

$$\lim_{(x,y)\to(a,b)} f(x,y) \;=\; L\,, \quad \text{and} \quad \lim_{(x,y)\to(a,b)} g(x,y) \;=\; M\,,$$

then

(a) $\displaystyle\lim_{(x,y)\to(a,b)} (f(x,y) + g(x,y)) \;=\; L + M\,,$

(b) $$\lim_{(x,y)\to(a,b)} f(x,y)g(x,y) = LM \text{ , and}$$

(c) $$\lim_{(x,y)\to(a,b)} \frac{f(x,y)}{g(x,y)} = \frac{L}{M} \text{ , provided } M \neq 0 \text{ .}$$

□

Continuity is also defined similarly to the one-dimensional case.

Definition 3.8. A function $f(x,y)$ is **continuous** at the point (a,b) if:

(a) f is defined at (a,b),

(b) $\lim_{(x,y)\to(a,b)} f(x,y)$ exists, and

(c) $$\lim_{(x,y)\to(a,b)} f(x,y) = f(a,b) \text{ .}$$

Corollary 3.9. If the function $f(x,y)$ is continuous at (a,b) then

$$\lim_{(x,y)\to(a,b)} f(x,y) = \lim_{x\to a}\lim_{y\to b} f(x,y) = \lim_{y\to b}\lim_{x\to a} f(x,y) \text{ .} \qquad \square \tag{3.9}$$

The following lemma is analogous to the one variable case.

Lemma 3.10. Let f , g and h be functions of two variables, such that:

(a) $f(x,y)$ and $g(x,y)$ are both continuous at the point (a,b), and

(b) $h(u,v)$ is continuous at the point $(f(a,b),g(a,b))$.

Then the composite function $h(f(x,y),g(x,y))$ is continuous at (a,b),

$$\lim_{(x,y)\to(a,b)} h(f(x,y),g(x,y)) = h(f(a,b),g(a,b)) \text{ .} \qquad \square \tag{3.10}$$

Remark 3.11. Let $f(x,y)$ be as in Lemma 3.10, and let $h(u)$ be a function of one variable, continuous at the point $f(a,b)$. Then the composite function $h(f(x,y))$ is continuous at (a,b) and

$$\lim_{(x,y)\to(a,b)} h(f(x,y)) = h(f(a,b)) \text{ .} \tag{3.11}$$

This result is a special case of Lemma 3.10.

Example 3.12. Consider the limit

$$\lim_{(x,y)\to(0,0)} \ln\sqrt{1-x^2-y^2} \text{ .}$$

The function $\ln\sqrt{1-x^2-y^2}$ is a composite $h(f(x,y)$ where

$$f(x,y) := \sqrt{1-x^2-y^2} \text{ is continuous at } (x,y) = (0,0)$$

and

$$h(u) := \ln u \text{ is continuous at } u = \lim_{(x,y)\to(0,0)} \sqrt{1-x^2-y^2} = 1 \text{ .}$$

Therefore $\lim_{(x,y)\to(0,0)} \ln\sqrt{1-x^2-y^2} = \ln 1 = 0$.

Definition 3.13. Let

$$z := f(x, y) , \tag{3.12}$$

and let α be any real number. Then the set

$$\{(x, y) : f(x, y) = \alpha\} \tag{3.13}$$

is called the **level curve** (or **contour**) of f.
Given a point (x, y, z), we interpret the coordinate z as altitude. The α–level curve is thus the set of all points (x, y) with the same altitude $f(x, y) = \alpha$. In particular, the α–level curve may be empty if there are no (x, y) with $f(x, y) = \alpha$.

For example, the "snapshots" of Example 3.2 are level curves, for constant t, of the light cone (3.2).

Level curves offer a convenient way to visualize the graphs of functions $f(x, y)$ of two variables. They are used in topographical maps, see e.g. Figure 6.1, where the level curves are obtained by slicing the hill at certain altitudes and projecting the intersections to a horizontal plane.

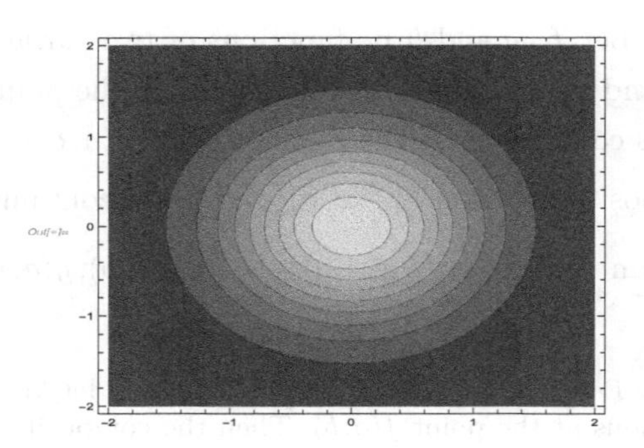

FIGURE 3.3: Level curves of a round hill.

Example 3.14. The paraboloid of revolution $z = x^2 + y^2$ has the circles

$$x^2 + y^2 = \alpha , \alpha \geq 0 ,$$

as its level curves. There are no level curves for $\alpha < 0$.

Mathematica session 3.2. *In this session we show how to use* **ContourPlot**.

ContourPlot[Cos[x * y], {x, −3, 3}, {y, −3, 3}]

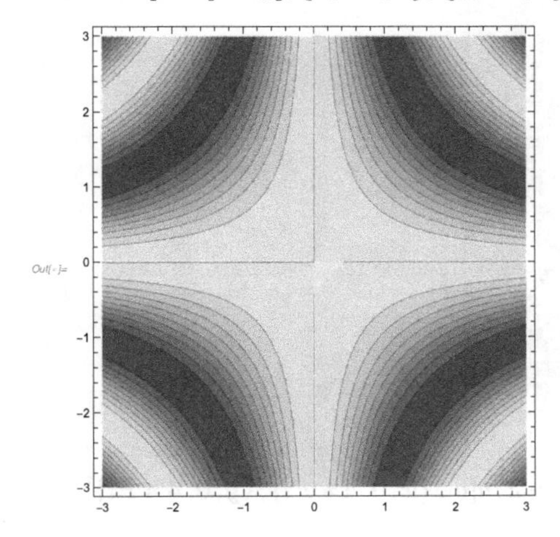

ContourPlot[Sin[x * y], {x, −3, 3}, {y, −3, 3}]

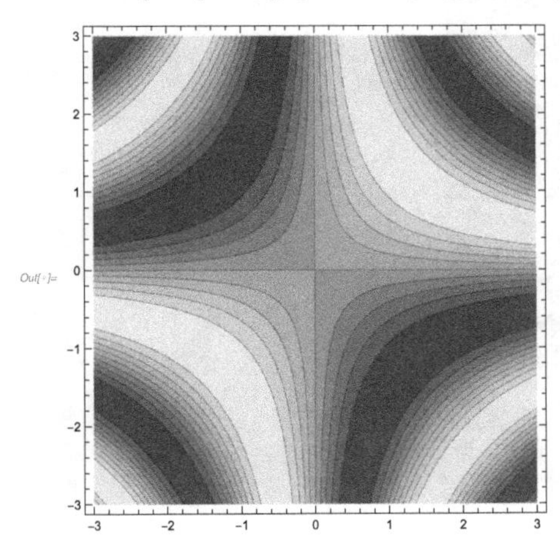

ContourPlot[Cos[x] + Cos[y], {x, 0, 4 * Pi}, {y, 0, 4 * Pi}]

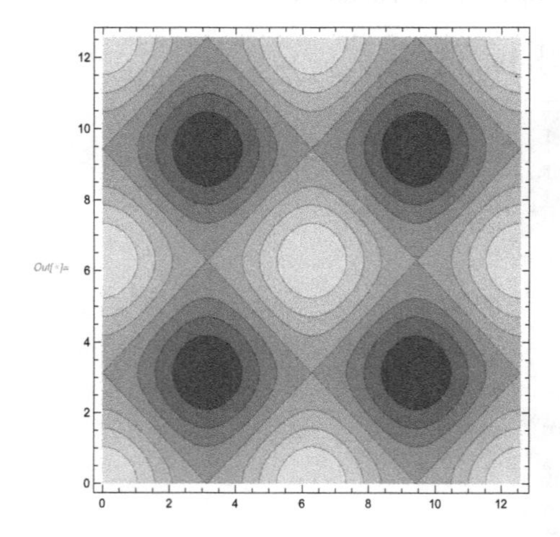

PICTURES/ContourPlot[Sin[x] + Sin[y], {x, 0, 4 * Pi}, {y, 0, 4 * Pi}]

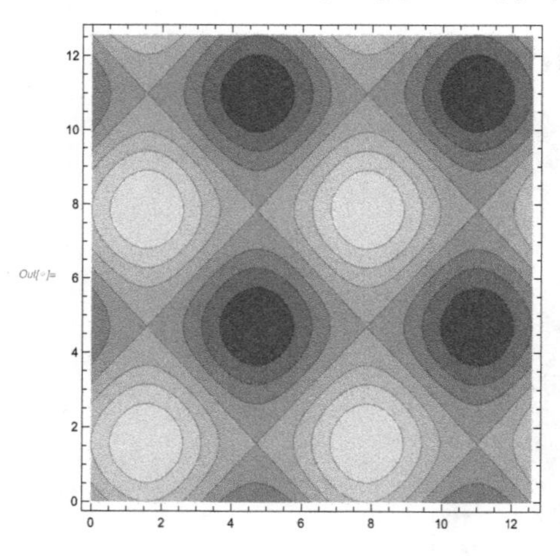

ContourPlot[x * y/(x^2 + y^2), {x, −3, 3}, {y, −3, 3}]

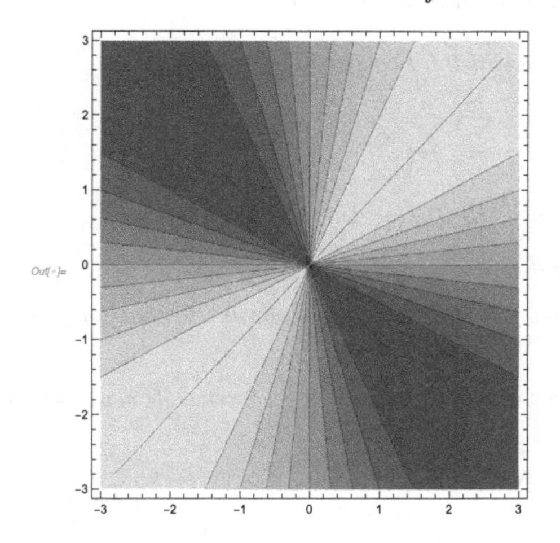

♦

Exercise 3.15. *Let* $f(x,y) := \sqrt{x^2-y^2}$. *Find the following values:*

(a) $f(\sqrt{3},1)$ (b) $f(5a,3a)$ (c) $f(a+b,a-b)$

(d) $f(1,1)$ (e) $f(c,c)$ (f) $f(u^2+v^2,2uv)$

Exercise 3.16. *Let* $f(x) := \frac{e^x-e^{-x}}{e^x+e^{-x}}$. *Evaluate and simplify*

(a) $f(x+y)$ (b) $f(x-y)$ (c) $f(x/y)$

Exercise 3.17. *If* $F(u,v) := u^3v+uv^3$, *find* $F(u(x,y),v(x,y))$ *if* u *and* v *are given as below.*

(a) $u := \sqrt{xy}\ ,\quad v := \sqrt{x-y}$ (b) $u := x-y\ ,\quad v := x+y$

(c) $u := \sin x\ ,\quad v := \cos x$ (d) $u := \sec x\ ,\quad v := \tan x$

Exercise 3.18. *For the following functions sketch the domains of definition. Indicate at which boundary points of these domains the functions are defined.*

(a) $f(x,y) := \sqrt{1-x^2-y^2}$ (b) $f(x,y) := \dfrac{x^2+y^2}{1-x^2-y^2}$

(c) $f(x,y) := \dfrac{x+y}{x-y}$ (d) $f(x,y) := \dfrac{x-y}{x+y}$

(e) $f(x,y) := e^{\frac{1}{xy}}$ (f) $f(x,y) := e^{\frac{1}{x^2+y^2}}$

(g) $f(x,y) := \sqrt{\dfrac{x-y}{x+y}}$ (h) $f(x,y) := \sqrt{\dfrac{x+y}{x-y}}$

(i) $f(x,y) := \sin\left(\dfrac{x+y}{x-y}\right)$ (j) $f(x,y) := \tan\left(\dfrac{x+y}{x-y}\right)$

(k) $f(x,y) := \cot\left(\dfrac{x-y}{x+y}\right)$ (l) $f(x,y) := \ln\left(\sin(|xy|)\right)$

Exercise 3.19. *Prove Lemma 3.7.*

Exercise 3.20. *Show: A polynomial in the two variables x and y*

$$p(x,y) \ := \ \sum_{j=1}^{m} \sum_{k=1}^{n} a_{jk}\, x^j\, y^k$$

is continuous at all points $(x,y) \in \mathbb{R}^2$. Hint: Use Lemma 3.7(a)–(b).

Exercise 3.21. *Prove Corollary 3.9.*

Exercise 3.22. *Prove Lemma 3.10. Hint: The proof is similar to that for functions of one variable.*

Exercise 3.23. *Use* MATHEMATICA *to plot several level curves $f(x,y) = \alpha$ for the following functions*

(a) $x^2 + 4y^2$ (b) $\sqrt{x^2+y^2+1}$ (c) $x^2 - y^2$

(d) $|x - y|$ (e) $\sqrt{x^2+y^2-1}$ (f) $\sqrt{\dfrac{x-y}{x+y}}$

3.2 Partial derivatives

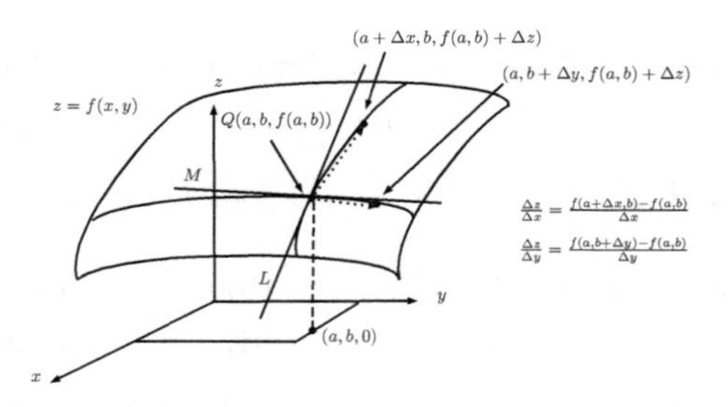

FIGURE 3.4: Partial derivative.

Consider a function

$$z \ := \ f(x,y)\ , \tag{3.14}$$

of two variables (x,y) , defined at a point $P(a,b)$ and in the neighborhood of P . The intersection of GRAPH f with the (vertical) plane $y = b$ gives a curve in that plane,

$$z \ = \ \varphi(x) \ := \ f(x,b)\ , \tag{3.15}$$

whose tangent at the point $Q(a,b,f(a,b))$ is denoted by L, see Figure 3.4. The slope of L is the derivative $\varphi'(a)$, which we call the **partial derivative** of f w.r.t. x at the point (a,b) , denoted by

$$\frac{\partial f}{\partial x}(a,b) \ := \ \lim_{\Delta x \to 0} \frac{f(a+\Delta x,b) - f(a,b)}{\Delta x}\ . \tag{3.16}$$

The RHS of (3.16) summarizes the above two steps:

(1) fix $y = b$ to get $f(x, b)$, and

(2) differentiate $f(x, b)$ at the point $x = a$.

Alternatively, we can fix $x = a$ to get a curve

$$z = \psi(y) := f(a, y) \,, \quad \text{in the plane } x = a \,, \tag{3.17}$$

with a tangent line M at the point Q, see Figure 3.4. The slope of M is the derivative $\psi'(b)$, called the **partial derivative** of f w.r.t. y at the point (a, b) , and denoted by

$$\frac{\partial f}{\partial y}(a, b) := \lim_{\Delta y \to 0} \frac{f(a, b + \Delta y) - f(a, b)}{\Delta y} \,. \tag{3.18}$$

The function f is called **(partially) differentiable w.r.t.** x [y] at the point (a, b) if the partial derivative $\frac{\partial f}{\partial x}(a, b)$ [$\frac{\partial f}{\partial y}(a, b)$] exists.

If we vary the point (a, b), the partial derivatives $\frac{\partial f}{\partial x}(a, b)$ and $\frac{\partial f}{\partial y}(a, b)$ give two new functions, defined next.

Definition 3.24. The **partial derivative** of f with respect to x is the function:

$$\frac{\partial f}{\partial x} := \lim_{\Delta x \to 0} \frac{f(x + \Delta x, y) - f(x, y)}{\Delta x} \tag{3.19}$$

and with respect to y is the function:

$$\frac{\partial f}{\partial y} := \lim_{\Delta y \to 0} \frac{f(x, y + \Delta y) - f(x, y)}{\Delta y} \,. \tag{3.20}$$

There are several notations for these partial derivatives:

$$\begin{aligned} \frac{\partial f}{\partial x} &= \frac{\partial}{\partial x} f = \frac{\partial z}{\partial x} = f_x(x, y) = D_x f(x, y) = D_1 f(x, y) \,, \text{ and} \\ \frac{\partial f}{\partial y} &= \frac{\partial}{\partial y} f = \frac{\partial z}{\partial y} = f_y(x, y) = D_y f(x, y) = D_2 f(x, y) \,, \end{aligned}$$

where D_x [D_y] denotes differentiation w.r.t. x [y], and D_1 [D_2] is derivative w.r.t. the first [second] argument. The partial derivatives of f at a point (a, b) are denoted by:

$$\begin{aligned} \frac{\partial f}{\partial x}(a, b) &= f_x(a, b) = D_x f(a, b) = D_1 f(a, b) \,, \\ \frac{\partial f}{\partial y}(a, b) &= f_y(a, b) = D_y f(a, b) = D_2 f(a, b) \,. \end{aligned}$$

Example 3.25. Let

$$f(x, y) := (x^2 + y^2) \sin(x - y) \,. \tag{3.21}$$

When computing the partial derivative w.r.t. x we treat y as a constant. Using the product rule,

$$\begin{aligned}\frac{\partial}{\partial x} f &= (x^2+y^2)\frac{\partial}{\partial x}\sin(x-y) + \sin(x-y)\frac{\partial}{\partial x}(x^2+y^2) \\ &= (x^2+y^2)\cos(x-y) + 2x\sin(x-y)\ .\end{aligned}$$

Similarly, x is considered constant when computing:

$$\begin{aligned}\frac{\partial}{\partial y} f &= (x^2+y^2)\frac{\partial}{\partial y}\sin(x-y) + \sin(x-y)\frac{\partial}{\partial y}(x^2+y^2) \\ &= -(x^2+y^2)\cos(x-y) + 2y\ \sin(x-y)\ .\end{aligned}$$

Partial derivatives are computed exactly as derivatives, so the differentiation rules of single variable calculus may apply. This is illustrated in the following

Example 3.26. Let

$$f(x,y) := \tan\left(\frac{xy}{x^2+y^2}\right) .$$

The partial derivative w.r.t. x is computed, using the chain rule

$$f_x(x,y) = \sec^2\left(\frac{xy}{x^2+y^2}\right)\frac{\partial}{\partial x}\left(\frac{xy}{x^2+y^2}\right) = \sec^2\left(\frac{xy}{x^2+y^2}\right)\frac{y(y^2-x^2)}{(x^2+y^2)^2} .$$

Since $f(x,y)$ is symmetric in x and y, we obtain the partial derivative w.r.t. y without computation,

$$f_y(x,y) = \sec^2\left(\frac{xy}{x^2+y^2}\right)\frac{x(x^2-y^2)}{(x^2+y^2)^2} .$$

△

Example 3.27. Let $z = f(x,y)$ where $x := u(t)$, $y := v(t)$ are functions of the single variable t . Therefore z is also a function of the single variable t ,

$$z = z(t) := f(u(t)\,,\, v(t))\ .$$

The derivative $z'(t)$ is then given by the chain rule

$$\frac{dz}{dt} = \frac{d}{dt}f\left(u(t),v(t)\right) = f_x\left(u(t),v(t)\right)\frac{du}{dt} + f_y\left(u(t),v(t)\right)\frac{dv}{dt}\ , \tag{3.22}$$

provided all derivatives exist. This can be abbreviated as

$$\frac{dz}{dt} = \frac{\partial z}{\partial x}\frac{dx}{dt} + \frac{\partial z}{\partial y}\frac{dy}{dt}\ . \tag{3.23}$$

For example, let $z = f(x,y) := \sin(x-y)$ where $x(t) := t^3$ and $y(t) := t^{-3}$. Then

$$\frac{dz}{dt} = \frac{\partial z}{\partial x}\frac{dx}{dt} + \frac{\partial z}{\partial y}\frac{dy}{dt}$$

$$= 3t^2\cos\left(t^3-t^{-3}\right) + 3t^{-4}\cos\left(t^3-t^{-3}\right) = 3\left(t^2+t^{-4}\right)\cos\left(t^3-t^{-3}\right)$$

For functions $f(x,y,z,\ldots)$ of three (or more) variables, the partial derivatives

$$\frac{\partial}{\partial x} f(x,y,z,\ldots) \,, \quad \frac{\partial}{\partial y} f(x,y,z,\ldots) \,, \quad \frac{\partial}{\partial z} f(x,y,z,\ldots) \,, \quad \ldots$$

are computed as ordinary derivatives w.r.t. a single variable, when all other variables are fixed constant.

Example 3.28. The partial derivatives of

$$f(x,y,z) \;:=\; \sqrt{x}\,\sin z\, e^{x+y+z}$$

are

$$f_x(x,y,z) \;=\; \frac{1}{2}\,x^{-1/2}\,\sin z\, e^{x+y+z} + \sqrt{x}\,\sin z\, e^{x+y+z} \,,$$

$$f_y(x,y,z) \;=\; \sqrt{x}\,\sin z\, e^{x+y+z} \,,$$

and

$$f_z(x,y,z) \;=\; \sqrt{x}\,\cos z\, e^{x+y+z} + \sqrt{x}\,\sin z\, e^{x+y+z} \,.$$

△

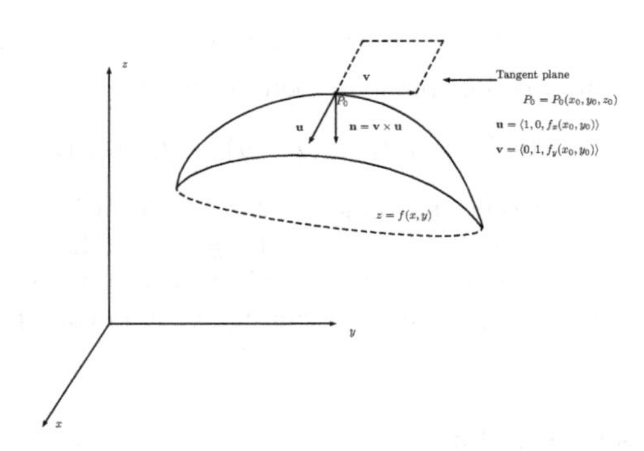

FIGURE 3.5: Illustration of the tangent plane.

Given a function f of one variable, the line tangent to the graph $y = f(x)$ at a point $(a, f(a))$ is given, in terms of the derivative $f'(a)$ as

$$y \;=\; f(a) + f'(a)(x-a) \,.$$

Similarly, partial derivatives are used to give the tangent plane to a graph of a function of two variables.

Definition 3.29. Let the function $f(x,y)$ be partially differentiable at (a,b), with partial derivatives $f_x(a,b)$ and $f_y(a,b)$. The **tangent plane** of the graph $z = f(x,y)$ at the point $(a,b,f(a,b))$ is the plane

$$z \;=\; f(a,b) + f_x(a,b)(x-a) + f_y(a,b)(y-b) \,. \tag{3.24}$$

We justify this definition as follows. Consider the curves made from $z \;=\; \varphi(x) \;:=\; f(x,b)$,i.e., the intersection of $z = f(x,y)$ and $y = b$, $z = \psi(y) := f(a,y)$ the intersection of $z = f(x,y)$ and $x = a$, and their tangent lines L and M at the point $Q(a, b, f(a, b))$, see Figures 3.4 and 3.5. We show now that the tangent plane of Definition 3.29 is PLANE $\{L, M\}$, the plane determined by the lines L and M .

The slope of the tangent line L (in the plane $y = b$) is the partial derivative $f_x(a, b)$. Therefore the vector

$$\mathbf{u} := \mathbf{i} + f_x(a, b)\,\mathbf{k}\,, \tag{3.25}$$

points in the direction of L, see Figure 3.5. Similarly, the vector

$$\mathbf{v} := \mathbf{j} + f_y(a, b)\,\mathbf{k}\,, \tag{3.26}$$

points in the direction of the tangent line M . The cross product

$$\mathbf{n} := \mathbf{u} \times \mathbf{v} = -f_x(a, b)\,\mathbf{i} - f_y(a, b)\,\mathbf{j} + \mathbf{k} = < -f_x(a, b), -f_y(a, b), 1 > \tag{3.27}$$

is perpendicular to both $\mathbf{u}$ and $\mathbf{v}$, and is therefore a normal to PLANE $\{L, M\}$.

A plane is determined by two lines, or alternatively, by a point and a normal. The plane is thus determined by the point $(a, b, f(a, b))$ and the normal $\mathbf{n}$. Its equation is

$$< x - a, y - b, z - f(a, b) > \cdot < -f_x(a, b)\,, -f_y(a, b)\,, 1 >= 0, \tag{3.28}$$

which simplifies to the tangent plane of Definition 3.29,

$$z = f(a, b) + f_x(a, b)(x - a) + f_y(a, b)(y - b). \tag{3.29}$$

Our goal here is to construct the tangent plane to an arbitrary surface which can be written in the canonical form $z = f(x, y)$. In order to do this we need to compute two tangent vectors to a surface. By intersecting the surface with the planes $\Pi_1 := \{x = x_0\}$, and $\Pi_2 := \{y = y_0\}$ we can compute in each plane respectively the tangent vectors $\vec{i} + \vec{k}\frac{\partial f}{\partial x}$, and $\vec{j} + \vec{k}\frac{\partial f}{\partial y}$.

Mathematica session 3.3. *First we clear any previous usage of the notation* $f(x, y)$.

```
Clear[f, z]
```

Next we introduce two vectors which are normal to the surface described by $z = f(x, y)$.

```
U = {1, 0, D[f[x, y], x]}
```

$\{1, 0, f^{(1,0)}[x, y]\}$

```
V = {0, 1, D[f[x, y], y]}
```

$\{0, 1, f^{(0,1)}[x, y]\}$

The normal to the surface is perpendicular to both the tangent vectors **U** *and* **V**.

```
n = V × U
```

$\left\{f^{(1,0)}[x,y], f^{(0,1)}[x,y], -1\right\}$

We now want to "normalize" this vector, i.e. make its length 1; *hence, we first calculate it length and then divide the normal vector by its length.*

```
myNorm[w_]:=Simplify[Sqrt[w.w]]
```

The following vector has length 1.

```
{x, y, z}.{x, y, z}
```

$x^2 + y^2 + z^2$

Using the above steps we now introduce a MATHEMATICA *function, written as a* **Module** *which will produce the normal to the surface given in the explicit form* $z = f(x,z)$.

```
surfaceNormal[f_]:=Module[{u, s}, u = {D[f, x], D[f, y], -1};
s = myNorm[u];
Simplify[u/s]]
```

```
surfaceNormal[x + y + 1]
```

$\left\{\frac{1}{\sqrt{3}}, \frac{1}{\sqrt{3}}, -\frac{1}{\sqrt{3}}\right\}$

```
surfaceNormal[2 - x^2 - y^2]
```

$\left\{-\frac{2x}{\sqrt{1+4x^2+4y^2}}, -\frac{2y}{\sqrt{1+4x^2+4y^2}}, -\frac{1}{\sqrt{1+4x^2+4y^2}}\right\}$

In the case the function is written in non-explicit form, i.e. $F(x,y,z) = 0$, *we use the* MATHEMATICA *function given below*

```
surfaceNormalNonexplicit[f_]:=Module[{u, s}, u = {D[f, x], D[f, y], D[f, z]};
s = myNorm[u];
Simplify[u/s]]
```

```
surfaceNormalNonexplicit[x^2 + y^2 + z^2 - 9]
```

$\left\{\frac{x}{\sqrt{x^2+y^2+z^2}}, \frac{y}{\sqrt{x^2+y^2+z^2}}, \frac{z}{\sqrt{x^2+y^2+z^2}}\right\}$

♦

Exercises

Exercise 3.30. *Compute the partial derivatives f_x and f_y of the following functions*

(a) $f(x,y) := exp\{x^{4-xy^3}\}$ (b) $f(x,y) := \tanh(x-y)$

(c) $f(x,y) := \ln(x^2-y^2)$ (d) $f(x,y) := x^2 \sin(y+x^2)$

Exercise 3.31. *Determine the tangent planes to the following surfaces, at the point $(2,1,z)$ on the surface.*

(a) $z := exp\{x^{4-xy^3}\}$ (b) $z := \tanh(x-y)$

(c) $z := \ln(x^2-y^2)$ (d) $z := x^2 \sin(y+x^2)$

Exercise 3.32. *The ellipsoid $x^2+3y^2+2z^2 = 9$ and the sphere $x^2+y^2+z^2 = 6$ intersect at the point $(-2,1,1)$. Determine the angle their tangent planes make at this point.*

Exercise 3.33.

(a) *Verify that (3.27) gives $\mathbf{u}\times\mathbf{v}$ for $\mathbf{u}$ and $\mathbf{v}$ as in (3.25) and (3.26).*

(b) *Verify that (3.28) gives the plane through $(a,b,f(a,b))$ with normal $\mathbf{n}$ as*

in (3.27).

Exercise 3.34. *[Euler rule for homogeneous functions] A function $f(x,y)$ such that*

$$f(\lambda x\,,\,\lambda y) \;=\; \lambda^n f(x\,,\,y)\;, \quad \text{for all } x\,,\,y\,,\,\lambda\,, \tag{3.30}$$

is called **homogeneous** *of the* n^{th}–**order**. *Prove: If f is n^{th}–order homogeneous and differentiable then*

$$x\,\frac{\partial f}{\partial x} \;+\; y\,\frac{\partial f}{\partial y} \;=\; n\,f\;. \tag{3.31}$$

This result is due to Euler[2].
Hint: Differentiate both sides of (3.30) w.r.t λ, and set $\lambda := 1$.

Exercise 3.35. *Show that the function $f(x,y) := a\,x^n + b\,y^n + c\,x^m\,y^{n-m}$ is homogeneous of order n, for all constants a, b, c and m. Then verify that (3.31) holds for this function.*

3.3 Gas thermodynamics

A region in space is occupied by some gas (e.g. air, oxygen, nitrogen, etc.). At any time we can measure the gas

[2] Leonhard Euler [1707–1783]

volume	V	given in units of volume, say [$\mathtt{cm}^3$],
pressure	p	with units of force/area, say [$\mathtt{N/cm}^2$], and
temperature	T	measured in **Kelvin degrees** °K.

A **Kelvin degree** °K is the same as a **Celsius degree** °C (or **centigrade**), but the two scales have different origins: $0^\circ\,\mathrm{K} = -273^\circ\,\mathrm{C}$ and $0^\circ\,\mathrm{C} = 273^\circ\,\mathrm{K}$.

The volume changes when the gas is compressed, or allowed to expand. The temperature and pressure are caused by the movements of the gas molecules. If a gas is cooled to a sufficiently low temperature, its molecules will slow and the gas will liquify, i.e. change to a liquid state[3]. We assume here that gases stay gaseous, i.e. the temperature does not fall too low.

The gas volume V, pressure p and temperature T are not independent, but are related by an equation

$$F(p, V, T) = 0 , \tag{3.32}$$

called a **state equation** of the gas. Knowing any two of these variables determines the value of the third, i.e. the implicit equation (3.104) can be solved for any of the variables p, V and T in terms of the other two. In particular, we may express the pressure p as a function

$$p = p(V, T) . \tag{3.33}$$

If V is fixed, i.e. the gas is contained in a closed vessel of volume V, and its temperature is changed (the gas is heated or cooled), how does its pressure change? In other words, how does p change with T? The answer is through the partial derivative

$$\left(\frac{\partial p}{\partial T}\right)_V ,$$

where the subscript V denotes the fact that V is constant. Similarly, we may solve (3.104) for the volume V

$$V = V(p, T) . \tag{3.34}$$

If the pressure is held constant (see Exercise 3.40), and we change the gas temperature, how does its volume change? The rate of change of V w.r.t. T, with constant p, is measured by the partial derivative

$$\left(\frac{\partial V}{\partial T}\right)_p .$$

Example 3.36 (Ideal gas). An **ideal gas** is one where the interactions between the gas molecules are ignored. The state equation of an ideal gas is

$$pV = RT , \tag{3.35}$$

[3]The difference between the gaseous state and liquid state is roughly that a liquid cannot be compressed.

where R is the **gas constant**, depending on the specific gas and its amount present[4]. This is called the **Boyle**[5] (or **Mariotte**[6]) **law**. For an ideal gas we have

$$\left(\frac{\partial p}{\partial T}\right)_V = \frac{R}{V}\,,\ \left(\frac{\partial p}{\partial V}\right)_T = -\frac{RT}{V^2}\,,\ \left(\frac{\partial T}{\partial p}\right)_V = \frac{V}{R}\,,\ \left(\frac{\partial V}{\partial p}\right)_T = -\frac{RT}{p^2}\,,\ \text{etc.}$$

Example 3.37 (van der Waals gas)**.** This is the case of a non–ideal gas, with the state equation (called the **van der Waals**[7] **law**)

$$\left(p+\frac{\alpha}{V^2}\right)(V-\beta) = RT\,, \tag{3.36}$$

where the term $\frac{\alpha}{V^2}$ measures the interaction between molecules[8] and β represents the (finite) volume of the molecules themselves. The Boyle–Mariotte law (3.61) of the ideal gas is the special case $\alpha = \beta = 0$. Solving (3.36) for p we get

$$p(V,T) = \frac{RT}{V-\beta} - \frac{\alpha}{V^2}\,, \tag{3.37}$$

and the partial derivatives

$$\left(\frac{\partial p}{\partial T}\right)_V = \frac{R}{V-\beta}\,,\quad \left(\frac{\partial p}{\partial V}\right)_T = -\frac{RT}{(V-\beta)^2}+\frac{2\alpha}{V^3}\,,\quad \text{etc.} \tag{3.38}$$

Example 3.38 (Virial state equation)**.** This state equation is

$$\frac{pV}{RT} = 1 + \frac{B(T)}{V} + \frac{C(T)}{V^2} + \frac{D(T)}{V^3} + \cdots$$

where $B(T)$, $C(T)$, $D(T)$, ... are called the **second, third, fourth ... virial coefficients**[9]. For simplicity, we consider only the second coefficient, which for most gases is approximated by

$$B(T) = \beta - \frac{\alpha}{RT}$$

where α , β are the constants in the van der Waals equation (3.36). This gives the **virial equation**

$$p(V,T) = \frac{RT}{V} + \frac{\beta\, RT - \alpha}{V^2}\,. \tag{3.39}$$

Indeed, solving (3.36) we get

$$\begin{aligned}\frac{pV}{RT} &= 1+\frac{\beta p}{RT} - \left(\frac{\alpha}{RT}\right)\frac{1}{V} + \left(\frac{\alpha\beta}{RT}\right)\frac{1}{V^2}\\ &\approx 1+\frac{\beta p}{RT} - \left(\frac{\alpha}{RT}\right)\frac{1}{V} \quad \text{ignoring the term with } \frac{1}{V^2}\\ &\approx 1+\beta\frac{1}{V} - \left(\frac{\alpha}{RT}\right)\frac{1}{V} \quad \text{since } pV \approx RT\,,\ \text{by (3.61)}\,.\end{aligned}$$

[4]In chemistry texts this equation may appear as $pV = NRT$ or $pV = nRT$, where n or N denote the amount of gas. Here we fix the units so as to get a simple state equation.

[5]Robert Boyle [1627–1691].

[6]Edme Mariotte [1620–1684].

[7]Johannes Diederik van der Waals [1837–1923].

[8]Which increases as the gas is compressed, i.e. when V decreases.

[9]See e.g. F.C. Andrews, *Thermodynamics: Principles and Applications*, Wiley–Interscience, 1971.

Exercises

Exercise 3.39. *Compute the partial derivatives*

(a) $\left(\frac{\partial p}{\partial V}\right)_T$ (b) $\left(\frac{\partial V}{\partial p}\right)_T$

(c) $\left(\frac{\partial T}{\partial V}\right)_p$ (d) $\left(\frac{\partial p}{\partial T}\right)_V$

for the virial equation of state (3.120).

Exercise 3.40. *A process where the gas pressure is constant is called* **isobaric**. *How do you keep a gas in constant pressure? Describe a simple experiment for doing this.*
Hint: Consider gas inside a vertical cylinder with an open top, which is sealed by a piston free to move up or down. Let A be the area of the piston, and W its weight (which can be adjusted by additional weights on top of the piston). The piston may drop initially, compressing the gas until a pressure $p = W/A$ is reached, at which point the piston will come to a stop. What happens when you heat, or cool, the gas inside the cylinder?

3.4 Higher order partial derivatives

The partial derivatives of $f(x,y)$ are themselves functions $f_x(x,y)$ and $f_y(x,y)$ of (x,y). Their partial derivatives, if they exist, are called the **second order partial derivatives** of f.

There are four such partial derivatives, denoted by

$$f_{xx} = (f_x)_x = \frac{\partial^2 f}{\partial x^2} \quad , \quad f_{xy} = (f_x)_y = \frac{\partial^2 f}{\partial y \partial x} \,, \tag{3.40a}$$

$$f_{yx} = (f_y)_x = \frac{\partial^2 f}{\partial x \partial y} \quad , \quad f_{yy} = (f_y)_y = \frac{\partial^2 f}{\partial y^2} \,. \tag{3.40b}$$

The partial derivatives f_{xy}, f_{yx} are called **mixed**. Partial derivatives of higher orders are defined similarly:

$$f_{xxx}(x,y) := \frac{\partial f_{xx}(x,y)}{\partial x} \quad , \quad f_{yxx}(x,y) := \frac{\partial f_{xx}(x,y)}{\partial y} \,, \tag{3.41a}$$

$$f_{yxy}(x,y) := \frac{\partial f_{xy}(x,y)}{\partial y} \quad , \quad f_{xxy}(x,y) := \frac{\partial f_{xy}(x,y)}{\partial x} \,, \quad \text{etc.} \tag{3.41b}$$

Example 3.41. Let $$f(x,y) := \frac{1}{\sqrt{x^2+y^2}} \,.$$

The first order partial derivatives are

$$f_x = \frac{-x}{(x^2+y^2)^{\frac{3}{2}}} \quad , \quad f_y = \frac{-y}{(x^2+y^2)^{\frac{3}{2}}} \,,$$

and their partial derivatives are the second order partial derivatives of f

$$f_{xx} = \frac{3x^2}{(x^2+y^2)^{\frac{5}{2}}} - \frac{1}{(x^2+y^2)^{\frac{3}{2}}} \;,\quad f_{xy} = \frac{3xy}{(x^2+y^2)^{\frac{5}{2}}}$$
$$f_{yx} = \frac{3xy}{(x^2+y^2)^{\frac{5}{2}}} \;,\quad f_{yy} = \frac{3y^2}{(x^2+y^2)^{\frac{5}{2}}} - \frac{1}{(x^2+y^2)^{\frac{3}{2}}} \;.$$

Note that the mixed derivatives are equal: $f_{xy} = f_{yx}$. ○

Example 3.42. *In general $f_{xy} \neq f_{yx}$, i.e. the second order mixed derivatives are not necessarily equal. This is illustrated by*

$$f(x,y) := \begin{cases} \frac{xy(x^2-y^2)}{x^2+y^2} & \text{if } (x,y) \neq (0,0) \\ 0 & \text{if } (x,y) = (0,0). \end{cases} .$$

We compute the first order derivative

$$f_x(x,y) = \frac{y(x^4+4x^2y^2-y^4)}{(x^2+y^2)^2}, \quad \text{when } (x,y) \neq (0,0).$$

At the point $(0,0)$ we can compute $f_x(0,0) = 0$, using Definition 3.16 directly. Therefore

$$\begin{aligned} f_x(0,y) &= -y \,, && \textit{for all values of } y \,. \\ \textit{Differentiating w.r.t. } y \textit{ we get} \quad f_{xy}(0,y) &= -1 \,, && \textit{for all values of } y \,, \\ \textit{in particular,} \quad f_{xy}(0,0) &= -1 \,. \end{aligned}$$

We now differentiate in reverse order. First:

$$f_y(x,y) = \begin{cases} \frac{x(x^4-4x^2y^2-y^4)}{(x^2+y^2)^2} & \text{if } (x,y) \neq (0,0) \,, \\ 0 & \text{if } (x,y) = (0,0) \,. \end{cases}$$

Therefore

$$\begin{aligned} f_y(x,0) &= x \,, && \textit{for all values of } x \,, \\ \textit{and, by differentiating w.r.t. } x, \quad f_{yx}(x,0) &= 1 \,, \\ \textit{showing that} \quad f_{xy}(0,0) &\neq f_{yx}(0,0) \,. \end{aligned}$$

○

It can be shown that if the mixed derivatives f_{xy} and f_{yx} are **continuous functions** of x and y, then they are equal. A similar result holds for functions $f(x,y,z)$ of three variables. There the continuity of mixed derivatives (as functions of x , y , z) guarantees that

$$\begin{aligned} f_{xy}(x,y,z) &= f_{yx}(x,y,z) \,, \\ f_{xz}(x,y,z) &= f_{zx}(x,y,z) \,, \text{ etc.} \end{aligned}$$

We assume from now on that mixed derivatives are equal:
$f_{xy} = f_{yx} \,, \quad f_{xyz} = f_{xzy} = f_{yxz} = f_{yzx} = f_{zxy} = f_{zyx}$, etc.

Mathematica session 3.4. *The* n^{th}*-order derivative of* f *w.r.t.* x *is computed by* `D[f,{x,n}]` *. Higher order partial derivatives are computed similarly. For example let us introduce the function* $f(x,y)$ *as* MATHEMATICA *lets us do some calculations by the chain rule. For example, if we consider taking the derivative of* $f(x,y,z)$ *whose derivative is* **known** *to* MATHEMATICA*, for example* $\sin(x+z)$*, and* z *is an implicitly given but* **arbitrary** *function* MATHEMATICA *uses the chain rule to compute* $\frac{\partial f}{\partial x}$.

However, if both functions, $F(x,y,z)$ *and* $z = f(x,y)$ *are both unspecified,* MATHEMATICA *can not know the chain rule for this case. Consequently, we need to write a procedure to handle this situation.*

```
D[x^2 * y + Sin[x * y], x]
```

$2xy + y\,Cos[xy]$

We can write the mixed derivative in MATHEMATICA *as*

```
D[x^2 * y + Sin[x * y], x, y]
```

$2x + Cos[xy] - xy\,Sin[xy]$

A higher order derivative, in this case the sixth order is written using the syntax

```
D[x^2 * y + Sin[x * y], {x, 6}]
```

$-y^6\,Sin[xy]$

```
partialDerivatives[f_, n_, m_]:=
If[0 < n&&0 < m, D[D[f, {x, n}], {y, m}],
If[0 < n&&m = 0, D[f, {x, n}], If[m < 0&&n = 0, D[f, {y, m}]]]]
```

```
partialDerivatives[Sin[x * y], 3, 2]
```

$-6y\,Cos[xy] + x^2y^3\,Cos[xy] + 6xy^2\,Sin[xy]$

Here is a MATHEMATICA *function which computes the mixed partial derivative* $\frac{\partial^{n+m} f(x,y)}{\partial x^n \partial y^m}$

```
partialDerivatives[f_, n_, m_]:=
If[0<n&&0<m, D[D[f, {x, n}], {y, m}], If[0<n&&m==0, D[f, {x, n}],
D[f, {y, m}]]]
```

```
partialDerivatives[x^5y^3, 1, 2]
```

$30x^4y$

```
partialDerivatives[x^2y + Sin[x * y], 3, 2]
```

$-6y\,Cos[xy] + x^2y^3\,Cos[xy] + 6xy^2\,Sin[xy]$

Finally we write a MATHEMATICA *function for computing all derivatives of mixed order* $1 \geq N$.

```
derivesOrderN[f_, N_]:=Do[Print[D[D[f, {x, i}], {y, N - i}]], {i, 0, N}]
```

```
derivesOrderN[Sin[x * y], 9]
```

$x^9\,Cos[xy]$

$x^8y\,Cos[xy] + 8x^7\,Sin[xy]$

$-42x^5\,Cos[xy] + x^7y^2\,Cos[xy] + 14x^6y\,Sin[xy]$

$-90x^4y\,Cos[xy] + x^6y^3\,Cos[xy] - 120x^3\,Sin[xy] + 18x^5y^2\,Sin[xy]$

$120x\,Cos[xy] - 120x^3y^2\,Cos[xy] + x^5y^4\,Cos[xy] - 240x^2y\,Sin[xy] + 20x^4y^3\,Sin[xy]$

$120y\,Cos[xy] - 120x^2y^3\,Cos[xy] + x^4y^5\,Cos[xy] - 240xy^2\,Sin[xy] + 20x^3y^4\,Sin[xy]$

$-90xy^4\,Cos[xy] + x^3y^6\,Cos[xy] - 120y^3\,Sin[xy] + 18x^2y^5\,Sin[xy]$

$-42y^5\,Cos[xy] + x^2y^7\,Cos[xy] + 14xy^6\,Sin[xy]$

$xy^8\,Cos[xy] + 8y^7\,Sin[xy]$

$y^9\,Cos[xy]$

♦

Example 3.43. Recall the van der Waals equation of state, see Example 3.37,

$$p(V,T) = \frac{RT}{V-\beta} - \frac{\alpha}{V^2}, \tag{3.42}$$

and the partial derivatives

$$\left(\frac{\partial p}{\partial T}\right)_V = \frac{R}{V-\beta}, \quad \left(\frac{\partial p}{\partial V}\right)_T = -\frac{RT}{(V-\beta)^2} + \frac{2\alpha}{V^3}. \tag{3.43}$$

The second order partial derivatives are then computed as

$$\frac{\partial^2 p}{\partial T^2} = 0 \quad , \quad \frac{\partial^2 p}{\partial V \partial T} = -\frac{R}{(V-\beta)^2},$$

$$\frac{\partial^2 p}{\partial T \partial V} = -\frac{R}{(V-\beta)^2} \quad , \quad \frac{\partial^2 p}{\partial V^2} = \frac{2RT}{(V-\beta)^3} - \frac{6\alpha}{V^4}.$$

Exercises

Exercise 3.44. *Express the following derivatives in subscript notation.*

(a) $\dfrac{\partial^4 f}{\partial x^2 \partial y^2}$ (b) $\dfrac{\partial^5 f}{\partial y^2 \partial x^3}$ (c) $\dfrac{\partial^4 f}{\partial y^4}$

(d) $\dfrac{\partial^4 f}{\partial x^x \partial y^2 \partial x}$ (e) $\dfrac{\partial^3 f}{\partial x \partial y^2}$ (f) $\dfrac{\partial^4 f}{\partial x \partial y^2 \partial z}$

Exercise 3.45. *Express the following derivatives in ∂ notation.*

(a) F_{xxyx} (b) F_{xyz} (c) F_{zxy}

(d) F_{yyyy} (e) F_{xyxy} (f) F_{xxyxx}

Exercise 3.46. *Check whether $f_{xy} = f_{yx}$ in the examples below*

(a) $f(x,y) := \sqrt{\dfrac{x^2-y^2}{x^2+y^2}}$ (b) $f(x,y) := (x+y)^7$

(c) $f(x,y) := x^{100} + 5x^{\pi}y^{22} + x\ln(xy)$ (d) $f(x,y) := \sin\left(x^2+y^2\right)$

(e) $f(x,y) := \begin{cases} \frac{x^2-y^2}{x^2+y^2} & \text{if } (x,y) \neq (0,0) \\ 0 & \text{otherwise} \end{cases}$

(f) $f(x,y) := \begin{cases} \frac{xy}{x^2+y^2} & \text{if } (x,y) \neq (0,0) \\ 0 & \text{otherwise} \end{cases}$

(g) $f(x,y) := \begin{cases} \frac{x^3-y^3}{x^2+y^2} & \text{if } (x,y) \neq (0,0) \\ 0 & \text{otherwise} \end{cases}$

(h) $f(x,y) := \begin{cases} \frac{x^2y^2}{x^4+y^4} & \text{if } (x,y) \neq (0,0) \\ 0 & \text{otherwise} \end{cases}$

Exercise 3.47. *Show that the following functions satisfy the equation*

$$\frac{\partial^2 f}{\partial x^2} + \frac{\partial^2 f}{\partial y^2} = 0.$$

(a)	$f := x^2 - y^2$	(b)	$f := xy$	(c)	$f := x^3 - 3xy^2$
(d)	$f := y^3 - 3x^2y$	(e)	$f := \ln(x^2 + y^2)$	(f)	$f := e^{-y}\cos x$
(g)	$f := \text{atan}\left(\frac{2xy}{x^2 - y^2}\right)$	(h)	$f := \sin x \cosh y$	(i)	$f := \cos x \sinh y$

3.5 Differentials

Let $f(x)$ be a function of a single variable. As x changes from x_0 to $x_0 + \Delta x$, the function changes by

$$\Delta f \;:=\; f(x_0 + \Delta x) - f(x_0)$$

a **difference**, which is approximated by the **differential**

$$df \;:=\; f'(x_0)\,\Delta x\;,$$

measured along the tangent line. For a function $f(x, y)$ of two variables, the difference can also be approximated by a differential, measured along the tangent plane. Let

$$z \;=\; f(x, y) \tag{3.44}$$

be a function of two variables, and let these variables change from (x_0, y_0) to $(x_0 + \Delta x, y_0 + \Delta y)$. The corresponding change in f is the (actual) difference

$$\Delta f \;:=\; f(x_0 + \Delta x, y_0 + \Delta y) - f(x_0, y_0)\;.$$

The tangent plane of the surface (3.95) at the point (x_0, y_0) is

$$z \;=\; f(x_0, y_0) + f_x(x_0, y_0)(x - x_0) + f_y(x_0, y_0)(y - y_0)\;, \quad \text{see (3.29)}\;,$$

and the difference, measured along the tangent plane, is

$$(f(x_0, y_0) + f_x(x_0, y_0)\Delta x + f_y(x_0, y_0)\Delta y) - f(x_0, y_0) \;=\; f_x(x_0, y_0)\Delta x + f_y(x_0, y_0)\Delta y\;,$$

which is the subject of the next definition.

Definition 3.48. The **differential** of $f(x, y)$ **at the point** (x_0, y_0) (where f is differentiable) is defined as

$$df \;:=\; f_x(x_0, y_0)\,dx \;+\; f_y(x_0, y_0)\,dy\;. \tag{3.45}$$

For fixed (x_0, y_0) the differential df is a linear function of the differences dx and dy. We show next that the differential can be used to approximate the actual difference Δf .

Lemma 3.49. Let the partial derivatives f_x and f_y be continuous in a neighborhood of the point (x_0, y_0) , and let

$$\Delta f \;:=\; f(x_0 + \Delta x, y_0 + \Delta y) - f(x_0, y_0) \tag{3.46a}$$

$$df \;:=\; f_x(x_0, y_0)\,\Delta x \;+\; f_y(x_0, y_0)\,\Delta y\;. \tag{3.46b}$$

Then

$$\Delta f \;=\; df \;+\; \varepsilon_1\,\Delta x \;+\; \varepsilon_2\,\Delta y \tag{3.47}$$

where

$$\lim_{\Delta x \to 0} \varepsilon_1 \;=\; 0\;, \quad \lim_{\Delta y \to 0} \varepsilon_2 \;=\; 0\;. \tag{3.48}$$

Proof. We assume for convenience that Δx and Δy are positive. The difference (3.46a) can be written as

$$\Delta f = (f(x_0+\Delta x, y_0 + \Delta y) - f(x_0 + \Delta x, y_0)) + (f(x_0+\Delta x, y_0) - f(x_0, y_0)) \ ,$$

and, using the Mean Value Theorem,

$$\Delta f \ = \ f_y(x_0 + \Delta x, \eta)\Delta y \ + \ f_x(\xi, y_0)\Delta x \ , \tag{3.49}$$

where $\eta \in (y_0, y_0 + \Delta y)$, and $\xi \in (x_0, x_0 + \Delta x)$. As the coefficients of Δy and Δx are continuous they can be approximated by their values at the point (x_0, y_0). To this end we write

$$\begin{aligned} f_y(x_0 + \Delta x, \eta) &= f_y(x_0, y_0) + (f_y(x_0 + \Delta x, \eta) - f_y(x_0, y_0)) \\ &:= f_y(x_0, y_0) + \varepsilon_2 \ , \\ f_x(\xi, y_0) &= f_x(x_0, y_0) + (f_x(\xi, y_0) - f_x(x_0, y_0)) := f_x(x_0, y_0) + \varepsilon_1 \ . \\ \therefore \quad \Delta f &= (f_x(x_0, y_0) + \varepsilon_1) \ \Delta x \ + \ (f_y(x_0, y_0) + \varepsilon_2) \ \Delta y \\ &= df \ + \ \varepsilon_1 \Delta x \ + \ \varepsilon_2 \Delta y \ , \end{aligned}$$

and (3.48) follows from the continuity of $f_x(x, y)$ and $f_y(x, y)$, and the fact that ξ and η are bounded above and below by $x_0 < \xi < x_0 + \Delta x$ and $y_0 < \eta < y_0 + \Delta y$.

The proof that both ε_1 and ε_2 tend to zero with Δx and Δy, for general signs of Δx and Δy, is left for Exercise 3.62 below. □

Equations (3.47)–(3.48) justify using the differential df to approximate the difference Δf, and in turn, to approximate $f(x_0+\Delta x, y_0+\Delta y)$, if the function f and its partial derivatives f_x and f_y are known at (x_0, y_0),

$$\begin{aligned} f(x_0 + \Delta x, y_0 + \Delta y) &= f(x_0, y_0) + \Delta f \\ &\approx f(x_0, y_0) + df \\ &= f(x_0, y_0) + f_x(x_0, y_0)\Delta x + f_y(x_0, y_0)\Delta y \ . \end{aligned} \tag{3.50}$$

Example 3.50 (Approximation of a function value). It is required to evaluate the function

$$f(x, y) \ := \ \sin\left(\frac{\pi}{8}(x^2 + y^2)\right)$$

at $(x, y) = (1.01,\ 1.03)$. At the nearby point $(x_0, y_0) = (1,\ 1)$ the value is $f(1, 1) = \sin\left(\frac{\pi}{8}(1 + 1)\right) = \frac{1}{\sqrt{2}}$.
The differential is, by (3.46b),

$$\begin{aligned} df &= \frac{\partial f}{\partial x}\, dx \ + \ \frac{\partial f}{\partial y}\, dy \\ &= \frac{\pi x}{4} \cos\left(\frac{\pi}{8}(x^2 + y^2)\right) dx \ + \ \frac{\pi y}{4} \cos\left(\frac{\pi}{8}(x^2 + y^2)\right) dy \ , \end{aligned}$$

and at $(x_0, y_0) = (1,\ 1)$,

$$df \ = \ \frac{\pi}{4} \cos\left(\frac{\pi}{4}\right) dx \ + \ \frac{\pi}{4} \cos\left(\frac{\pi}{4}\right) dy \ .$$

Therefore, by (3.50),

$$\begin{aligned}\sin\left(\frac{\pi}{8}([1.01]^2+[1.03]^2)\right) &\approx \frac{1}{\sqrt{2}}+\frac{\pi}{4}\cos\left(\frac{\pi}{4}\right)(.01)+\frac{\pi}{4}\cos\left(\frac{\pi}{4}\right)(.03)\\ &= \frac{1}{\sqrt{2}}+\frac{\pi}{4}\frac{1}{\sqrt{2}}0.01+\frac{\pi}{4}\frac{1}{\sqrt{2}}0.03\\ &= \frac{1+(0.01+0.03)\pi/4}{\sqrt{2}} \approx 0.729321...,\end{aligned}$$

as compared with the value $f(1.01,\, 1.03) = 0.729237\ldots$

Example 3.51 (Volume differential in ideal gas). An ideal gas, represented by

$$pV = RT\,, \tag{3.51}$$

changes from $A(p_1, T_1)$ to $B(p_2, T_2)$, two points sufficiently close in the pT–plane. It is required to compute the volume change, approximated by the differential

$$\begin{aligned}dV &= \frac{\partial V}{\partial T}\,dT + \frac{\partial V}{\partial p}\,dp\\ &= \frac{R}{p}\,dT - \frac{RT}{p^2}\,dp\,,\ \text{since } V = R\,\frac{T}{p}\,.\end{aligned} \tag{3.52}$$

See Figure 3.6. We consider two cases:
(a) The change is along the line AB. Along that line

$$dT = \frac{T_2-T_1}{p_2-p_1}\,dp\,, \tag{3.53a}$$

$$p^2 \approx p_1\,p_2\,,\quad \text{since } A,\, B \text{ are close}\,. \tag{3.53b}$$

Substituting (3.53a) in (3.60) we obtain

$$\begin{aligned}dV &= \left(\frac{R}{p}\left(\frac{T_2-T_1}{p_2-p_1}\right) - \frac{RT}{p^2}\right)dp\\ &\approx \frac{R\,(T_2p_1-p_2T_1)}{(p_2-p_1)\,p_1\,p_2}\,dp,\quad \text{using (3.53b)}\ ,\ T\approx\tfrac{T_1+T_2}{2} \text{ and } p\approx\tfrac{p_1+p_2}{2}\,,\\ &= R\left(\frac{T_2}{p_2}-\frac{T_1}{p_1}\right) = V_2-V_1\,,\quad \text{by (3.61)}\,.\end{aligned}$$

(b) Isothermal change (constant T) followed by isobaric change (constant p),

$$\begin{aligned}\text{Isothermal}: T = T_1 \quad \therefore\ dV &= -\frac{R\,T_1}{p^2}\,dp\,,\quad \text{by (3.60)}\,,\\ \text{isobaric}: p = p_2 \quad \therefore\ dV &= \frac{R}{p_2}\,dT\,.\\ \therefore\ \text{The total } dV &= -\frac{R\,T_1}{p^2}\,dp + \frac{R}{p_2}\,dT\\ &\approx R\left(\frac{T_2}{p_2}-\frac{T_1}{p_1}\right),\ \text{by (3.53b) and (3.61)},\\ &= V_2 - V_1\,.\end{aligned}$$

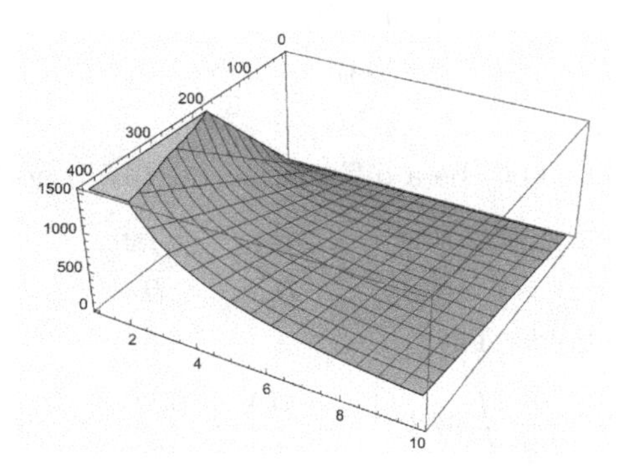

FIGURE 3.6: Pressure–temperature, with volume as a dependent variable.

We see that in both cases we get the same answer

$$dV \approx \Delta V = V_2 - V_1 .$$

Definition 3.52 (Exact differential)**.** Let $f(x, y)$ be a differentiable function of two variables, and let dx and dy be two additional variables, called the **differentials** of x and y respectively. Then the function df (of the four variables x, y, dx, dy)

$$df := \frac{\partial f}{\partial x}\, dx + \frac{\partial f}{\partial y}\, dy \tag{3.54}$$

is called the **exact** (or **complete**) **differential** of f. The terms $\frac{\partial f}{\partial x}\, dx$ and $\frac{\partial f}{\partial y}\, dy$ are called **partial differentials** of f.

An exact differential is a function of the form

$$M(x, y)\, dx + N(x, y)\, dy , \quad \text{where } M , N \text{ are functions of } (x, y) .$$

A natural question here is: For which functions M, N is $M\, dx + N\, dy$ an exact differential, i.e. there exists a differentiable function $f(x, y)$ such that

$$M(x, y) = \frac{\partial f}{\partial x} , \quad N(x, y) = \frac{\partial f}{\partial y} . \tag{3.55}$$

The answer follows.

Theorem 3.53. Let the functions $M(x,y)$, $N(x,y)$ be differentiable. Then the function

$$M(x,y)\,dx \;+\; N(x,y)\,dy \tag{3.56}$$

is an exact differential if, and only if,

$$\frac{\partial M}{\partial y} \;=\; \frac{\partial N}{\partial x}\,. \tag{3.57}$$

Proof. Only if: Let (3.56) be a differential, i.e. (3.55) holds for some f. Then

$$\frac{\partial^2 f}{\partial x\,\partial y} \;=\; \frac{\partial M}{\partial y} \;=\; \frac{\partial N}{\partial x}\,.$$

If: Assume (3.57) holds. Then

$$\frac{\partial}{\partial x}\left\{N \;-\; \frac{\partial}{\partial y}\int M dx\right\} \;=\; \frac{\partial N}{\partial x} - \frac{\partial M}{\partial y} \;=\; 0\,, \quad \text{by (3.57)}\,,$$

and the function f defined by

$$f \;:=\; \int M dx \;+\; \int\left\{N - \frac{\partial}{\partial y}\int M dx\right\} dy \;+\; \text{constant} \tag{3.58}$$

satisfies

$$\frac{\partial f}{\partial x} \;=\; M\,,\; \frac{\partial f}{\partial y} \;=\; N\,,$$

proving that (3.56) is a differential. □

Remark 3.54. *The function (3.58) is not the only function f satisfying (3.55). Another function can be defined, by symmetry,*

$$f \;:=\; \int N dy \;+\; \int\left\{M - \frac{\partial}{\partial x}\int N dy\right\} dx \;+\; \textit{constant}\,. \tag{3.59}$$

Example 3.55. The function (of the variables p, T, dp, dT)

$$\frac{R}{p}\,dT \;-\; \frac{RT}{p^2}\,dp \tag{3.60}$$

is an exact differential, since

$$\frac{\partial}{\partial p}\left(\frac{R}{p}\right) \;=\; \frac{\partial}{\partial T}\left(-\frac{RT}{p^2}\right) \;=\; -\frac{R}{p^2}\,.$$

Indeed, (3.60) is the differential dV of the volume of an ideal gas

$$V \;=\; \frac{RT}{p}\,. \tag{3.61}$$

see Exercise 3.63. The physical meaning of the fact that dV is an exact differential: As the ideal gas changes from $A(p_1, T_1)$ to $B(p_2, T_2)$, its volume change

$$dV \;=\; V_2 - V_1 \;=\; \frac{RT_2}{p_2} - \frac{RT_1}{p_1}$$

depends only on the endpoints A and B, not on the route taken from A to B, see illustration in Example 3.51.

Example 3.56. The mechanical work done by a gas expanding by a differential volume dV is

$$dW := p\,dV\,. \tag{3.62}$$

For an ideal gas this becomes, using (3.60),

$$dW = R\,dT - \frac{RT}{p}\,dp\,. \tag{3.63}$$

The function (3.63) violates the exactness condition (3.57)

$$\frac{\partial}{\partial p} R = 0 \neq \frac{\partial}{\partial T}\left(-\frac{RT}{p}\right) = -\frac{R}{p}\,,$$

therefore dW is not an exact differential. The physical meaning of the fact that dW is not exact: The mechanical work done by the gas as it changes from $A(p_1, T_1)$ to $B(p_2, T_2)$ depends on the route taken from A to B and not just on A and B, see Exercise 3.64.

Example 3.57. In § 3.3, the thermodynamic system was represented using the variables (V, p, T). Other variables can be used instead of (V, p, T), for example:

volume	V	given in units of volume, say [cm^3],
internal energy	U	with units of energy, say [Ncm], and
entropy	S	measured in [cal/°K]

Let Q be the **heat** (measured in [cal]) contained in the system. The differential of heat, dQ, absorbed or emitted by a process is defined as

$$dQ = T\,dS \tag{3.64}$$

which can be used to define the entropy S. From physics we know that

$$dU := dQ - dW = T\,dS - p\,dV \tag{3.65}$$

is an exact differential. This implies that there is a differentiable function $U(S, V)$ such that

$$\frac{\partial U}{\partial S} = T\,, \quad \frac{\partial U}{\partial V} = -p\,, \tag{3.66}$$

and by the exactness criterion (3.57),

$$\frac{\partial T}{\partial V} = -\frac{\partial p}{\partial S}\,.$$

This is the first of the **Maxwell relations**[10], see Exercise 3.72.

Example 3.58. The internal energy $U(S, V)$ has the differential

$$dU = T\,dS - p\,dV$$

where T and p can be determined by differentiating U, namely

$$T = \left(\frac{\partial U}{\partial S}\right)_V\,, \quad p = -\left(\frac{\partial U}{\partial V}\right)_S\,. \tag{3.67}$$

[10] James Clerk Maxwell [1831–1879].

A reasonable approximation of the internal energy for a free gas[11] is given by

$$U = aV^{-\frac{2}{3}}e^{\frac{2S}{3R}} , \qquad (3.68)$$

where the gas constant R is measured, for example in Joules $(\frac{1}{K^\circ}\frac{1}{\text{mole}})$. Using (3.67) we compute

$$p = -\left(\frac{\partial U}{\partial V}\right)_S = \frac{2}{3}\frac{U}{V}, \quad \text{and} \quad T = \left(\frac{\partial U}{\partial S}\right)_V = \frac{2}{3}\frac{U}{R}.$$

We thus get the **caloric equation of state**

$$U = \frac{3}{2}RT ,$$

and by combining the above identities we establish

$$p = \frac{RT}{V} ,$$

the equation of state of **ideal gas**.

Exercises

Exercise 3.59. *Approximate the change in $f(x, y)$ as (x,y) changes from (x_0, y_0) to (x_1, y_1). Compare your results with the true value of $f(x_1, y_1)$.*

(a) $f(x, y) := 1/x + 1/y$; $(x_0, y_0) = (1, 2)$, $(x_1, y_1) = (2, 2.5)$

(b) $f(x, y) := \sec(x/y)$; $(x_0, y_0) = (0, 2)$, $(x_1, y_1) = (\pi/2, 10)$

(c) $f(x, y) := \text{atan}\sqrt{x^2 + y^2}$; $(x_0, y_0) = (0, 0)$, $(x_1, y_1) = (\pi/30, -\pi/20)$

Exercise 3.60. *Approximate the change in $f(x, y, z)$ as (x, y, z) changes from (x_0, y_0, z_0) to (x_1, y_1, z_1). Compare your results with the true value of $f(x_1, y_1, z_1)$.*

(a) $f(x, y, z) := 1/x^2 + 1/y + z^2; (x_0, y_0, z_0) = (1, 1, 0), (x_1, y_1, z_1) = (1, 1.2, 0.5)$

(b) $f(x, y, z) := \cos\left(\frac{x}{y^2 + z}\right); (x_0, y_0, z_0) = (0, 2, 0), (x_1, y_1, z_1) = (\frac{\pi}{10}, 2.2, 0.01)$

Exercise 3.61. *The percent error in measuring a quantity f, PE(f) is given by the following formula*

$$PE(f) := \frac{\Delta f}{f} \times 100\% ,$$

where Δf is the **error** *in measuring f. Use differentials to estimate the percent error of $f(x_1, y_1, z_1)$ from the errors in measuring x, y, and z. Notice that if the errors are small enough we may get good results by using differentials.*

(a) $f(x, y, z) := xyz^3$; *the percent errors of x, y, and z are* 1%, 0.2%, *and* 2%.

(b) $f(x, y, z) := \frac{z}{xy^2}$; *the percent errors of x, y, and z are* 0.5%, 1%, *and* 0.1%.

(c) $f(x, y, z) := \frac{1}{x} + \frac{1}{y} + \frac{1}{z}$; *the percent errors of x, y, z are* 0.1%, 0.2%, 0.1%.

[11] *Thermodynamics: Principles and Applications*, F.C. Andrews, Wiley Interscience (1971).

Exercise 3.62. *Show that the terms represented by ε_1, and ε_2 in equation (3.47) tend to zero as Δx, Δy tend to zero irrespective of their signs.*

Exercise 3.63. *Use (3.58) or (3.59) to construct the function*

$$V = \frac{RT}{p} \quad \text{from its differential} \quad \frac{R}{p}\,dT - \frac{RT}{p^2}\,dp\,, \quad \text{see Example 3.55.}$$

Exercise 3.64. *Approximate the work dW done by the gas in Example 3.51 as it changes from $A(p_1, T_1)$ to $B(p_2, T_2)$:*

(a) *along the line AB*

(b) *along $T = T_1$ until (p_2, T_1) , then along $p = p_2$ to (p_2, T_2).*

Hint: Since the points A and B are close, in the pT–plane, we can use the approximations

$$p \approx \frac{p_1 + p_2}{2}\,, \quad T \approx \frac{T_1 + T_2}{2}\,.$$

Exercise 3.65. *Determine which of the following expressions is an exact differential, in which case, determine a function of which it is a differential.*

(a) $2dx - 3dy$

(b) $xdx + ydy$

(c) $ydx + xdy$

(d) $\dfrac{xdx + ydy}{\sqrt{x^2 + y^2}}$

(e) $\dfrac{1}{y}dx - \dfrac{x}{y^2}dy$

(f) $e^{-xy}\,(ydx + xdy)$

Exercise 3.66. *Find a function f whose exact differential is:*

$$\left(x - \frac{y}{x^2 + y^2}\right) dx + \left(y + \frac{x}{x^2 + y^2}\right) dy\,.$$

Exercise 3.67. *Show that*

$$\frac{dy + f(\frac{y}{x})dx}{y + xf(\frac{y}{x})}$$

is an exact differential.

Exercise 3.68. *The* **Helmholtz**[12] **free energy** *function A also has an exact differential*

$$dA := -S\,dT - p\,dV\,. \tag{3.69}$$

Let A be given by $\quad A := -\dfrac{\alpha}{V} - RT\ln(V - \beta) + f(T)\,,$

where α, β are constants, and f is an arbitrary function of T. Show that this gas has the van der Waals equation of state

$$p = \frac{RT}{V - \beta} - \frac{\alpha}{V^2}\,. \tag{3.70}$$

Exercise 3.69. *The* **Gibbs**[13] **free energy** *function G also has an exact differential*

$$dG = -S\,dT + V\,dp\,.$$

Let G be given by $\quad G = RT\ln p + A + Bp + \dfrac{C}{2}p^2,$

where A, B, C are constants. Find the equation of state for this gas.

---

[12] Ludwig Ferdinand von Helmholtz [1821–1894].

[13] Josiah Willard Gibbs [1839–1903].

Exercise 3.70. *The* **enthalpy** *differential is defined as*

$$dH \;:=\; dU \;+\; p\,dV \;=\; T\,dS \;+\; V\,dp\,. \tag{3.71}$$

Assuming this differential is exact, prove the following relations

$$T \;=\; \frac{\partial H}{\partial S}\;,\quad V \;=\; \frac{\partial H}{\partial p}\,.$$

Exercise 3.71. *The* **free energy** *is defined by*

$$F \;:=\; U \;-\; T\,S \;+\; p\,V\,.$$

Its differential

$$\begin{aligned} dF &= dU - T\,dS - S\,dT + p\,dV + V\,dp \\ &= -S\,dT + V\,dp\;, \quad \text{by (3.71)} \end{aligned}$$

is exact. Show that this implies

$$S \;=\; -\frac{\partial F}{\partial T}\;,\quad V \;=\; \frac{\partial F}{\partial p}\,.$$

Exercise 3.72. *In the following expressions, T, p, S, V are the usual thermodynamic variables temperature, pressure, entropy, and volume respectively. Verify the following* **Maxwell relations** *by making use of the fact that certain thermodynamic variables have exact differentials. Refer to Example 3.57 in the text for the procedure.*

$$\left(\frac{\partial T}{\partial V}\right)_S \;=\; -\left(\frac{\partial p}{\partial S}\right)_V\;,\quad \left(\frac{\partial S}{\partial V}\right)_T \;=\; \left(\frac{\partial p}{\partial T}\right)_V\;, \tag{3.72a}$$

$$\left(\frac{\partial T}{\partial p}\right)_S \;=\; \left(\frac{\partial V}{\partial S}\right)_p\;,\quad \left(\frac{\partial S}{\partial p}\right)_T \;=\; -\left(\frac{\partial V}{\partial T}\right)_p\,. \tag{3.72b}$$

3.6 The chain rule for several variables

Theorem 3.73 (Chain rule: ordinary derivative)**.** Let $z \;:=\; f(x,y)$ have continuous first partial derivatives for $(x,y) \in \mathcal{D} \subset \mathbb{R}^2$. Moreover, let $x = x(t)$, $y = y(t)$ be differentiable functions of $t \in [a,b]$, such that $(x(t),y(t)) \in \mathcal{D}$ for $t \in [a,b]$. Then

$$\frac{df}{dt} \;=\; \frac{\partial f}{\partial x}\,\frac{dx}{dt} \;+\; \frac{\partial f}{\partial y}\,\frac{dy}{dt}\;,\quad t \in [a,b]\,. \tag{3.73}$$

Proof. As t changes from t_0 to $t_0 + \Delta t$, the corresponding change in z is

$$\Delta z \;=\; f_x\,(u(t_0),v(t_0))\,\Delta x + f_y\,(u(t_0),v(t_0))\,\Delta y + \varepsilon_1 \Delta x + \varepsilon_2 \Delta y\;,$$

where Δx and Δy are the corresponding changes in x and y respectively, and ε_1 and ε_2 tend to zero as $\Delta x \to 0$ and $\Delta y \to 0$, see Lemma 3.49. Dividing Δz by Δt and taking the limit as $\Delta t \to 0$,

$$\frac{dz}{dt} = \lim_{\Delta t \to 0} \frac{\Delta z}{\Delta t} = f_x\left(u(t_0), v(t_0)\right) \lim_{\Delta t \to 0} \frac{\Delta x}{\Delta t} + f_y\left(u(t_0), v(t_0)\right) \lim_{\Delta t \to 0} \frac{\Delta y}{\Delta t}$$

proving (3.73). □

This result extends to functions of n variables which are themselves functions of the **independent variable** t. Let

$$z := f\left(x_1(t), x_2(t), x_3(t), \ldots, x_n(t)\right),$$

then $\frac{dz}{dt}$ can be computed using the chain rule

$$\frac{dz}{dt} = \frac{\partial f}{\partial x_1}\frac{dx_1}{dt} + \frac{\partial f}{\partial x_2}\frac{dx_2}{dt} + \cdots + \frac{\partial f}{\partial x_n}\frac{dx_n}{dt}. \tag{3.74}$$

The proof of the above result is similar to the case of two variables and we leave it for Exercise 3.78.

Theorem 3.74 (Chain rule: partial derivatives). Let $w := f(x, y, z)$ have continuous first partial derivatives in $\mathcal{G} \subset \mathbb{R}^3$. Furthermore, let x, y, z themselves be represented as differentiable functions of the variables s and t. Then the partial derivatives $\frac{\partial w}{\partial s}$ and $\frac{\partial w}{\partial t}$ may be computed by the following rules

$$\frac{\partial w}{\partial s} = \frac{\partial f}{\partial x}\frac{\partial x}{\partial s} + \frac{\partial f}{\partial y}\frac{\partial y}{\partial s} + \frac{\partial f}{\partial z}\frac{\partial z}{\partial s}, \tag{3.75a}$$

$$\text{and} \quad \frac{\partial w}{\partial t} = \frac{\partial f}{\partial x}\frac{\partial x}{\partial t} + \frac{\partial f}{\partial y}\frac{\partial y}{\partial t} + \frac{\partial f}{\partial z}\frac{\partial z}{\partial t}. \tag{3.75b}$$

Proof. The proof of this result is similar to the previous theorem, and is left for Exercise 3.77. □

Example 3.75. Sometimes it is convenient to change from the Cartesian coordinates ((x, y), or (x, y, z)) to another coordinate system, which takes advantage of a symmetry in a given problem. For example, if there is **radial symmetry** it is convenient to use **polar coordinates**.

$$\begin{aligned} x &= r\cos\theta \\ y &= r\sin\theta \end{aligned} \tag{3.76}$$

Let $u := f(x, y)$ be a continuously differentiable function of the variables x and y. It is required to compute the partial derivatives

$$\frac{\partial u}{\partial r} \quad \text{and} \quad \frac{\partial u}{\partial \theta}.$$

First we compute the partial derivatives

$$\frac{\partial x}{\partial r} = \cos\theta \ , \quad \frac{\partial y}{\partial r} = \sin\theta \ , \tag{3.77a}$$

$$\frac{\partial x}{\partial \theta} = -r\sin\theta \ , \quad \frac{\partial y}{\partial \theta} = r\cos\theta \ , \tag{3.77b}$$

then using the chain rule,

$$\frac{\partial w}{\partial r} = \frac{\partial w}{\partial x}\frac{\partial x}{\partial r} + \frac{\partial w}{\partial y}\frac{\partial y}{\partial r} = \frac{\partial w}{\partial x}\cos\theta + \frac{\partial w}{\partial y}\sin\theta \ , \tag{3.78a}$$

$$\frac{\partial w}{\partial \theta} = \frac{\partial w}{\partial x}\frac{\partial x}{\partial \theta} + \frac{\partial w}{\partial y}\frac{\partial y}{\partial \theta} = -r\frac{\partial w}{\partial x}\sin\theta + r\frac{\partial w}{\partial y}\cos\theta \ . \tag{3.78b}$$

Example 3.76. Consider the function $w = f(u, v, x, y)$ where u and v are functions of x and y, i.e. $u = g(x, y)$ and $v = h(x, y)$. Then

$$\frac{\partial w}{\partial x} = \frac{\partial f}{\partial x} + \frac{\partial w}{\partial u}\frac{\partial g}{\partial x} + \frac{\partial w}{\partial v}\frac{\partial h}{\partial x} \ ,$$

taking into account the fact that u and v are also functions of x. Similarly we compute

$$\frac{\partial w}{\partial y} = \frac{\partial f}{\partial y} + \frac{\partial w}{\partial u}\frac{\partial g}{\partial y} + \frac{\partial w}{\partial v}\frac{\partial h}{\partial y} \ .$$

Exercises

Exercise 3.77. *Prove the partial derivative form of the chain rule.*

Exercise 3.78. *Prove the n-variable version of the chain rule equation (3.74).*

Exercise 3.79. *Find* $\frac{\partial u}{\partial r}$ *and* $\frac{\partial u}{\partial s}$ *for*

(a) $u := x^2 + y^2$ *where* $\begin{cases} x = 2r + 3s + 4 \\ y = -r + 5s - 27 \end{cases}$

(b) $u := \dfrac{x - y}{1 + xy}$ *where* $\begin{cases} x = \tan(2r + s) \\ y = \cot(r^2 s) \end{cases}$

Exercise 3.80. *Find*

$$\frac{du}{d\theta} \quad \textit{for} \quad u := xyz \quad \textit{where} \quad \begin{cases} x = a\cos\theta \\ y = a\sin\theta \\ z = b\theta \end{cases}$$

Exercise 3.81. *Find* $\frac{\partial u}{\partial \xi}$, $\frac{\partial u}{\partial \eta}$, $\frac{\partial u}{\partial \zeta}$ *for* u*, a function of* x , y , z*, where*

$$\begin{cases} x = & a_1\xi + b_1\eta + c_1\zeta \\ y = & a_2\xi + b_2\eta + c_2\zeta \\ z = & a_3\xi + b_3\eta + c_3\zeta \end{cases}$$

Exercise 3.82. *Show that the change of variables* $\begin{cases} x = r\cos\theta \\ y = r\sin\theta \end{cases}$ *transforms the equation*

$$\left(\frac{\partial u}{\partial x}\right)^2 + \left(\frac{\partial u}{\partial y}\right)^2 = 0 \quad \textit{into} \quad \left(\frac{\partial u}{\partial r}\right)^2 + \frac{1}{r^2}\left(\frac{\partial u}{\partial \theta}\right)^2 = 0 \ .$$

Exercise 3.83. *If $f(\frac{x}{y})$ is differentiable show that it satisfies the partial differential equation*

$$x\,\frac{\partial f}{\partial x} + y\,\frac{\partial f}{\partial y} = 0\,.$$

Exercise 3.84. *Let u and v satisfy the* **Cauchy**[14]**–Riemann**[15] **equations**

$$\frac{\partial u}{\partial x} = \frac{\partial v}{\partial y}\,, \quad \text{and} \quad \frac{\partial u}{\partial y} = -\frac{\partial v}{\partial x}\,. \tag{3.79}$$

Show that

$$\frac{\partial u}{\partial r} = \frac{1}{r}\frac{\partial v}{\partial \theta}\,, \quad \text{and} \quad \frac{1}{r}\frac{\partial u}{\partial \theta} = -\frac{\partial v}{\partial r}\,.$$

Exercise 3.85. *Let u and v satisfy the equations (3.79) and let $w(x,y)$ be* **any** *third function. Show that*

$$\frac{\partial^2 w}{\partial x^2} + \frac{\partial^2 w}{\partial y^2} = \left(\frac{\partial^2 w}{\partial u^2} + \frac{\partial^2 w}{\partial v^2}\right)\left(\left(\frac{\partial w}{\partial x}\right)^2 + \left(\frac{\partial w}{\partial y}\right)^2\right)\,.$$

Exercise 3.86. *Let $f(x,y,z)$ have two continuous derivatives, and change from Cartesian coordinates to spherical coordinates:*

$$x := \rho\sin\varphi\cos\theta\,, \quad y := \rho\sin\varphi\sin\theta\,, \quad z := \rho\cos\varphi\,. \tag{3.80}$$

Prove the identity:

$$\frac{\partial^2 f}{\partial x^2} + \frac{\partial^2 f}{\partial y^2} + \frac{\partial^2 f}{\partial z^2} = \frac{\partial^2 f}{\partial \rho^2} + \frac{1}{\rho^2}\frac{\partial^2 f}{\partial \varphi^2} + \frac{1}{\rho^2\sin^2\varphi}\frac{\partial^2 f}{\partial \theta^2} + \frac{2}{\rho}\frac{\partial f}{\partial \rho} + \frac{\cot\varphi}{\rho^2}\frac{\partial f}{\partial \varphi}\,. \tag{3.81}$$

Exercise 3.87. *Let $f(x,y,z)$ have two continuous derivatives, and change from Cartesian coordinates to cylindrical coordinates:*

$$x := r\cos\theta\,, \quad y := r\sin\theta\,, \quad z := z\,. \tag{3.82}$$

Prove the identity

$$\frac{\partial^2 f}{\partial x^2} + \frac{\partial^2 f}{\partial y^2} + \frac{\partial^2 f}{\partial z^2} = \frac{\partial^2 f}{\partial r^2} + \frac{1}{r}\frac{\partial f}{\partial r} + \frac{1}{r^2}\frac{\partial^2 f}{\partial \theta^2} + \frac{\partial^2 f}{\partial z^2}\,. \tag{3.83}$$

Exercise 3.88. *Show that the twice differentiable functions $f(x+t)$ and $g(x-t)$ satisfy the* **partial differential equation**

$$\frac{\partial^2 u}{\partial x^2} = \frac{\partial^2 u}{\partial t^2}\,.$$

Exercise 3.89. *Let u and v satisfy the equations (3.79), and change to polar coordinates*

$$\begin{aligned} x &= r\cos\theta \\ y &= r\sin\theta \end{aligned} \tag{3.84}$$

Show that both u and v satisfy the identity (here w stands for either u or v)

$$\frac{\partial^2 w}{\partial x^2} + \frac{\partial^2 w}{\partial y^2} = \frac{\partial^2 w}{\partial r^2} + \frac{1}{r^2}\frac{\partial^2 w}{\partial \theta^2}\,. \tag{3.85}$$

[14] Augustin-Louis Cauchy [1789–1857]
[15] Bernhard Riemann [1826–1866]

3.7 The implicit function theorem

This section is optional. It can be skipped at first reading. The material covered here is not used elsewhere in this book.

Consider an implicit relation

$$F(x, y) = 0 , \tag{3.86}$$

and its **graph**, GRAPH(3.86) := $\{(x, y) : F(x, y) = 0\}$. This implicit equation can be solved for y

$$y := f(x) , \tag{3.87}$$

if, and only if, GRAPH(3.86) satisfies the vertical line test. Substituting $y = f(x)$ in (3.86) we then get an identity

$$F(x, f(x)) \equiv 0 , \tag{3.88}$$

which holds for all x in the domain of f.

Often it suffices to find a local solution of (3.86) at a point (x_0, y_0) on GRAPH(3.86). I.e. it is required to find a solution $y = f(x)$ such that (3.92) holds in a neighborhood of (x_0, y_0). Differentiating (3.92) we get

$$F_x + F_y \frac{dy}{dx} = 0 \tag{3.89a}$$

$$\therefore \frac{dy}{dx} = -\frac{F_x}{F_y} \tag{3.89b}$$

Assuming the partial derivatives F_x and F_y are continuous, a local solution $y = f(x)$ exists at (x_0, y_0) if

$$F_y(x_0, y_0) \neq 0 . \tag{3.90}$$

Indeed, from (3.89b) we get the slope of the local solution $y = f(x)$, which is finite by (3.91). The solution can then be approximated, near x_0, by

$$f(x_0 + dx) \approx y_0 + f'(x_0)\, dx = y_0 - \frac{F_x(x_0, y_0)}{F_y(x_0, y_0)}\, dx .$$

The finiteness of the slope $f'(x)$ is analogous to the vertical line test for GRAPH(3.86).

The following theorem, stated without proof, guarantees also that the local solution is continuously differentiable at x_0.

Theorem 3.90 (The Implicit Function Theorem–I). Let the function $F(x, y)$ have continuous partial derivatives in a neighborhood of the point (x_0, y_0), and let $F(x_0, y_0) = 0$. If

$$F_y(x_0, y_0) \neq 0 , \tag{3.91}$$

then there is a function $y = f(x)$ which satisfies:

$$F(x, f(x)) \equiv 0 , \tag{3.92}$$

for all x in some neighborhood of x_0. Moreover, the derivative $f'(x)$ is continuous in a neighborhood of x_0. □

An analogous result holds for implicit equations in three variables

$$F(x, y, z) = 0 \,. \tag{3.93}$$

Let the partial derivatives F_x , F_y , F_z be continuous in a neighborhood of a point (x_0, y_0, z_0) where $F(x_0, y_0, z_0) = 0$. If one of the partial derivatives is nonzero at (x_0, y_0, z_0), say

$$F_z(x_0, y_0, z_0) \neq 0 \,, \tag{3.94}$$

then it is possible to solve (3.99) for z to obtain a local solution

$$z = f(x, y) \tag{3.95}$$

satisfying

$$F(x, y, f(x, y)) = 0 \,, \tag{3.96}$$

in some neighborhood of (x_0, y_0, z_0). This fact is now stated without proof.

Theorem 3.91 (The Implicit Function Theorem–II)**.** Let the function $F(x, y, z)$ have continuous partial derivatives in a neighborhood of the point (x_0, y_0, z_0), and let $F(x_0, y_0, z_0) = 0$. If

$$F_z(x_0, y_0, z_0) \neq 0 \,, \tag{3.97}$$

then there is a function $z = f(x, y)$ which satisfies:

$$F(x, y, f(x, y)) = 0 \,, \tag{3.98}$$

for all (x, y) in some neighborhood of (x_0, y_0). Moreover, the function f has continuous partial derivatives in some neighborhood of (x_0, y_0). □

The proofs of the Implicit Function Theorems 3.90–3.91 belong to a more advanced course.

Exercises

Exercise 3.92. *Use Theorem 3.91 to check whether the following implicit equations can be solved for z at the point $(x_0, y_0, z_0) := (0, 0, 0)$. If possible, find the solution.*

(a) $F(x, y, z) := x^3 + 5xy^2 + \sin z$ (b) $F(x, y, z) := \exp(x + y + z)$

(c) $F(x, y, z) := (x^3 - y^3 + z^3)(x^2 + y^2 + z^2)$ (d) $F(x, y, z) := \sin(x^2 + y^2 + z^2)$

3.8 Implicit differentiation

Consider an implicit equation

$$F(x, y, z) = 0 \,, \tag{3.99}$$

satisfied by a point $P(x_0, y_0, z_0)$. We assume that F and its partial derivatives are continuous in a neighborhood of P, and at least one of the partial derivatives does not vanish at P, say

$$F_z(x_0, y_0, z_0) \neq 0 .$$

Then solving for z we get a local solution

$$z = f(x, y) .$$

We consider z a function of two **independent variables**, x and y. Differentiating $F(x, y, z) = 0$ partially with respect to x we get, using the chain rule and $\frac{\partial y}{\partial x} = \frac{\partial x}{\partial y} = 0$ (since x and y are independent):

$$\frac{\partial F}{\partial x} + \frac{\partial F}{\partial z}\frac{\partial z}{\partial x} = 0 ; \tag{3.100}$$

similarly, differentiation w.r.t. y yields

$$\frac{\partial F}{\partial y} + \frac{\partial F}{\partial z}\frac{\partial z}{\partial y} = 0 . \tag{3.101}$$

We solve to get the implicit differentiation formulæ

$$\frac{\partial z}{\partial x} = -\frac{F_x}{F_z}, \quad \frac{\partial z}{\partial y} = -\frac{F_y}{F_z}. \tag{3.102}$$

Example 3.93. Consider the implicit equation

$$x^2 + y^2 + z^2 = 1 ,$$

whose graph is the unit sphere. Solving for z we get two branches

$$z := +\sqrt{1 - x^2 - y^2} \quad \text{and} \quad z := -\sqrt{1 - x^2 - y^2} .$$

The partial derivatives of the **positive** branch (the upper hemisphere) are computed directly as:

$$\frac{\partial z}{\partial x} = -\frac{x}{\sqrt{1 - x^2 - y^2}} , \quad \frac{\partial z}{\partial y} = -\frac{y}{\sqrt{1 - x^2 - y^2}} .$$

The partial derivatives of the **negative branch** are seen to be the negatives of these.

The same partial derivatives are now computed indirectly, using the implicit differentiation (3.102):

$$2x + 2z\frac{\partial z}{\partial x} = 0 \quad \text{and} \quad 2y + 2z\frac{\partial z}{\partial y} = 0 ,$$

which we solve for the partial derivatives to get

$$\frac{\partial z}{\partial x} = -\frac{x}{z} \quad \text{and} \quad \frac{\partial z}{\partial y} = -\frac{y}{z} .$$

Substituting z (the positive, or negative, branch) we get the correct partial derivatives for that branch. △

Implicit differentiation can be used also to compute higher order derivatives. Differentiating equations (3.100)–(3.101) with respect to x and with respect to y leads to just three new equations,

$$\frac{\partial^2 F}{\partial x^2} + 2\frac{\partial^2 F}{\partial x\,\partial z}\frac{\partial z}{\partial x} + \frac{\partial^2 F}{\partial z^2}\left(\frac{\partial z}{\partial x}\right)^2 + \frac{\partial F}{\partial z}\frac{\partial^2 z}{\partial x^2} = 0 \quad (3.103a)$$

$$\frac{\partial^2 F}{\partial x\,\partial y} + \frac{\partial^2 F}{\partial x\,\partial z}\frac{\partial z}{\partial y} + \frac{\partial^2 F}{\partial y\,\partial z}\frac{\partial z}{\partial x} + \frac{\partial^2 F}{\partial z^2}\frac{\partial z}{\partial x}\frac{\partial z}{\partial y} + \frac{\partial F}{\partial z}\frac{\partial^2 z}{\partial x\,\partial y} = 0 \quad (3.103b)$$

$$\frac{\partial^2 F}{\partial y^2} + 2\frac{\partial^2 F}{\partial y\,\partial z}\frac{\partial z}{\partial y} + \frac{\partial^2 F}{\partial z^2}\left(\frac{\partial z}{\partial y}\right)^2 + \frac{\partial F}{\partial z}\frac{\partial^2 z}{\partial y^2} = 0 \quad (3.103c)$$

because

$$\frac{\partial(3.100)}{\partial y} = \frac{\partial(3.101)}{\partial x} = (3.103b)\ .$$

These equations can be solved for the second order partial derivatives

$$\frac{\partial^2 z}{\partial x^2}\ ,\ \frac{\partial^2 z}{\partial x\,\partial y}\ ,\ \frac{\partial^2 z}{\partial y^2}$$

after computing all the other partial derivatives in (3.103).

Example 3.94. Consider a **state equation**

$$F(p,V,T) = 0\ , \quad (3.104)$$

relating the pressure p, volume V and temperature T of a gas. The differential of F is

$$dF = \left(\frac{\partial F}{\partial p}\right)_{V,T} dp + \left(\frac{\partial F}{\partial V}\right)_{p,T} dV + \left(\frac{\partial F}{\partial T}\right)_{p,V} dT = 0\ , \quad (3.105)$$

where subscripts denote the variables which are held constant when taking the derivative. If pressure is held fixed, i.e. $dp = 0$, then we obtain from (3.105)

$$\left(\frac{\partial V}{\partial T}\right)_p = -\frac{(\partial F/\partial T)_{p,V}}{(\partial F/\partial V)_{p,T}}\ . \quad (3.106a)$$

Similarly, if $dV = 0$

$$\left(\frac{\partial T}{\partial p}\right)_V = -\frac{(\partial F/\partial p)_{T,V}}{(\partial F/\partial T)_{p,V}}\ , \quad (3.106b)$$

and for $dT = 0$

$$\left(\frac{\partial p}{\partial V}\right)_T = -\frac{(\partial F/\partial V)_{p,T}}{(\partial F/\partial p)_{T,V}}\ . \quad (3.106c)$$

Three different products may be obtained by multiplying these equations together, one of which is

$$\left(\frac{\partial p}{\partial V}\right)_T \left(\frac{\partial V}{\partial T}\right)_p = -\left(\frac{\partial p}{\partial T}\right)_V\ . \quad (3.107)$$

See Exercise 3.105 the other two identities.

Example 3.95 (Change of Independent Variables)**.** In thermodynamics it is sometimes convenient to use variables other than (p,V,T) in the state equation (3.104). We show here how to compute the partial derivatives w.r.t. the new variables.

For example, let the pressure p be given as a function of the **internal energy** U and the **volume** V. Then clearly

$$dp = \left(\frac{\partial p}{\partial U}\right)_V dU + \left(\frac{\partial p}{\partial V}\right)_U dV . \tag{3.108}$$

Suppose further that the internal energy is given as a function of temperature and volume,

$$U = U(T,V) .$$

Then

$$dU = \left(\frac{\partial U}{\partial T}\right)_V dT + \left(\frac{\partial U}{\partial V}\right)_T dV . \tag{3.109}$$

Substituting dU from (3.109) into (3.108) we obtain

$$dp = \left(\frac{\partial p}{\partial U}\right)_V \left(\frac{\partial U}{\partial T}\right)_V dT + \left[\left(\frac{\partial p}{\partial V}\right)_U + \left(\frac{\partial p}{\partial U}\right)_V \left(\frac{\partial U}{\partial V}\right)_T\right] dV . \tag{3.110}$$

On the other hand, substituting $U = U(T,V)$ into $p = p(U,V)$ we obtain a new function $p = p(T,V)$[16] with differential

$$dp = \left(\frac{\partial p}{\partial T}\right)_V dT + \left(\frac{\partial p}{\partial V}\right)_T dV . \tag{3.111}$$

Comparing coefficients in (3.110) and (3.111) we conclude

$$\left(\frac{\partial p}{\partial T}\right)_V = \left(\frac{\partial p}{\partial U}\right)_V \left(\frac{\partial U}{\partial T}\right)_V , \tag{3.112a}$$

$$\left(\frac{\partial p}{\partial V}\right)_T = \left(\frac{\partial p}{\partial V}\right)_U + \left(\frac{\partial p}{\partial U}\right)_V \left(\frac{\partial U}{\partial V}\right)_T . \tag{3.112b}$$

We apply this technique for a specific example, of the **virial equation of state**

$$p(T,V) = \frac{RT}{V} + \frac{\beta\, RT - \alpha}{V^2} . \tag{3.113}$$

Let the internal energy be given by

$$U(T,V) = \frac{5}{2} RT + \alpha \left(\frac{1}{V} - \frac{1}{V_0}\right) . \tag{3.114}$$

Here α, β and V_0 are constants. From (3.121) we can compute directly

$$\left(\frac{\partial U}{\partial V}\right)_T = -\frac{\alpha}{v^2} . \tag{3.115}$$

The computation of $\left(\frac{\partial U}{\partial V}\right)_p$ is more complicated. We use the identity

$$\left(\frac{\partial U}{\partial V}\right)_p = \left(\frac{\partial U}{\partial V}\right)_T + \left(\frac{\partial U}{\partial T}\right)_V \left(\frac{\partial T}{\partial V}\right)_p . \tag{3.116}$$

[16]Here we abuse mathematical notation, as physicists sometimes do, by writing $p = p(T,V)$ and $p = p(U,V)$ as if these are the same function.

We now compute

$$\left(\frac{\partial U}{\partial T}\right)_V = \frac{5}{2}R\,,$$

and use

$$\left(\frac{\partial T}{\partial V}\right)_p = \frac{\left(\frac{\partial p}{\partial V}\right)_T}{\left(\frac{\partial p}{\partial T}\right)_V}$$

to compute $\left(\frac{\partial T}{\partial V}\right)_p$. We have

$$\begin{aligned}\left(\frac{\partial p}{\partial V}\right)_T &= -\frac{2RT}{V^3\left(\beta-\frac{\alpha}{RT}\right) - \frac{RT}{V^2}} \\ \text{and} \quad \left(\frac{\partial p}{\partial T}\right)_V &= \frac{R}{V}\left(1+\frac{\beta}{V}\right). \\ \therefore \quad \left(\frac{\partial T}{\partial V}\right)_p &= \frac{T\left(\frac{2}{V}(\beta-\frac{\alpha}{RT})-1\right)}{V+\beta}\,,\end{aligned}$$

and the desired result is (3.117)

$$\left(\frac{\partial U}{\partial V}\right)_p = -\frac{\alpha}{V^2} + \left(\frac{5R}{2}\right)\frac{T\left(\frac{2}{V}(\beta-\frac{\alpha}{RT})-1\right)}{V+\beta}\,. \quad (3.118)$$

This is a rather long problem, but it illustrates well the usefulness of the procedure. Try on the other hand to solve (3.120) for V and substitute this into (3.121) to compute the derivative directly. See Exercise 3.109 in this regard.

Exercises

Exercise 3.96. *Compute the first order partial derivatives of z with respect to x and y of the functions below using implicit differentiation.*

(a) $1 = e^{x^4+z^2-xy^3}$ (b) $z = \tanh(x-y+z^2)$

(c) $z + x = \ln(x^2+y^2+z^2)$ (d) $1 = x^2\sin(y+x^2+z^2)$

Exercise 3.97. *Compute implicitly the first order partial derivatives with respect to x and y of z in the exercises below.*

(a) $e^{x^4-zy^3} = 0$ (b) $z^2 = \tanh(x-zy)$

(c) $z = \ln(x^2-y^2)$ (d) $x^2\sin(y+x^2z) = 1$

Exercise 3.98. *Compute implicitly the second order derivatives of z in the functions below.*

(a) $x^3+5zy^2+z^2 = 1$ (b) $(x+y+z)^7 = z$

(c) $\dfrac{x^3-y^3+z}{x^2+y^2+z^2} = 1$ (d) $\dfrac{\sin(x^2+y^2)+\cos(z)}{x^4+y^4+z^4} = 1$

Exercise 3.99. *Compute the formulas for the third derivative of y with respect to x where an implicit function is defined by $F(x,y) = 0$.*

Exercise 3.100. *Compute the formulas for the third derivatives of z with respect to x and y where z is defined as an implicit function $F(x,y,z) = 0$.*

Exercise 3.101. *Suppose a curve $\mathcal{C}$ is represented as the intersection of the two surfaces $\mathcal{S}_1$ and $\mathcal{S}_2$. Moreover, let the surface $\mathcal{S}_j$ be defined by the implicit relation $F_j(x, y, z) = 0$, (j=1,2). Compute a formula for the curvature of $\mathcal{C}$.*

Exercise 3.102. *Compute the partial derivatives $\frac{\partial z}{\partial x}$ and $\frac{\partial z}{\partial y}$ for the following surfaces. Use the Implicit Function Theorem to state where you may solve for z. Where can you solve for x and where can you solve for y?*

(a) $\dfrac{x^2}{4} + \dfrac{y^2}{1} + \dfrac{z^2}{9} = 1$ (b) $\dfrac{x^2}{2} - \dfrac{y^2}{1} + \dfrac{z^2}{9} = 1$

(c) $\dfrac{x^2}{1} + \dfrac{y^2}{1} - \dfrac{z^2}{9} = 1$ (d) $\dfrac{x^2}{9} - \dfrac{y^2}{1} - \dfrac{z^2}{4} = 1$

(e) $-\dfrac{x^2}{4} - \dfrac{y^2}{9} + z^2 = 0$ (f) $\dfrac{x^2}{9} + \dfrac{y^2}{4} = z$

Exercise 3.103. *Compute implicitly the first order partial derivatives with respect to x and y of z in the exercises below.*

(a) $e^{x^4 - zy^3} = 0$ (b) $z^2 = \tanh(x - zy)$

(c) $z = \ln(x^2 - y^2)$ (d) $x^2 \sin(y + x^2 z) = 1$

Exercise 3.104. *Consider the surfaces $F(x, y, z,) = 0$ below. Compute the derivatives $\frac{\partial F}{\partial x}$, $\frac{\partial F}{\partial y}$, $\frac{\partial F}{\partial z}$. Where are these derivatives defined? Where are the derivatives $\frac{\partial z}{\partial x}$ and $\frac{\partial z}{\partial y}$ defined? Compute the expressions*

$$-\frac{\partial F}{\partial x} / \frac{\partial F}{\partial z}$$

and compare these with $\frac{\partial z}{\partial x}$. What do you conclude? Repeat the process for

$$-\frac{\partial F}{\partial y} / \frac{\partial F}{\partial z}$$

and $\frac{\partial z}{\partial y}$ and draw a conclusion.

(a) $x^2 + y^2 - 4z = 0$ (b) $16x + y^2 - 4z^2 = 0$

(c) $9x^2 + 4y^2 + z^2 = 36$ (d) $2x^2 + y + 4z^2 + 1 = 0$

(e) $x = \dfrac{y^2}{4} + \dfrac{z^2}{3}$

Exercise 3.105. *Given an equation of state*

$$F(p, V, T) \;=\; 0\,,$$

derive, similarly to (3.107), the identities

$$\left(\frac{\partial p}{\partial V}\right)_T \left(\frac{\partial T}{\partial p}\right)_V \left(\frac{\partial V}{\partial T}\right)_p \;=\; -1\,, \tag{3.119a}$$

$$\left(\frac{\partial V}{\partial T}\right)_p \left(\frac{\partial T}{\partial p}\right)_V \left(\frac{\partial p}{\partial V}\right)_T \;=\; -1. \tag{3.119b}$$

Exercise 3.106. *An* **ideal gas** *has an equation of state of the form* $F(p, V, T) := pV - RT$*, where* RT *is a* **universal constant** *and depends on the units used. Compute the partial derivatives* $\left(\frac{\partial p}{\partial V}\right)_T$*,* $\left(\frac{\partial V}{\partial T}\right)_p$*, and* $\left(\frac{\partial T}{\partial p}\right)_V$*. Verify the identities (3.119a)–(3.119b) above for the case of the ideal gas.*

Exercise 3.107. *Repeat the previous problem for the van der Waals gas (3.36). gas.*

Exercise 3.108. *Let the pressure of a gas be given by the virial equation of state*

$$p(T, V) \;=\; \frac{RT}{V} \;+\; \frac{\beta\, RT - \alpha}{V^2}\,. \tag{3.120}$$

and the internal energy be given by

$$U(T, V) \;=\; \frac{5}{2}\,RT \;+\; \alpha\left(\frac{1}{V} - \frac{1}{V_0}\right)\,. \tag{3.121}$$

Show that

$$\left(\frac{\partial U}{\partial T}\right)_p \;=\; \frac{5R}{2} \;-\; \frac{\alpha(V + \beta)}{V^2 T}\,. \tag{3.122}$$

Exercise 3.109. *Using* MATHEMATICA *compute*

(a) $\left(\frac{\partial U}{\partial T}\right)_p$ (b) $\left(\frac{\partial U}{\partial V}\right)_p$

by first solving for V *and substituting into the internal energy as given in the previous exercise. Compare the answer of (a) to (3.122).*

3.9 Jacobians

Sometimes it is desirable to change variables, say move from the variables (x, y) to new variables (u, v). In this connection we need to study certain matrices, called **Jacobi matrices** or **Jacobians**, and their determinants.

Let the variables (x, y) be expressed in terms of the variables (u, v) as follows

$$x \;=\; f(u, v) \tag{3.123a}$$

$$y \;=\; g(u, v) \tag{3.123b}$$

This is a transformation $\mathbf{T}$ taking points (u, v) into points (x, y),

$$(x, y) \;=\; \mathbf{T}(u, v)\,. \tag{3.124}$$

We assume that the transformation $\mathbf{T}$ maps a region $\mathbf{U}$ in the uv–plane into a region $\mathbf{X}$ in the xy–plane and is one-to-one, i.e. $\mathbf{T}$ gives exactly one $(x, y) \in \mathbf{X}$ for every $(u, v) \in \mathbf{U}$. The functions f and g are assumed to have continuous first partial derivatives in $\mathbf{U}$.

Definition 3.110. Let $f, g, \mathbf{U}$ and $\mathbf{X}$ be as above. The **Jacobi**[17] **matrix** (or

[17] KARL GUSTAV JACOB JACOBI [1804–1851].

Jacobian) of $\mathbf{T}$ is the 2×2 matrix of partial derivatives

$$J_{\mathbf{T}} := \begin{pmatrix} \dfrac{\partial f}{\partial u} & \dfrac{\partial f}{\partial v} \\ \dfrac{\partial g}{\partial u} & \dfrac{\partial g}{\partial v} \end{pmatrix}, \quad \text{or} \quad \begin{pmatrix} f_u & f_v \\ g_u & g_v \end{pmatrix}, \quad \text{or} \quad \begin{pmatrix} x_u & x_v \\ y_u & y_v \end{pmatrix}. \tag{3.125}$$

Its determinant is denoted by

$$\frac{\partial(f,g)}{\partial(u,v)} := \mathbf{det} \begin{pmatrix} \dfrac{\partial f}{\partial u} & \dfrac{\partial f}{\partial v} \\ \dfrac{\partial g}{\partial u} & \dfrac{\partial g}{\partial v} \end{pmatrix}, \quad \text{also denoted} \quad \frac{\partial(x,y)}{\partial(u,v)}. \tag{3.126}$$

The elements of the Jacobian (5.58) are functions of (u, v). Therefore the Jacobian itself is a function of (u, v),

$$J_{\mathbf{T}}(u,v) = \begin{pmatrix} f_u(u,v) & f_v(u,v) \\ g_u(u,v) & g_v(u,v) \end{pmatrix}.$$

The determinant of the Jacobian, $\mathbf{det}\, J_{\mathbf{T}}(u, v)$, is also a function of (u, v).

Since the mapping $\mathbf{T}$ is one-to-one, each point $(x, y) \in \mathbf{X}$ corresponds to a unique point $(u, v) \in \mathbf{U}$. Then it is possible to go back from $\mathbf{X}$ to $\mathbf{U}$ using the **inverse mapping** $\mathbf{T}^{-1}$,

$$(u,v) = \mathbf{T}^{-1}(x,y) \tag{3.127}$$

taking points $(x, y) \in \mathbf{X}$ into points $(u, v) \in \mathbf{U}$. We recall that $\mathbf{T}^{-1}$ is also one-to-one, i.e. any point $(x, y) \in \mathbf{X}$ is mapped into a unique point $(u, v) \in \mathbf{U}$, given by (5.57). We write this inverse transformation in detail as

$$u = F(x,y) \tag{3.128a}$$

$$v = G(x,y) \tag{3.128b}$$

Substituting the inverse mapping (3.128) into the original mapping (3.123), we get

$$x = f(F(x,y), G(x,y)) \tag{3.129a}$$

$$y = g(F(x,y), G(x,y)) \tag{3.129b}$$

Differentiating the equations (3.129) partially w.r.t. x we obtain

$$1 = \frac{\partial f}{\partial u}\frac{\partial F}{\partial x} + \frac{\partial f}{\partial v}\frac{\partial G}{\partial x} \quad \text{or} \quad \frac{\partial x}{\partial u}\frac{\partial u}{\partial x} + \frac{\partial x}{\partial v}\frac{\partial v}{\partial x} = 1 \tag{3.130a}$$

$$0 = \frac{\partial g}{\partial u}\frac{\partial F}{\partial x} + \frac{\partial g}{\partial v}\frac{\partial G}{\partial x} \quad \text{or} \quad \frac{\partial y}{\partial u}\frac{\partial u}{\partial x} + \frac{\partial y}{\partial v}\frac{\partial v}{\partial x} = 0 \tag{3.130b}$$

Similarly, partial differentiation of (3.129) w.r.t. y gives

$$0 = \frac{\partial f}{\partial u}\frac{\partial F}{\partial y} + \frac{\partial f}{\partial v}\frac{\partial G}{\partial y} \quad \text{or} \quad \frac{\partial x}{\partial u}\frac{\partial u}{\partial y} + \frac{\partial x}{\partial v}\frac{\partial v}{\partial y} = 0 \quad (3.131a)$$

$$1 = \frac{\partial g}{\partial u}\frac{\partial F}{\partial y} + \frac{\partial g}{\partial v}\frac{\partial G}{\partial y} \quad \text{or} \quad \frac{\partial y}{\partial u}\frac{\partial u}{\partial y} + \frac{\partial y}{\partial v}\frac{\partial v}{\partial y} = 1 \quad (3.131b)$$

These identities can be combined in a matrix multiplication

$$\begin{pmatrix} \frac{\partial f}{\partial u} & \frac{\partial f}{\partial v} \\ \frac{\partial g}{\partial u} & \frac{\partial g}{\partial v} \end{pmatrix} \begin{pmatrix} \frac{\partial F}{\partial x} & \frac{\partial F}{\partial y} \\ \frac{\partial G}{\partial x} & \frac{\partial G}{\partial y} \end{pmatrix} = \begin{pmatrix} 1 & 0 \\ 0 & 1 \end{pmatrix} . \quad (3.132)$$

Therefore the Jacobian $J_{\mathbf{T}}$ is nonsingular, and its inverse is

$$J_{\mathbf{T}}^{-1} = \begin{pmatrix} \frac{\partial f}{\partial u} & \frac{\partial f}{\partial v} \\ \frac{\partial g}{\partial u} & \frac{\partial g}{\partial v} \end{pmatrix}^{-1} = \begin{pmatrix} \frac{\partial F}{\partial x} & \frac{\partial F}{\partial y} \\ \frac{\partial G}{\partial x} & \frac{\partial G}{\partial y} \end{pmatrix} = J_{\mathbf{T}^{-1}} , \quad (3.133)$$

the **Jacobi matrix** of the inverse transformation $\mathbf{T}^{-1}$. We see that the assumption that $\mathbf{T}$ is one-to-one implies that the Jacobian $J_{\mathbf{T}}$ is nonsingular or equivalently, that its determinant $\partial(f,g)/\partial(u,v)$ is nonzero. We can express (3.133) as

the Jacobian of $\mathbf{T}^{-1}$ is the inverse of the Jacobian of $\mathbf{T}$.

Taking determinants of (3.132) we get,

$$\mathbf{det}\left(J_{\mathbf{T}} J_{\mathbf{T}}^{-1}\right) = \mathbf{det}\left(J_{\mathbf{T}}\right) \mathbf{det}\left(J_{\mathbf{T}}^{-1}\right) = \mathbf{det}\begin{pmatrix} 1 & 0 \\ 0 & 1 \end{pmatrix} = 1 ,$$

so the determinants of the Jacobians of $\mathbf{T}$ and $\mathbf{T}^{-1}$ satisfy the important identity

$$\frac{\partial(x,y)}{\partial(u,v)}\frac{\partial(u,v)}{\partial(x,y)} = 1 . \quad (3.134)$$

Example 3.111 (A linear transformation)**.** The transformation $\mathbf{T}$ is **linear** if it is defined by

$$< x , y > = T < u , v > , \quad (3.135)$$

where T is a given 2×2 matrix. The Jacobian of $\mathbf{T}$ is then the same as the matrix T. If this matrix is nonsingular, then $\mathbf{T}$ maps the whole xy–plane into the whole uv–plane, and the inverse transformation $\mathbf{T}^{-1}$ is given by

$$< u , v > = T^{-1} < x , y > ; , \quad (3.136)$$

for all (x, y) . For example, let $(x, y) = \mathbf{T}(u, v)$ be defined by

$$\begin{matrix} x & = & u & + & v \\ y & = & u & - & v \end{matrix} \quad \text{or} \quad < x , y > = T < u , v > \quad \text{where} \quad T = \begin{pmatrix} 1 & 1 \\ 1 & -1 \end{pmatrix} .$$

In particular, $\mathbf{T}$ maps the unit square $\mathbf{U}$ to the parallelogram $\mathbf{X}$.

The Jacobian is

$$J_{\mathbf{T}} = \begin{pmatrix} x_u & x_v \\ y_u & y_v \end{pmatrix} = \begin{pmatrix} 1 & 1 \\ 1 & -1 \end{pmatrix} = T\,.$$

The mapping $\mathbf{T}$ has an inverse mapping

$$\begin{aligned} u &= \tfrac{x}{2} + \tfrac{y}{2} \\ v &= \tfrac{x}{2} - \tfrac{y}{2} \end{aligned} \quad \text{or} \quad < u, v >= \begin{pmatrix} \frac{1}{2} & \frac{1}{2} \\ \frac{1}{2} & -\frac{1}{2} \end{pmatrix} < u, v >= T^{-1} < x, y > \,.$$

Example 3.112. The identity (3.133) allows computing the partial derivatives $\frac{\partial F}{\partial x}$, $\frac{\partial F}{\partial y}$, $\frac{\partial G}{\partial x}$ and $\frac{\partial G}{\partial y}$ indirectly, by computing the inverse of the Jacobian $J_{\mathbf{T}}$. This is an algebraic method, which sometimes is easier than the direct, analytical omputation of these derivatives.

Consider for example the mapping $(x, y) = \mathbf{T}(r, \theta)$ from the polar coordinates (r, θ) to the Cartesian coordinates (x, y) ,

$$\begin{aligned} x &= r \cos \theta \\ y &= r \sin \theta \end{aligned}$$

The Jacobi matrix $J_{\mathbf{T}}$ is

$$J_{\mathbf{T}} = \begin{pmatrix} \cos\theta & -r \sin\theta \\ \sin\theta & r \cos\theta \end{pmatrix} \tag{3.137}$$

and is nonsingular if $r \neq 0$ (i.e. everywhere except at the origin). The inverse of $J_{\mathbf{T}}$ is

$$\begin{pmatrix} \cos\theta & -r \sin\theta \\ \sin\theta & r \cos\theta \end{pmatrix}^{-1} = \frac{1}{r} \begin{pmatrix} r \cos\theta & r \sin\theta \\ -\sin\theta & \cos\theta \end{pmatrix}$$

from which we read the partial derivatives

$$\begin{aligned} \frac{\partial r}{\partial x} &= \cos\theta & \qquad \frac{\partial r}{\partial y} &= \sin\theta \\ \frac{\partial \theta}{\partial x} &= -\frac{\sin\theta}{r} & \qquad \frac{\partial \theta}{\partial y} &= \frac{\cos\theta}{r} \end{aligned}$$

The direct computation of these partial derivatives requires using the inverse mapping $(r, \theta) = \mathbf{T}^{-1}(x, y)$ given by

$$\begin{aligned} r &= \sqrt{x^2 + y^2} \\ \theta &= \operatorname{atan} \frac{y}{x} \end{aligned}$$

from which we get

$$\begin{aligned} \frac{\partial r}{\partial x} &= \frac{x}{\sqrt{x^2+y^2}} = \frac{r \cos\,\theta}{r} & \qquad \frac{\partial r}{\partial y} &= \frac{y}{\sqrt{x^2+y^2}} = \frac{r \sin\theta}{r} \\ \frac{\partial \theta}{\partial x} &= -\frac{y}{x^2+y^2} = -\frac{r \sin\theta}{r^2} & \qquad \frac{\partial \theta}{\partial y} &= \frac{1/x}{1+(y/x)^2} = \frac{r \cos\theta}{r^2} \end{aligned}$$

in agreement with the above.

The determinant rules apply to the determinant of the Jacobian. For example, switching two rows, or two columns, reverses the sign

$$\frac{\partial(x,y)}{\partial(u,v)} = -\frac{\partial(y,x)}{\partial(u,v)}, \quad \frac{\partial(x,y)}{\partial(u,v)} = -\frac{\partial(x,y)}{\partial(v,u)}.$$

A determinant with two equal rows, or columns, is zero

$$\frac{\partial(x,x)}{\partial(u,v)} = 0, \quad \frac{\partial(x,y)}{\partial(u,u)} = 0.$$

Consider now the case where in addition to the transformation $(x, y) = \mathbf{T}(u, v)$ given by (3.123a)–(3.123b), there is a transformation $(u, v) = \mathbf{S}(\alpha, \beta)$ given as follows

$$u = u(\alpha, \beta) \tag{3.138a}$$

$$v = v(\alpha, \beta) \tag{3.138b}$$

Thus x and y are functions of (α, β),

$$x = f(u(\alpha,\beta), v(\alpha,\beta)) \tag{3.139a}$$

$$y = g(u(\alpha,\beta), v(\alpha,\beta)) \tag{3.139b}$$

The partial derivatives of x and y with respect to α and β can be computed by the chain rule

$$\begin{aligned}
\frac{\partial x}{\partial \alpha} &= \frac{\partial f}{\partial u}\frac{\partial u}{\partial \alpha} + \frac{\partial f}{\partial v}\frac{\partial v}{\partial \alpha} \\
\frac{\partial x}{\partial \beta} &= \frac{\partial f}{\partial u}\frac{\partial u}{\partial \beta} + \frac{\partial f}{\partial v}\frac{\partial v}{\partial \beta} \\
\frac{\partial y}{\partial \alpha} &= \frac{\partial g}{\partial u}\frac{\partial u}{\partial \alpha} + \frac{\partial g}{\partial v}\frac{\partial v}{\partial \alpha} \\
\frac{\partial y}{\partial \beta} &= \frac{\partial g}{\partial u}\frac{\partial u}{\partial \beta} + \frac{\partial g}{\partial v}\frac{\partial v}{\partial \beta}
\end{aligned}$$

giving the Jacobi matrix of the composition $\mathbf{T}(\mathbf{S})$ as the product of the Jacobi matrices of $\mathbf{T}$ and $\mathbf{S}$. This is summarized in the next theorem.

Mathematica session 3.5. *Note that the Jacobian matrix of the inverse transform is the inverse matrix of the Jacbian of the original transform. We check this by multiplying the two matrices together to obtain the identity matrix. In* MATHEMATICA *"." indicates matrix mutiplication.*

```
jacobian[exprs_, vars_]:=Outer[D, exprs, vars]
```

```
jacobian[{r * Cos[theta] * Sin[phi], r * Sin[theta] * Sin[phi], r * Cos[phi]}, {r,
theta, phi}]//MatrixForm
```

$$\begin{pmatrix} Cos[theta]Sin[phi] & -r\,Sin[phi]Sin[theta] & r\,Cos[phi]Cos[theta] \\ Sin[phi]Sin[theta] & r\,Cos[theta]Sin[phi] & r\,Cos[phi]Sin[theta] \\ Cos[phi] & 0 & -r\,Sin[phi] \end{pmatrix}$$

```
A = {U1[x1, x2, x3], U2[x1, x2, x3], U3[x1, x2, x3]}
```

{U1[x1, x2, x3], U2[x1, x2, x3], U3[x1, x2, x3]}

```
B = {x1, x2, x3}
```

{x1, x2, x3}

```
jacobian[A, B]//MatrixForm
```

$$\begin{pmatrix} U1^{(1,0,0)}[x1, x2, x3] & U1^{(0,1,0)}[x1, x2, x3] & U1^{(0,0,1)}[x1, x2, x3] \\ U2^{(1,0,0)}[x1, x2, x3] & U2^{(0,1,0)}[x1, x2, x3] & U2^{(0,0,1)}[x1, x2, x3] \\ U3^{(1,0,0)}[x1, x2, x3] & U3^{(0,1,0)}[x1, x2, x3] & U3^{(0,0,1)}[x1, x2, x3] \end{pmatrix}$$

```
jacobiandeterminant[exprs_, vars_]:=Simplify[Det[Outer[D, exprs, vars]]]

jacobiandeterminant[{r * Cos[theta], r * Sin[theta], z}, {r, theta, z}]
```

r

```
x = Cosh[mu] * Cos[nu] * Cos[phi];
y = Cosh[mu] * Cos[nu] * Sin[phi];
z = Sinh[mu] * Sin[nu]
```

Sin[nu]Sinh[mu]

Prolate Spheroidal Coordinates. These are not one-to-one as shown below

jacobiandeterminant[{Cosh[μ] * Cos[ν] * Cos[φ], Cosh[μ] * Cos[ν] * Sin[φ], Sinh[μ] * Sin[ν]}, {μ, nu, φ}]

0

Instead we use the alternate transform which is, with the exception of certain values, one-to-one.

jacobiandeterminant[{Sqr[(sigma^2 − 1)(1 − tau^2)] * Cos[phi], Sqrt[(sigma^2 − 1)(1 − tau^2)] * Sin[phi], sigma * tau}, {sigma, tau, phi}]

$((sigma^2 - tau^2)(-Sin[phi]^2 Sqr[-(-1 + sigma^2)(-1 + tau^2)] + 2(-1 + sigma^2)(-1 + tau^2) Cos[phi]^2 Sqr'[-(-1 + sigma^2)(-1 + tau^2)]))/ (\sqrt{-(-1 + sigma^2)(-1 + tau^2)})$

♦

Theorem 3.113. The Jacobi matrices of the transformations **T** , **S** and **T**(**S**) are related by

$$J_{\mathbf{T}(\mathbf{S})} = J_{\mathbf{T}}\, J_{\mathbf{S}}\,, \quad \text{or} \quad \begin{pmatrix} \dfrac{\partial x}{\partial \alpha} & \dfrac{\partial x}{\partial \beta} \\ \dfrac{\partial y}{\partial \alpha} & \dfrac{\partial y}{\partial \beta} \end{pmatrix} = \begin{pmatrix} \dfrac{\partial x}{\partial u} & \dfrac{\partial x}{\partial v} \\ \dfrac{\partial y}{\partial u} & \dfrac{\partial y}{\partial v} \end{pmatrix} \begin{pmatrix} \dfrac{\partial u}{\partial \alpha} & \dfrac{\partial u}{\partial \beta} \\ \dfrac{\partial v}{\partial \alpha} & \dfrac{\partial v}{\partial \beta} \end{pmatrix}, \tag{3.140}$$

and the determinants of the Jacobians satisfy

$$\frac{\partial(x, y)}{\partial(\alpha, \beta)} = \frac{\partial(x, y)}{\partial(u, v)} \frac{\partial(u, v)}{\partial(\alpha, \beta)}\,. \tag{3.141}$$

□

Mathematica session 3.6. *In this* MATHEMATICA *section we investigate inversion schemes.*

jacobianInverse[F_, X_]:=Module[{j, n }, j = Inverse[jacobian[F, X]]]

jacobian[{x^2, y^2, z^2}, {x, y, z}]

$\{\{2x, 0, 0\}, \{0, 2y, 0\}, \{0, 0, 2z\}\}$

jacobianInverse[{x^2, y^2, z^2}, {x, y, z}]

$\left\{\left\{\frac{1}{2x},0,0\right\},\left\{0,\frac{1}{2y},0\right\},\left\{0,0,\frac{1}{2z}\right\}\right\}$

Det[%]

$\frac{1}{8xyz}$

We now introduce a function that will be helpful when we want to solve equations using an iteration method. First we try it in one-dimension and then proceed to several dimensions. Basically the function is applied to the same input-output recursively.

```
iterates[f_, x0_, n_]:=Module[{k},
For[k = 0; v = x0, k < n, k++, v = N[f/.{x → v}]];
Print[N[v]]]
```

We iterate sin(x) at x = 1 4 times.

iterates[Sin[x], 1, 4]

0.627572

An alternate procedure is given by the MATHEMATICA *command* **Nest**. *We illustrate how it functions below.*

Nest[Sin, x, 3]

Sin[Sin[Sin[x]]]

N[Nest[Sin, x, 10]/.{x → 1}]

0.462958

$Sin[Sin[Sin[1]]]$

We now illustrate how to use a **Do** *statement. This will be useful to us later.*

For[i = 1; t = x, i^2 < 10, i++, t = t^2 + i; Print[t]]

$1 + x^2$
$2 + (1 + x^2)^2$
$3 + \left(2 + (1 + x^2)^2\right)^2$

N[Nest[Sin, x, 100]/.{x → 1}]

0.168852

N[Nest[Sin, x, 1000]/.{x → 1}]

0.054593

size[i_]:=N[Nest[Sin, x, i]/.{x → 1}]

size[100]

0.168852

size[1000]

0.054593

size[10000]

0.0173136

size[12000]

size[12000]

We now introduce a **module** *that applies Newton's method for solving an equa-*

tion in one variable. The method will then be extended to systems of equations in several variables below.

```
myNewton[f_, x0_, n_]:=Module[{g}, g = x - f/D[f, x];
N[iterates[g, x0, n]]]
```

```
myNewton[x^2 - 2, 4, 6]
```

1.41421

```
myNewton[x^2 - 2, 4, 7]
```

1.41421

We now introduce a three-dimensional iteration scheme.

```
threeDimIterates[F_, X_, X0_, n_]:=Module[{k},
For[k = 0; {v = X0}, k < n, k++,
{v[[1]], v[[2]], v[[3]]} = F/.{X[[1]] → v[[1]], X[[2]] → v[[2]], X[[3]] → v[[3]]}];
Print[N[{v[[1]], v[[2]], v[[3]]}]]]
```

```
threeDimIterates[{Sin[x], Tan[x], Cos[x]}, {x, y, z}, {4, 4, 4}, 6]
```

{−0.529721, −0.624544, 0.848172}

```
threeDimIterates[{Sin[x], Tan[x], Cos[x]}, {x, y, z}, {4, 4, 4}, 20]
```

{−0.344902, −0.36745, 0.938639}

```
threeDimIterates[{Sin[x], Tan[x], Cos[x]}, {x, y, z}, {4, 4, 4}, 100]
```

{−0.168162, −0.170591, 0.985759}

```
threeDimIterates[{Sin[x * y], Tan[x * z] * Cos[y], Cos[z * y]}, {x, y, z}, {4, 4, 4},
4]
```

{0.000838998, 0.0150894, 0.998674}

Now we are in a position to formulate a three-dimensional Newton iteration scheme.

```
myThreeDimensionalNewton[F_, X_, X0_, n_]:=Module[{G, J}, J =
jacobianInverse[F, X]; G = X − J.F; N[threeDimIterates[G, X, X0, n]]]
```

```
myThreeDimensionalNewton[{x^2 − 2, y^2 − 3, z^2 − 4}, {x, y, z}, {4, 4, 4}, 6]
{1.41421, 1.73205, 2.}
```

```
myThreeDimensionalNewton[{x^2 * y + y^2 * z + 2x * y * z, x^2 − y^2
−z^2, x * y * x^2}, {x, y, z}, {4, 4, 4}, 6]
{0.559077, 0.576556, −0.225385}
```

♦

The transformation

$$(x, y) = \mathbf{T}(u, v) \tag{3.142}$$

takes points in the uv–plane into points in the xy–plane, which we represent as vectors $\mathbf{r}(u, v)$. It is necessary to find how areas are changed by going from (u, v) to $\mathbf{r}(u, v)$. In particular consider the area element $du\,dv$ shown in Figure 3.7. It is mapped into a (curved) area element in the xy–plane, shown

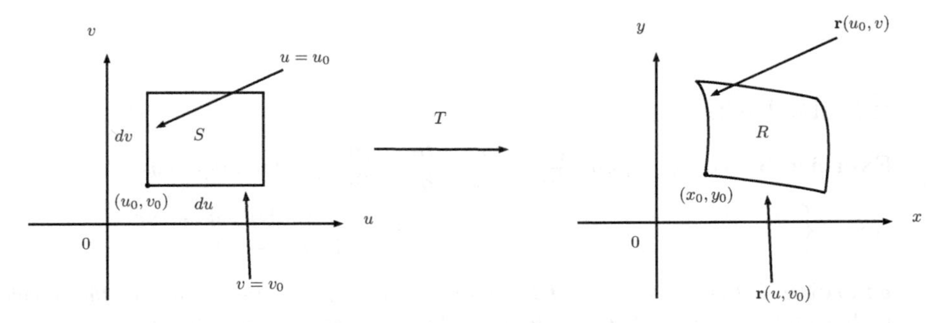

FIGURE 3.7: Corresponding area elements in the uv–plane and in the xy–plane.

in Figure 3.7 . It may be approximated by the parallelogram spanned by the vectors

$$\begin{aligned}
\mathbf{r}(u_0 + du, v_0) - \mathbf{r}(u_0, v_0) &\approx \frac{\partial \mathbf{r}}{\partial u}\, du \\
\mathbf{r}(u_0, v_0 + dv) - \mathbf{r}(u_0, v_0) &\approx \frac{\partial \mathbf{r}}{\partial v}\, dv
\end{aligned}$$

and so its area is approximated by

$$\begin{aligned} dA &= \left|\frac{\partial \mathbf{r}}{\partial u} \times \frac{\partial \mathbf{r}}{\partial v}\right| du\,dv = \left|\mathbf{det}\begin{pmatrix} \dfrac{\partial x}{\partial u} & \dfrac{\partial x}{\partial v} \\ \dfrac{\partial y}{\partial u} & \dfrac{\partial y}{\partial v} \end{pmatrix}\right| du\,dv \\ &= \left|\frac{\partial(x,y)}{\partial(u,v)}\right| du\,dv\ . \end{aligned} \tag{3.143}$$

We can write this as

$$\frac{dA}{du\,dv} = \left|\frac{\partial(x,y)}{\partial(u,v)}\right| ,$$

or, in words: at any point (u_0, v_0), the absolute value of the determinant of the Jacobian, $|\mathbf{det}\, J_{\mathbf{T}}(u_0, v_0)|$, is the ratio of corresponding area elements.

Example 3.114. Consider again the transformation

$$\begin{aligned} x &= r\cos\theta \\ y &= r\sin\theta \end{aligned}$$

mapping the polar coordinates (r, θ) into (x, y). The determinant of the Jacobian is, by Example 3.112,

$$\frac{\partial(x,y)}{\partial(r,\theta)} = \mathbf{det}\begin{pmatrix} \cos\theta & -r\sin\theta \\ \sin\theta & r\cos\theta \end{pmatrix} = r\ ,$$

Therefore, by (5.61),

$$dA = r\,dr\,d\theta\ . \tag{3.144}$$

Exercises

Exercise 3.115. *Compute* $\dfrac{\partial u}{\partial x}, \dfrac{\partial u}{\partial y}, \dfrac{\partial v}{\partial x}, \dfrac{\partial v}{\partial y}$ *for the mappings*

(a) $\left\{\begin{array}{lll} x & := & u + v \\ y & := & u^2 + v^2 \end{array}\right.$ (b) $\left\{\begin{array}{lll} x & := & u + v\,e^u \\ y & := & v \end{array}\right.$

Exercise 3.116. *For each of the following transformations compute the Jacobi matrix and its inverse. Specify where the transformation is one–to–one.*

(a) $x := u + v\ ,\quad y := u - v$ (b) $x := u\,v\ ,\quad y := \dfrac{u}{v}$

(c) $x := u\,v\ ,\quad y := u^2 - v^2$ (d) $x := u^2 + v^2\ ,\quad y := u^2 - v^2$

(e) $x := \dfrac{u}{u^2 + v^2}\ ,\quad y := \dfrac{v}{u^2 + v^2}$

Chapter 4

Directional Derivatives and Extremum Problems

4.1 Directional derivatives and gradients

Let $f(x, y)$ be a function of two variables, $P(x_0, y_0)$ a point where f is differentiable. The partial derivative $\frac{\partial f}{\partial x}(x_0, y_0)$ measures the rate of change of f with respect to distance measured along the direction of the x–axis, at the point P. Similarly, the partial derivative $\frac{\partial f}{\partial y}(x_0, y_0)$ is the rate of change of f along the direction of the y–axis, at the point P. In this section we study **directional derivatives** which measure the rate of change of the function f along an arbitrary direction.

Definition 4.1. Let $\mathbf{u} := (u_1, u_2)$ be a given vector. The **directional derivative** of f in the direction $\mathbf{u}$ at $P(x_0, y_0)$ is defined as

$$D_{\mathbf{u}} f(x_0, y_0) := \lim_{t \to 0} \frac{f(x_0 + t u_1\,,\, y_0 + t u_2) \;-\; f(x_0, y_0)}{t} \,. \tag{4.1}$$

The partial derivatives of f at (x_0, y_0) are special cases of the directional derivatives defined above. Indeed, the partial derivative $\frac{\partial f}{\partial x}$ is the directional derivative of f in the direction of the unit vector $\mathbf{i} = < 1, 0 >$,

$$\frac{\partial f}{\partial x}(x_0, y_0) \;=\; \lim_{t \to 0} \frac{f(x_0 + t, y_0) \;-\; f(x_0, y_0)}{t},$$

where this is (3.19) with t used instead of Δx. For $D_{\mathbf{i}} f(x_0, y_0)$ we operate similarly, i.e.

$$\frac{\partial f}{\partial y}(x_0, y_0) \;=\; D_{\mathbf{j}} f(x_0, y_0),$$

where the derivative in the direction of $\mathbf{j} = < 0, 1 >$ is used.

Lemma 4.2. Let $\mathbf{u} = < u_1, u_2 >$ and let $\lambda > 0$. The derivative of f at (x_0, y_0) in the direction of $\lambda\,\mathbf{u} = < \lambda u_1, \lambda u_2 >$ is

$$D_{\lambda\,\mathbf{u}} f(x_0, y_0) \;=\; \lambda\, D_{\mathbf{u}} f(x_0, y_0) \,. \tag{4.2}$$

Proof:

$$D_{\lambda\,\mathbf{u}} f(x_0, y_0) \;=\; \lim_{t \to 0} \frac{f(x_0 + t\lambda\, u_1\,,\, y_0 + t\lambda\, u_2) \;-\; f(x_0, y_0)}{t}$$

$$= \lambda \lim_{t \to 0} \frac{f(x_0 + t\lambda u_1 \, , \, y_0 + t\lambda u_2) \; - \; f(x_0, y_0)}{t\lambda}$$

$$= \lambda \lim_{\tau \to 0} \frac{f(x_0 + \tau u_1 \, , \, y_0 + \tau u_2) \; - \; f(x_0, y_0)}{\tau} \, , \quad \text{where } \tau := t\lambda$$

$$= \lambda \, D_{\mathbf{u}} f(x_0, y_0) \, .$$

□

Example 4.3. We use Definition 4.1 to compute the directional derivative of $f(x,y) := x^2 - y^2$ at the point $P(1,1)$ in the direction $\mathbf{u} = -\frac{1}{\sqrt{2}}\mathbf{i} + \frac{1}{\sqrt{2}}\mathbf{j}$.

$$\begin{aligned} D_{\mathbf{u}} f(1,1) &= \frac{1}{\sqrt{2}} D_{\sqrt{2}\mathbf{u}} f(1,1) \, , \quad \text{by (4.2)} \\ &= \frac{1}{\sqrt{2}} D_{(-1,1)} \, f(1,1) \\ &= \frac{1}{\sqrt{2}} \lim_{t \to 0} \frac{((1-t)^2 - (1+t)^2) \; - \; (1^2 - 1^2)}{t} \\ &= \frac{1}{\sqrt{2}} \lim_{t \to 0} \frac{-4t}{t} \; = \; -2\sqrt{2} \, . \end{aligned}$$

If the directional derivative $D_{\mathbf{u}} \, f(x_0, y_0)$ is known, then Lemma 4.2 allows computing the directional derivative for all vectors in the direction of $\mathbf{u}$. Without loss of generality, we can assume that $\mathbf{u} = (u_1, u_2)$ is a unit vector, i.e. $\|\mathbf{u}\| = \sqrt{u_1^2 + u_2^2} = 1$.

If the partial derivatives $\frac{\partial f}{\partial x}(x_0, y_0)$ and $\frac{\partial f}{\partial y}(x_0, y_0)$ are known, it is possible to compute the directional derivative, in any direction, without taking the limit (4.28). This is shown in the next theorem.

Theorem 4.4. Let $f(x,y)$ have continuous partial derivatives at (x_0, y_0), and let $\mathbf{u}$ be a unit vector, making an angle θ with the positive x–axis. Then

$$D_{\mathbf{u}} \, f(x_0, y_0) \; = \; \cos\theta \, \frac{\partial f}{\partial x}(x_0, y_0) \; + \; \sin\theta \, \frac{\partial f}{\partial y}(x_0, y_0) \, . \tag{4.3}$$

Proof:
We rotate the coordinate axes by the angle θ, to obtain new axes $(\overline{x}, \overline{y})$, with the positive $\overline{x}$–axis parallel to $\mathbf{u}$, see Figure 4.1. The coordinates (x,y) are related to the new coordinates $(\overline{x}, \overline{y})$ by

$$(T_1) \qquad \begin{aligned} x &= \bar{x} \cos\theta - \bar{y} \sin\theta \, , \\ y &= \bar{x} \sin\theta + \bar{y} \cos\theta \, . \end{aligned} \tag{4.4}$$

The directional derivative $D_{\mathbf{u}} \, f(x_0, y_0)$ is the same as the partial derivative of f with respect to $\overline{x}$,

$$\begin{aligned} D_{\mathbf{u}} \, f(x_0, y_0) &= \frac{\partial f}{\partial \overline{x}}(x_0, y_0) \\ &= \frac{\partial f}{\partial x}(x_0, y_0) \, \frac{\partial x}{\partial \overline{x}} \; + \; \frac{\partial f}{\partial y}(x_0, y_0) \, \frac{\partial y}{\partial \overline{x}} \, , \quad \text{using the chain rule} \\ &= \frac{\partial f}{\partial x}(x_0, y_0) \cos\theta \; + \; \frac{\partial f}{\partial y}(x_0, y_0) \sin\theta \, , \quad \text{using} \quad (T1) - (T_2) \, . \end{aligned}$$

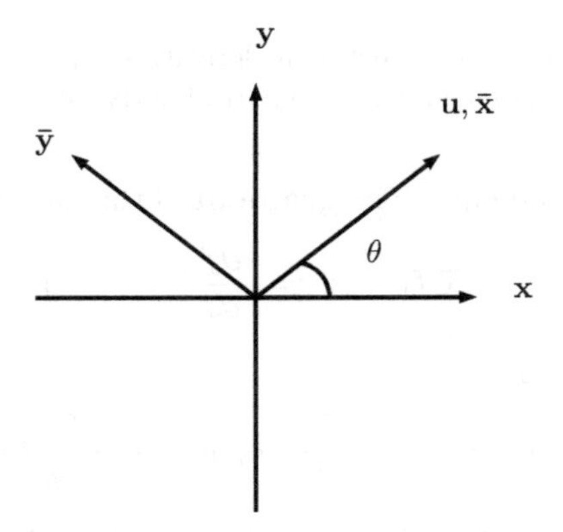

FIGURE 4.1: A rotation of the Cartesian coordinates in $\mathbb{R}^2$ by θ.

□

A convenient way to express (4.3) for arbitrary (i.e. not necessarily unit) vectors $\mathbf{u} =< u_1, u_2 >$ is

$$D_{\mathbf{u}}\, f \;=\; \|\mathbf{u}\| \left(\frac{\partial f}{\partial x} \cos\alpha \;+\; \frac{\partial f}{\partial y} \cos\beta \right) \tag{4.5}$$

$$\;=\; \frac{\partial f}{\partial x}\, u_1 \;+\; \frac{\partial f}{\partial y}\, u_2 \tag{4.6}$$

where:

$$\|\mathbf{u}\| \;:=\; \sqrt{u_1^2 + u_2^2} \quad \text{is the } \textbf{norm} \text{ (or length) of } \mathbf{u}\,,$$

$$\alpha \;:=\; \theta\;,\;\beta\;:=\;\frac{\pi}{2} - \theta \quad \text{are the } \textbf{direction angles} \text{ of } \mathbf{u}\,,$$

$$\frac{u_1}{\|\mathbf{u}\|} \;=\; \cos\alpha\;,\;\frac{u_2}{\|\mathbf{u}\|} \;=\; \cos\beta \quad \text{are its } \textbf{direction cosines}\,,$$

see Figure 4.1. Equation (4.5) can be written as

$$D_{<\cos\alpha,\cos\beta>} f \;=\; \frac{\partial f}{\partial x} \cos\alpha \;+\; \frac{\partial f}{\partial y} \cos\beta\,. \tag{4.7}$$

Example 4.5. We compute again the directional derivative of $f(x, y) := x^2 - y^2$ at $(1, 1)$ in the direction $\mathbf{u} := -\frac{1}{\sqrt{2}}\mathbf{i} + \frac{1}{\sqrt{2}}\mathbf{j}$.

$$\begin{aligned} D_{\mathbf{u}} f(1,1) \;&=\; \frac{\partial f}{\partial x}(1,1)\left(-\frac{1}{\sqrt{2}}\right) \;+\; \frac{\partial f}{\partial y}(1,1)\left(\frac{1}{\sqrt{2}}\right)\,, \quad \text{using (4.6)} \\ &=\; 2\left(-\frac{1}{\sqrt{2}}\right) \;+\; (-2)\left(\frac{1}{\sqrt{2}}\right) \\ &=\; -\frac{4}{\sqrt{2}}\,, \quad \text{in agreement with Example 4.3.} \end{aligned}$$

Equation (4.6) gives the directional derivative $D_{\mathbf{u}}f$ as the inner product of $\mathbf{u}$ and a vector whose components are the partial derivatives of f. This vector deserves a special name:

Definition 4.6 (Gradient). The **gradient** of the function $f(x, y)$ is the vector

$$\nabla f(x,y) \ := \ \frac{\partial f}{\partial x}\,\mathbf{i} \ + \ \frac{\partial f}{\partial y}\,\mathbf{j} \tag{4.8}$$

and at a point (x_0, y_0),

$$\nabla f(x_0,y_0) \ := \ \mathbf{i}\,\frac{\partial f}{\partial x}\,(x_0,y_0) \ + \ \mathbf{j}\,\frac{\partial f}{\partial y}\,(x_0,y_0)\,. \tag{4.9}$$

The symbol ∇ is pronounced **del** (short form of **delta**) or **nabla**[1].

Equation (4.6) is thus written as

$$D_{\mathbf{u}}f(x_0,y_0) \ = \ \mathbf{u} \ \cdot \ \nabla f(x_0,y_0)\,. \tag{4.10}$$

For a function $f(x, y, z)$, of three variables, we define the gradient analogously,

$$\nabla f(x,y,z) \ := \ \frac{\partial f}{\partial x}\,\mathbf{i} \ + \ \frac{\partial f}{\partial y}\,\mathbf{j} \ + \ \frac{\partial f}{\partial z}\,\mathbf{k} \tag{4.11}$$

and the directional derivative of f at (x_0, y_0, z_0) in the direction of $\mathbf{u} = u_1\mathbf{i} + u_2\mathbf{j} + u_3\mathbf{k}$ is, in analogy with (4.30),

$$D_{\mathbf{u}}f(x_0,y_0,z_0) \ = \ \mathbf{u} \ \cdot \ \nabla f(x_0,y_0,z_0) \ =$$

$$u_1\,\frac{\partial f}{\partial x}(x_0,y_0,z_0) \ + \ u_2\,\frac{\partial f}{\partial x}(x_0,y_0,z_0) \ + \ u_3\,\frac{\partial f}{\partial y}(x_0,y_0,z_0). \tag{4.12}$$

Example 4.7. Compute the directional derivative of

$$f(x,y,z) \ := \ \frac{1}{\sqrt{x^2+y^2+z^2}}\,, \quad \text{at}(1,2,-1)\text{along the unit vector}$$

$$\mathbf{u} \ := \ \frac{1}{\sqrt{3}}\,\mathbf{i} - \frac{1}{\sqrt{3}}\,\mathbf{j} + \frac{1}{\sqrt{3}}\,\mathbf{k}\,.$$

We compute first the gradient $\nabla f(x, y, z) = f_x\,\mathbf{i} + f_y\,\mathbf{j} + f_z\,\mathbf{k}$ where

$$f_x \ = \ \frac{-x}{\left(\sqrt{x^2+y^2+z^2}\right)^3}\,, \ f_y \ = \ \frac{-y}{\left(\sqrt{x^2+y^2+z^2}\right)^3}\,, \ f_z \ = \ \frac{-z}{\left(\sqrt{x^2+y^2+z^2}\right)^3}\,.$$

At the point $(1, 2, -1)$ we have

$$\nabla f(1,2,-1) \ = \ \frac{-1}{(\sqrt{6})^3}\,\mathbf{i} \ - \ \frac{2}{(\sqrt{6})^3}\,\mathbf{j} \ + \ \frac{1}{(\sqrt{6})^3}\,\mathbf{k}\,.$$

Therefore, by (4.12

$$D_{\mathbf{u}}f(1,2,-1) = \mathbf{u}\cdot\nabla f(1,2,-1) = \frac{-1}{\sqrt{3}}\,\frac{1}{(\sqrt{6})^3} + \frac{1}{\sqrt{3}}\,\frac{2}{(\sqrt{6})^3} + \frac{1}{\sqrt{3}}\,\frac{1}{(\sqrt{6})^3} \ = \ \frac{2}{\sqrt{3}\cdot\sqrt{6}^3}\,.$$

[1] **Nabla** is the Greek word for harp, a musical instrument with shape similar to ∇.

Mathematica session 4.1. $R = Sqrt[x\wedge 2 + y\wedge 2 + z\wedge 2]$

$\sqrt{x^2 + y^2 + z^2}$

Grad[%, {x, y, z}]

$\left\{\frac{x}{\sqrt{x^2+y^2+z^2}}, \frac{y}{\sqrt{x^2+y^2+z^2}}, \frac{z}{\sqrt{x^2+y^2+z^2}}\right\}$

Grad[(x * y) * z/(x^2 + y^2 + z^2), {x, y, z}]

$\left\{-\frac{2x^2yz}{(x^2+y^2+z^2)^2} + \frac{yz}{x^2+y^2+z^2}, -\frac{2xy^2z}{(x^2+y^2+z^2)^2} + \frac{xz}{x^2+y^2+z^2}, -\frac{2xyz^2}{(x^2+y^2+z^2)^2} + \frac{xy}{x^2+y^2+z^2}\right\}$

Grad[1/Sqrt[x^2 + y^2 + z^2], {x, y, z}]

$\left\{-\frac{x}{(x^2+y^2+z^2)^{3/2}}, -\frac{y}{(x^2+y^2+z^2)^{3/2}}, -\frac{z}{(x^2+y^2+z^2)^{3/2}}\right\}$

Grad[1/R^2, {x, y, z}]

$\left\{-\frac{2x}{(x^2+y^2+z^2)^2}, -\frac{2y}{(x^2+y^2+z^2)^2}, -\frac{2z}{(x^2+y^2+z^2)^2}\right\}$

v:={x, y, z}

Grad[1/R^2, v]

$\left\{-\frac{2x}{(x^2+y^2+z^2)^2}, -\frac{2y}{(x^2+y^2+z^2)^2}, -\frac{2z}{(x^2+y^2+z^2)^2}\right\}$

Grad[f[r, theta, z], {r, theta, z}, "Cylindrical"]

$\left\{f^{(1,0,0)}[r, theta, z], \frac{f^{(0,1,0)}[r, theta, z]}{r}, f^{(0,0,1)}[r, theta, z]\right\}$

Grad[r * Cos[theta] * z^2, {r, theta, z}, "Cylindrical"]

$\{z^2\, Cos[theta], -z^2\, Sin[theta], 2rz\, Cos[theta]\}$

w:={r, theta, phi}

Grad[r^2 * Cos[theta] * Sin[phi], w, "Spherical"]

$\{2r\, Cos[theta]\, Sin[phi], -r\, Sin[phi]\, Sin[theta], r\, Cos[phi]\, Cot[theta]\}$

Grad[r * Cos[theta] * z^2, {r, theta, z}, "Cylindrical"]

$\{z^2\,Cos[theta], -z^2\,Sin[theta], 2rz\,Cos[theta]\}$

Grad[r^2 * Cos[theta] * Sin[phi], w, "Spherical"]

$\{2r\,Cos[theta]\,Sin[phi], -r\,Sin[phi]\,Sin[theta], r\,Cos[phi]\,Cot[theta]\}$

♦

An important property of the gradient $\nabla f(x_0, y_0)$ is that its direction is that along which the directional derivative $D_{\mathbf{u}}f(x_0, y_0)$ is maximal, i.e. $\nabla f(x_0, y_0)$ points in the direction where the rate of change of the function is the greatest.

Lemma 4.8. Let $f(x, y)$ be differentiable at (x_0, y_0) where $\nabla f(x_0, y_0) \neq \mathbf{0}$. Then the maximum of the directional derivative $D_{\mathbf{u}}f(x_0, y_0)$ over all unit vectors $\mathbf{u}$ is

$$\max_{\|\mathbf{u}\|=1} D_{\mathbf{u}}f(x_0, y_0) = \|\nabla f(x_0, y_0)\| \tag{4.13}$$

and it occurs in the direction of $\nabla f(x_0, y_0)$.

Proof:
Let $\mathbf{u}$ be an arbitrary unit vector, and let θ be the angle between $\mathbf{u}$ and the gradient $\nabla f(x_0, y_0)$. Then

$$\begin{aligned} D_{\mathbf{u}}f(x_0, y_0) &= \mathbf{u} \cdot \nabla f(x_0, y_0)\,, \quad \text{by (4.30)} \\ &= \|\mathbf{u}\|\ \|\nabla f(x_0, y_0)\|\ \cos\theta \\ &= \|\nabla f(x_0, y_0)\|\ \cos\theta\,, \quad \text{since } \|\mathbf{u}\| = 1 \end{aligned} \tag{4.14}$$

is maximal when $\cos\theta = 1$, i.e. when $\theta = 0$. □

Remark 4.9. From (4.14) we conclude that the directional derivative $D_{\mathbf{u}}f(x_0, y_0)$ is minimal in the direction of $-\nabla f(x_0, y_0)$. This means that the smallest rate of change of f is against the direction of the gradient ∇f.

Example 4.10. We compute the maximal rate of change of the function $f(x, y) := e^y \cosh x$ at the point $(-1, 0)$, and the direction $\mathbf{u}$ where this maximum occurs. First we compute the gradient

$$\nabla f(x, y) = f_x\,\mathbf{i} + f_y\,\mathbf{j} = (e^y \sinh x)\,\mathbf{i} + (e^y \cosh x)\,\mathbf{j}\,,$$

and at the point $(-1, 0)$,

$$\nabla f(-1, 0) = (e^{-1} \sinh 0)\,\mathbf{i} + (e^{-1} \cosh 0)\,\mathbf{j} = e^{-1}\mathbf{j}\,.$$

By (4.13), the maximum value of $D_{\mathbf{u}}f(-1, 0)$ is e^{-1}, in the direction of $\mathbf{j}$. △

We recall the definition

$$\{(x,y) \,:\, f(x,y) \;=\; \alpha\}$$

of the α–level curve of the function f. Let $\mathcal{C}$ be the level curve passing through the point (x_0, y_0), i.e.

$$\mathcal{C} \;=\; \{(x,y) \,:\, f(x,y) \;=\; \alpha_0\} \quad \text{where} \quad \alpha_0 \;=\; f(x_0, y_0)\,.$$

We parameterize $\mathcal{C}$ with respect to the arclength s as

$$x := x(s), y \;:=\; y(s)\,, \quad \text{such that} \quad f(x(s), y(s)) \equiv \alpha_0 \quad \text{and} \quad x(0) = x_0\,,\, y(0) = y_0\,.$$

Differentiating $f(x(s), y(s)) \equiv \alpha_0$ with respect to arc-length s we get

$$\begin{aligned} f_x(x(s), y(s))\, x'(s) \;+\; f_y(x(s), y(s))\, y'(s) \;&=\; 0\,, \\ \text{or} \quad \nabla f(x(s), y(s)) \;\cdot\; \left(x'(s)\,\mathbf{i} \;+\; y'(s)\,\mathbf{j}\right) \;&=\; 0\,, \end{aligned}$$

where $x'(s)\,\mathbf{i} + y'(s)\,\mathbf{j}$ is tangent to the curve $\mathcal{C}$ at $(x(s), y(s))$. In particular, for $s = 0$ we conclude that the $\nabla f(x_0, y_0)$ is perpendicular to the tangent $x'(0)\,\mathbf{i} \;+\; y'(0)\,\mathbf{j}$ of the level curve $\mathcal{C}$. We have thus proved:

Corollary 4.11. If $\nabla f(x_0, y_0) \neq \mathbf{0}$ then the gradient $\nabla f(x_0, y_0)$ is perpendicular at (x_0, y_0) to the level curve[2]

$$\{(x,y) \,:\, f(x,y) \;=\; f(x_0, y_0)\}\,.$$

□

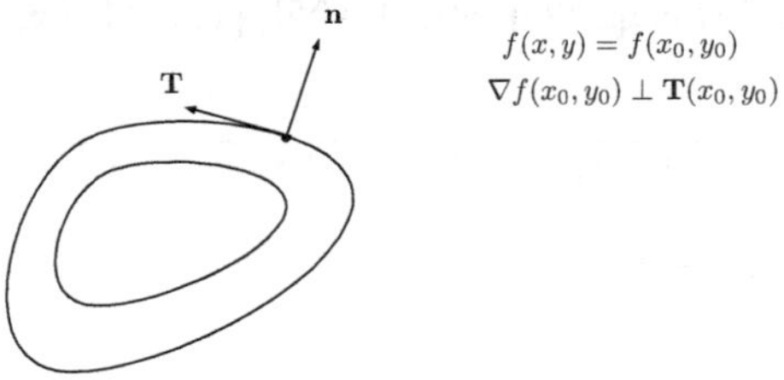

FIGURE 4.2: Illustration of Corollary 4.11.

Example 4.12. Consider a long charged wire which we take as the z–axis. The electrostatic potential in the xy–plane, due to the charged wire, is given by

$$V(x,y) \;=\; \frac{1}{\sqrt{x^2+y^2}}\,.$$

The level curve of V through the point $(2, -1)$ is

$$\begin{aligned} \mathcal{C} \;&=\; \{(x,y) : \frac{1}{\sqrt{x^2+y^2}} \;=\; V(2,-1) \;=\; \frac{1}{\sqrt{5}}\}\,, \\ &=\; \{(x,y) : x^2 + y^2 \;=\; 5\}\,. \end{aligned}$$

---

[2]See Figure 4.2.

The gradient of V is

$$\nabla V(x,y) = -\frac{x\,\mathbf{i} + y\,\mathbf{j}}{\left(\sqrt{x^2 + y^2}\right)^3}\,, \quad \text{and at } (2,-1)\,, \quad \nabla V(2,-1) = -\frac{2\,\mathbf{i} - \mathbf{j}}{5^{\frac{3}{2}}}$$

which points towards the origin, and is thus perpendicular to the level curve $\mathcal{C}$ at $(2,-1)$.

Mathematica session 4.2. $\{x, y\}$

```
directalDerivative[f_, a_, v_]:=Module[{alpha}, alpha=a/Sqrt[a.a]; Grad[f, v].alpha]
```

```
directalDerivative[x^2 - y^2, {-1, 1}, {x, y}]
```

$-\sqrt{2}x - \sqrt{2}y$

```
directalDerivative[x^2 + y^2 + z^2, {1, 1, 1}, {x, y, z}]
```

$\frac{2x}{\sqrt{3}} + \frac{2y}{\sqrt{3}} + \frac{2z}{\sqrt{3}}$

```
directalDerivative[1/Sqrt[x^2 + y^2 + z^2], {1, 1, 1}, {x, y, z}]
```

$-\frac{x}{\sqrt{3}(x^2+y^2+z^2)^{3/2}} - \frac{y}{\sqrt{3}(x^2+y^2+z^2)^{3/2}} - \frac{z}{\sqrt{3}(x^2+y^2+z^2)^{3/2}}$

```
%/.{x → 1, y → 0, z → 0}
```

$-\frac{1}{\sqrt{3}}$

♦

Exercises

Exercise 4.13. *For the following functions $f(x, y)$, points (x_0, y_0) and directions $\mathbf{u}$ compute the directional derivative $D_{\mathbf{u}}f(x_0, y_0)$ using (i) the direct definition (4.28), and (ii) partial derivatives, as in (4.6).*

(a) $f(x,y) := x^2 - y^2\,, \quad (x_0, y_0) = (1,1)\,, \quad \mathbf{u} = \frac{1}{\sqrt{2}}\mathbf{i} - \frac{1}{\sqrt{2}}\mathbf{j}$

(b) $f(x,y) := xy\,, \quad (x_0, y_0) = (0,0)\,, \quad \mathbf{u}$ *arbitrary*

(c) $f(x,y) := xy\,, \quad (x_0, y_0) = (\varepsilon, \varepsilon)\,, \quad \mathbf{u} = \frac{1}{\sqrt{2}}\mathbf{i} + \frac{1}{\sqrt{2}}\mathbf{j}$, *for arbitrary* ε

4.2 A Taylor theorem for functions of two variables

As in the case of functions of one variable, it is useful if a function $f(x, y)$ can be approximated by polynomials in x and y, in the neighborhood of a point (x_0, y_0), say

$$f(x, y) \approx \sum a_{mn} (x - x_0)^m (y - y_0)^n . \tag{4.15}$$

We seek an analogue of the one variable Taylor Theorem for two variables. For simplicity assume that $f(x, y)$ has arbitrarily many derivatives of the form $\frac{\partial^{j+k} f}{\partial x^j \partial y^k}$ at the point (x_0, y_0). Denote $\Delta x := x - x_0$, $\Delta y := y - y_0$ and define

$$\Phi(t) := f(x_0 + t\Delta x, y_0 + t\Delta y) , \quad \text{for } t \in [0, 1] . \tag{4.16}$$

The Maclaurin form of the Taylor Theorem with $x_0 = 0$) gives

$$\Phi(1) = \Phi(0) + \Phi'(0) + \cdots + \frac{1}{n!}\Phi^{(n)}(0) + \frac{1}{(n+1)!}\Phi^{(n+1)}(\theta) , \quad \text{for some } \theta \text{ in}[0, 1] , \tag{4.17}$$

which we can use to get a Taylor polynomial approximation for $f(x, y)$ near (x_0, y_0). In particular, for $n = 0$ we get

$$f(x, y) = f(x_0, y_0) + \Delta x(f_x(x_0 + \theta\Delta x, y_0 + \theta\Delta y)) + \Delta y(f_y(x_0 + \theta\Delta x, y_0 + \theta\Delta y)) , \tag{4.18}$$

for some $\theta \in [0, 1]$. For $n = 1$ we get similarly

$$\begin{aligned} f(x, y) &= f(x_0, y_0) + \Delta x f_x(x_0, y_0) + \Delta y f_y(x_0, y_0) \\ &+ \frac{1}{2}\left((\Delta x)^2 f_{xx}(X, Y) + 2\Delta x\Delta y f_{xy}(X, Y) + (\Delta y)^2 f_{yy}(X, Y)\right) \end{aligned} \tag{4.19}$$

where $X = x_0 + \theta\Delta x$ and $Y = y_0 + \theta\Delta y$. We state the result for general n without proof:

Theorem 4.14. If $f(x, y)$ has partial derivatives of orders $n + 1$ at the point (x_0, y_0) then it may be expanded as a truncated power series

$$f(x, y) = \sum_{j+k \leq n} \frac{\partial^{j+k} f(x_0, y_0)}{\partial x^j \, \partial y^k} \frac{(x - x_0)^j (y - y_0)^k}{j!k!} + R_{n+1}(x, y) , \tag{4.20}$$

where $R_{n+1}(x, y)$ is the remainder term

$$R_{n+1}(x, y) := \sum_{j+k=n+1} \frac{\partial^{j+k} f(x_0 + \theta\Delta x, y_0 + \theta\Delta y)}{\partial x^j \, \partial y^k} \frac{(x - x_0)^j (y - y_0)^k}{j!k!} , \tag{4.21}$$

for some $\theta \in [0, 1]$. □

Remark 4.15. The polynomial expansion of the truncated Taylor series is a good approximation provided the remainder is small. If the remainder tends to zero then the function may be expanded as an infinite Taylor series. For this to happen it is necessary, but not sufficient, for f to have partial derivatives of all orders. In this case we write

$$f(x,y) \;=\; \sum_{j+k\,=\,0}^{\infty} \frac{\partial^{j+k} f(x_0,y_0)}{\partial x^j\,\partial y^k} \left(\frac{(x-x_0)^j\,(y-y_0)^k}{j!\,k!}\right) . \tag{4.22}$$

The remainder term in the Taylor series may be used to estimate the error incurred when truncating the series. Suppose we approximate a function f by just the linear terms,

$$f(x,y) \;\approx\; f(x_0,y_0) + (f_x(x_0,y_0))\,\Delta x + (f_y(x_0,y_0))\,\Delta y \; ,$$

then the remainder is given by

$$R_2(x,y) \;:=\; \frac{1}{2}\left(f_{xx}(X,Y)\,(\Delta x)^2 + 2\,f_{xy}(X,Y)\,\Delta x\,\Delta y + f_{yy}(X,Y)\,(\Delta y)^2\right)$$
$$\text{where } X := x_0 + \theta\Delta x\,,\; Y \;:=\; y_0 + \theta\Delta y\;.$$

$$\therefore \text{If} \quad M \;:=\; \max_{\substack{x_0-\Delta x \,\le\, x \,\le\, x_0+\Delta x \\ y_0-\Delta y \,\le\, y \,\le\, y_0+\Delta y}} \{|f_{xx}(x,y)|\,,|f_{xy}(x,y)|,|f_{yy}(x,y)|\}$$

$$\text{then} \quad |R_2(x,y)| \;\le\; \frac{M}{2}\,(|\Delta x| + |\Delta y|)^2 \; .$$

The magnitude $E := |R_2(x,y)|$ is the error in approximating a function by its linear terms; hence, an estimate of this error is given by

$$E \le \frac{M}{2}\,(|\Delta x| + |\Delta y|)^2 \tag{4.23}$$

A similar formula holds when we approximate a function by its TAYLOR polynomial of degree n, namely

$$f(x,y) \approx \sum_{j+k\le n} \frac{\partial^{j+k} f(x_0,y_0)}{\partial x^j\,\partial y^k}\,\frac{(x-x_0)^j\,(y-y_0)^k}{j!k!}.$$

Then as

$$E_{n+1}(x,y) \;\le\; \sum_{j+k=n+1} \left|\frac{\partial^{j+k} f(x_0+\theta\Delta x, y_0+\theta\Delta y)}{\partial x^j\,\partial y^k}\right| \frac{|\Delta x|^j\,|\Delta y|^k}{j!k!} \,,$$

we have the estimate

$$E_{n+1}(x,y) \;\le\; M\,\frac{|\Delta x \,+\, \Delta y|^{n+1}}{(n+1)!}, \tag{4.24}$$

where M is the maximum of the derivatives

$$\left|\frac{\partial^{j+k} f(x,y)}{\partial x^j\,\partial y^k}\right|$$

over the set $[x_0-\Delta x, x_0+\Delta x,] \times [y_0-\Delta y, y_0+\Delta y,]$.

It is useful to have a list of derivatives in many applications. One is in constructing a Taylor polynomial to approximate a particular function. We discuss this in detail later in this chapter; however, let us formally define the Taylor polynomial[3] $\mathbf{t}_{n,m}f[x,y]$ for a function $f(x,y)$ as follows

$$\mathbf{t}_{n,m}[x,y] \;:=\; f(0,0) + \sum_{j=1}^{n}\sum_{k=1}^{m} \frac{(\partial^{j+k}f)(0,0)}{\partial^j\partial^k}\frac{x^j y^k}{j!k!}.$$

Mathematica session 4.3. *The* MATHEMATICA *function* **Series** *calculates the Taylor polynomial of a function of several variables, which we take, in this case to be x and y.* [4] *We choose to take the Taylor polynomial about the point (x_0, y_0) and take orders up to 2 in x and y*

Series[f[x, y], {x, x0, 2}, {y, y0, 2}]

$$\left(f[x_0,y_0] + f^{(0,1)}[x_0,y_0](y-y_0) + \frac{1}{2}f^{(0,2)}[x_0,y_0](y-y_0)^2 + O[y-y_0]^3\right)$$
$$+\left(f^{(1,0)}[x_0,y_0], + f^{(1,1)}[x_0,y_0](y-y_0) + \frac{1}{2}f^{(1,2)}[x_0,y_0](y-y_0)^2 + O[y-y_0]^3\right)\cdot$$
$$(x-x_0)+\left(\frac{1}{2}f^{(2,0)}[x_0,y_0] + \frac{1}{2}f^{(2,1)}[x_0,y_0](y-y_0)^2(y-y_0) + \frac{1}{4}f^{(2,2)}[x_0,y_0](y-y_0)^2\right.$$
$$\left. + O[y-y_0]^3\right)(x-x_0)^2 + O[x-x_0]^3$$

Next we try to take a Taylor series of $\sin(xy)$ *about the point* $(0,0)$ *up to orders 6 in both variables.*

Series[Sin[x * y], {x, 0, 6}, {y, 0, 6}]

$$\left(y + O[y]^7\right)x + \left(-\frac{y^3}{6} + O[y]^7\right)x^3 + \left(\frac{y^5}{120} + O[y]^7\right)x^5 + O[x]^7$$

To make this resemble the standard form of the Taylor series we apply the MATHEMATICA *command* **Normal**

Normal[%]

$$xy - \frac{x^3y^3}{6} + \frac{x^5y^5}{120}$$

We next expand a rational function of two variables about $(0,0)$*; however,*

[3] Taylor polynomials are constructed with regard to a **center** and the order of the polynomial. Here we are picking the **center** be the origin. More about this later!

[4] We recall that all built in MATHEMATICA functions begin with a capital.

the polynomial terms are not listed as one might expect, namely in increasing powers of x and y.

```
Series[1/(1 − x^2 − y^2), {x, 0, 6}, {y, 0, 6}]
```

$(1+y^2+y^4+y^6+O[y]^7)+(1+2y^2+3y^4+4y^6+O[y]^7)x^2+(1+3y^2+6y^4+10y^6+O[y]^7)x^4+(1+4y^2+10y^4+20y^6+O[y]^7)x^6+O[x]^7$

```
tp = Normal[%]
```

$1+y^2+y^4+y^6+x^2\left(1+2y^2+3y^4+4y^6\right)+x^4\left(1+3y^2+6y^4+10y^6\right)+x^6\left(1+4y^2+10y^4+20y^6\right)$

```
N[Integrate[tp, {x, 0, 1}, {y, 0, 1}]]
```

4.57515

```
N[Integrate[Series[1/(1−x^2−y^2), {x, 1/4, 2}, {y, 1/4, 2}],{x, 0, 1}, {y, 0, 1}]]
```

2.29478

This is a rather poor approximation. How many terms in the Taylor approximation are needed to get three decimals of accuracy? We now turn to writing our own command to compute the Taylor polynomial. To this end we first construct a function which evaluates the mixed derivatives occurring in the Taylor approximation at the point we are expanding the series.

```
derivesValues[f_, N_, P_]:=
D[D[f/(N[[1]]! * N[[2]]!), {x, N[[1]]}], {y, N[[2]]}]/.{x → P[[1]], y → P[[2]]}
```

```
derivesValues[Sin[x * y], {5, 5}, {0, 0}]
```

$\frac{1}{120}$

```
derivesValues[Sin[x * y], {7, 7}, {0, 0}]
```

$-\frac{1}{5040}$

Can you tell why does the function evaluate to 0 here?

```
derivesValues[Sin[x * y], {5, 4}, {0, 0}]
```

0

```
derivesValues[Sin[x * y], {4, 4}, {0, 0}]
```

0

```
derivesValues[Cos[x * y], {4, 4}, {0, 0}]
```

$\frac{1}{24}$

Finally we produce a function that gives a Taylor polynomial. Are there alternate forms that might produce more presentable outputs? Try **taylorPolynom** *on some other functions of two variables.*

```
taylorPolynom[f_, N_, P_]:=
Sum[Sum[derivesValues[f,{i, j}, {P[[1]],P[[2]]}]*(x−P[[1]])^i*(y−P[[2]])^(j),
{i, 0, N[[1]]}], {j, 0, N[[2]]}]
```

We obtain the terms of order less than or equal to 5 *in the expansion of* $\sin(xy)$.

```
taylorPolynom[Sin[x * y], {5, 5}, {0, 0}]
```

$xy - \frac{x^3y^3}{6} + \frac{x^5y^5}{120}$

♦

Example 4.16. We next find the error in approximating $f := e^{x+y}$ by its linear TAYLOR polynomial

$$e^{x+y} = 1 + x + y.$$

Then since $f_{xx} = f_{xy} = f_{yy} = e^{x+y}$, on the set $[-0.1, 0.1] \times [-0.1, 0.1]$ the error in this approximation is bounded by

$$E_2(x, y) \leq \frac{(0.1 + 0.1)^2}{2!} \max_{|x|\leq 0.1, |y|\leq 0.1} \left(e^{x+y}\right) \leq \frac{1}{200} e^{0.2} \approx .$$

Example 4.17. Any power series expansion of the form

$$\sum_{j+k=0}^{\infty} a_{jk} \frac{(x-x_0)^j (y-y_0)^k}{j!k!}, \tag{4.25}$$

where the a_{jk} are real constants, and which converges in a domain $[x_0-\alpha, x_0+\alpha] \times [y_0-\beta, y_0+\beta]$ represents there a function $f(x,y)$, whose TAYLOR series coincides with this power series. This fact is very useful when computing TAYLOR series by hand computation. Suppose we wish to find the TAYLOR series of

$$f(x,y) := \frac{1}{1-x^2-y^2}$$

about the origin. Recall the geometric series

$$\frac{1}{1-r} = \sum_{k=0}^{\infty} r^k, \text{ which converges for } r < 1.$$

By setting $r = x^2+y^2$ in the geometric series we obtain

$$\frac{1}{1-x^2-y^2} = \sum_{k=0}^{\infty} \left(x^2+y^2\right)^k .$$

Using the binomial expansion we have

$$\left(x^2+y^2\right)^k = \sum_{j=0}^{k} \frac{k!}{j!(k-j)!} x^{2j} y^{2(k-j)} \quad ;$$

hence, a power series expansion is obtained in the form

$$\begin{aligned} f(x,y) &= \sum_{k=0}^{\infty} \sum_{j=0}^{k} \frac{k!}{j!(k-j)!} x^{2j}\, y^{2(k-j)} \\ &= 1 + x^2 + y^2 + x^4 + 2\,x^2 y^2 + y^4 + \text{ higher order terms}. \end{aligned}$$

Let us try this once more. Suppose we wish to expand $f(x,y) := e^{x+y}$ about the origin. Recall the expansion for e^x, namely

$$e^x = \sum_{k=0}^{\infty} \frac{x^k}{k!}.$$

Then as an expansion

$$\begin{aligned} f(x,y) &:= \sum_{k=0}^{\infty} \frac{(x+y)^k}{k!} = \sum_{k=0}^{\infty} \sum_{j=0}^{k} \frac{k!}{j!(k-j)!} x^j\, y^{k-j} \\ &= 1 + x + y + \frac{1}{2}\left(x^2 + 2xy + y^2\right) \text{ plus third order terms.} \end{aligned}$$

EXERCISES

Exercise 4.18. *Using the* TAYLOR *expansion theorem obtain by hand calculations the following functions as truncated series with center at the origin. Obtain all second order terms and compare your results using* MATHEMATICA . *Use* MATHEMATICA *to compute the errors in this order truncation if the x and y intervals are both* $[-0.1, 0.1]$.

(a) $e^x \sin y$ (b) $e^x \cos y$

(c) $\dfrac{1}{x+y-1}$ (d) e^{x+y}

(e) $\ln(4+2x+2y)$ (f) $\dfrac{x+2}{y-1}$

Exercise 4.19. *By hand calculation find a power series expansion, about the origin, of the following functions. Do not use the* TAYLOR *theorem to find the expansions. Check your results using the* MATHEMATICA *command* `taylor`.

(a) $\dfrac{1}{1+x^2+y^2}$ (b) $\dfrac{1+2x+3y}{1-x^2+y}$

(c) $\dfrac{x^2+y^2}{1-x-y+2xy}$ (d) $\sin(x^2+y^2)$

(e) $\ln(1+x+y)$ (f) $ye^{1+x}+xe^{1+y}$

(g) $\cos(x+y)$ (h) $(x+y)\tan(x+y)$

(i) $\dfrac{x}{2^{x-y}}$ (j) $\dfrac{1}{1+\sin(x+y)}$

(k) $\cos x \sin y$ (l) $\dfrac{\sec^2(x+y)}{2x+3y+1}$

Exercise 4.20. *A similar* TAYLOR *expansion holds for functions of three (and more) variables. For three variables we have the truncated* TAYLOR *expansion*

$$f(x,y) = \sum_{i+j+k\leq n} \frac{\partial^{i+j+k} f(x_0,y_0)}{\partial x^i\,\partial y^j\,\partial z^k} \frac{(x-x_0)^i\,(y-y_0)^j\,(z-z_0)^k}{i!j!k!} + R_{n+1}(x,y,z)\,. \tag{4.26}$$

Using this expression compute the second-order TAYLOR *polynomials of the following functions and check your results with* MATHEMATICA .

(a) $e^{x+y}\sin(y+z)$ (b) e^{x+y+z}

(c) $\dfrac{1}{x+y+z-1}$ (d) $e^{x^2+y^2+z^2}$

(e) $\ln(1+x+y+z)$ (f) $\dfrac{x+2y}{y+z-1}$

Exercise 4.21. *Write the Taylor expansion (4.20) and the remainder (4.21) for* $n=2$.

Exercise 4.22. *Let $\Phi(t)$ be defined by (4.16). Show that the general term of (4.20) corresponding to $\frac{\Phi^{(n)}(0)}{n!}$ can be written as*

$$\frac{1}{n!}\left(\Delta x\frac{\partial}{\partial x}+\Delta y\frac{\partial}{\partial y}\right)^n f(x,y)\bigg|_{x=x_0,y=y_0}.$$

4.3 Unconstrained extremum problems

We are already acquainted with extremum problems for functions of one variable from our course on one variable calculus. In this chapter we obtain similar results for functions of several variables. These results are easy to visualize for functions of two variables. We therefore state most of the definitions and results in two dimensions; the extensions to three or more dimensions are straightforward.

We first clarify what it means to be **near** a given point $P(x_0, y_0)$. A **neighborhood** of the point P, denoted $\mathcal{N}(P)$, usually means either

$$\begin{array}{rl} \text{the } \textbf{disk} \text{ centered at } P: & \{(x,y) : (x-x_0)^2 + (y-y_0)^2 \le r^2\} \,, \\ \text{or the } \textbf{square} \text{ centered at } P: & \{(x,y) : \max\{|x-x_0|,\, |y-y_0|\} \le r\} \end{array}$$

for some $r > 0$. If r is important, we denote the neighborhood by $\mathcal{N}(P, r)$. An arbitrary neighborhood usually means a neighborhood $\mathcal{N}(P, r)$ of arbitrarily small r.

Definition 4.23. Let $P(x_0, y_0)$ be a point in a domain $\mathcal{D}$ in $\mathbb{R}^2$. Then P is:
(a) an **interior point** of $\mathcal{D}$ if $\mathcal{D}$ contains some neighborhood of P ,
(b) a **boundary point** of $\mathcal{D}$ if not (a), i.e. if every neighborhood of P contains points not in $\mathcal{D}$.
Let f be defined in $\mathcal{D}$. Then f has at P:
(c) a **local maximum** if $f(x, y) \le f(x_0, y_0)$ for all (x, y) in some neighborhood of P,
(d) a **local minimum** if $f(x, y) \ge f(x_0, y_0)$ for all (x, y) in some neighborhood of P,
(e) a **local extremum** if either (a) or (b),
(f) a **global maximum** if $f(x, y) \le f(x_0, y_0)$ for all (x, y) in $\mathcal{D}$,
(g) a **global minimum** if $f(x, y) \ge f(x_0, y_0)$ for all (x, y) in $\mathcal{D}$,
(h) a **global extremum** if either (d) or (e).
Moreover, the point $P(x_0, y_0)$ is
(i) **stationary** if f is differentiable at P, and

$$\nabla f(x_0, y_0) = \mathbf{0} \,, \tag{4.27}$$

(j) **critical** if either stationary or if f is not differentiable at P.

These concepts are easily understood in terms of the graph

$$z = f(x, y) \,.$$

The local maxima are the altitudes z of mountain "peaks", and the local minima are the bottoms of wells, see e.g. Figure 4.3. A global maximum is the altitude of highest the mountain–top; a global minimum is the altitude of the lowest point on the graph. A stationary point is where the tangent plane exists and is horizontal.

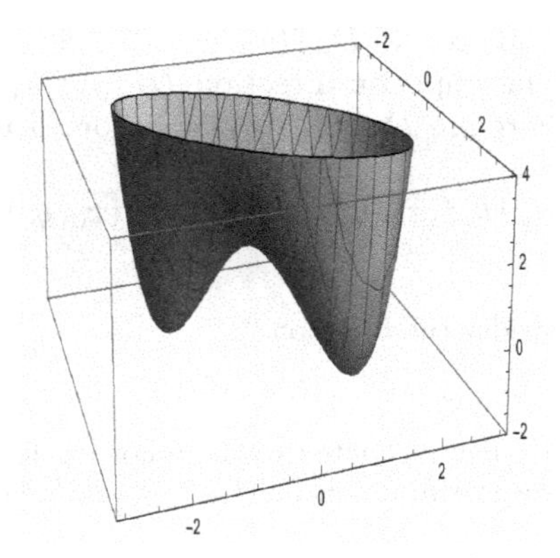

FIGURE 4.3: An example of a surface having two local minima and a saddle point.

We recall from the calculus of one variable that a necessary condition for f to have an extremum at x^* is for x^* to be critical. The first derivative was used in the **First Derivative Test** to test whether a function has a maximum or a minimum at x^*.

We generalize these results to functions of two variables.

Theorem 4.24. Let the function $f(x, y)$ be continuous at the point (x_0, y_0), and let $\mathcal{U}$ be the set of directions for which the directional derivatives $D_{\mathbf{u}}f(x_0, y_0)$ exist. If
(a) f has a local maximum at (x_0, y_0) then $D_{\mathbf{u}}f(x_0, y_0) \le 0$ for all $\mathbf{u} \in \mathcal{U}$,
(b) f has a local minimum at (x_0, y_0) then $D_{\mathbf{u}}f(x_0, y_0) \ge 0$ for all $\mathbf{u} \in \mathcal{U}$.

Proof:
(a) Let f have a maximum at (x_0, y_0), and let $\mathbf{u} \in \mathcal{U}$. Then the directional derivative

$$\lim_{t \to 0} \frac{f(x_0 + tu_1\, ,\, y_0 + tu_2) \;-\; f(x_0, y_0)}{t} \tag{4.28}$$

is nonpositive since $f(x_0 + tu_1, y_0 + tu_2) \le f(x_0, y_0)$ for t sufficiently small.
(b) is similarly proved. □

Corollary 4.25. Let the function f be differentiable at (x_0, y_0).
If f has a local extremum (maximum or minimum) at (x_0, y_0), then

$$\nabla f(x_0, y_0) \;=\; \mathbf{0}\,. \tag{4.29}$$

In other words, if f has a local extremum at a point P, then P is a critical point.

Proof:
Let f have a maximum at (x_0, y_0). By the first-order test it follows that

$D_{\mathbf{u}}f(x_0, y_0) \leq 0$ for all $\mathbf{u}$. If $D_{\mathbf{u}}f(x_0, y_0) < 0$ for some $\mathbf{u}$ then the directional derivative in the opposite direction $D_{\{-\mathbf{u}\}}f(x_0, y_0) > 0$ contradicting Theorem 4.24(a). Therefore $D_{\mathbf{u}}f(x_0, y_0) = 0$ for all $\mathbf{u}$, and $\nabla f(x_0, y_0) = \mathbf{0}$ by

$$D_{\mathbf{u}}f(x_0, y_0) = \mathbf{u} \cdot \nabla f(x_0, y_0). \tag{4.30}$$

□

Example 4.26. Consider the function [5]

$$f(x, y) := (y - x^2)(y - 2x^2),$$

and the minimum of f when restricted to the following lines

(a) the x–axis: here the function is $f(x, 0) = 2x^4$ which has its minimum at $x = 0$,

(b) the y–axis: on this line the function becomes $f(0, y) = y^2$, and its minimum is at $y = 0$, and

(c) any line $y = m\,x$ with finite, nonzero, m: we define the restriction of f to this line by

$$F(x) := f(x, m\,x) = (m\,x - x^2)(m\,x - 2\,x^2);$$

then $F'(0) = 0$, $F''(0) = 2\,m^2$ and therefore F has a minimum at $x = 0$.

We conclude that f has a minimum at the origin along any line which passes through the origin. However, f does not have a minimum at the origin, because arbitrarily close to the origin it has both positive and negative values, so $f(0, 0) = 0$ cannot be an extremum.

Note that the gradient $\nabla f(0, 0) = \mathbf{0}$, showing that the necessary condition (4.29) is not sufficient. △

Example 4.27. Consider the hyperbolic paraboloid

$$f(x, y) = x^2 - y^2.$$

The origin $(0, 0)$ is a stationary point of f , but is not a local extremum. Indeed f has at $(0, 0)$:

(a) a minimum along the x–axis where $f = f(x, 0) = x^2$, and

(b) a maximum along the y–axis where $f = f(0, y) = -y^2$. As suggested by this picture, the point $(0, 0)$ is called a **saddle point** of f. △

Example 4.28. The function

$$f(x, y) := 1 - (x^2 + y^2)^{\frac{1}{3}}$$

has a global maximum at the origin. Plot $f(x, y)$ with MATHEMATICA to verify this. See Figure 4.6. *The first partial derivatives*

$$\frac{\partial f}{\partial x} = -\frac{2}{3}\frac{x}{(x^2 + y^2)^{\frac{2}{3}}}, \quad \frac{\partial f}{\partial y} = -\frac{2}{3}\frac{y}{(x^2 + y^2)^{\frac{2}{3}}}.$$

blow-up at the origin, showing that the derivatives need not exist at an extremum. △

[5]See Figures 4.4 and ?? .

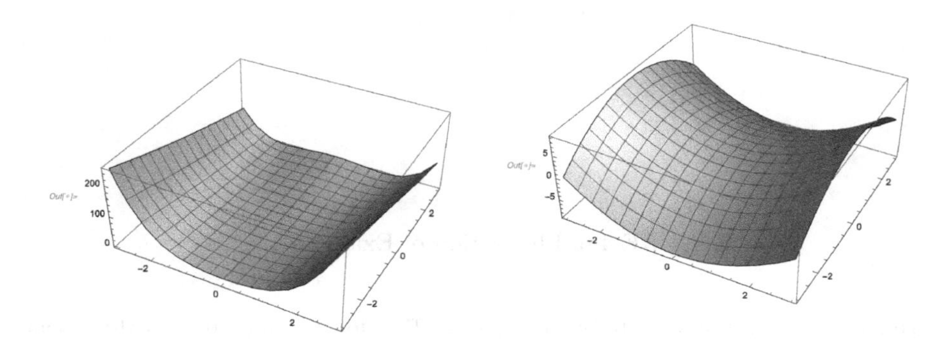

FIGURE 4.4: Illustration of Example 4.26

FIGURE 4.5: Illustration of Example 4.27

We see in Definition 4.23 that **local extrema** are defined for a function f and a point P; and **global extrema** also depend on the set $\mathcal{D}$. If f has a global extremum at a point P on the boundary of $\mathcal{D}$, then this point is not covered by the necessary condition in Theorem 4.24 and Corollary 4.25. In other words, if $P(x_0, y_0)$ is a boundary point of $\mathcal{D}$ where f has a global extremum, then it is possible for $\nabla f(x_0, y_0) \neq \mathbf{0}$.

Example 4.29. Let $\mathcal{D}$ be the set points inside and on the ellipse

$$\mathcal{D} := \left\{ (x, y) : \frac{x^2}{2} + y^2 \leq 1 \right\},$$

and consider the local and global extrema, in $\mathcal{D}$, of the function

$$f(x, y) := \sqrt{2\,x^2 + y^2}\,.$$

First we find the critical points. The stationary points are found by solving

$$\frac{\partial f}{\partial x} = \frac{2\,x}{\sqrt{2\,x^2 + y^2}} = 0\,, \quad \frac{\partial f}{\partial y} = \frac{y}{\sqrt{2\,x^2 + y^2}} = 0\,.$$

these never vanish simultaneously, so f has no stationary points. The origin $(0, 0)$ is a critical point (the partial derivatives f_x and f_y are undefined at $(0, 0)$). Since

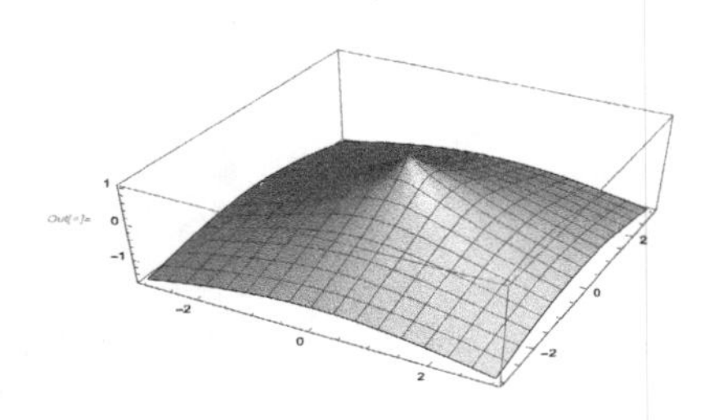

FIGURE 4.6: Illustration of Example 4.28.

$f(0,0) = 0$ and $f(x,y) \geq 0$ for all $(x,y) \in \mathcal{D}$ it follows that $(0,0)$ is the global minimum.

Since f is continuous it does have a global maximum on $\mathcal{D}$. The global maximum must then occur on the boundary as the only interior candidate was $(0,0)$ which corresponds to a global minimum. On the boundary of $\mathcal{D}$ the function f takes on the values

$$\sqrt{4 - 3\,y^2}\,, \quad -1 \leq y \leq 1\,.$$

Therefore, the global maxima of f on $\mathcal{D}$ occur at the boundary points where $y = 0$, i.e. the two points $(\pm\sqrt{2}, 0)$.

Plot the boundary of $\mathcal{D}$, and the contour map of f, and verify the above conclusions. △

A more complicated situation is where the boundary of $\mathcal{D}$ consists of two parts or more, where different parameterizations can be used. This is illustrated in the following:

Example 4.30. Let $\mathcal{D}$ be the triangular domain bounded by the three straight lines

$$y = x\,, \quad y = 0 \quad \text{and} \quad y = -x + 2\,.$$

It is required to find the extremal values of the function

$$f(x,y) := x^2\,y^2 + x + y$$

on $\mathcal{D}$. The function f is differentiable throughout $\mathcal{D}$, and its partial derivatives are

$$\frac{\partial f}{\partial x} = 2\,x\,y^2 + 1\,, \quad \frac{\partial f}{\partial y} = 2\,y\,x^2 + 1\,.$$

We find the stationary points by solving $f_x = f_y = 0$ to obtain $x = y = -2^{-\frac{1}{3}}$. However, this point is outside the domain $\mathcal{D}$, showing that f has no local extrema which are interior points of $\mathcal{D}$. We must therefore examine the behavior of f at the boundary points of $\mathcal{D}$. Since the boundary is made of three line segments, they must

be considered separately:
(a) Along the segment on the line $y = x$: Here $f = f(y, y) = y^4 + 2y$ where $0 \le y \le 1$ (check!), and f has a minimum at $y = 0$ and a maximum at $y = 1$. Therefore $f(0, 0) = 0$ is the minimum, and $f(1, 1) = 3$ is the maximum on this segment.
(b) Along the segment on the line $y = ,0$: Here $f = f(x, 0) = x$ where $0 \le x \le 2$ (check!), and f has a minimum at $x = 0$ and a maximum at $x = 2$. The minimum on this segment is $f(0, 0) = 0$, and the maximum is $f(2, 0) = 2$. Note that the point $(0, 0)$ is also the minimizer in part (a).
(c) Along the segment on the line $y = -x + 2$: Here $f = f(x, -x + 2) = x^4 - 4x^3 + 4x^2 + 2$ and $1 \le x \le 2$ (check!). The function $g(x) := x^4 - 4x^3 + 4x^2 + 2$ has derivative $g'(x) = 4x^3 - 12x^2 + 8x = 4x(x - 1)(x - 2)$. The stationary points are $x = 0$, $x = 1$ and $x = 2$. The point $(0, 2)$ is outside $\mathcal{D}$, and the other two points, $(1, 1)$ and $(2, 0)$ were encountered before.

In summary, the function f has on $\mathcal{D}$ a global maximum at $(1, 1)$: $f(1, 1) = 3$, and a global minimum at $(0, 0)$: $f(0, 0) = 0$. △

Example 4.31. Let $\mathcal{D}$ be the annulus

$$\mathcal{D} := \left\{(x, y) : \frac{1}{2} \le \sqrt{x^2 + y^2} \le 1\right\}$$

and consider the function

$$f(x, y) := \frac{1}{x\,y + 1} .$$

It is required to find the global extrema of f in $\mathcal{D}$.

We first compute the partial derivatives

$$\frac{\partial f}{\partial x} = -\frac{y}{(x\,y + 1)^2} , \quad \frac{\partial f}{\partial y} = -\frac{x}{(x\,y + 1)^2} .$$

The stationary point $(x, y) = (0, 0)$ is outside $\mathcal{D}$. Other critical points are along the hyperbolas $x\,y = -1$, which do not intersect $\mathcal{D}$ (check!). We conclude that the global extrema are on the boundary of $\mathcal{D}$. The boundary is made of two parts, which are analyzed separately:
(a) The outer circle $\mathcal{C}_1$: $\sqrt{x^2 + y^2} = 1$. We parametrize this curve by

$$x = \cos\theta , \quad y = \sin\theta , \quad \text{where} \quad 0 \le \theta \le 2\pi ,$$

and the function f restricted to $\mathcal{C}_1$ is

$$g(\theta) := f(\cos\theta, \sin\theta) = \frac{1}{1 + \cos\theta\,\sin\theta} .$$

To find the extrema of $g(\theta)$ on the interval $[0, 2\pi]$ we compute

$$\frac{d\,g(\theta)}{d\theta} = \frac{-4\,\cos 2\theta}{(2 + \sin 2\theta)^2}$$

whose denominator is always positive, and the numerator is zero for the four points $\pi/4$, $3\pi/4$, $5\pi/4$, $7\pi/4$ in $[0, 2\pi]$. These correspond to the four points on $\mathcal{C}_1$:

$$A\left(\frac{1}{\sqrt{2}}, \frac{1}{\sqrt{2}}\right) , \quad B\left(\frac{1}{\sqrt{2}}, -\frac{1}{\sqrt{2}}\right) , \quad C\left(-\frac{1}{\sqrt{2}}, -\frac{1}{\sqrt{2}}\right) , \quad D\left(-\frac{1}{\sqrt{2}}, \frac{1}{\sqrt{2}}\right) ,$$

Examining the function $f(x,y) = 1/(x\,y+1)$ we conclude that minimum occurs if x and y have the same sign; maximum if x and y have opposite signs (why?). Therefore A and C with $f(A) = f(C) = \frac{2}{3}$ are candidates for global minima, and B and D, with $f(B) = f(D) = 2$ are candidates for global maxima.
(b) The inner circle $\mathcal{C}_2$: $\sqrt{x^2+y^2} = \frac{1}{2}$. This curve is parametrized by

$$x = \frac{1}{2}\cos\theta\ ,\quad y = \frac{1}{2}\sin\theta \quad \text{where} \quad 0 \le \theta \le 2\,\pi\ ,$$

and the function f restricted to $\mathcal{C}_2$ is

$$h(\theta) := f(\frac{1}{2}\cos\theta, \frac{1}{2}\sin\theta) = \frac{8}{8+\sin 2\,\theta}\ .$$

Proceeding as above we get four stationary points of h on $\mathcal{C}_2$,

$$E\left(\frac{1}{2\sqrt{2}},\frac{1}{2\sqrt{2}}\right)\ ,\ F\left(\frac{1}{2\sqrt{2}},-\frac{1}{2\sqrt{2}}\right)\ ,\ G\left(-\frac{1}{2\sqrt{2}},-\frac{1}{2\sqrt{2}}\right)\ ,\ H\left(-\frac{1}{2\sqrt{2}},\frac{1}{2\sqrt{2}}\right)\ .$$

Computing f at these points, and comparing with the values at A, B, C, D we conclude that $f(B) = f(D) = 2$ are the global maxima, and $f(A) = f(C) = \frac{2}{3}$ are the global minima, of f at $\mathcal{D}$. △

We can summarize the procedure for finding global extrema as follows:

A scheme for finding the global extrema of f in $\mathcal{D}$	
Step1 :	Find all critical points of f in $\mathcal{D}$; stationary points found by solving $\nabla f(x,y) = \mathbf{0}$, and also points where f is not differentiable .
Step2 :	Determine which of the critical points are local maximizers or local minimizers .
Step3 :	Restrict f to the boundary $\mathcal{C}$ of $\mathcal{D}$, and compute its local maximizers and local minimizers on $\mathcal{C}$.
Step4 :	Compare the values of f in Steps 2 and 3, and determine the global extremizer(s) and extremal values.

Exercises

Exercise 4.32. *For the following f and $\mathcal{D}$, compute the global extrema of f on $\mathcal{D}$.*

(a) $f(x,y) := \dfrac{x^2-y^2}{x^2+y^2}\ ,\quad \mathcal{D} := \{(x,y) : \max\{|x|,|y|\} \le 1\}$

(b) $f(x,y) := (x^2+y^2)\sin(x^2+y^2)\ ,\quad \mathcal{D} := \{(x,y) : x^2+y^2 \le 2\,\pi\}$

(c) $f(x,y) := x^y\ ,\quad \mathcal{D} := \{(x,y) : 0 \le x,y \le 1\}$

4.4 Second derivative test for extrema

We saw in the previous section that a stationary point may be a local maximum, a local minimum or a saddle point (i.e. not an extremum at all). In this section we develop a sufficient condition for a stationary point of the function $f(x, y)$ to be a local maximum or a local minimum. This condition uses the second partial derivatives

$$\frac{\partial^2 f}{\partial x^2}\,, \quad \frac{\partial^2 f}{\partial x\, \partial y}\,, \quad \frac{\partial^2 f}{\partial y^2}\,,$$

and is analogous to the second derivative test in single-variable calculus. First we require some terminology:

Definition 4.33. Let A, B and C be constants. The function

$$f(x, y) \;:=\; A\,x^2 + 2\,B\,x\,y + C\,y^2 \tag{4.31}$$

is called a **quadratic form**. It is called:
(a) **definite** if $f(x, y) = 0$ only if $x = y = 0$, in particular,
(b) **positive definite** if $f(x, y) > 0$ for all nonzero (x, y),
(c) **negative definite** if $f(x, y) < 0$ for all nonzero (x, y),
(d) **indefinite** if neither (b) nor (c), i.e. if $f(x, y)$ has both positive values and negative values.
The **discriminant** of the quadratic form (4.31) is defined as[6]

$$\delta \;=\; B^2 - A\,C\,. \tag{4.32}$$

Remark 4.34. *(a) If $A\,x^2 + 2\,B\,x\,y + C\,y^2$ is:*
positive definite *then the point* $(0, 0)$ *is the global minimum,*
negative definite *then the point* $(0, 0)$ *is the global maximum,*
indefinite *then the point* $(0, 0)$ *is a saddle point.*
(b) The function (4.31) can be written, using vector–matrix notation, as

$$f(x, y) \;:=< x,\, y >^T \begin{pmatrix} A & B \\ B & C \end{pmatrix} < x,\, y >= (x, y) \begin{pmatrix} A & B \\ B & C \end{pmatrix} < x,\, y > . \tag{4.33}$$

The determinant of the 2×2 *matrix in (4.33) is*

$$\mathbf{det} \begin{pmatrix} A & B \\ B & C \end{pmatrix} \;=\; A\,C - B^2$$

the negative of the discriminant (4.32).

[6]Note that the discriminant of the general quadratic function

$$D_1 \qquad A\,x^2 + B\,x\,y + C\,y^2 + D\,x + E\,y + F \;=\; 0\,,$$

is defined as $B^2 - 4\,A\,C$, because the coefficient of $x\,y$ is B, not $2\,B$ as in (4.31).

Example 4.35. The quadratic forms

(a) $f(x,y) := x^2 + y^2$ (b) $f(x,y) := -x^2 - y^2$

(c) $f(x,y) := x^2 - y^2$

are

(a) positive definite (b) negative definite

(c) indefinite, respectively.

△

The test for local extrema is based on deciding if a quadratic form is positive definite or negative definite. As in Remark 4.34, the answer depends on the discriminant.

Lemma 4.36. Let A, B, and C be constants. Then a necessary and sufficient condition for

$$A\,x^2 + 2\,B\,x\,y + C\,y^2 \tag{4.34}$$

to be definite is that

$$A\,C - B^2 \;>\; 0\,, \tag{4.35}$$

in which case (4.34) is positive definite if $A > 0$ (or equivalently, if $C > 0$), negative definite if $A < 0$ (or equivalently, if $C < 0$).

Proof:
Sufficiency: Let (4.35) hold. Since $B^2 \geq 0$ it follows that both A and C are nonzero. Therefore (4.34) becomes

$$A\,x^2 + 2\,B\,x\,y + C\,y^2 \;=\; \frac{1}{A}\left[(A\,x + B\,y)^2 + \left(A\,C - B^2\right)\,y^2\right] \tag{4.36}$$

which is zero only if $A\,x + B\,y = 0$ and $y = 0$, i.e. only if $x = y = 0$.
Necessity: Let (4.34) be definite, and assume (4.35) does not hold, i.e.

$$A\,C - B^2 \;\leq\; 0\,. \tag{4.37}$$

If $A \neq 0$ then the form is definite provided

$$A\,x^2 + 2\,B\,x\,y + C\,y^2 \;=\; \frac{1}{A}\left[(A\,x + B\,y)^2 + \left(A\,C - B^2\right)\,y^2\right] \;\neq\; 0$$

for all nonzero x and y, which implies $A\,C - B^2 > 0$, contradicting (4.37). A similar contradiction of (4.37) is obtained if $C \neq 0$. Therefore $A = C = 0$, and B must be nonzero (otherwise (4.34) is identically zero). Then (4.34) is zero for $(0, 1)$, showing it is indefinite, a contradiction.
Finally, if (4.35) holds, then (4.36) shows that for nonzero (x, y), the sign of $A\,x^2 + 2\,B\,x\,y + C\,y^2$ is the same as the sign of A. □

Example 4.37. The discriminants of the quadratic forms in Example 4.35 are

(a) $\delta \;=\; -1$ (b) $\delta \;=\; -1$ (c) $\delta \;=\; 1$

in agreement with Lemma 4.36. Similarly, $x^2 + 4\,y^2$ is positive definite since $\delta = -4$, $A > 0$. $2\,x^2 - y^2$ is indefinite since $\delta = 2$. A quadratic form may be definite in some region, and indefinite in others. For example, the form

$$3\,x^2 + 7\,x\,y + 2\,y^2$$

is positive definite in the first quadrant. However, there are no simple conditions, such as (4.35), to determine a region where a quadratic form is definite. △

In the test for extrema the A, B, and C of the quadratic form are the second partial derivatives f_{xx}, f_{xy}, and f_{yy} evaluated at the critical points.

Theorem 4.38. Let (x_0, y_0) be a stationary point of f, and let f have continuous partial derivatives of the second order in some neighborhood of (x_0, y_0), with

$$\frac{\partial^2 f}{\partial x^2}\frac{\partial^2 f}{\partial y^2} - \left(\frac{\partial^2 f}{\partial x \partial y}\right)^2 > 0\,, \tag{4.38}$$

holding at (x_0, y_0). Then f has at (x_0, y_0):
(a) a local minimum if

$$\frac{\partial^2 f}{\partial x^2}(x_0, y_0) > 0\,,$$

(b) a local maximum if

$$\frac{\partial^2 f}{\partial x^2}(x_0, y_0) < 0\,.$$

Proof:
Let (x_0, y_0) satisfy the conditions of the theorem. Since $f(x, y)$ has continuous derivatives up to second order, its Taylor approximation at all points $(x_0 + h, y_0 + k)$ is, by (4.20)–(4.21),

$$f(x_0 + h, y_0 + k) =$$
$$f(x_0, y_0) + h\,\frac{\partial f}{\partial x}(x_0, y_0) + k\,\frac{\partial f}{\partial y}(x_0, y_0) + R_2(x, y)\,, \tag{4.39}$$

where

$$R_2(x, y) = \frac{1}{2}\,\left(A(\theta)\,h^2 + 2\,B(\theta)\,h\,k + C(\theta)\,k^2\right)\,,$$
$$A(\theta) := \frac{\partial^2 f}{\partial x^2}(x_0 + \theta\,h, y_0 + \theta\,k)\,,$$
$$B(\theta) := \frac{\partial^2 f}{\partial x\,\partial y}(x_0 + \theta\,h, y_0 + \theta\,k)\,,$$
$$C(\theta) := \frac{\partial^2 f}{\partial y^2}(x_0 + \theta\,h, y_0 + \theta\,k)\,,$$

for some $0 \le \theta \le 1$. Since (x_0, y_0) is stationary, the first derivatives in (4.39) are zero, and we get

$$f(x_0 + h, y_0 + k) = f(x_0, y_0) + \frac{1}{2}\,\left(A(\theta)\,h^2 + 2\,B(\theta)\,h\,k + C(\theta)\,k^2\right)\,.$$

Since the second partial derivatives are continuous near (x_0, y_0), it follows that

$$\begin{aligned}\text{SIGN}\left(A(\theta)\,C(\theta)-B^2(\theta)\right) &= \text{SIGN}\left(A(0)\,C(0)-B^2(0)\right)\\ &= \text{SIGN}\left(\frac{\partial^2 f}{\partial x^2}(x_0,y_0)\,\frac{\partial^2 f}{\partial y^2}(x_0,y_0)-\left(\frac{\partial^2 f}{\partial x\partial y}(x_0,y_0)\right)^2\right)\\ \text{and}\quad \text{SIGN}\,A(\theta) &= \text{SIGN}\,A(0) = \text{SIGN}\,\frac{\partial^2 f}{\partial x^2}(x_0,y_0)\end{aligned}$$

for all sufficiently small h, k. The proof follows then from Lemma 4.36 and Remark 4.34(a), with A, B, C given directly above. □

It is convenient to group the second partial derivatives, appearing in Theorem 4.38, in a matrix which deserves a special name:

Definition 4.39. Let $f(x,y)$ have second partial derivatives at the point $\mathbf{r} =< x_0, y_0 >$. The **Hesse matrix**[7] (or **Hessian**) of f at (x_0, y_0) is the matrix

$$H_f(\mathbf{r}) := \begin{pmatrix} \frac{\partial^2 f}{\partial x^2}(\mathbf{r}) & \frac{\partial^2 f}{\partial x\partial y}(\mathbf{r}) \\ \frac{\partial^2 f}{\partial y\partial x}(\mathbf{r}) & \frac{\partial^2 f}{\partial y^2}(\mathbf{r}) \end{pmatrix}. \tag{4.40}$$

Mathematica session 4.4. *Mathematica is able to work with the Hessian. See, for example, Figure 4.7 and Figure 4.8.*

```
D[f[x, y], {{x, y}}]
```

$\{f^{(1,0)}[x,y], f^{(0,1)}[x,y]\}$

```
hessian[f_]:=MatrixForm[D[f, {{x, y}}, {{x, y}}]]
```

```
hessian[(x^2 - y)/(x^2 + y^2 + 1)]
```

$$\begin{pmatrix} \frac{8x^2(x^2-y)}{(1+x^2+y^2)^3} - \frac{8x^2+2(x^2-y)}{(1+x^2+y^2)^2} + \frac{2}{1+x^2+y^2} & \frac{8x(x^2-y)y}{(1+x^2+y^2)^3} + \frac{2x}{(1+x^2+y^2)^2} - \frac{4xy}{(1+x^2+y^2)^2} \\ \frac{8x(x^2-y)y}{(1+x^2+y^2)^3} + \frac{2x}{(1+x^2+y^2)^2} - \frac{4xy}{(1+x^2+y^2)^2} & \frac{8(x^2-y)y^2}{(1+x^2+y^2)^3} - \frac{2(x^2-y)}{(1+x^2+y^2)^2} + \frac{4y}{(1+x^2+y^2)^2} \end{pmatrix}$$

```
Det[%]
```

$$-\frac{8x^2}{(1+x^2+y^2)^6} + \frac{8x^6}{(1+x^2+y^2)^6} + \frac{12y}{(1+x^2+y^2)^6} - \frac{12x^2y}{(1+x^2+y^2)^6} - \frac{28x^4y}{(1+x^2+y^2)^6} - \frac{4x^6y}{(1+x^2+y^2)^6} +$$
$$\frac{12y^2}{(1+x^2+y^2)^6} - \frac{12x^2y^2}{(1+x^2+y^2)^6} - \frac{12x^4y^2}{(1+x^2+y^2)^6} - \frac{4x^6y^2}{(1+x^2+y^2)^6} + \frac{20y^3}{(1+x^2+y^2)^6} - \frac{24x^2y^3}{(1+x^2+y^2)^6} -$$

[7] Ludwig Otto Hesse [1811–1874].

$$\frac{12x^4y^3}{(1+x^2+y^2)^6} + \frac{8y^4}{(1+x^2+y^2)^6} - \frac{24x^2y^4}{(1+x^2+y^2)^6} - \frac{8x^4y^4}{(1+x^2+y^2)^6} + \frac{4y^5}{(1+x^2+y^2)^6} - \frac{12x^2y^5}{(1+x^2+y^2)^6} - \frac{4y^6}{(1+x^2+y^2)^6} - \frac{4x^2y^6}{(1+x^2+y^2)^6} - \frac{4y^7}{(1+x^2+y^2)^6}$$

***Simplify*[%]**

$$-\frac{4\left(x^4\left(-2+y+y^2\right)+y\left(-3-3y-2y^2+y^3+y^4\right)+x^2\left(2+6y+4y^2+2y^3+y^4\right)\right)}{(1+x^2+y^2)^5}$$

Plot3D $\left[4\left(x^4\left(-2+y+y^2\right)+y\left(-3-3y-2y^2+y^3+y^4\right)+x^2\left(2+6y +4y^2+2y^3+y^4\right)\right), \{x,-2,2\}, \{y,-2,2\}\right]$

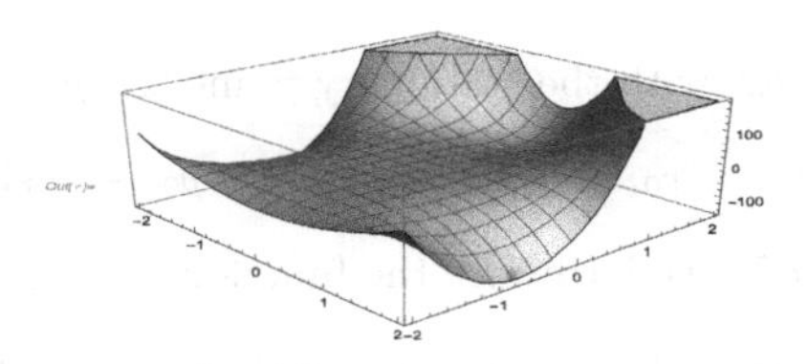

FIGURE 4.7: A plot showing the Hessian, determinant for the function $(x^2 - y)/(x^2 + y^2 + 1)$.

ContourPlot $\left[x^4\left(-2+y+y^2\right)+y\left(-3-3y-2y^2+y^3+y^4\right)+x^2\left(2+6y +4y^2+2y^3+y^4\right) == 0, \{x,-4,4\}, \{y,-4,4\}\right]$

***hessian*[x^2 + y^2]**

$$\begin{pmatrix} 2 & 0 \\ 0 & 2 \end{pmatrix}$$

♦

Remark 4.40. The conditions of Theorem 4.38 can be reduced to a single variable

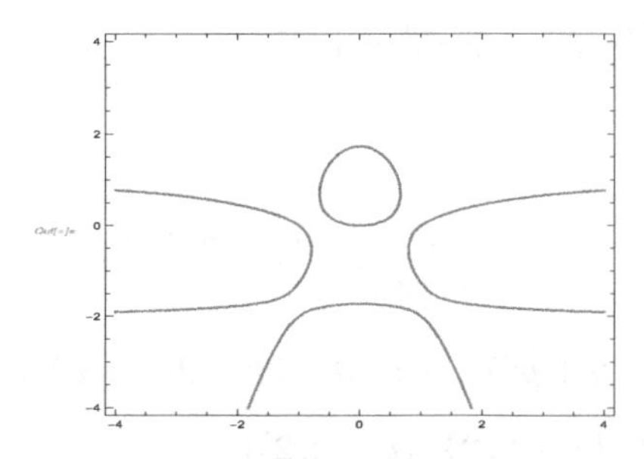

FIGURE 4.8: A contour plot showing the Hessian, determinant's zeros for the function $(x^2 - y)/(x^2 + y^2 + 1)$.

using polar coordinates[8] with the stationary point (x_0, y_0) as the origin,

$$x := x_0 + r \cos\theta , \quad y := y_0 + r \sin\theta .$$

For each fixed value of θ we introduce the function

$$F(r : \theta) := f(x_0 + r \cos\theta, y_0 + r \sin\theta) .$$

of the single variable r. Then

$$\frac{dF(r : \theta)}{dr}(0) = f_x(x_0, y_0) \cos\theta + f_y(x_0, y_0) \sin\theta . \tag{4.41}$$

If (x_0, y_0) is a stationary point of $f(x, y)$, then $\frac{dF(r:\theta)}{dr}(0) = 0$ and $r = 0$ is a stationary point of $F(r : \theta)$ for all θ. Conversely, if $r = 0$ is a stationary point of $F(r : \theta)$ for all θ, then $f_x(x_0, y_0) = f_y(x_0, y_0) = 0$, since $\cos\theta$ and $\sin\theta$ are never simultaneously zero. It follows that (x_0, y_0) is a stationary point of $f(x, y)$ if and only if $r = 0$ is a stationary point of $F(r : \theta)$ for all θ.

We can then determine if the stationary point corresponds to a maximum, a minimum or a saddle point, using the second derivative of $F(r : \theta)$ at $r = 0$,

$$\frac{d^2F(r : \theta)}{dr^2}(0) = f_{xx}(x_0, y_0) \cos^2\theta + 2f_{xy}(x_0, y_0) \cos\theta \sin\theta + f_{yy}(x_0, y_)) \sin^2\theta , \tag{4.42}$$

a quadratic form in $\cos\theta$, $\sin\theta$. From the second derivative test, Theorem 4.38, it follows that $F(r : \theta)$ has:

(i) a **local minimum** at $r = 0$ if the quadratic form (4.42) is **positive definite**,

(ii) a **local maximum** at $r = 0$ if the quadratic form (4.42) is **negative definite**,

and the rest follows from Lemma 4.36.

[8] This reduction was suggested by Leonard Gillman, "The second-partials test for local extrema of $f(x, y)$", *The American Mathematical Monthly* **101**(1994), 1004–1006.

Example 4.41. The stationary points of

$$f(x,y) \ := \ x^4 + y^4 + 2\,x^2 + 4\,y^2 \ .$$

are found by solving the system

$$\frac{\partial f}{\partial x} \ = \ 4\,x\,(x^2+1) \ = \ 0 \ \text{and} \ \frac{\partial f}{\partial y} \ = \ 4\,y\,(y^2+2) \ = \ 0$$

giving $(x_0, y_0) \ = \ (0,0)$ as the sole stationary point. At this point

$$\frac{\partial^2 f}{\partial x^2} \ = \ 4\,(3\,x^2+1) \ = \ 4 \ , \qquad \frac{\partial^2 f}{\partial x\,\partial y} \ = \ 0 \ , \qquad \frac{\partial^2 f}{\partial y^2} \ = \ 4\,(3\,y^2+2) \ = \ 8$$

and

$$\left(\frac{\partial^2 f}{\partial x\,\partial y}\right)^2 - \frac{\partial^2 f}{\partial x^2}\,\frac{\partial^2 f}{\partial y^2} \ = \ -32$$

showing that $(0,0)$ is a local minimum, by Theorem 4.38. △

Example 4.42. We determine the stationary points of the function

$$f(x,y) \ := \ x^4 + 4\,x\,y^3$$

by solving the system

$$\frac{\partial f}{\partial x} \ = \ 4\,x^3 + 4\,y^3 \ = \ 0$$

and

$$\frac{\partial f}{\partial y} \ = \ 12\,x\,y^2 \ = \ 0$$

giving $(x_0, y_0) = (0,0)$ as the only stationary point. The second derivative test of Theorem 4.38 fails, because all second derivatives vanish at $(0,0)$. We use again the trick of Remark 4.40, and change to polar coordinates around the stationary point $(0,0)$,

$$x \ := \ r\,\cos\theta \ , \quad y \ := \ r\,\sin\theta \ .$$

For each fixed θ we define the function

$$F(r:\theta) \ = \ f(r\,\cos\theta,\, r\,\sin\theta) \ = \ r^4\,\left(\cos^4\theta + 4\,\cos\theta\,\sin^3\theta\right)$$

and a new function

$$G(r:\theta) \ = \ \begin{cases} \dfrac{F(r:\theta)}{r^2} \ = \ r^2\,\left(\cos^4\theta + 4\,\cos\theta\,\sin^3\theta\right) & \text{if } r > 0 \\ 0 & \text{if } r = 0 \end{cases}$$

which has the same sign as $F(r:\theta)$ for all θ. Then

$$\frac{dG(r:\theta)}{dr}(0) \ = \ 0 \ , \qquad \frac{d^2G(r:\theta)}{dr^2}(0) \ = \ 2\,\left(\cos^4\theta + 4\,\cos\theta\,\sin^3\theta\right)$$

The second derivative changes sign as θ increases from 0 to $2\,\pi$, showing that the stationary point $(0,0)$ cannot be a local extremizer, i.e. it is a saddle point. △

Sometimes it is required to find extrema of a function $f(x,y)$ on a curve $\mathcal{C}$ parametrized by

$$\mathcal{C} := \{(x(t), y(t)) : t \in [0,T]\} .$$

Substituting $x = x(t)$ and $y = y(t)$ in $f(x,y)$, we get a function of the single variable t

$$F(t) := f(x(t), y(t))$$

and the problem reduces to finding the extreme values of $F(t)$ in the interval $[0,T]$, a one–dimensional problem solvable by the methods of single variable calculus.

Example 4.43. Find the extreme values of

$$f(x,y) := x^2 y + y^2 x \quad \text{on the curve} \quad \mathcal{C} := \{(\cos t, \sin t) : 0 \le t \le \pi\} .$$

Substituting $x := \cos t$, $y := \sin t$ in f we get

$$F(t) := \cos^2 t \, \sin t + \sin^2 \cos t .$$

The stationary points are the roots of

$$F'(t) = -2 \cos t \sin^2 t + \cos^3 t + 2 \sin t \cos^2 t - \sin^3 t = 0 ,$$

or

$$\xi^3 + 2\xi^2 - 2\xi - 1 = 0$$

where $\xi := \cot t$. The roots of this equation are

$$\xi = 1 , \quad \xi = \frac{-3 \pm \sqrt{5}}{2} .$$

Translating back to $t := \operatorname{acot} \xi$ we get the stationary points of $F(t)$,

$$t = \frac{\pi}{4} , \quad t = \pi - \operatorname{acot}\left(\frac{3 \pm \sqrt{5}}{2}\right) .$$

Exercises

Exercise 4.44. *Find the minimum value of the function $x^4 + y^4 + 4x - 16y + 9$.*

Exercise 4.45. *Complete the details of the necessity argument for the proof that a quadratic form is definite if $B^2 - AC < 0$.*

Exercise 4.46. *In the proof of the second derivative criteria, complete the proof of the sufficiency that f has a local maximum if $\frac{\partial^2 f}{\partial x^2} < 0$ and a local minimum if $\frac{\partial^2 f}{\partial x^2} > 0$. Hint: consider when the quadratic form remains positive under this condition and when it remains negative.*

Mathematica session 4.5. *In this session we use* MATHEMATICA *to find the extrema of a function of several variables. Let us first write a procedure to find the critical points of a function of two variables. The reader can easily modify this procedure to work for functions of three or more variables. To test whether the point* $(0,0)$ *turns out to be a maximum, minimum, or a saddle-point, we need to compute the discriminant and the second derivative with respect to* x *and check their signs. We write first a procedure to compute the discriminant.*

```
criticalPoints[f_]:=Module[{dfx, dfy, sol}, dfx = D[f, x]; dfy = D[f, y];
sol = Solve[{dfx==0, dfy==0}, {x, y}]]
```

```
criticalPoints[x^2 + 3 * x * y + y^2]
```

$\{\{x \to 0, y \to 0\}\}$

We now use the MATHEMATICA *command* **Part** *to pick out the right-hand sides of the output. This will be useful for the next module.*

```
{x -> 0, y -> 0}[[1, 2]]
```

0

```
{x -> 0, y -> 0}[[2, 2]]
```

0

```
discrim[f_, P_]:=Module[{d}, d = D[f, {x, 2}] * D[f, {y, 2}] - D[D[f, x], y]^2;
d/.{x -> P[[1]], y -> P[[2]]}]
```

```
check[f_, P_]:=If[discrim[f, P] > 0&& D[f, {x, 2}] > 0,
Print[" local maximum at critical point"], Print["false"]]
```

```
check[x^2 + 3 * x * y + y^2, {0, 0}]
```

false

```
fxx[f_, P_]:= D[f, {x, 2}]/.{x -> P[[1]], y -> P[[2]]}
```

test[f_, P_]:=If[discrim[f, P] > 0 > 0&&

fxx[f, P] > 0, Print[" local maximum at critical point"], Print["Not"]]

0

test[x^2 + 3 * x * y + y^2, {0, 0}]

Not

derivativeTest[f_, P_]:=If[discrim[f, P] > 0&&

fxx[f, P] > 0, Print[" local minimum at critical point"], If[discrim[f, P] > 0&&

fxx[f, P] < 0, Print[" local maximum at critical point"],

Print["Saddle Point at critical point"]]]

derivativeTest[x^2 + 3 * x * y + y^2, {0, 0}]

Saddle Point at critical point

criticalPoints[x^3 − y^3 + 2 * x * y − 6]

$\{\{x \to 0, y \to 0\}, \{x \to \frac{2}{3}, y \to -\frac{2}{3}\}, \{x \to -\frac{2}{3}(-1)^{1/3}, y \to -\frac{2}{3}(-1)^{2/3}\}, \{x \to \frac{2}{3}(-1)^{2/3}, y \to \frac{2}{3}(-1)^{1/3}\}\}$

N $\left[\{\{x \to 0, y \to 0\}, \{x \to \frac{2}{3}, y \to -\frac{2}{3}\}, \{x \to -\frac{2}{3}(-1)^{1/3}, y \to -\frac{2}{3}(-1)^{2/3}\}, \{x \to \frac{2}{3}(-1)^{2/3}, y \to \frac{2}{3}(-1)^{1/3}\}\}\right]$

$\{\{x \to 0., y \to 0.\}, \{x \to 0.666667, y \to -0.666667\}, \{x \to -0.333333 - 0.57735i, y \to 0.333333 - 0.57735i\}, \{x \to -0.333333 + 0.57735i, y \to 0.333333 + 0.57735i\}\}$

derivativeTest[x^3 − y^3 + 2 * x * y − 6, {0, 0}]

Saddle Point at critical point

derivativeTest[x^3 − y^3 + 2 * x * y − 6, {2/3, 2/3}]

Saddle Point at critical point

Plot3D[x^3 − y^3 + 2 * x * y − 6, {x, −1, 1}, {y, −1, 1}, AxesLabel → {x, y, z}]

criticalPoints[−x^4 − y^4 + 8 * x * y]

$\{\{x \to 0, y \to 0\}, \{x \to -\sqrt{2}, y \to -\sqrt{2}\}, \{x \to -i\sqrt{2}, y \to i\sqrt{2}\}, \{x \to i\sqrt{2}, y \to -i\sqrt{2}\}, \{x \to \sqrt{2}, y \to \sqrt{2}\}, \{x \to -(-1)^{1/4}\sqrt{2}, y \to -(-1)^{3/4}\sqrt{2}\}, \{x \to (-1)^{1/4}\sqrt{2}, y \to (-1)^{3/4}\sqrt{2}\}, \{x \to -(-1)^{3/4}\sqrt{2}, y \to -(-1)^{1/4}\sqrt{2}\}, \{x \to (-1)^{3/4}\sqrt{2}, y \to (-1)^{1/4}\sqrt{2}\}\}$

derivativeTest[4 * x * y − x^4 − y^4, {Sqrt[2], Sqrt[2]}]

local maximum at critical point

derivativeTest[4 * x * y − x^4 − y^4, {−Sqrt[2], −Sqrt[2]}]

local maximum at critical point

derivativeTest[4 * x * y − x^4 − y^4, {0, 0}]

Saddle Point at critical point

Plot3D[−x^4 − y^4 + 8 * x * y, {x, −3, 3}, {y, −3, 3}, AxesLabel → {x, y, z}]

Exercise 4.47. *Find the critical points of the following functions and determine whether they are local maxima, minima, or saddle points. For plots of the functions in exercises (c) and (d) see Figure 4.9 and Figure 4.10 respectively.*

(a) $f(x,y) := x\,y$ (b) $f(x,y) := x^2 + y^2$

(c) $f(x,y) := x^3 + 3\,x^2 - 2\,x\,y + 5\,y^2 - 4\,y^3$ (d) $f(x,y) := \dfrac{x-y}{x^2+y^2}$

(e) $f(x,y) := x^2 + 2\,x\,y + 5\,y^2 + 2\,x - 2\,y + 12$ (f) $f(x,y) := x^2 + 4\,y^2 - 2\,x\,y$

Exercise 4.48. *Find all maxima, minima, and saddle points of the functions listed below.*

(a) $f(x,y) := x^2 + 2xy + y^2 + 4x + 3y - 5$

(b) $f(x,y) := x^2 + 4xy + 4y^2 + 13x + 7y - 1$

(c) $f(x,y) := -x^2 + xy + -y^2 + 14x + 30y - 11$

(d) $f(x,y) := -2x^2 - 6xy + y^2 + 3x + 17y + 20$

(e) $f(x,y) := -3x^2 + xy - 2y^2 + 4x - 6$

(f) $f(x,y) := 4x^2 - 14xy + 3y^2 + 12x$

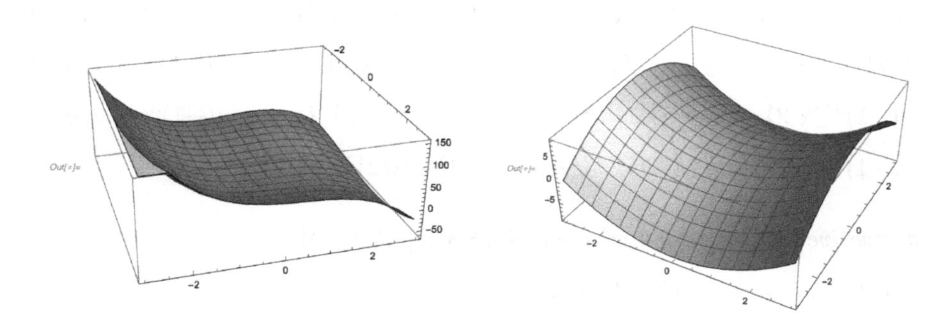

FIGURE 4.9: A plot showing the example (c) in 4.47.

FIGURE 4.10: A plot showing the example (d) in4.47.

(g) $f(x, y) := x^2 - 4xy - 8y^2 + x + 5y + 8$

(h) $f(x, y) := -x^2 - 2xy + 2y^2 + 8x + 3y + 21$

Exercise 4.49. *Find all maxima, minima, and saddle points of the functions listed below.*

(a) $f(x, y) := \dfrac{1}{1 + x^2 + y^2}$

(b) $f(x, y) := \dfrac{1}{\sqrt{1 + x^2 + y^2}}$

(c) $f(x, y) := \dfrac{1}{|x| + |y| + 1}$

(d) $f(x, y) := \dfrac{|x| - |y|}{|x| + |y|}$

(e) $f(x, y) := \displaystyle\int_{1-x^2}^{1-y^2} t\, dt$

(f) $f(x, y) := \displaystyle\int_{\frac{1}{1-x^2}}^{\frac{1}{1-y^2}} t^2\, dt$

(g) $f(x, y) := \sin(x + y)$

(h) $f(x, y) := \cos(x - y)$

Exercise 4.50. *A triangle has vertices at $(0, 1)$, $(-1, 0)$, and $(1, 0)$, and let $f(x, y)$ be the sum of distances of a point (x, y) from the above vertices. Find the minimum of $f(x, y)$, and where it occurs.*

Exercise 4.51. *Let two lines be given parametrically by $x = t, y = 2t-1, z = 3t+2$ and $x = 2s, y = -s + 4, z = s - 1$. Let $d^2(s, t)$ be the square of the distance between points on the two lines. Show that $d^2(s, t)$ has a minimum.*

Exercise 4.52. *Let $d^2(x,y)$ be the square of the distance of the point $(1,0,0)$ from a surface $z = f(x,y)$. Find and classify the critical points of $d^2(x,y)$ for the following surfaces.*

(a) $z = x^2 + y^2$ (b) $z = x^2 - y^2$ (c) $z = \sqrt{x^2 + y^2}$

Exercise 4.53. *Use* MATHEMATICA *to find and classify the critical points of the following functions.*

(a) $f(x,y) := x^3 - 3\,x\,y^2$ (b) $f(x,y) := x^3 - 3\,x^2\,y$ (c) $f(x,y) := x^3 - y^3$
(d) $f(x,y) := x\,y\,(x^2 - y^2)$ (e) $f(x,y) := x\,y\,(x^3 - y^3)$

Exercise 4.54. *Use the Newton iteration method to find the stationary points of the following functions.*

(a) $f(x,y) := x^4 + 3x^2y^2 + y^3 + 2x + 4y$ (b) $f(x,y) := x^5 + 12xy^2 + x^3 + 4y^2$
(c) $f(x,y) := y^4 + x^2y + 4y^2 + 10x + 4y$ (d) $f(x,y) := x^4 + 6xy + y^2 + 22x + y$
(e) $f(x,y) := x^4 + x^3 + 2y^3 + 4y$ (f) $f(x,y) := x^5 + 3x^3y^2 + y^4 + 12x^2 + 4y^2$

Classify the stationary points as maxima, minima or saddle–points.

Exercise 4.55. *Suppose there are three particles of mass m_k, $(k = 1,2,3)$ which are situated in the $x - y$ plane at the points (x_k, y_k), $(k = 1,2,3)$. The moment of inertia $I(x,y,z)$ of this system about the point $P = (x,y)$ is given by*

$$\begin{aligned} I(x,y,z) &= \frac{1}{2}\left[m_1\left((x-x_1)^2 + (y-y_1)^2\right) + m_2\left((x-x_2)^2 + (y-y_2)^2\right)\right. \\ &+ \left. m_3\left((x-x_3)^2 + (y-y_3)^2\right)\right] . \end{aligned}$$

Clearly as the point P recedes to infinitely distant points the moment of inertia grows without bound. Also the moment of inertia is bounded below by zero. Hence, there is a least value. This value must be achieved for some value of (x,y,z) which is a stationary point of I. Find those points where the minimum value of the moment of inertia is achieved, and find the value of the moment of inertia.

4.5 Constrained extremal problems and Lagrange multipliers

In §§ 4.3–4.4 we investigated the extrema of functions whose variable were free to vary in some given domain. Such problems are called **unconstrained.**

In this section we consider **constrained problems**, with an added condition, or **constraint**, on the variables. A typical problem is

$$\text{extremize}\, f(x,y,z) \quad \text{s.t.} \quad \varphi(x,y,z) = 0 \tag{4.43}$$

- **extremize** stands for **maximize** (abbreviated "max") or **minimize** (abbreviated "min"),

- **s.t.** is an abbreviation of "such that",
- $f(x, y, z)$ is called the **objective function**, and
- $\varphi(x, y, z)$ is called the **constraint function**.

Example 4.56. Find the closest point to the origin on the plane

$$x + y + z = 1 .$$

The distance of a point (x, y, z) to the origin is given by $\sqrt{x^2 + y^2 + z^2}$. It is minimized whenever its square

$$f(x, y, z) := x^2 + y^2 + z^2$$

is a minimum. The problem is therefore equivalent to:

$$\min f(x, y, z) := x^2 + y^2 + z^2 \quad \text{s.t.} \quad x + y + z = 1 .$$

We use the **constraint** $x + y + z = 1$ to eliminate one of the variables, say $z = 1 - x - y$, and substitute in f to get

$$F(x, y) := f(x, y, 1-x-y) = x^2+y^2+(1-x-y)^2 = 2\,(x^2 + y^2 + x\,y - x - y)+1 .$$

The stationary points are found by solving the system

$$(1) \quad \frac{\partial F}{\partial x} = 4\,x + 2\,y - 2 = 0 \qquad\qquad (2) \quad \frac{\partial F}{\partial y} = 4\,y + 2\,x - 2 = 0$$

to give $x = y = \frac{1}{3}$. Moreover,

$$\frac{\partial^2 F}{\partial x^2} = 4 , \quad \frac{\partial^2 F}{\partial x\,\partial y} = 2 , \quad \frac{\partial^2 F}{\partial y^2} = 4$$

and the stationary point is a local minimum, by Theorem 4.38.

The value of z corresponding to $x = y = \frac{1}{3}$ is

$$z = 1 - x - y = 1 - \frac{1}{3} - \frac{1}{3} = \frac{1}{3}$$

so the minimum is $(x, y, z) = (\frac{1}{3}, \frac{1}{3}, \frac{1}{3})$ where the distance to the origin is $\frac{1}{\sqrt{3}}$. △

Example 4.57 (Largest rectilinear parallelpiped in a sphere **).** It is required to place a rectangular parallelpiped of maximum volume inside a sphere of radius 10.

For convenience we center the sphere at the origin, and take the faces of the parallelepiped parallel to the coordinate planes, see Figure 4.11. The volume V of the parallelepiped is then given by $V = 8xyz$, where (x, y, z) is the vertex in the first octant. The coordinates $(x,\ y,\ z)$ are connected by the fact that this point lies on the sphere given by

$$x^2 + y^2 + z^2 = 100 .$$

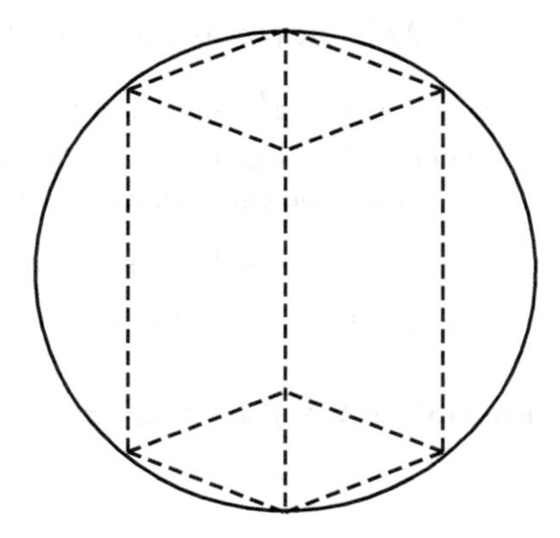

FIGURE 4.11: Illustrations of Example 4.57.

Consequently, we may eliminate one of the variables, say z, and substitute in the volume to obtain

$$V(x,y) \;=\; 8\,x\,y\,\sqrt{100-x^2-y^2}\,,$$

defined for all points (x,y) such that $x^2+y^2 \le 100$. If $x^2+y^2 \;=\; 100$ then $z=0$ and the volume is clearly zero. We have thus found a solution to the problem of finding the smallest volume! However, we are looking for the parallelepiped of largest volume. We **know** a solution must exist, indeed the size of this volume must be bounded above by the volume of the sphere, namely $\frac{4}{3}\pi 10^3$. To this end, we solve the system

$$(1)\quad \frac{\partial V}{\partial x} \;=\; 8y\sqrt{100-x^2-y^2} - \frac{8x^2y}{\sqrt{100-x^2-y^2}} \;=\; 0$$

$$(2)\quad \frac{\partial V}{\partial y} \;=\; 8x\sqrt{100-x^2-y^2} - \frac{8y^2x}{\sqrt{100-x^2-y^2}} \;=\; 0$$

These two equations simplify to

$$\text{(i)}\quad 2\,x^2+y^2 \;=\; 100 \qquad\qquad \text{(ii)}\quad x^2+2\,y^2 \;=\; 100$$

with the solutions $x \;=\; \pm\frac{10}{\sqrt{3}}$, $y \;=\; \pm\frac{10}{\sqrt{3}}$. The corresponding value of z is

$$z \;=\; \sqrt{100-x^2-y^2} \;=\; \sqrt{100-\left(\frac{10}{\sqrt{3}}\right)^2-\left(\frac{10}{\sqrt{3}}\right)^2} \;=\; \pm\frac{10}{\sqrt{3}}\,.$$

Since the point (x,y,z) is in the first octant, we only accept positive solution values for x, y, and z. The maximum volume is thereby given by $V_{\max} \;=\; \frac{8}{9}\sqrt{3}\,10^3$. △

Example 4.58. It is required to find the point closest to the origin on the paraboloid $z = 1 - x^2 - y^2$. As in the previous example, this problem is equivalent to

$$\min f(x, y, z) := x^2 + y^2 + z^2 \quad \text{s.t.} \quad z - 1 + x^2 + y^2 = 0 .$$

Using the constraint to eliminate the variable z, and substituting in the objective f we get a function $F(x, y)$, and the problem is equivalent to

$$\min F(x, y) := x^2 + y^2 + (x^2 + y^2 - 1)^2 \quad \text{in the domain} \quad x^2 + y^2 \leq 1 .$$

Let us use MATHEMATICA to complete this problem.

Mathematica session 4.6. $x\text{^}2 + y\text{^}2 + z\text{^}2/.\{z \to 1 - x\text{^}2 - y\text{^}2\}$

$x^2 + y^2 + (1 - x^2 - y^2)^2$

***Expand*[%]**

$1 - x^2 + x^4 - y^2 + 2x^2y^2 + y^4$

***criticalPoints*[*f*_]:=*Module*[{*dfx*, *dfy*, *sol*}, *dfx* = *D*[*f*, *x*]; *dfy* = *D*[*f*, *y*];**

***sol* = *Solve*[{*dfx*==0, *dfy*==0}, {*x*, *y*}]]**

criticalPoints $[1 - x^2 + x^4 - y^2 + 2x^2y^2 + y^4]$

$\left\{\left\{y \to -\frac{\sqrt{1-2x^2}}{\sqrt{2}}\right\}, \left\{y \to \frac{\sqrt{1-2x^2}}{\sqrt{2}}\right\}, \{x \to 0, y \to 0\}\right\}$

Expand $\left[1 - x^2 + x^4 - y^2 + 2x^2y^2 + y^4/.\left\{y \to -\frac{\sqrt{1-2x^2}}{\sqrt{2}}\right\}\right]$

$\frac{3}{4}$

So it would appear that the nearest points should be on the circle $x^2 + y^2 = 1/2$ *lying in the* $x - y$ *plane.*

♦

It is clear that $x = 0 = y$ does satisfy both equations. However, if $x \neq 0$ and $y \neq 0$, then these equations become through cancellation the single equation

$$2x^2 + 2y^2 - 1 = 0,$$

which means that all the points on the circle $x^2 + y^2 = \frac{1}{2}$ are stationary points. Since

$\delta(0,0) = -4$, and $F_{xx}(0,0) = -2$, we have a local maximum $(0,0)$. The expression for $\delta(x,y)$ factors into the form

$$\delta(x,y) := -48(x^2+y^2)^2 + 32(x^2+y^2) - 4;$$

hence, on the circle $x^2+y^2 = \frac{1}{2}$, $\delta = 0$, and the test is inconclusive. Let us try using polar coordinates. Since $x^2+y^2 = r^2$, the function $F(r\cos\theta, r\sin\theta)$ becomes

$$F(r) := r^2 + (r^2-1)^2.$$

Then $F'(r) = 4r^3 - 2r$ and $F''(r) = 12r^2 - 2$. Hence, as before the stationary points are $r = 0$ and $r = \frac{1}{\sqrt{2}}$. At $r = 0$ $F''(0) = -2$ which corresponds to a local maximum, and this agrees with our previous result. On the other hand $F(\frac{1}{\sqrt{2}}) = 2$ and this corresponds to a local minimum. If we had not made use of the polar coordinates the two variable second derivative test would not have furnished the necessary information to come to this conclusion. △

A constrained extremum problem

$$\text{extremize}\,!\, f(x,y,z) \quad \text{s.t.} \quad \varphi(x,y,z) \;=\; 0 \tag{4.44}$$

can be solved by using the constraint to eliminate one of the variables, say

$$z \;=\; z(x,y)$$

which substituted in the objective gives a function of two variables

$$F(x,y) \;:=\; f(x,y,z(x,y))\ , \tag{4.45}$$

reducing the problem to an unconstrained problem of two variables. This approach was illustrated in Examples 4.56–4.58.

The stationary points of F satisfy

$$\frac{\partial F}{\partial x} \;=\; \frac{\partial f}{\partial x} + \frac{\partial f}{\partial z}\frac{\partial z}{\partial x} \;=\; 0\ , \tag{4.46a}$$

$$\frac{\partial F}{\partial y} \;=\; \frac{\partial f}{\partial y} + \frac{\partial f}{\partial z}\frac{\partial z}{\partial y} \;=\; 0\ . \tag{4.46b}$$

where we differentiated F using the chain rule. On the other hand, since $\varphi(x,y,z) = 0$,

$$\frac{\partial \varphi}{\partial x} + \frac{\partial \varphi}{\partial z}\frac{\partial z}{\partial x} \;=\; 0 \tag{4.47a}$$

$$\frac{\partial \varphi}{\partial y} + \frac{\partial \varphi}{\partial z}\frac{\partial z}{\partial y} \;=\; 0\ . \tag{4.47b}$$

We can combine the above two pairs of equations to eliminate $\frac{\partial z}{\partial x}$ and $\frac{\partial z}{\partial y}$. We obtain

$$\frac{\partial f}{\partial x}\frac{\partial \varphi}{\partial z} - \frac{\partial f}{\partial z}\frac{\partial \varphi}{\partial x} \;=\; 0 \tag{4.48a}$$

$$\frac{\partial f}{\partial y}\frac{\partial \varphi}{\partial z} - \frac{\partial f}{\partial z}\frac{\partial \varphi}{\partial y} \;=\; 0 \tag{4.48b}$$

These two equations with the constraint $\varphi(x, y, z) = 0$, then serve to determine the stationary points.

The **Lagrange multiplier method** [9] is an alternative method of solving constrained extremum problems without eliminations/substitutions. It is based on the auxiliary function

$$L(x, y, z, \lambda) := f(x, y, z) - \lambda\, \varphi(x, y, z) \tag{4.49}$$

called the **Lagrange function.** . The function L (or **Lagrangian**) of the problem (4.44). Here λ is a parameter, called the **Lagrange multiplier** .

$$\frac{\partial L}{\partial x} = f_x - \lambda\, \varphi_x = 0\,, \tag{4.50a}$$

$$\frac{\partial L}{\partial y} = f_y - \lambda\, \varphi_y = 0\,, \tag{4.50b}$$

$$\frac{\partial L}{\partial z} = f_z - \lambda\, \varphi_z = 0\,, \tag{4.50c}$$

$$\frac{\partial L}{\partial \lambda} = -\varphi(x, y, z) = 0\,. \tag{4.50d}$$

If these equations are to hold simultaneously, then solving (4.50c) for λ we get

$$\lambda = \frac{f_z}{\varphi_z} \tag{4.51}$$

and substitute in (4.50a)–(4.50b) to get

$$\begin{aligned} f_x\, \varphi_z - f_z\, \varphi_x &= 0\,, \\ f_y\, \varphi_z - f_z\, \varphi_y &= 0\,, \end{aligned}$$

which is the pair (4.48a)–(4.48b) obtained earlier. The equation (4.50d) is just $\varphi(x, y, z) = 0$, which together with (4.48a)–(4.48b) determine the stationary points.

Remark 4.59. Since $\varphi = 0$ is equivalent to $-\varphi = 0$, we can write the Lagrangian alternatively as

$$L(x, y, z, \lambda) := f(x, y, z) + \lambda\, \varphi(x, y, z) \tag{4.52}$$

giving the same stationary points (x, y, z) as we get by using the Lagrangian (4.49).

Example 4.60. We use the Lagrange Multiplier Method to solve the problem in Example 4.56. The Lagrangian is

$$L(x, y, z, \lambda) := x^2 + y^2 + z^2 - \lambda\, (x + y + z - 1)\ .$$

Computing the first partial derivatives, and equating to zero, we get

[9] Joseph Louis Lagrange, [1736 – 1813].

$$(1)\quad \frac{\partial L}{\partial x} = 2x - \lambda = 0 \qquad (2)\quad \frac{\partial L}{\partial y} = 2y - \lambda = 0$$

$$(3)\quad \frac{\partial L}{\partial z} = 2z - \lambda = 0 \qquad (4)\quad \frac{\partial L}{\partial \lambda} = -(x + y + z - 1) = 0$$

Solving for λ we obtain $\lambda = 2x = 2y = 2z$, showing that $x = y = z$. Substituting $y = x$ and $z = x$ in equation (4), which is the constraint, we get $3x = 1$. We conclude that $\left(\frac{1}{3}, \frac{1}{3}, \frac{1}{3}\right)$ is the nearest point to the origin on the plane, in agreement with the answer in Example 4.56 obtained by the previous method. △

Example 4.61. We repeat the problem in Example 4.58. The Lagrangian here is

$$L(x, y, z, \lambda) := x^2 + y^2 + z^2 - \lambda\left(z - 1 + x^2 + y^2\right) .$$

Computing the first partial derivatives, and equating to zero, we get

$$(1)\quad \frac{\partial L}{\partial x} = 2x(1 - \lambda) = 0 \qquad (2)\quad \frac{\partial L}{\partial y} = 2y(1 - \lambda) = 0$$

$$(3)\quad \frac{\partial L}{\partial z} = 2z - \lambda = 0 \qquad (4)\quad \frac{\partial L}{\partial \lambda} = -(z - 1 + x^2 + y^2) = 0$$

The value $\lambda = 1$ agrees with equations (1) and (2). Substituting $\lambda = 1$ in equation (3) we get $z = \frac{1}{2}$. Equation (4) then gives $x^2 + y^2 = \frac{1}{2}$.

Alternatively, $x = y = 0$ is an acceptable solution of equations (1) and (2), giving $z = 1$ in equation (4).

Example 4.62. It is required to find the stationary points of $f := x\,y\,z$ subject to $\varphi(x, y, z) := x + y + z - 1 = 0$. The Lagrangian is

$$L(x, y, z, \lambda) := x\,y\,z + \lambda\,(x + y + z - 1) .$$

Equating the first partial derivatives to zero we get

$$(1)\quad \frac{\partial L}{\partial x} = y\,z + \lambda = 0 \qquad (2)\quad \frac{\partial L}{\partial y} = x\,z + \lambda = 0$$

$$(3)\quad \frac{\partial L}{\partial z} = x\,y + \lambda = 0 \qquad (4)\quad \frac{\partial L}{\partial \lambda} = x + y + z - 1 = 0$$

Equation (3) gives $\lambda = -xy$. Substituting this in equations (1)–(2) we get

$$y\,z - x\,y = 0 , \quad x\,z - x\,y = 0 , \quad x + y + z = 1 ,$$

having four solutions $(x_1, y_1, z_1) = \left(\frac{1}{3}, \frac{1}{3}, \frac{1}{3}\right)$, $(x_2, y_2, z_2) = (1, 0, 0)$, $(x_3, y_3, z_3) = (0, 1, 0)$ and $(x_4, y_4, z_4) = (0, 0, 1)$. Of these, (x_1, y_1, z_1) is a local maximum, the other three solutions are saddle points.

Note that the values of $f := x\,y\,z$ on the plane $\varphi(x, y, z) = 0$ are unbounded. For example, the points $(-\alpha, -\alpha, 2\,\alpha + 1)$ satisfy the constraint for all real α, and $f(-\alpha, -\alpha, 2\,\alpha + 1) = \alpha^2(2\,\alpha + 1)$ which $\to \pm\infty$ with α. △

Example 4.63. We find the stationary points of $x^2 + y^2 + 4z^2$ subject to $x^2 + y^2 + z^2 = 1$. The corresponding Lagrangian is

$$L(x, y, z, \lambda) := x^2 + y^2 + 4\,z^2 + \lambda\left(x^2 + y^2 + z^2 - 1\right) .$$

Equating the first partial derivatives of L to zero we get four equations

$$\begin{array}{llll} (1) & 2x + \lambda 2x = 0 & (2) & 2y + \lambda 2y = 0 \\ (3) & 8z + \lambda 2z = 0 & (4) & x^2 + y^2 + z^2 = 1 \end{array}$$

Therefore $\lambda = -4$ giving three equations

$$\text{(i)} \quad 2x - 8x = 0 \qquad \text{(ii)} \quad 2y - 8y = 0 \qquad \text{(iii)} \quad x^2 + y^2 + z^2 = 1$$

determining the stationary points $(0, 0, \pm 1)$. △

Example 4.64 (Largest cone in a sphere)**.** It is required to determine the right circular cone of largest volume which can be inscribed in the sphere $x^2+y^2+z^2 = 100$. Recall that the volume of an arbitrary circular cone with base radius b and height h is

$$V = \frac{\pi}{3} h b^2 .$$

We can simplify the analysis by taking into account the symmetry about the cone's

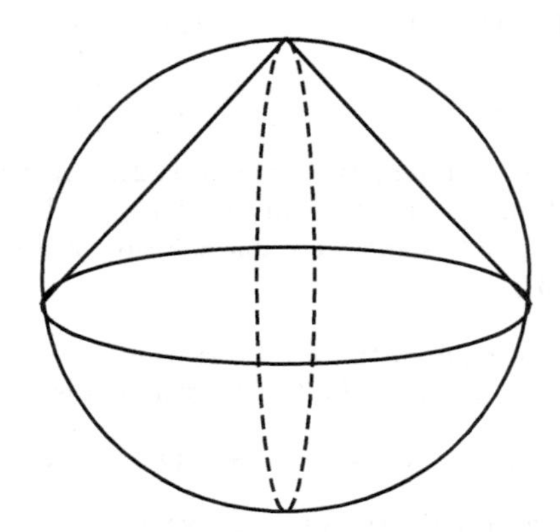

FIGURE 4.12: Right cone inscribed in sphere.

axis. This suggests use of the cylindrical coordinates (r, φ, z), in terms of which the sphere becomes $r^2 + z^2 = 100$. Because of the axial symmetry we can suppress the angular variable and view the situation in the r and z coordinates. In the cross-sectional view the cone in the sphere appears as a triangle in a disk, with the vertex at the north pole $(0, 10)$. The base of the cone is represented by the line segment joining the points $(-b, -z)$ and $(b, -z)$, see Figure 4.12. The height of the triangle (cone) is $h = 10 + z$. We therefore maximize

$$V := \frac{\pi}{3} b^2 (10 + z) ,$$

subject to the constraint

$$z^2 + b^2 = 100 .$$

The corresponding Lagrangian is

$$L(b, z, \lambda) := \frac{\pi}{3} b^2 (10 + z) + \lambda (z^2 + b^2 - 100) ,$$

giving the following equations,

$$(1) \quad \frac{\partial L}{\partial b} = \frac{2\pi}{3} b(10+z) + 2\lambda b = 0 \qquad (2) \quad \frac{\partial L}{\partial z} = \frac{\pi}{3} b^2 + 2\lambda z = 0$$

$$(3) \quad \frac{\partial L}{\partial \lambda} = z^2 + b^2 - 100 = 0$$

From equations (1)–(2) we get

$$\lambda = -\frac{\pi}{3}(10+z) = -\frac{\pi b^2}{6z} .$$

Substituting this in equation (3) we get $3z^2 + 20z - 100 = 0$, with two roots $z = \frac{10}{3}$ and $z = -10$. The second root gives the minimal volume (why?). The first root corresponds to $b = \frac{20\sqrt{2}}{3}$. The maximum volume of a right cone, inscribed in a sphere of radius 10, is therefore:

$$V = \frac{\pi}{3}\left(\frac{20\sqrt{2}}{3}\right)^2 \left(10 + \frac{10}{3}\right) = \frac{32000\pi}{81} . \qquad \triangle$$

A justification of the Lagrange method is contained in the next theorem.

Theorem 4.65. Let $f(x, y, z)$ and $\varphi(x, y, z)$ be differentiable functions, and let (x_0, y_0, z_0) be a local extremizer of $f(x, y, z)$ subject to $\varphi(x, y, z) = 0$. Then the gradients of f and φ are collinear at (x_0, y_0, z_0), i.e. there is a real number λ such that

$$\nabla f(x_0, y_0, z_0) = \lambda \nabla \varphi(x_0, y_0, z_0) . \tag{4.53}$$

Proof:
Let (x_0, y_0, z_0) be a point on the surface $\varphi = 0$, where f has a local extremum relative to the values f takes on the surface. Our result depends on the following claim. If $\mathcal{C} := \{x = x(t), y = y(t)\}$ is a differentiable curve on the surface $\varphi = 0$ passing through (x_0, y_0, z_0), then the gradient of f is orthogonal to the tangent to $\mathcal{C}$ at (x_0, y_0, z_0). Since the gradient of f is orthogonal to <u>all</u> curves lying on the surface and passing through the extremal point, ∇f is orthogonal to the tangent plane to $\varphi = 0$ at the extremal point. Hence, ∇f is collinear to the normal to the surface φ, which is $\nabla \varphi$. We have still to demonstrate that ∇f is orthogonal to any differentiable curve $\mathcal{C}$. On $\mathcal{C}$ f takes on the values $f(x(t), y(t), z(t))$; hence, by the chain rule

$$\frac{df}{dt} = f_x \frac{dx}{dt} + f_y \frac{dy}{dt} + f_z \frac{dz}{dt} = \nabla f \cdot \frac{d\mathbf{r}}{dt} .$$

Now, since $\frac{df}{dt} = 0$ at an extremal point, the claim follows. □

Mathematica session 4.7. *We want to find the minimal distance of a point on a surface to the origin. For this purpose we enter the square of the distance to an arbitrary point* (x, y, z)*. The surface is the circular paraboloid,* *which becomes our constraint. We enter the derivatives of the auxiliary function* $F(x, y, z, \lambda) := f(x, y, z, \lambda) + g(x, y, z, \lambda)$ *as the system of simultaneous equations to be solved. To find the location of the possible extrema we find the critical points by solving the system of equations for* x, y, z, *and* λ. [10]

```
f = x^2 + y^2 + z^2
```

$x^2 + y^2 + z^2$

```
g = z - 16 + x^2 + y^2
```

$-16 + x^2 + y^2 + z$

```
{2x + 2λx, 2y + 2λy, λ + 2z}
```

$\{2x + 2\lambda x, 2y + 2y\lambda, 2z + \lambda\}$

```
Solve[{2x + 2λx == 0, 2y + 2λy == 0, λ + 2z == 0, z - 16 + x^2+
y^2 == 0}, {x, y, z, λ}]
```

Solve : Equations may not give solutions for all "solve" variables.

$\left\{\left\{y \to -\frac{\sqrt{31-2x^2}}{\sqrt{2}}, z \to \frac{1}{2}, \lambda \to -1\right\}, \left\{y \to \frac{\sqrt{31-2x^2}}{\sqrt{2}}, z \to \frac{1}{2}, \lambda \to -1\right\},\right.$
$\left.\{x \to 0, y \to 0, z \to 16, \lambda \to -32\}\right\}$

```
checkMagnitude[f_, g_, n_]:=
Module[{F, root}, F = f - λ * g;
root = Solve[{D[F, x] == 0, D[F, y] == 0, D[F, z] == 0, g == 0}, {x, y, z, λ}];
f /.root[[n]]]
```

```
Simplify[checkMagnitude[x^2 + y^2 + z^2, z - 16 + x^2 + y^2, 1]]
```

Solve : Equations may not give solutions for all "solve" variables.

$\frac{63}{4}$

---

[10] To enter a greek letter in mathematica use esc greek letter esc, for example escape lambda escape for λ.

Simplify[checkMagnitude[x^2 + y^2 + z^2, z − 16 + x^2 + y^2, 2]]

Solve : Equations may not give solutions for all "solve" variables.

$\frac{63}{4}$

Simplify[checkMagnitude[x^2 + y^2 + z^2, z − 16 + x^2 + y^2, 3]]

Solve : Equations may not give solutions for all "solve" variables.

256

Solve[Grad[x^2 + y^2 + z^2 + λ ∗ (x + y + z − 1), {x, y, z, λ}] == {0, 0, 0, 0},

{x, y, z, λ}]

$\{\{x \to \frac{1}{3}, y \to \frac{1}{3}, z \to \frac{1}{3}, \lambda \to -\frac{2}{3}\}\}$

♦

In general, when solving for critical points, some may not have meaning for us. In particular, we should in many cases test for the complex roots and discard these as we are usually solving for real solutions. The Lagrange Multipliers Method extends easily to more than three variables. Consider the extremum problem in n variables,

$$\text{extremize}\, f(x_1, x_2, \ldots, x_n) \quad \text{s.t.} \quad \varphi(x_1, x_2, \ldots, x_n) \;=\; 0\,. \tag{4.54}$$

In analogy with the case of three variables we define the **Lagrangian**,

$$L(x_1, x_2, \ldots, x_n, \lambda) \;:=\; f(x_1, x_2, \ldots, x_n) - \lambda\, \varphi(x_1, x_2, \ldots, x_n)\,. \tag{4.55}$$

Computing the partial derivatives, and equating to zero, we get the system of $n+1$ equations in $n+1$ variables,

$$\begin{aligned}
\frac{\partial L}{\partial x_1} &= \frac{\partial f}{\partial x_1} - \lambda\,\frac{\partial \varphi}{\partial x_1} = 0 \\
\frac{\partial L}{\partial x_2} &= \frac{\partial f}{\partial x_2} - \lambda\,\frac{\partial \varphi}{\partial x_2} = 0 \\
\cdots &= \cdots \\
\frac{\partial L}{\partial x_n} &= \frac{\partial f}{\partial x_n} - \lambda\,\frac{\partial \varphi}{\partial x_n} = 0 \\
\frac{\partial L}{\partial \lambda} &= -\varphi(x_1, x_2, \ldots, x_n) = 0
\end{aligned} \tag{4.56}$$

Eliminating λ, say

$$\lambda \;=\; \frac{\partial f}{\partial x_n} \Big/ \frac{\partial \varphi}{\partial x_n}$$

and substituting in the first $n-1$ equations, we get a system in n equations for the n variables $x_1, x_2, \ldots, x_n$.

EXERCISES

Exercise 4.66. *What are the extreme values of $f = xy$ on the ellipse $\frac{x^2}{4} + y^2 = 1$?*

Exercise 4.67. *Find the dimensions of the right circular cone of smallest surface area whose volume is given by 20π.*

Exercise 4.68. *Suppose we wish to design a box in the form of a rectangular parallelepiped which is to contain a certain fixed volume V_0 of material. However, to reduce the cost of manufacturing the box we wish to use the minimum amount of material. Find the dimensions of such a box with minimum surface area.*

Exercise 4.69. *Design prerequisites require that the surface of a rectangular box be fixed. Find the dimensions of such a box of maximum volume.*

Exercise 4.70. *Suppose a scientific farmer wishes to design a silo whose lateral sides are cylindrical, and whose roof is in the shape of a cone. If the volume of the silo is fixed find the most economical dimensions.*

Exercise 4.71. *Imagine that the farmer has a traditionally minded brother who has made his silo in the customary form, that is with cylindrical sides and a semi-spherical roof. What are the most economical dimensions for such a silo if it contains the same fixed volume as in the previous problem?*

Exercise 4.72. *Find the maximum volume of rectangular parallelepiped inscribed in the ellipsoid*

$$\frac{x^2}{a^2} + \frac{y^2}{b^2} + \frac{z^2}{c^2} = 1.$$

Exercise 4.73. *Find the minimum distance of the the surface $x^2 - \frac{y^2}{4} + z^2 = 1$ from the origin.*

Exercise 4.74. *How close does the line $y = 2x+3$ come to the ellipse $\frac{x^2}{4} + y^2 = 1$? Hint: Find the distance of an arbitrary point to the line. Then restrict the point to lie on the ellipse.*

Exercise 4.75. *Find how close the parabola $y = x^2$ is to the parabola $x = 4 + y^2$.*

Exercise 4.76. *How close does the plane $z = x + y + 1$ come to the paraboloid $x = z^2 + y^2$?*

Exercise 4.77. *Find the rectangular parallelpiped of given surface area and maximum volume using Lagrange Multipliers.*

Exercise 4.78. *Suppose a tank is to be built which is to be vertically cylindrical with the upper end open. In order to minimize cost determine the dimensions if the volume of the cylinder is to be fixed.*

Exercise 4.79. *The same problem as above, but for a rectangular tank.*

Exercise 4.80. *Develop the method of Lagrange multipliers for functions of two variables, that is seek the extrema of $f(x, y)$ subject to the condition $\varphi(x, y) = 0$. Check your answers against the one variable theory.*

Exercise 4.81. *Find the smallest right truncated cone in which the sphere $x^2 + y^2 + z^2 = 1$ can be inscribed.*

Exercise 4.82. *Find the rectangular parallelpiped of largest volume which can be inscribed in the ellipsoid*

$$\frac{x^2}{4} + \frac{y^2}{9} + \frac{z^2}{4} = 1.$$

Exercise 4.83. *Find the right pyramid, with square base, which has the largest volume that can be inscribed in the sphere $x^2 + y^2 + z^2 = 1$.*

4.6 Several constraints

We consider now an extremal problem which has more than one constraint. Suppose it is required to find the minimum (or maximum) of the function

$$w = F(x, y, u, v) \tag{4.57}$$

subject to the two constraints

$$G(x, y, u, v) = 0, \quad H(x, y, u, v) = 0. \tag{4.58}$$

Now in the instance we can solve the second two equations for u and v, say as $u = f(x, y)$ and $v = g(x, y)$, these may be substituted in the function to be minimized and we proceed to minimize

$$F(x, y, f(x, y), g(x, y)) \tag{4.59}$$

as a function of x and y. Since the minima occur at the stationary points we are led to the pair of equations

$$F_x + F_u f_x + F_v g_x = 0, \qquad F_y + F_u f_y + F_v g_y = 0. \tag{4.60}$$

Now since we really have not solved for f and for g, we we must determine the derivatives f_x, f_y, g_x, and g_y using implicit differentiation of the two constraining equations by which f and g were supposedly uniquely determined. Implicit differentiation of the constraining equations produces the system

$$\begin{cases} G_x + G_u f_x + G_v g_x = 0, & G_y + G_u f_y + G_v g_y = 0 \\ H_x + H_u f_x + H_v g_x = 0, & H_y + H_u f_y + H_v g_y = 0, \end{cases} . \tag{4.61}$$

The system of the three equations for f_x and g_x in (4.60), (4.61) has the following matrix form

$$\begin{pmatrix} F_x & F_u & F_v \\ G_x & G_u & G_v \\ H_x & H_u & H_v \end{pmatrix} \begin{pmatrix} 1 \\ f_x \\ g_x \end{pmatrix} = \begin{pmatrix} 0 \\ 0 \\ 0 \end{pmatrix} \tag{4.62}$$

In order to have a solution, the coefficient matrix must be non-invertible because otherwise we will have the contradiction $(1, f_x, g_x) = (0, 0, 0)$. Therefore, we must have

$$\begin{vmatrix} F_x & F_u & F_v \\ G_x & G_u & G_v \\ H_x & H_u & H_v \end{vmatrix} = 0, \tag{4.63}$$

Similarly, it must be true that

$$\begin{vmatrix} F_y & F_u & F_v \\ G_y & G_u & G_v \\ H_y & H_u & H_v \end{vmatrix} = 0. \tag{4.64}$$

Moreover, we may use these two equations (4.63), (4.64) along with the pair (4.58) to solve for the stationary point (x^*, y^*, u^*, v^*). Let us illustrate this procedure with an example.

Example 4.84. For simplicity, we consider a function F of only three variables. Such a situation occurs when we want to find the minimum distance from the origin of those points lying both on the ellipsoid $x^2 + \frac{y^2}{4} + z^2 = 1$ and the plane $x+y+z = 1$. According to our procedure we set

$$F(x, y, z) := x^2 + y^2 + z^2,$$

$$G(x, y, z) := x^2 + \frac{y^2}{4} + z^2 - 1, \text{ and } H(x, y, z) := x + y + z - 1.$$

Since in the present case there are only three variables x,y,z in contrast to the method outlined above which had four, we notice that if we could solve $G(x, y, z) = 0$ and $H(x, y, z) = 0$ simultaneously for $y = f(x)$ and $z = g(x)$, then this would lead to the system

$$\begin{cases} F_x + F_y f'(x) + F_z g'(x) & = 0 \\ G_x + G_y f'(x) + G_z g'(x) & = 0 \\ H_x + H_y f'(x) + H_z g'(x) & = 0, \end{cases}$$

which means that we must have the determinant of the coefficients vanish, that is

$$\begin{vmatrix} F_x & F_y & F_z \\ G_x & G_y & G_z \\ H_x & H_y & H_z, \end{vmatrix} = 0. \tag{4.65}$$

The equation (4.65) combined with the two constraints $G(x, y, z) = 0$ and

$H(x, y, z) = 0$ may then be used to find the stationary point. For the problem at hand we compute the determinantal entries as

$$\begin{array}{lll} F_x = 2x, & F_y = 2y, & F_z = 2z, \\ G_x = 2x, & G_y = \frac{y}{2}, & G_z = 2z, \\ H_x = 1, & H_y = 1 & H_z = 1; \end{array} \tag{4.66}$$

hence, the determinantal equation becomes

$$\begin{vmatrix} 2x & 2y & 2z \\ 2x & \frac{y}{2} & 2z \\ 1 & 1 & 1, \end{vmatrix} \equiv 3y(z - x) = 0,$$

that is, either $y = 0$ or $z = x$. Let us consider case (i), namely that $y = 0$. Then $G := x^2 + z^2 - 1$ and $H := x + z - 1$, and setting $G = 0 = H$ we have $z = 1 - x$ which substituted into $G = 0$ yields

$$x^2 + (1 - x)^2 = 1.$$

We then get the two roots $x = 0, 1$. So for case (i) we have two stationary points

$$(i)\ (x, y, z) = (0, 0, 1), \text{ and } (x, y, z) = (1, 0, 0).$$

We now turn to case (ii), namely that $z = x$. The equations $G = 0$, and $H = 0$ then become

$$2x^2 + \frac{1}{4}y^2 = 1, \text{ and } 2x + y = 1.$$

Solving the second equation for $y = 1 - 2x$ and substituting this into the first equation leads to

$$2x^2 + \frac{1}{4}(1 - 4x + 4x^2) = 1,$$

which has the solutions $x = \frac{1 \pm \sqrt{10}}{6}$. Consequently, we get the pair of stationary points for case (ii)

$$(ii)\ (x, y, z) = \left(\frac{1 \pm \sqrt{10}}{6}, \frac{2 \mp \sqrt{10}}{3}, \frac{1 \pm \sqrt{10}}{6} \right).$$

To see which points correspond to the minimum distance evaluate $x^2 + y^2 + z^2$ at these points and notice that the minimum occurs at $(0, 0, 1)$ and $(1, 0, 0)$.
$\triangle$

We now show how to extend the method of Lagrange multipliers to handle multiple constraints. The Lagrange multiplier method for our problem with two constraints is then formulated using the **Lagrangian**

$$L(x, y, u, v, \lambda) := F(x, y, u, v) + \lambda G(x, y, u, v) + \mu H(x, y, u, v) \tag{4.67}$$

where λ and μ are constants, to be determined later, called the **Lagrange multipliers**. The stationary points of the Lagrangian function L are determined by setting its partial derivatives $\frac{\partial L}{\partial x}, \ldots, \frac{\partial L}{\partial \lambda}, \frac{\partial L}{\partial \mu}$ equal to zero. The first four equations give the system

$$\left\{ \begin{array}{ll} F_x + \lambda G_x + \mu H_x = 0, & F_y + \lambda G_y + \mu H_y = 0 \\ F_u + \lambda G_u + \mu H_u = 0, & F_v + \lambda G_v + \mu H_v = 0 \end{array} \right. \tag{4.68}$$

We solve the last two equations for λ and μ,

$$\lambda = \frac{F_u H_v - F_v H_u}{G_u H_v - G_v H_u}, \quad \mu = \frac{F_u G_v - F_v G_u}{G_u H_v - G_v H_u},$$

and substitute in the first two equations to get

$$\begin{aligned} F_x (G_u H_v - G_v H_u) + G_x (F_v H_u - F_u H_v) + H_x (F_u G_v - F_v G_u) &= 0\,, \\ F_y (G_u H_v - G_v H_u) + G_y (F_v H_u - F_u H_v) + H_y (F_u G_v - F_v G_u) &= 0\,. \end{aligned}$$

which are exactly the determinantal equations (4.60),(4.61) given above. So we conclude that this extension of the Lagrange multiplier trick to several constraints gives exactly the same stationary points as we would get by solving and substituting back into the function to be minimized (maximized).

Example 4.85. It is required to find the points, lying on the intersection of the plane $x + 2y + 3z = 6$ and the cylinder $x^2 + z^2 = 1$, which are closest and farthest from the origin. Since there are two constraints, the Lagrangian function has two multipliers, λ and μ,

$$L(x, y, z, \lambda, \mu) \;:=\; x^2 + y^2 + z^2 - \lambda\,(x + 2y + 3z - 6) - \mu\,\left(x^2 + z^2 - 1\right)\,.$$

Then

(1) $L_x = 2\,x - \lambda - 2\,x\,\mu = 0$ (2) $L_y = 2\,y - 2\,\lambda = 0$

(3) $L_z = 2\,z - 3\,\lambda - 2\,z\,\mu = 0$ (4) $L_\lambda = -(x + 2\,y - 3\,z - 6) = 0$

(5) $L_\mu = -(x^2 + z^2 - 1) = 0$

Equation (2) yields $\lambda = y$. Substituting this in equations (1) and (3), and solving for μ we get

$$\mu = \frac{2x - y}{2x} = \frac{2z - 3y}{2z}\,,$$

this leads to $2zy = 6xy$ which <u>if</u> $y \neq 0$ becomes $z = 3x$. Substituting this into equations (4)–(5) (the constraints), we get the system of two equations

(i) $x + 2\,y - 9\,x - 6 = 0$ (ii) $10\,x^2 = 1$

giving $x = \pm\frac{1}{\sqrt{10}}$ and $y = 3 \pm \frac{4}{\sqrt{10}}$. If, on the other hand, $y = 0$ we must solve the simultaneous equations

(iii) $x - 3\,z - 6 = 0$ (iv) $x^2 + z^2 = 1$

which have only complex solutions. △

Mathematica session 4.8. *We show here how* MATHEMATICA *can be used to solve problems with two constraints. We first consider the problem of finding the nearest and farthest point from the origin which lies on the intersection of the paraboloid* $z = 16 - x^2 - y^2$ *with the plane* $x + y + z = 1$*. We introduce the distance squared as the function* f*, and the two* *constraints as the functions g and h. We solve the equations that arise in the two Lagrange multiplier method using the* MATHEMATICA *command* **NSolve[equations,variables]**. *We should check if the x, y, z roots are real, as only these are of significance in $\mathbb{R}^3$. For this there is the alternate command* **NSolve[equations,variables,Real]** *We wish to find the maxima of the function $f(x, y, z)$ subject to the fact that (x, y, z) lies on the intersection of $g(x, y, z) = 0$ and $h(x, y, z) = 0$. First we input the function f which we wish to extremize*

```
f = x^2 + y^2 + z^2
```

$x^2 + y^2 + z^2$

Next we input the two functions acting as constraints

```
g = z - 16 + x^2 + y^2
```

$-16 + x^2 + y^2 + z$

```
h = x + y + z
```

$x + y + z$

We now set up the gradient of the function $F(x, y, z) = f - \lambda g - \mu h$. Term-wise these are

```
eq1 = D[f - λ * g - μ * h, x]
```

$2x - 2x\lambda - \mu$

```
eq2 = D[f - λ * g - μ * h, y]
```

$2y - 2y\lambda - \mu$

```
eq3 = D[f - λ * g - μ * h, z]
```

$2z - \lambda - \mu$

eq4* = *g

$-16 + x^2 + y^2 + z$

eq5* = *h

$x + y + z$

Now we use the components of the gradient plus the constraints and solve for x, y, z, λ, μ *using* **NSolve**.

***NSolve*[{*eq1* == 0, *eq2* == 0, *eq3* == 0, *eq4* == 0, *eq5* == 0}, {*x*, *y*, *z*, λ, μ}]**

$\{\{x \to 4.53113, y \to 0., z \to -4.53113, \lambda \to 1., \mu \to 0.\}, \{x \to -3.53113, y \to 0., z \to 3.53113, \lambda \to 1., \mu \to 0.\}, \{x \to 1.69165, y \to -3.38329, z \to 1.69165, \lambda \to 2., \mu \to -3.38329\}, \{x \to -1.89165, y \to 3.78329, z \to -1.89165, \lambda \to 2., \mu \to 3.78329\}\}$

These are the critical points. We must now test these to see which corresponds to a maximum or minimum value of the function f.

***f*/.{*x* → "*4.53113*", *y* → "*0.*", *z* → −"*4.53113*"}**

41.0623

***f*/.{*x* → −"*3.53113*", *y* → "*0.*", *z* → "*3.53113*", λ → "*1.*",**
μ → "*0.*"}

24.9377

***f*/.{*x* → "*1.69165*", *y* → −"*3.38329*", *z* → "*1.69165*",**
λ → "*2.*", μ → −"*3.38329*"}

17.17

$\boldsymbol{f/.\{x \to -\text{“1.89165”}, y \to \text{“3.78329”}, z \to -\text{“1.89165”},}$
$\boldsymbol{\lambda \to \text{“2.”}, \mu \to \text{“3.78329”}\}}$

21.47

It appears that the first root returns a maximum value for f; whereas, the third root corresponds to a minimum value. We repeat this procedure with a new set of constraints. To do this let us **Clear** *the previous values for g, h.*

Clear[{g, h}]

$\boldsymbol{g = x\text{^}2 + y\text{^}2/4 + z\text{^}2 - 1}$

$-1 + x^2 + \frac{y^2}{4} + z^2$

$\boldsymbol{h = x + y + z - 1}$

$-1 + x + y + z$

$\boldsymbol{eq1 = D[f - \lambda * g - \mu * h, x]}$

$2x - 2x\lambda - \mu$

$\boldsymbol{eq2 = D[f - \lambda * g - \mu * h, y]}$

$2y - \frac{y\lambda}{2} - \mu$

$\boldsymbol{eq3 = D[f - \lambda * g - \mu * h, z]}$

$2z - 2z\lambda - \mu$

We again solve for the values of x, y, z, λ, μ which lead to extrema.

$\boldsymbol{NSolve[\{eq1 == 0, eq2 == 0, eq3 == 0, g == 0, h == 0\}, \{x, y, z, \lambda, \mu\}]}$

$\{\{x \to 0.693713, y \to -0.387426, z \to 0.693713, \lambda \to 1.36754, \mu \to -0.509941\}, \{x \to 1., y \to 0., z \to 0., \lambda \to 1., \mu \to 0.\}, \{x \to -0.36038, y \to$

$1.72076, z \to -0.36038, \lambda \to 2.63246, \mu \to 1.17661\}, \{x \to 0., y \to 0., z \to 1., \lambda \to 1., \mu \to 0.\}\}$

f/.{x → "0.693713", y → −"0.387426", z → "0.693713"}

1.11257

f/.{x → "1.", y → "0.", z → "0."}

1.

f/.{x → −"0.36038", y → "1.72076", z → −"0.36038"}

3.22076

f/.{x → "0.", y → "0.", z → "1."}

1.

Again the third root leads to a maximum. Which root corresponds to a minimum?

Exercises

Exercise 4.86. *Find the extreme values of $f(x, y, z) := x^2 - 3xz + z^2$ on the intersection of the planes $x - y = 1$ and $y + 2z = 2$.*

Exercise 4.87. *Find the extreme values of $f(x, y, z) := 2x^2 - 3y^2 - z^2$ on the intersection of the planes $x - 2y = 0$ and $y + z = 0$.*

Exercise 4.88. *Find the points of the circle, at the intersection of the sphere $x^2 + y^2 + z^2 = 1$ and the plane $x + 2y - z = 0$, where the function $f(x, y, z) := 2\,x^2 + 4\,y^2 + z^2$ attains its greatest and least values.*

Exercise 4.89. *Design a rectangular tank to have the largest volume under the constraints that its cost of production, to be explained below, and its lateral surface area is fixed. The cost of production is $2.00 [ft]$^{-2}$ for the base material, $1.50 [ft]$^{-2}$ for the lateral material, and $1.00[ft]$^{-2}$ for the top material. Suppose that the amount of funds allotted to build the tank is $250.00 for the materials, and the lateral area is to be 60 [ft]^2.*

Exercise 4.90. *Design a cylindrical tank to have the largest volume under the constraints that its cost of production, to be explained below, and its lateral surface area is fixed. The cost of production is $2.00 [ft]$^{-2}$ for the base material, $1.50 [ft]$^{-2}$ for the lateral material, and $1.00[ft]$^{-2}$ for the top material. Suppose that the amount of funds allotted to build the tank is $250.00 for the materials, and the area of the base is to be 20 [ft]^2.*

Exercise 4.91. *It is desired to build a tank of cylindrical shape, having hemispherical caps as its two ends. The builder wishes to have the largest volume tank possible under certain cost constraints. The material used for the cylindrical sides costs $1.50* `[ft]`$^{-2}$*; whereas, the material used to form the caps costs $2.50* `[ft]`$^{-2}$. *Suppose that the amount of funds allotted to build the tank is $1500.00 for the materials, and the total length of the tank is to be 20* `[ft]`.

Exercise 4.92. *The method of* LAGRANGE *multipliers for functions of $n+p$ variables. (Assume that the number of constraints is less than n.*

$$F(x_1, \ldots, x_n, u_1, \ldots, u_p) \tag{4.69}$$

with p constraints

$$H^1(x_1, \ldots, x_n, u_1, \ldots, u_p) = 0, \ldots, H^p(x_1, \ldots, x_n, u_1, \ldots, u_p) = 0 \tag{4.70}$$

may be developed along the same lines as the two constraint case discussed in the text. Consider the function

$$w(x_1, \ldots, x_n, u_1, \ldots, u_p) :=$$

$$F(x_1, \ldots, x_n, u_1, \ldots, u_p) + \sum_1^p \lambda_i H^i(x_1, \ldots, x_n, u_1, \ldots, u_p).$$

and show that the equations you get are equivalent to

$$\begin{vmatrix} F_{x_k} & F_{x_{n-p+1}} & \ldots & F_{x_n} \\ H^1_{x_k} & H^1_{x_{n-p+1}} & \ldots & H^1_{x_n} \\ \ldots & \ldots & \ldots & \ldots \\ H^p_{x_k} & H^p_{x_{n-p+1}} & \ldots & H^p_{x_n} \end{vmatrix} = 0. \tag{4.71}$$

4.7 Least squares

Sometimes when one has collected data from a scientific experiment there is a need to display this information graphically. For example an engineer might record the time it takes to fill a mold cavity with polymer versus the temperature of the polymer at the injection gate. He may wish to make a graph of this data so that his assistant may adjust the temperatures in order to fill the required number of molds in a given time. In order to allow the assistant to calibrate the controls it is convenient if a straight line can be drawn through the data. Now if all points lie on a straight line this is trivial; the usual situation is that the points do not lie in a line and we want to get a **best fit** to the data. What do we mean by a best fit? To answer this, first we define a way of measuring how close a fit actually is, and then try to minimize this quantity. **Closeness of fit** can be defined in many ways. For example, if we have n points (x_i, y_i), $i = 1, 2, \ldots, n$ through which we might wish to pass the straight line $y = Ax + B$ as closely as possible we might measure the closeness of fit as

$$e(A, B) := \sum_{i=1}^{n} |Ax_i + B - y_i| \tag{4.72}$$

which is the sum of the vertical distances of the points (x_i, y_i) from the line. This seems to be a reasonable measure of closeness of the data to the straight line; however, it is not so convenient to work with as the following measure

$$E(A, B) := \sum_{i=1}^{n} (Ax_i + B - y_i)^2 \, . \tag{4.73}$$

which uses the sum of the squares of the vertical distance. In equation (4.73) we are to determine the constants A and B in order to make $E(A, B)$ a minimum. This means that we must seek the stationary points of the function $E(A, B)$, and these are obtained by solving

$$\frac{\partial E(A, B)}{\partial A} := 2 \sum_{i=1}^{n} (Ax_i + B - y_i)\, x_i = 0$$

and

$$\frac{\partial E(A, B)}{\partial B} := 2 \sum_{i=1}^{n} (Ax_i + B - y_i) = 0.$$

This is a pair of linear equations for the unknown constants A and B. We may solve such an equation easily with MATHEMATICA . Note first, however, that this system may be written in the more convenient form

$$A \sum_{i=1}^{n} x_i^2 + B \sum_{i=0}^{n} x_i = \sum_{i=0}^{n} x_i y_i, \tag{4.74}$$

and

$$A \sum_{i=1}^{n} x_i + B n = \sum_{i=0}^{n} y_i, \tag{4.75}$$

Mathematica session 4.9. *Let us solve the problem algebraically with* MATHEMATICA. *We will write a procedure to do this. The procedure will have variables* **XY**, *an array of points, and* **n** *the number of points in our data set.*

Before presenting the procedure we show how MATHEMATICA *uses* **Part** *to extract subarrays from arrays. Let us illustrate this with the simple array.*

```
XY = {{1, 2}, {3, 4}, {5, 6}, {7, 8}}
```

```
{{1, 2}, {3, 4}, {5, 6}, {7, 8}}
```

To extract the first and third subarrays we use the **Part** *command*

XY[[{1, 3}]]

{{1, 2}, {5, 6}}

To extract the first terms in these two arrays, we use

XY[[{1, 3}, 1]]

{1, 5}

To extract two terms from the first and third subsets we use

XY[[{1, 3}, {1, 2}]]

{{1, 2}, {5, 6}}

To extract the first two terms from the fourth and first subarrays

XY[[{4, 1}]]

{{7, 8}, {1, 2}}

To extract the first term from the first subarray

XY[[{1}, 1]]

{1}

and to extract the second term in the first subarray

XY[[{1}, 2]]

{2}

Now we are in a position to write a procedure for the least squares method.

leastSquares[XY_, n_]:=

Solve[

```
{A * Sum[XY[[{i}, 1]]^2, {i, 1, n}] + B * Sum[XY[[{i}, 1]], {i, 1, n}]-
Sum[XY[[{i}, 1]] * XY[[{i}, 2]], {i, 1, n}] == 0,
A * Sum[XY[[{i}, 1]], {i, 1, n}] + n * B - Sum[XY[[{i}, 2]], {i, 1, n}] == 0},
{A, B}]
```

We test this method on the simple array introduced above.

```
leastSquares[{{1, 2}, {3, 4}, {5, 6}, {7, 8}}, 4]
```

$\{\{A \to 1, B \to 1\}\}$

We obtain as expected a straight line.

```
XY = {{1, 0}, {2, 1}, {3, 5}, {15, 11}}
```

We try another array for **XY**

{{1, 0}, {2, 1}, {3, 5}, {15, 11}}

```
leastSquares[XY, 4]
```

$\{\{A \to \frac{371}{515}, B \to \frac{241}{515}\}\}$

To see the straight line approximation to this last data set we plot both the straight line and the data on the same graph. Note how we do this by defining the plots, but not executing this by using a semicolon at the end of the statement. The use of the command **Show** *allows us to plot both data sets. We see this Plot in Figure 4.13.*

```
P1 = Plot[371/515 * x + 241/515, {x, 0, 18}];

P2 = ListPlot[{{1, 0}, {2, 1}, {3, 5}, {15, 11}}];

Show[P1, P2]
```

♦

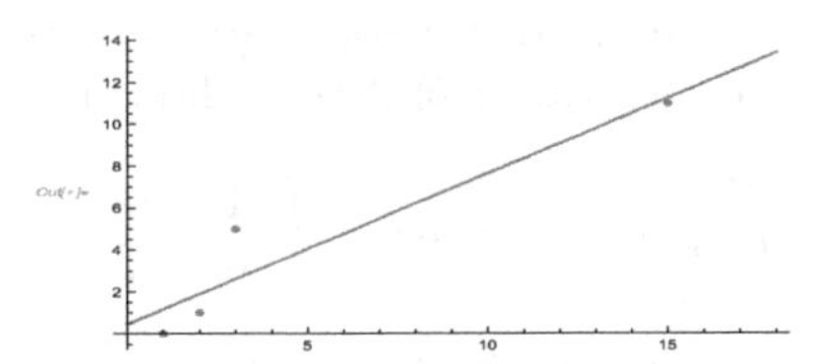

FIGURE 4.13: This plot shows the least squares approximation and also the points use to create the approximation.

Another type of LEAST SQUARES APPROXIMATION occurs when we want to approximate a given function on a finite interval $[0, L]$ by a **linear combination** of a given set of functions. For example, we may wish to approximate a continuous function f by the set of functions x^k, $k = 0, 1, 2, \ldots, n$. In this case we use a linear combination

$$f(x) \approx \sum_{k=0}^{n} c_k x^k \tag{4.76}$$

as an approximation, where the c_k are constants to be determined. Another family of functions might be the sine functions $\sin\left(\frac{k\pi x}{L}\right)$ $k = 0, 1, 2, \ldots, n$, and in this instance we use the linear combination

$$f(x) \approx \sum_{k=0}^{n} c_k \sin\left(\frac{k\pi x}{L}\right). \tag{4.77}$$

What should we use as our measure of how good the approximation is? One commonly used measure is the **least squares measure** defined by

$$E(c_1, c_2, \ldots, c_n) := \int_a^b |f(x) - \sum_{k=0}^{n} c_k \varphi^k(x)|^2 dx, \tag{4.78}$$

where $\varphi^k(x)$ $(k = 0, 1, \ldots, n)$ stand for the set of functions used in the approximation. The constants c_k $(k = 0, 1, \ldots, n)$ are to be chosen so as to minimize the integral E. The minima of functions of several variables occur at the stationary points, and these are given by

$$\frac{\partial E}{\partial c_k} := -2 \int_a^b \left(f(x) - c_0\varphi^0(x) - \ldots - c_n\varphi^n(x)\right) \varphi^k(x) dx = 0, \tag{4.79}$$

where $k = 0, 1, \ldots, n$. This produces a linear system for the constants c_k

$$c_0 \int_a^b \varphi^0 \varphi^k dx + \ldots + c_n \int_a^b \varphi^n \varphi^k dx = \int_a^b f \varphi^k dx. \tag{4.80}$$

Certain functions turn out to be better for approximations than other ones, primarily because of their **orthonormal** properties.

Definition 4.93. The family of functions φ^k, $k = 0, 1, \ldots, n$ are **mutually orthonormal** over the interval $[a, b]$ if the following equalities hold

$$\int_a^b \varphi^j(x)\varphi^k(x)\,dx = \begin{cases} 1 & \text{if } j = k \\ 0 & \text{if } j \neq k \end{cases} \tag{4.81}$$

Remark 4.94. Notice that if the functions $\varphi^k(x)$ are orthogonal then the equation (4.80) reduce to an explicit formula for the coefficients c_k

$$c_k \int_a^b \varphi^k\varphi^k dx = \int_a^b f\varphi^k\,dx \quad k = 0, 1, \ldots \quad . \tag{4.82}$$

We have seen orthogonal functions earlier in the book. Recall, for example, the identities

$$\int_0^\pi \sin(jx)\sin(kx)\,dx = \begin{cases} \frac{\pi}{2} & \text{if } j = k \neq 0 \\ 0 & \text{if } j \neq k \end{cases} . \tag{4.83}$$

Hence, the functions $\sqrt{\frac{2}{\pi}}\sin(kx)$, $(k = 1, 2, \ldots, n)$ form a mutually orthonormal family over the interval $[0, \pi]$. For the interval $[0, L]$ a simple change of integration parameter shows that the $\sqrt{\frac{2}{L}}\sin\left(\frac{j\pi x}{L}\right)$, $j = (1, 2, \ldots, n)$ also form a mutually orthonormal family

$$\int_0^L \sin\frac{j\pi x}{L}\sin\frac{k\pi x}{L}\,dx = \begin{cases} \frac{L}{2} & \text{if } j = k \\ 0 & \text{if } j \neq k \end{cases} . \tag{4.84}$$

Another such family is $\sqrt{\frac{1}{L}}, \sqrt{\frac{2}{L}\cos\left(\frac{j\pi x}{L}\right), j = 1, 2, \ldots, n\}}$ since

$$\int_0^L \cos\left(\frac{j\pi x}{L}\right)\cos\left(\frac{k\pi x}{L}\right)dx = \begin{cases} L & \text{if } j = k = 0 \\ \frac{L}{2} & \text{if } j = k \neq 0 \\ 0 & \text{if } j \neq k \end{cases} . \tag{4.85}$$

There are many other such families, a few of which the reader will meet in the exercises.

Exercises

Exercise 4.95. *Find the best least squares approximation of the function* $f(x) \equiv 1$ *on the interval* $[0, 1]$ *using the orthonormal set of functions* $\sqrt{\frac{2}{\pi}}\sin(k\pi x)$, $k = 1, 2, \ldots, 4$.

Exercise 4.96. *Find the best least squares approximation of the function*

$$f(x) \equiv \begin{cases} 1 & \text{on the interval } -1 < x < 0 \\ 0 & \text{on the interval } 0 < x < 1 \end{cases}$$

using the orthonormal set of functions $\{\sqrt{\frac{1}{2L}}, \frac{1}{L}\cos(k\pi x), k = 1, 2, 3, 4, L = 1.$

Exercise 4.97. *Find the best least squares approximation of the function*

$$f(x) \equiv \begin{cases} -1 & \text{on the interval} -1 < x < 0 \\ 1 & \text{on the interval } 0 < x < 1 \end{cases}$$

using the orthonormal set of functions $\sqrt{\frac{1}{L}}\sin(\frac{k\pi x}{L})$, $k = 1, 2, 3, 4$, *where* $L = 1$.

Exercise 4.98. *Find the best least squares approximation of the function*

$$f(x) \equiv \begin{cases} 1 & \text{on the interval } 0 < x < s \\ 0 & \text{on the interval } s < x < 2-s \\ 1 & \text{on the interval } 2-s < x < 2. \end{cases}$$

using the orthonormal set of functions $\sqrt{\frac{1}{2}}, \cos(\frac{k\pi x}{2}), k = 1, 2, 3, 4\}$.

Exercise 4.99. *The family of functions*

$$\frac{1}{2\sqrt{\pi}}, \frac{1}{\sqrt{\pi}}\cos x, \frac{1}{\sqrt{\pi}}\sin x, \frac{1}{\sqrt{\pi}}\cos 2x, \frac{1}{\sqrt{\pi}}\sin 2x, \ldots$$

is an orthonormal family over the interval $[-\pi, \pi]$. *Approximate the function* $f(x) = x^2 - \pi^2$ *using the first five functions in the family.*

Exercise 4.100. *Write a* MATHEMATICA *function for computing the coefficients of a least squares approximation of a continuous function with regard to an orthogonal family of functions. Check some of the hand calculations suggested by the earlier exercises.*

Chapter 5

Multiple Integrals

5.1 Introduction

A **double integral** may be written as

$$\iint_{\mathcal{A}} f(x,y)\,dx\,dy$$

where $\mathcal{A}$ is a 2-dimensional domain in the xy–plane, and f is a function of x and y. To see how such integrals arise, consider for example the 3–dimensional region $\mathcal{D}$ bounded above by the graph of f, below by $\mathcal{A}$, and on the side by vertical lines, see Figure 5.1.

The volume V of this region can be computed as follows: Let the domain $\mathcal{A}$ be partitioned into small rectangles, such as the rectangle $R_k := \Delta x_k \times \Delta y_k$ in Figure 5.1. Consider now the volume of that part of the region $\mathcal{D}$ which has R_k as its base. This volume can be approximated as the volume of the rectangular prism

$$f(\xi_k, \eta_k)\,\Delta x_k\,\Delta y_k\ ,$$

where (ξ_k, η_k) is any point in R_k, see Figure 5.2. Therefore the volume V is approximated by the Riemann sum

$$V \approx \sum_k f(\xi_k, \eta_k)\,\Delta x_k\,\Delta y_k\ ,$$

and the approximation is better as the rectangles R_k get smaller. If the limit of the Riemann sum exists, as the area of the rectangles R_k approach zero, then the volume V is written as the **double integral**,

$$V = \iint_{\mathcal{A}} f(x,y)\,dx\,dy\ . \tag{5.1}$$

Similarly, a **triple integral** may be written as

$$\iiint_{\mathcal{D}} f(x,y,z)\,dx\,dy\,dz\ ,$$

where $\mathcal{D}$ is a 3–dimensional region in the (x, y, z)–space, and f is a function of x, y and z. This integral, for example, expresses the mass of a 3–dimensional body occupying the region Q, if the specific density at a point (x, y, z) is given by $f(x, y, z)$.

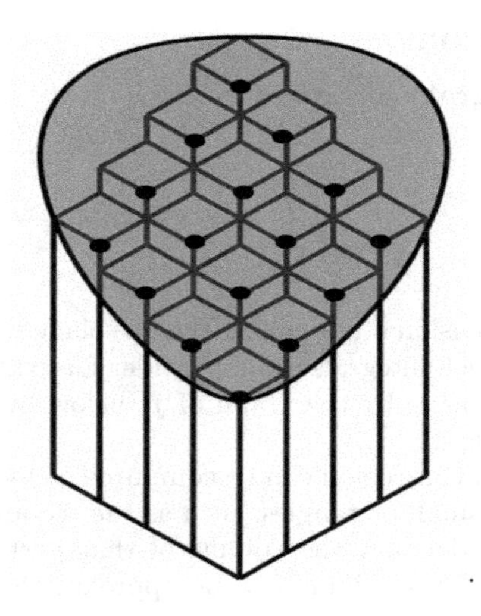

FIGURE 5.1: Approximating the volume under the surface $z = F(x, y)$, over the portion of the plane $z = 0$, as shown in Figure 5.2, using rectangular parallelpipeds.

5.2 Iterated double integrals

The integrand here is a function $f(x, y)$ of two variables. We first integrate w.r.t. one variable, say

$$\int f(x,y)\,dy\ , \quad (\text{the } \textbf{inner integration})\ , \tag{5.2}$$

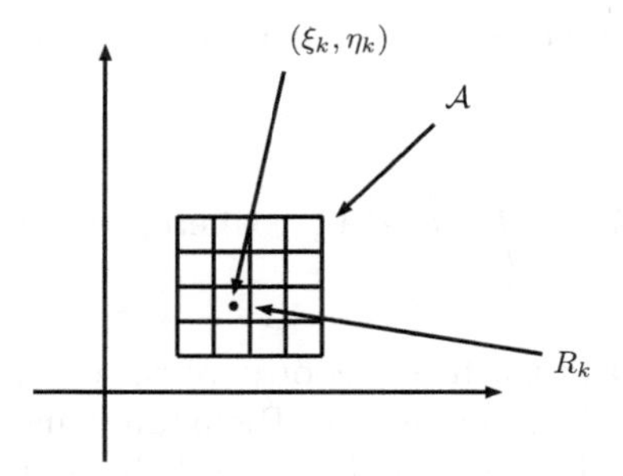

FIGURE 5.2: Partition of the projection of the surface $z = F(x, y)$, shown in Figure 5.1, onto the plane $z = 0$.

treating the other variable (here x) as a constant, and then integrate the result w.r.t. the other variable

$$\int \left(\int f(x,y)\,dy \right) dx \ , \quad \text{(the \textbf{outer integration})} \ . \tag{5.3}$$

When it is clear which integration is done first, or when the order does not matter, we can drop the brackets in (5.3). In particular, both

$$\int\int f(x,y)\,dy\,dx \quad \text{and} \quad \int dx \int f(x,y)\,dy$$

mean the same as (5.3), with y integrated first.

Area computations, such as using a vertical slicing method (5.2) or a horizontal slicing method, naturally give rise to iterated integrals with appropriate limits of integration. To see this, let us complicate the vertical slicing method

$$A \ = \ \int_a^b (g_2(x) - g_1(x))\,dx \ ,$$

by writing

$$g_2(x) - g_1(x) \ = \ \int_{g_1(x)}^{g_2(x)} dy \ . \tag{5.4}$$

Therefore (5.2) becomes an iterated integral,

$$A \ = \int_a^b \int_{g_1(x)}^{g_2(x)} dy\,dx \ , \tag{5.5}$$

where the y–integration is done first, treating x as constant, to obtain (5.4), which is then integrated from $x := a$ to $x := b$. The iterated integral (5.5) can be written as

$$A = \int_a^b \int_{g_1(x)}^{g_2(x)} dA \,, \quad \text{where} \quad dA = dx\,dy \tag{5.6}$$

is the **area element** in Cartesian coordinates, see Figure 5.3(a). The iterated integral (5.5) is therefore the limit of Riemann sums of area elements.

We can analogously write the horizontal slicing method

A_h $$A = \int_c^d (h_2(y) - h_1(y))\, dy \,,$$

as an iterated integral,

$$\int_c^d (h_2(y) - h_1(y))\, dy = \int_c^d \int_{h_1(y)}^{h_2(y)} dx\, dy \,, \tag{5.7}$$

where now the x–integration is carried out first (treating y as constant), then comes the outer integration, w.r.t y.

An area A calculated by polar coordinates,

A_r $$A = \tfrac{1}{2} \int_{\theta_1}^{\theta_2} \left(f^2(\theta) - g^2(\theta)\right) d\theta \,,$$

can also be written as an iterated integral

$$A = \int_{\theta_1}^{\theta_2} \int_{g(\theta)}^{f(\theta)} r\, dr\, d\theta \,, \tag{5.8}$$

where the inner integral

$$\int_{g(\theta)}^{f(\theta)} r\, dr = \left.\frac{r^2}{2}\right|_{g(\theta)}^{f(\theta)} = \frac{1}{2}\left(f^2(\theta) - g^2(\theta)\right)$$

is evaluated first.

The iterated integral (5.8) can be written as

$$A = \int_{\theta_1}^{\theta_2} \int_{g(\theta)}^{f(\theta)} r\, dr\, d\theta = \int_{\theta_1}^{\theta_2} \int_{g(\theta)}^{f(\theta)} dA \,, \tag{5.9}$$

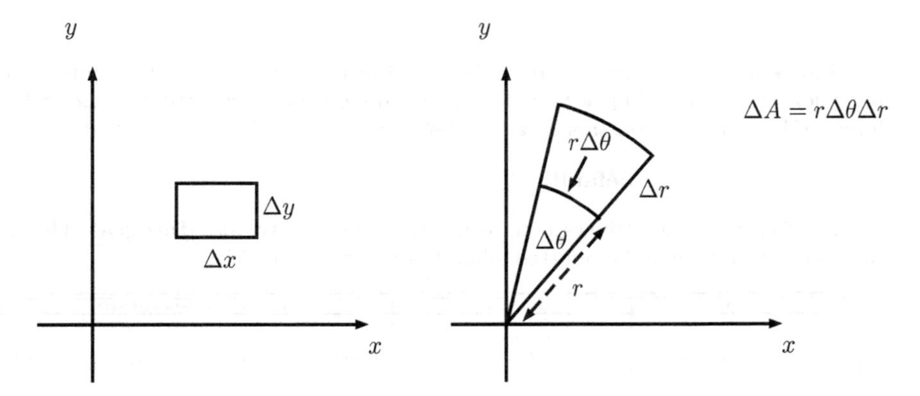

FIGURE 5.3: Area elements in Cartesian and polar coordinates.

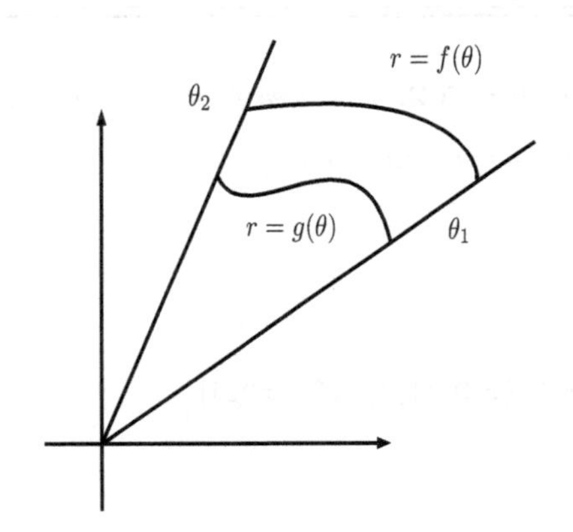

FIGURE 5.4: Polar integration.

where the area element in polar coordinates is

A_p $$dA = r\,dr\,d\theta\;,$$ see e.g. Figure 5.3.

We can compute the iterated integrals (5.5) and (5.7), by using the MATHEMATICA function.

Mathematica session 5.1. ***iteratedIntegral[default_, S_, T_]:=***

Integrate[default, {S[[1]], S[[2]], S[[3]]}, {T[[1]], T[[2]], T[[3]]}]

♦

Here • `s1` is the **inner integration variable**; `s2, s3` are its limits of integration (functions of `t`), • `t1` is the **outer integration variable**; `t2, t3` its limits of integration (constants), • and `default1` is the default setting [1], defined as

default:=1

The following table shows how to use the function **iteratedIntegral**. The parameter `default1` may be omitted when the function is called.

method	integral	the function iteratedIntegral
vertical slicing method	$\int_a^b \int_{g_1(x)}^{g_2(x)} 1\, dy\, dx$	**iteratedIntegral**[{x, a, b} , {y, $g_1(x)$, $g_2(x)$}]
horizontal slicing method	$\int_c^d \int_{h_1(y)}^{h_2(y)} 1\, dx\, dy$	**iteratedIntegral**[y, c, d, x, $h_1(y)$, $h_2(y)$]]

Mathematica session 5.2. *The area A of the region* $\mathbb{D} := \{(x, y) : x^3 \leq y \leq x^2, 0 \leq x \leq 1\}$ *is computed by vertical slicing,*

***default* = 1**

1

***iteratedIntegral*[*default*, {*x*, 0, 1}, {*y*, *x*^3, *x*^2}]**

$\frac{1}{12}$

***iteratedIntegral*[*default*, {*x*, 0, π/4}, {*y*, *Sin*[*x*], *Cos*[*x*]}]**

$-1 + \sqrt{2}$

***iteratedIntegral*[*default*, {*y*, 0, 1/*Sqrt*[2]}, {*x*, 0, *ArcSin*[*y*]}]+**
***iteratedIntegral*[*default*, {*y*, 1/*Sqrt*[2], 1}, {*x*, 0, *ArcCos*[*y*]}]**

$-1 - \frac{-4+\pi}{4\sqrt{2}} + \frac{4+\pi}{4\sqrt{2}}$

***Simplify*[%]**

$-1 + \sqrt{2}$

[1] The parameter `default1` allows using **iteratedIntegral** in other applications, see MATHEMATICA–Session 5.4.

```
areaPolar[default_, theta_, R_]:=
Integrate[r * default, {theta[[1]], theta[[2]], theta[[3]]}, {R[[1]], R[[2]], R[[3]]}]
```

To calculate the area inside the cardioid $r = 1 + \cos(\theta)$ we input

```
areaPolar[1, {theta, 0, 2 * Pi}, {r, 0, 1 + Cos[theta]}]
```

$\frac{3\pi}{2}$

```
iteratedIntegral[r, {theta, 0, 2 * Pi}, {r, 0, 1 + Cos[theta]}]
```

iteratedIntegral[r, {theta, 0, 2π}, {r, 0, 1 + Cos[theta]}]

Similarly, the area inside the nth revolution of the spiral $r = a\theta$ is obtained by

```
areaPolar[1, {theta, 2 * (n − 1) * Pi, 2 * n * Pi}, {r, 0, a * theta}]
```

$\frac{4}{3}a^2\left(1 - 3n + 3n^2\right)\pi^3$

Compute the area enclosed by the lemniscate

$$r^2 = 2a^2 \cos 2\theta\ .$$

Solution: This area is four times the value of

```
areaPolar[1, {theta, 0, Pi/4}, {r, 0, a * Sqrt[2 * Cos[2 * theta]]}]
```

$\frac{a^2}{2}$

♦

Exercises

Exercise 5.1. *Graph each of the following curves, and compute the area inside it. Assume a is positive.*

(a) $r = a\ \sin\theta$ (b) $r = a\ \sin 2\theta$ (c) $r = a\ \sin 3\theta$

(d) $r^2 = 8\cos 2\theta$ (e) $r = 3\ \cos\theta$ (f) $r = a\ \cos 3\theta$

Exercise 5.2. *Evaluate the iterated integrals by hand calculation and check your results with* MATHEMATICA

(a) $\int_0^1\int_0^3 (xy + 4)dx\,dy$ (b) $\int_2^3\int_{-1}^2 (xy^2 + 4x^2y)dx\,dy$ (c) $\int_{-1}^1\int_{-1}^1 (x^2+y^2)dy\,dx$

(d) $\displaystyle\int_0^{\ln 2}\int_{\ln 2}^{\ln 3} e^{x+y}dx\,dy$ (e) $\displaystyle\int_0^1\int_0^1 xy\sqrt{x^2+y^2}dx\,dy$ (f) $\displaystyle\int_0^\infty\int_0^\infty e^{-x-y}dx\,dy$

(g) $\displaystyle\int_0^\infty\int_0^\infty xye^{-x^2-y^2}dx\,dy$ (h) $\displaystyle\int_0^\pi\int_0^3(\sin x+\sin y)dx\,dy$ (i) $\displaystyle\int_0^\pi\int_0^\pi(\sin x\cos y)dx\,dy$

Exercise 5.3. *Evaluate the iterated integrals by hand calculation and check your results with* MATHEMATICA

(a) $\displaystyle\int_0^1\int_0^1 \frac{1}{(x+y)^2}dx\,dy$ (b) $\displaystyle\int_0^{\pi/4}\int_0^1 y\sec^2(xy)dx\,dy$ (c) $\displaystyle\int_0^\pi\int_0^1 x\cos(xy)dy\,dx$

(d) $\displaystyle\int_1^{e^2}\int_0^1 \frac{x^2}{x^3+y}dx\,dy$ (e) $\displaystyle\int_0^2\int_0^1 xe^{xy}dy\,dx$ (f) $\displaystyle\int_0^1\int_0^1(x\sin y+y\cos x)dy\,dx$

(g) $\displaystyle\int_0^\pi\int_0^1(1+x-y)dy\,dx$ (h) $\displaystyle\int_{-1}^1\int_0^1 \frac{x}{(xy+1)^{\frac{3}{2}}}dy\,dx$ (i) $\displaystyle\int_0^1\int_0^1 x\sinh(xy)dy\,dx$

5.3 Double integrals

In the calculus of one variable, a single integral was defined as the limit of Riemann sums, using finer and finer partitions of the interval $[a,b]$. The double integral (5.1), over a 2–dimensional region D , is obtained in a similar fashion. We first show how to partition D . Let $\mathcal{A}$ be a **bounded domain** in the xy–plane, meaning that $\mathcal{A}$ is contained in some finite disk or rectangle. Assume that $\mathcal{A}$ is

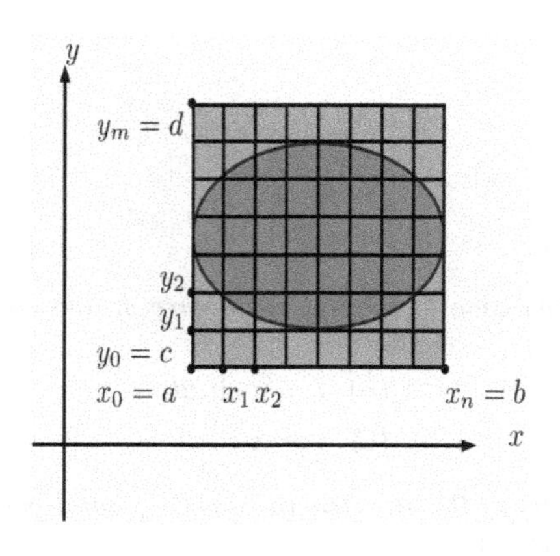

FIGURE 5.5: Partitioning a two–dimensional domain.

contained in a rectangle $\mathcal{R} := [a, b] \times [c, d]$, let

$$\mathcal{P}_{[a,b]} := \{x_0 = a < x_1 < \cdots < x_i < x_{i+1} < \cdots < x_n = b\} \text{ be a partition of } [a, b] , \tag{5.10}$$

and let

$$\mathcal{P}_{[c,d]} := \{y_0 = c < y_1 < \cdots < y_j < y_{j+1} < \cdots < y_m = d\} \text{ be a partition of } [c, d] . \tag{5.11}$$

The $n+1$ vertical lines $x := x_i$, and the $m+1$ horizontal lines $y := y_j$ partition the rectangle R into $m \times n$ smaller rectangles, see Figure (5.5). We denote this partition of $\mathcal{R}$ by $\mathcal{P}_{[a,b]} \times \mathcal{P}_{[c,d]}$. Some of these $m \times n$ rectangles lie inside $\mathcal{A}$, see the shaded area in Figure (5.5). We denote the number of such rectangles by $N(n, m)$.

Definition 5.4 (Riemann sum). Let $R_1, R_2, \ldots, R_{N(n,m)}$ be the rectangles of $\mathcal{P}_{[a,b]} \times \mathcal{P}_{[c,d]}$ which are contained in $\mathcal{A}$. The set

$$\mathcal{P}_{\mathcal{A}} := \{R_1, R_2, \ldots, R_{N(n,m)}\} \tag{5.12}$$

is called an **internal partition** of the domain $\mathcal{A}$. Denoting the area of the kth rectangle R_k by ΔA_k , we define the **norm** of the partition $\mathcal{P}_{\mathcal{A}}$ as the maximal area of its rectangles,

$$\|\mathcal{P}_{\mathcal{A}}\| := \max_{k=1,2,\ldots,N(n,m)} \Delta A_k . \tag{5.13}$$

Let the function $f(x, y)$ be defined and continuous on $\mathcal{A}$, and let $\mathcal{P}_{\mathcal{A}}$ be an internal partition as above. A **Riemann sum** of f on $\mathcal{A}$ is defined as

$$S_{\mathcal{P}_{\mathcal{A}}}(f, \mathcal{A}) := \sum_{k=1}^{N(n,m)} f(\xi_k, \eta_k)\, \Delta A_k , \tag{5.14}$$

where (ξ_k, η_k) is an arbitrary point in the kth rectangle.

Mathematica session 5.3. *In this* MATHEMATICA *session we evaluate some Riemann sums that approximate double integrals. We check the accuracy by analytical performing the integrations.*

```
riemannSumTwo[f_, X_, Y_, n_, m_]:=Module[{g}, xmin = X[[2]];
xmax = X[[3]];
ymin = Y[[2]];
ymax = Y[[3]];
g = f/.{x → xmin + (i/n) * (xmax − xmin), y → ymin + (j/n) * (ymax − ymin)};
Sum[g * (1/n)^2 * (xmax − xmin) * (ymax − ymin), {i, 0, n}, {j, 0, m}]]

riemannSumTwo[x^2 * y * 3, {x, 0, 2}, {y, 0, 2}, 4, 3]
```

$\frac{135}{8}$

```
N[riemannSumTwo[x^2 * y * 3, {x, 0, 2}, {y, 0, 2}, 10, 10]]
```

20.328

```
N[riemannSumTwo[x^2 * y * 3, {x, 0, 2}, {y, 0, 2}, 100, 100]]
```

16.4032

```
N[riemannSumTwo[x^2 * y * 3, {x, 0, 2}, {y, 0, 2}, 200, 200]]
```

16.2008

```
Integrate[x^2 * y * 3, {x, 0, 2}, {y, 0, 2}]
```

16

```
riemannSumTwo[Sin[x * y], {x, 0, Pi/2}, {y, 0, Pi/2}, 6, 6];
N[%]
```

1.47215

```
riemannSumTwo[Sin[x * y], {x, 0, Pi/2}, {y, 0, Pi/2}, 20, 20];
N[%]
```

1.27267

```
riemannSumTwo[Sin[x * y], {x, 0, Pi/2}, {y, 0, Pi/2}, 100, 100];
N[%]
```

1.20192

```
N[riemannSumTwo[Sin[x * y], {x, 0, Pi/2}, {y, 0, Pi/2}, 500, 500]]
```

1.18769

```
Integrate[Sin[x * y], {x, 0, Pi/2}, {y, 0, Pi/2}]
```

$EulerGamma - CosIntegral\left[\frac{\pi^2}{4}\right] + 2Log\left[\frac{\pi}{2}\right]$

N[%]

1.18412

♦

As in the definite integral, we expect the sums (5.14) to have a limit if the norms $\|\mathcal{P}_{\mathcal{A}}\|$ tend to zero. If this is the case, we are in business:

Definition 5.5 (Double integral). The function f is called **integrable** in $\mathcal{A}$ if the limit

$$\lim_{\|\mathcal{P}_{\mathcal{A}}\|\to 0} S_{\mathcal{P}_{\mathcal{A}}}(f, \mathcal{A})$$

exists, in which case the (**double**) **integral** of f on D is that limit,

$$\iint\limits_{\mathcal{A}} f(x,y)\,dx\,dy := \lim_{\|\mathcal{P}_{\mathcal{A}}\|\to 0} \sum_{k=1}^{N(n,m)} f(\xi_k, \eta_k)\Delta A_k \,. \tag{5.15}$$

The double integral (5.15) is also denoted by $\iint\limits_{\mathcal{A}} f(x,y)\,dA$ or just $\iint\limits_{\mathcal{A}} f\,dA$.

For the double integral (5.15) to exist, both the integrand f and the domain $\mathcal{A}$ must be well-behaved, for example, f continuous and D having a relatively smooth boundary. The situation is simpler for single integrals,

$$\int_a^b f(x)\,dx$$

where for bounded $[a, b]$, the continuity of f suffices, as was shown in the one-variable calculus.

Example 5.6 (Rectangular domain). If the domain $\mathcal{A}$ is itself a rectangle, say $\mathcal{A} = [a, b] \times [c, d]$, then the double integral (5.15) can be written as an iterated integral

$$\iint\limits_{[a,b]\times[c,d]} f(x,y)\,dx\,dy = \int_a^b \left(\int_c^d f(x,y)dy\right) dx = \int_c^d \left(\int_a^b f(x,y)dx\right) dy \,. \tag{5.16}$$

This is so because for $\mathcal{A} = [a, b] \times [c, d]$ we can use the partition[2] $\mathcal{P}_{[a,b]} \times \mathcal{P}_{[c,d]}$, so that the Riemann sum (5.14) becomes the double sum

$$\sum_{i=1}^{m} \sum_{j=1}^{n} f(\xi_i, \eta_j)\,\Delta x_i\,\Delta y_j \,.$$

[2]No need here for the internal partition $\mathcal{P}_D$.

If we carry first the summation on j , then on i , we get

$$\sum_{i=1}^{m}\left(\sum_{j=1}^{n} f(\xi_i,\eta_j)\,\Delta y_j\right)\Delta x_i\ , \quad \text{and in the limit} \quad \int_a^b\left(\int_c^d f(x,y)dy\right)dx\ . \tag{5.17}$$

Summing first on i , then on j , gives

$$\sum_{j=1}^{n}\left(\sum_{i=1}^{m} f(\xi_i,\eta_j)\,\Delta x_i\right)\Delta y_j\ , \quad \text{and in the limit} \quad \int_c^d\left(\int_a^b f(x,y)dx\right)dy\ . \tag{5.18}$$

Example 5.7. To illustrate (5.16), consider the integral of e^{x-3y} over the rectangle $\mathcal{A}\ :=\ [0,1]\times[0,2]$,

$$\iint\limits_{[0,1]\times[0,2]} e^{x-3y}\,dx\,dy\ .$$

Integrating first on y , then on x , we get

$$\begin{aligned}
&\int_0^1\left(\int_0^2 e^{x-3y}\,dy\right)dx \\
&\qquad = \int_0^1 e^x\left(\int_0^2 e^{-3y}\,dy\right)dx\ , \\
&\qquad\quad \text{as } e^x \text{ is constant in the inner integral,} \\
&\qquad = \int_0^1 e^x\left(-\frac{e^{-3y}}{3}\Big|_0^2\right)dx = \frac{\left(e^0-e^{-6}\right)}{3}\left(\ e^x\ \Big|_0^1\right) \\
&\qquad = \frac{1}{3}\left(1-e^{-6}\right)(e-1)\ .
\end{aligned}$$

Similarly, integrating first on x , then on y ,

$$\int_0^2\left(\int_0^1 e^{x-3y}\,dy\right)dx\ =\ \int_0^2 e^{-3y}\left(\int_0^1 e^x\,dx\right)dy\ =$$

$$\int_0^2 e^{-3y}\left(e^1-e^0\right)dy\ =\ \frac{1}{3}\,(e-1)\left(1-e^{-6}\right)\ .$$

There are other regions for which double integrals reduce to iterated integrals. These are the "vertically simple" and "horizontally simple" regions.
A **vertically simple** (or y–**simple**) **region** is a connected region whose intersection with any vertical line is either an interval (possibly a single point) or is empty[3]. This is illustrated in Figure (5.6),(5.7), where the region is bounded

- above by a curve $y = f_2(x)$
- below by a curve $y = f_1(x)$
- on the left by the vertical line $x = a$
- on the right by the vertical line $x{=}b$

[3]In other words, a connected region D is y–simple if its intersections with vertical lines have no holes.

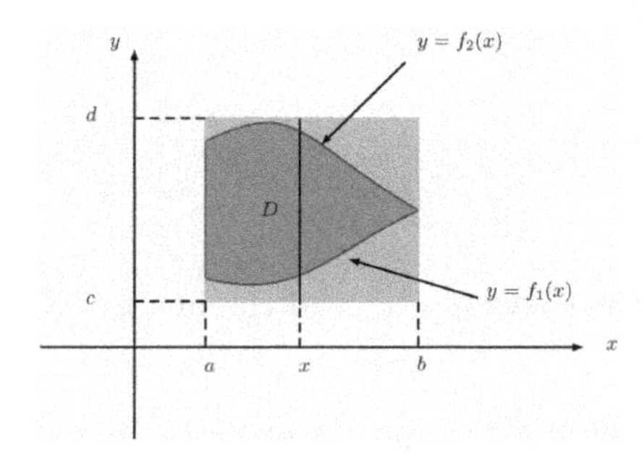

FIGURE 5.6: A vertically simple region.

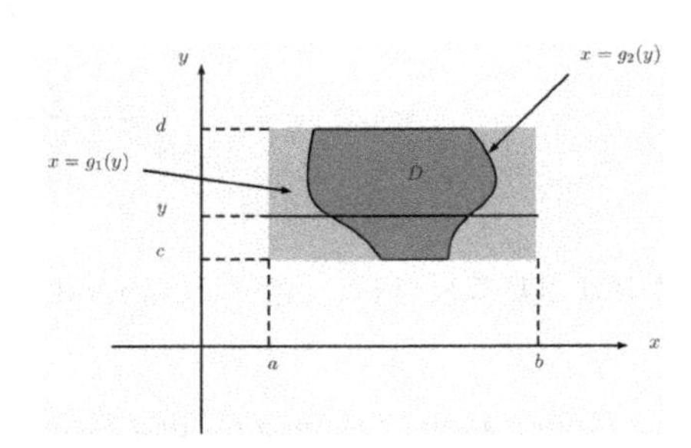

FIGURE 5.7: A horizontally simple region.

possibly $a = -\infty$ or $b = \infty$. If $\mathcal{A}$ is such a region, an integral over $\mathcal{A}$ can be replaced by an iterated integral

$$\iint_{\mathcal{A}} f(x,y)dA = \int_a^b \left(\int_{f_1(x)}^{f_2(x)} f(x,y)dy \right) dx \, . \tag{5.19}$$

A **horizontally simple** (or x–**simple**) **region,** is defined analogously for horizontal lines. An example is the region shown in Figure 5.6, which is bounded

- on the right by a curve $x = g_2(y)$
- on the left by a curve $x = g_1(y)$
- below by the horizontal line $y = c$
- above by the horizontal line $y = d$

where possibly $c = -\infty$ or $d = \infty$. For the horizontally simple region we replace the double integral by

$$\iint_{\mathcal{A}} f(x,y)\, dA = \int_c^d \left(\int_{g_1(y)}^{g_2(y)} f(x,y)\, dx \right) dy \, . \tag{5.20}$$

In MATHEMATICA 5.2 we calculated areas using iterated integrals. We now use MATHEMATICA to compute general iterated integrals.

Mathematica session 5.4. *The function* **iteratedIntegral[default,s,s1,s2, t,t1,t2]** *of* MATHEMATICA*–Session 5.2 can be used to compute the iterated integral*

$$\int_a^b \left(\int_{g_1(x)}^{g_2(x)} f(x,y)dy \right) dx \quad \textit{as} \quad \textbf{iteratedIntegral}[f(x,y), \{y, g_1(x), g_2(x)\}, \{x, a, b\}]$$

Note that the integrand $f(x,y)$ is used for the parameter `default1` *in the definition of* **iteratedIntegral**.

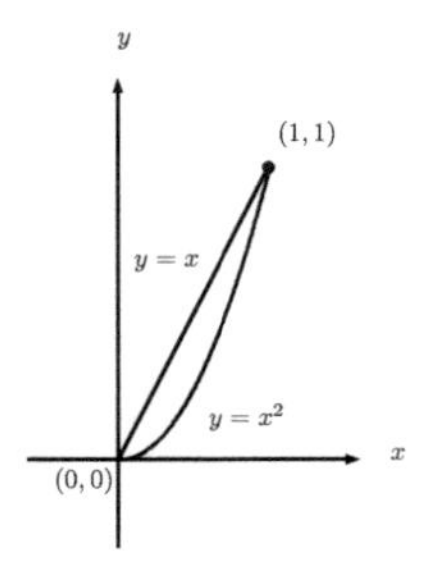

FIGURE 5.8: The region bounded by the curve $y = x^2$ and the line $y = x$.

For example, the integral of $f(x, y) := xy$ *over the region bounded by the curve* $y = x^2$ *and the line* $y = x$*, see Figure 5.8,*

$$\int_0^1 \left(\int_{x^2}^{x} xy\, dy \right) dx \ , \tag{5.21}$$

is computed by the procedure `iteratedIntegral` *which can also be used to compute integrals over* x*–simple regions (where the* x*–integration is carried out first)*

$$\int_c^d \left(\int_{h_1(y)}^{h_2(y)} f(x, y)\, dx \right) dy \qquad as \qquad \texttt{iter\_int}(x, h_1(y), h_2(y), y, c, d, f(x, y))$$

Thus, the integral (5.21) can be computed by **iteratedIntegral[default_, S_, T_]:=**

Integrate[default, {S[[1]], S[[2]], S[[3]]}, {T[[1]], T[[2]], T[[3]]}]

iteratedIntegral[x * y, {x, 0, 1}, {y, x^2, x}]

$\frac{1}{24}$

♦

Mathematica session 5.5. *Another example: To compute the integral of* $f(x, y) := (x + y)^2$ *over the region bounded by the parabola* $x = y^2$ *and the line* $y = x - 2$ *,*

$$\int_{-1}^{2} \left(\int_{y^2}^{y+2} (x + y)^2\, dx \right) dy$$

we input the expression

iteratedIntegral[(x + y)^2, {y, −1, 2}, {x, y^2, y + 2}]

$\frac{4131}{140}$

```
N[iteratedIntegral[E^(-x^2 - y^2), {x, -1, 1}, {y, -Sqrt[1 - x^2], Sqrt[1 - x^2]}]]
```

1.98587

```
areaPolar[default_, theta_, R_]:=
Integrate[r * default, {theta[[1]], theta[[2]], theta[[3]]}, {R[[1]], R[[2]], R[[3]]}]
```

```
areaPolar[E^(-r^2), {θ, 0, 2 * π}, {r, 0, 1}]
```

$\frac{(-1+e)\pi}{e}$

```
N[%]
```

1.98587

```
iteratedIntegral[12 + y - x^2, {x, 0, 1}, {y, x^2, Sqrt[x]}]
```

$\frac{569}{140}$

```
iteratedIntegral[r * r, {θ, -π/2, π/2}, {r, 0, 2 * Cos[θ]}]
```

$\frac{32}{9}$

♦

We consider now double integrals

$$\iint_{\mathcal{A}} f(x,y)\,dx\,dy$$

in the special case where $\mathcal{A}$ is **radially simple** (or r–**simple**), i.e. its intersections with radial lines are intervals or empty. Figure 5.4 shows an r–simple, bounded region.

- on the outside by a curve $r = f(\theta)$
- on the inside by a curve $r = g(\theta)$
- and the two radial lines $\theta := \theta_1$
- and $\theta := \theta_2$

Using polar coordinates, this integral becomes an iterated integral

$$\iint_{\mathcal{A}} f(x,y)\,dx\,dy = \int_{\theta_1}^{\theta_2} \int_{g(\theta)}^{f(\theta)} f(r\cos\theta, r\sin\theta)\,r\,dr\,d\theta \tag{5.22}$$

where we used again the fact $dA = r\,dr\,d\theta$.

Example 5.8. The integral

$$\iint_{\{(x,y)\,:\,x^2+y^2\le 1\}} e^{-x^2-y^2}\,dx\,dy \tag{5.23}$$

can be computed, using Cartesian coordinates, by using numerical integration, i.e. we use `N` before `integratedIntegral`.[4]

Mathematica session 5.6.

```
N[iteratedIntegral[E^(-x^2 - y^2), {x, -1, 1}, {y, -Sqrt[1 - x^2], Sqrt[1 - x^2]}]]
```

1.98587

A more natural way to compute (5.23) is by using polar coordinates. The domain of integration, the unit disk

$$\{(x,y) : x^2 + y^2 \le 1\} \qquad \textit{is rewritten as} \qquad \{(r,\theta) : 0 \le r \le 1\ ,\ 0 \le \theta \le 2\pi\}\ .$$

Then by (5.22),

$$\iint_{\{(x,y)\,:\,x^2+y^2\le 1\}} e^{-x^2-y^2}\,dx\,dy \qquad \textit{is rewritten as} \qquad \int_0^{2\pi}\int_0^1 e^{-r^2}\,r\,dr\,d\theta$$

which can be computed using the function `areaPolar` *of* MATHEMATICA*–Session 5.2.*

$\frac{(-1+e)\pi}{e}$

```
N[%]
```

1.98587

♦

Exercises

Exercise 5.9. *Integrate $f(x,y) := (x+y)^2$ over the regions, bounded by the following curves:*

(a) *The line $y = x$ and the parabola $y = x^2$*

(b) *The lines $y = x$, $y = -x$ and $y = 1$.* (c) *The parabolas $y = x^2$ and $x = y^2$.*

(d) *The line $y = x$ and the cubic parabola $y = x^3$, where $x \ge 0$.*

[4] MATHEMATICA did not evaluate this algebraically. It gave another integral in terms of the `erf` function.

Exercise 5.10. *In each of the problems listed evaluate the iterated integral, then evaluate again reversing the order of integration (sketching the domain of integration will help you to determine the new limits of integration.)*

(a) $\int_0^a dx \int_0^b e^{x+y}\, dy$

(b) $\int_0^1 dy \int_{y^2}^1 (x^2+y^2)\, dx$

(c) $\int_0^a dy \int_0^{\sqrt{a^2-y^2}} (x+y)^3\, dx$

(d) $\int_{\frac{1}{3}}^{\frac{2}{3}} dy \int_{y^2}^{\sqrt{y}} (xy+1)\, dx$

(e) $\int_0^{2\pi} d\theta \int_0^1 r\sqrt{1-r^2}\, dr$

(f) $\int_1^2 dr \int_0^{\frac{\pi}{4}} r(1+\cos\theta)\, d\theta$

(g) $\int_{\frac{-1}{4}}^{\frac{1}{4}} d\theta \int_0^{1+\cos\theta} r\, dr$

(h) $\int_1^{\sqrt{2}} dr \int_{\text{acos}\,\frac{1}{r}}^{\frac{\pi}{4}} r\,\sin\theta\, d\theta$

(i) $\int_{-\sqrt{2}}^{\sqrt{2}} \int_{x^2}^{x^2} ye^x\, dydx$

(j) $\int_0^{\pi/4} \left(\int_0^{\pi/2} \frac{\sec^2 y}{\tan^2 y + \cos^2 x}\, dy \right) dx$

(k) $\int_{\frac{-2}{\sqrt{3}}}^{\frac{2}{\sqrt{3}}} \int_{x^2}^{2\sqrt{1-\frac{x^2}{2}}} x^2 y dydx$

Exercise 5.11. *In each part are listed a function, and graphs which bound an area. Try to calculate each integral by using both vertical and horizontal slicing. Are some problems better done by choosing one method over the other? Compare your answers.*

(a) *The function* $f = x^4$ *the parabola* $y^2 = x$, *and the line* $y = 2x - 1$.

(b) *The function* $f = x^2 + y^2$, *and the parabolas* $y = 6x - x^2$, $y = x^2 - 2x$.

(c) *The function* $f = xy$, *the hyperbola* $y = \frac{1}{x}$, *the y-axis, and the lines* $y = 1$, $y = 2$.

(d) *The function* $f := x - y$, *the parabola* $y = \frac{1}{3}x^2$, *and the straight line* $y = 3$.

(e) *The function* $f := \sqrt{x^2+1}$, *and the lines* $y = x$, $x = 1$.

(f) *The function* $f := e^{xy}$, *the curve* $xy = 2$, *the* $x - axis$ *and the lines* $x = 1$, $x = e^4$.

(g) *The function* $f := 4x + 6y$, *and the ellipse* $\frac{x^2}{4} + \frac{y^2}{9} = 1$.

Exercise 5.12. *Compute the integral of*

$$f(x,y) = \frac{1}{x^2+y^2}$$

over the annulus $\{(x,y) : 1 \leq x^2 + y^2 \leq 4\}$, *using*

(a) *Cartesian coordinates* (b) *polar coordinates*

Exercise 5.13. *Sketch the region over which the iterated integral*

$$\int_{-\pi/2}^{\pi/2} d\theta \int_0^{2a\cos\theta} f(r\cos\theta,\, r\sin\theta)\, r\, dr$$

is taken.

5.4 Volume by double integration

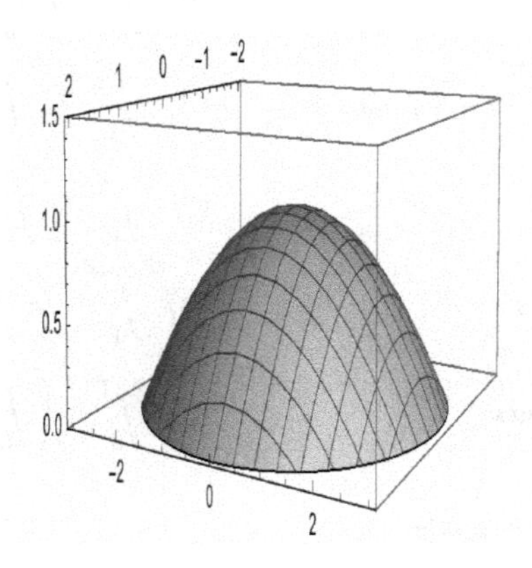

FIGURE 5.9: Illustration of Example 5.14).

Let $\mathcal{A}$ be a region in the xy–plane, $f(x, y)$ a function which is nonnegative in $\mathcal{A}$. Consider the 3–dimensional region $\mathcal{D}$ which is part of the vertical cylinder, with base $\mathcal{A}$, bounded above by the graph $z = f(x, y)$, see Figure 5.9. The volume of $\mathcal{D}$ is given by the integral

$$\iint\limits_{\mathcal{A}} f(x, y)\, dx\, dy$$

which, if an iterated integral, can be computed using the function `iteratedIntegral` of MATHEMATICA–Session 5.5.

Example 5.14. *It is required to compute the volume of the solid bounded by the paraboloid*

$$z = 1 - \frac{x^2}{4} - \frac{y^2}{9}$$

and the xy–plane. Because of symmetry we can compute the volume in the first octant, and multiply by 4. We therefore perform the area integral over the first quadrant of the ellipse

$$S := \left\{ (x, y) : \frac{x^2}{4} + \frac{y^2}{9} = 1 \ , \ x \geq 0 \ , \ y \geq 0 \right\} ,$$

obtaining the iterated integral

$$\frac{V}{4} = \int_0^2 \int_0^{\frac{3}{2}\sqrt{4-x^2}} \left(1 - \frac{x^2}{4} - \frac{y^2}{9}\right) dy\, dx$$

which can be computed using the MATHEMATICA *function* **iteratedIntegral** *(of* MATHEMATICA*-Session 5.2)*

```
iteratedIntegral[(1 - x^2/4 - y^2/9), {x, 0, 2}, {y, 0, (3/2) * Sqrt[4 - x^2]}]
```

$\frac{3\pi}{4}$

The required volume is therefore 3π .

Example 5.15. Consider the region D in the xy–plane bounded by the two parabolas $x = y^2$ and $y = x^2$, see Figure 5.10. It is required to compute the volume of the cylinder bounded below by D and bounded above by the surface $z = 12 + y - x^2$. The volume integral becomes

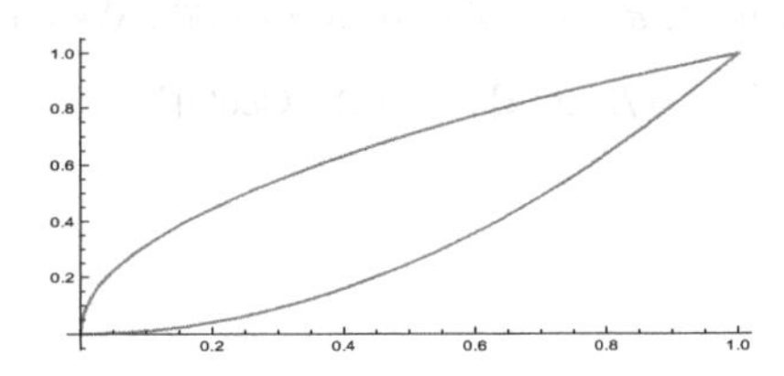

FIGURE 5.10: Illustration of Example 5.15.

$$V = \int_0^1 \int_{x^2}^{\sqrt{x}} \left(12 + y - x^2\right)\, dy\, dx$$

which is **iteratedIntegral[default_, S_, T_]:=**

```
Integrate[default, {S[[1]], S[[2]], S[[3]]}, {T[[1]], T[[2]], T[[3]]}]
```

```
iteratedIntegral[12 + y - x^2, {x, 0, 1}, {y, x^2, Sqrt[x]}]
```

$\frac{569}{140}$

computed using the MATHEMATICA function **iteratedIntegral** (of MATHEMATICA–Session 5.2)

Example 5.16. Compute the volume inside the cylinder $x^2 + y^2 = 2x$, lying above the xy–plane and below the cone $z = \sqrt{x^2 + y^2}$. This volume, using Cartesian coordinates, is

$$V = \iint_D \sqrt{x^2 + y^2}\, dx\, dy \tag{5.24}$$

where D is the disk bounded by $x^2 + y^2 = 2x$. It is however easier to compute using polar coordinates. Here, the cylinder

$$\{(x, y) : x^2 + y^2 = 2x\} \quad \text{becomes} \quad \{(r, \theta) : r = 2 \cos\theta\ ,\ -\frac{\pi}{2} \leq \theta \leq \frac{\pi}{2}\}$$

and the cone becomes $z = r$. Therefore the volume (5.24) is

$$V = \int_0^{2 \cos\theta} \int_{-\pi/2}^{\pi/2} (r)\,(r\, dr\, d\theta)$$

where the area element $dx\, dy$ is replaced by $r\, dr\, d\theta$. We compute this by inputting

```
iteratedIntegral[r^2, {θ, -π/2, π/2}, {r, 0, 2 * Cos[θ]}]
```

$\frac{32}{9}$

Exercises

Exercise 5.17. *Compute the volumes of the following solids by double integration.*

(a) *The tetrahedron bounded by the coordinate planes and the plane* $x+y+z=1$.

(b) *The tetrahedron bounded by the coordinate planes and the plane* $3x+2y+4z=1$.

(c) *The solid in the second octant bounded above by the paraboloid* $z = x^2 + y^2$, *bounded on the side by the planes* $x := -1$, $y := 2$ *and below by* $z := -1$

(d) *The solid lying between* $z = x + 1$ *and* $z = 3 + y$ *and contained in the cylinder* $x^2 + y^2 = 1$.

(e) *The solid enclosed between* $y = x$, $z = 0$, *and* $x + y = 1$

(f) *The solid enclosed between* $x = y^3$, $z = 1$, *and* $2x + y = 1$

(g) *The solid lying in the third octant bounded above by* $z = e^{x+y}$ *and laterally by the plane* $2x + 3y + 1 = 0$.

(h) *The solid bounded above by* $z = 4 - x^2 - y^2$ *and below by* $z = 2$.

(i) *The solid bounded by the paraboloid* $z = x^2 + y^2$ *and the cylinder* $x^2 + y^2 = 9$.

(j) *The solid, in the first octant, bounded laterally by the cylinder* $y := 1 - x^3$, *above by* $z := x^3$, *and below by the plane* $z := 0$

(k) *The solid lying in the intersection of* $x^2 + y^2 = 1$ *and* $y^2 + \frac{z^2}{4} = 1$

(l) *The solid bounded above by* $Ax + By + Cz = 1$, *with* $A, B, C > 0$ *and laterally bounded by* $\frac{x^2}{a^2} + \frac{y^2}{b^2} = 1$

Exercise 5.18. *Compute the volume (5.24) using Cartesian coordinates and Polar coordinates.*

Exercise 5.19. *A round hole of radius* 1 *is bored through the solid bounded by the paraboloid*

$$z = 1 - \frac{x^2}{4} - \frac{y^2}{9}$$

and the xy–plane. The axis of the hole coincides with the z-axis. Find the volume of the material removed.

Exercise 5.20. *A cylinder parallel to the z-axis has the cross section bounded by a single loop of the curve given (in polar coordinates) by $r = a\cos n\theta$. Find the volume of the region internal to this cylinder, truncated above and below by the sphere $x^2 + y^2 + z^2 = 2az$.*

5.5 Centroids and moments of inertia

In the calculus of one variable we computed centroids of simple plates and bodies, using single integrals. In general, such computations require multiple integrals. Consider for example a plate S in the xy–plane, with variable density δ (that is $\delta = \delta(x, y)$). See Figure 5.11 Let us form a partition of the

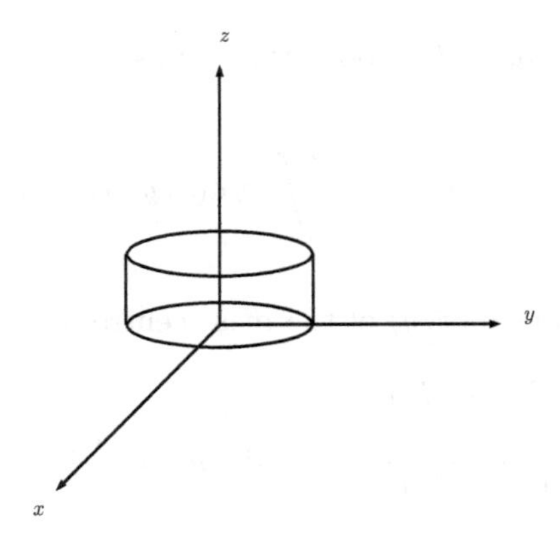

FIGURE 5.11: A disk width a variable density $\delta(x, y)$.

5.11, namely $\mathcal{P}$. An element of the partition is R_k, whose mass m_k may be approximated by

$$m_k \approx \delta(\xi_k, \eta_k)\,\Delta x_k\,\Delta y_k \ ,$$

the product of density (at some point (ξ_k, η_k) of R_k) and area $\Delta x_k \, \Delta y_k$. The mass of the plate S is therefore approximated by the Riemann sum

$$M \approx \sum_k m_k \approx \sum_k \delta(\xi_k, \eta_k) \, \Delta x_k \, \Delta y_k$$

and in the limit, as $\|\mathcal{P}\| \to 0$,

$$M = \iint_S \delta(x,y) \, dA \, . \tag{5.25}$$

The moment of a small element R_k, of the solid, about the y–axis is approximately $m_k \, \xi_k$, see Figure 5.12 where we approximate the small region as a point mass. Then the mass moment of the whole plate about the y–axis is approximated by summing over k ,

$$M_y \approx \sum_k m_k \, \xi_k \approx \sum_k \xi_k \, \delta(\xi_k, \eta_k) \, \Delta x_k \, \Delta y_k \, ,$$

leading to the integral

$$M_y = \iint_S x \, \delta(x,y) \, dA \, . \tag{5.26}$$

Similarly, the moment of R_k about the x–axis is

$$M_x = \iint_S y \, \delta(x,y) \, dx \, dy \, . \tag{5.27}$$

Finally, the coordinates $(\overline{x}, \overline{y})$ of the mass centroid are given by

$$\overline{x} = \frac{M_y}{M} = \frac{\iint_S x \, \delta(x,y) \, dx \, dy}{\iint_S \delta(x,y) \, dx \, dy} \quad \text{and} \quad \overline{y} = \frac{M_x}{M} = \frac{\iint_S y \, \delta(x,y) \, dx \, dy}{\iint_S \delta(x,y) \, dx \, dy} \, . \tag{5.28}$$

Mathematica session 5.7. Area and mass centroids *The coordinates $(\overline{x}, \overline{y})$ of (5.28) are computed simultaneously using the* MATHEMATICA *function*

twoCentroid[default_, S_, T_]:=

Module[{num, denom},

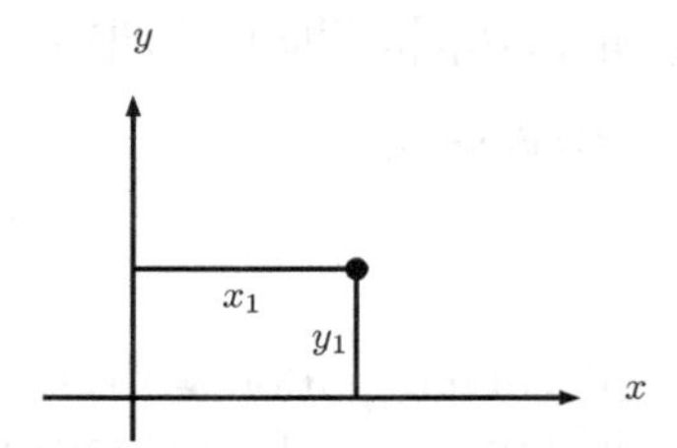

FIGURE 5.12: A mass and the moments $m \cdot x_1$ and $m \cdot y_1$.

```
num = Integrate[default * {x, y}, {S[[1]], S[[2]], S[[3]]}, {T[[1]], T[[2]], T[[3]]}];
denom = Integrate[1, {S[[1]], S[[2]], S[[3]]}, {T[[1]], T[[2]], T[[3]]}]; num/denom]
```

For three-dimensional centroids we use

```
averageValue[default_, R_, S_, T_]:=
Module[{num, denom},
num = Integrate[default, {R[[1]], R[[2]], R[[3]]}, {S[[1]], S[[2]], S[[3]]},
{T[[1]], T[[2]], T[[3]]}];
denom = Integrate[1, {R[[1]], R[[2]], R[[3]]}, {S[[1]], S[[2]], S[[3]]},
{T[[1]], T[[2]], T[[3]]}]; num/denom]

averageValue[x * y * z, {z, 0, 2}, {y, 0, 2}, {x, 0, 2}]
```

1

```
averageValue[x * Sin[y], {z, 0, 1}, {y, 0, Pi}, {x, 0, 1}]
```

$\frac{1}{\pi}$

```
threeCentroid[default_, R_, S_, T_]:=
Module[{num, denom},
num = Integrate[default * {x, y, z}, {R[[1]], R[[2]], R[[3]]}, {S[[1]], S[[2]], S[[3]]},
{T[[1]], T[[2]], T[[3]]}];
```

***denom* = *Integrate*[1, {*R*[[1]], *R*[[2]], *R*[[3]]}, {*S*[[1]], *S*[[2]], *S*[[3]]},**

{*T*[[1]], *T*[[2]], *T*[[3]]}]; *num*/*denom*]

♦

• `S2, S3` are its limits of integration, • `T` is the **outer integration variable** (the remaining variable); `T2, T3` its limits of integration, • `default1` is the default parameter, originally assigned the value 1.

The new feature here is the expression $[x, y]$, a vector with components x and y .

If we use the function in the form **threeCentroid**[$default, R, S, T$], we assume the density $\delta(x, y) \equiv 1$, so that (5.28) gives the area centroid. For a general density $\delta(x, y)$, we call the function as **threeCentroid** [$default, R, S, T$], with δ replacing `default`. Then the numerator **threeCentroid**[$default, R, S, T$] computes simultaneously the two mass moments (the numerators in (5.28)),

$$\iint x\,\delta(x, y)\,dx\,dy \qquad \text{and} \qquad \iint y\,\delta(x, y)\,dx\,dy\ ,$$

and the denominator is the mass in question, which is the denominator in (5.28).

To compute the centroid of the semi-sphere of unit density we input

Mathematica session 5.8.

***threeCentroid*[1, {*x*, −2, 2}, {*y*, −*Sqrt*[4 − *x*^2], *Sqrt*[4 − *x*^2]}, {*z*, 0, 4 − *x*^2− *y*^2}]**

We next compute the centroid with a density $\delta(x, y, z) = xyz$

$\{0, 0, \frac{4}{3}\}$

***threeCentroid*[*x* ∗ *y* ∗ *z*, {*x*, −2, 2}, {*y*, −*Sqrt*[4 − *x*^2], *Sqrt*[4 − *x*^2]},**

{*z*, 0, 4 − *x*^2 − *y*^2}]

$\{0, 0, 0\}$

***threeCentroid*[1 + *x*^2 + *y*^2 + *z*^2, {*x*, −2, 2}, {*y*, −*Sqrt*[4 − *x*^2],**

{*Sqrt*[4 − *x*^2]}, *z*, 0, 4 − *x*^2 − *y*^2}]

$\{0, 0, \frac{136}{15}\}$

```
Integrate[1, {x, -2, 2}, {y, -3 * Sqrt[1 - x^2/4], 3 * Sqrt[1 - x^2/4]},
{z, 0, 4 * Sqrt[1 - x^2/4 - y^2/9]}]
```

16π

♦

A mass m at a the point (x, y) makes a **moment** about the y–axis of

$$M_y := m\,x\,, \tag{5.29}$$

the product of the mass and its lever. The **second moment** is defined as the product of the mass and the **square** of its lever. Thus a mass m at (x, y) has a second moment of $m\,x^2$ about the y–axis. The second moments are called **moments of inertia**, and denoted by I . In particular, I_y and I_x are the moments of inertia about the y–axis and x–axis respectively.

For the plate S in Figure 5.11 we approximate I_y as a Riemann sum

$$I_y \approx \sum_k m_k\,\xi_k^2 \approx \sum_k \xi_k^2\,\delta(\xi_k, \eta_k)\,\Delta x_k\,\Delta y_k\,,$$

leading to the integral

$$I_y = \iint_S x^2\,\delta(x, y)\,dA\,. \tag{5.30a}$$

$$\text{Similarly} \quad I_x = \iint_S y^2\,\delta(x, y)\,dA\,. \tag{5.30b}$$

Exercises

Exercise 5.21. *Find the centroid of a lobe of the lemniscate* $r^2 = a^2 \cos 2\theta$.

Exercise 5.22. *Compute the x and y centroids for the following planar objects. Check your results using the function* `twoCentroid`.

(a) *The region in the first quadrant of the ellipse* $x^2 + y^2 = 1$

(b) *The region in the first quadrant of the ellipse* $\frac{x^2}{a^2} + \frac{y^2}{b^2} = 1$

(c) *The triangle with the vertices* $(0, 0)$, $(2, 0)$, *and* $(1, 1)$

(d) *The L-shaped region with vertices* $(0, 0)$, $(a, 0)$, $(a, 1)$, $(1, 1)$, $(1, b)$, *and* $(0, b)$, *where a and b are greater than* 1.

(e) *The region bounded by the circles* $x^2 + y^2 = a^2$, $(x-1)^2 + y^2 = a^2$, *where* $1 < a$

(f) *The region bounded by the x axis and one period of the cycloid* $x = a(t - \sin t$, $y = a(1 - \cos t)$

(g) *The T-shaped region with vertices* $(-1, 0)$, $(1, 0)$, $(1, h)$, (w, h), $(w, h + 1)$, $(-w, h + 1)$, $(-w, h)$, $(-1, h)$

Exercise 5.23. *Compute the centers of mass of the following composite regions having the given mass densities.*

(a) *The upper semicircle of* $x^2 + y^2 = a^2$ *with constant mass density* δ_1 *joined to the lower semicircle with mass density* δ_2

(b) *The circles* $x^2 + y^2 = a^2$ *and* $(x - 2a)^2 + y^2 = a^2$, *where one circle has mass density* δ_1 *and the other* δ_2

(c) *The region made up by joining the upper part of the ellipse* $\frac{x^2}{a^2} + \frac{y^2}{b^2} = 1$ *with the lower part of the circle* $x^2 + y^2 = a^2$
Assume that the mass density of the ellipse is δ_1 *and that of the circle is* δ_2

(d) *The region made up by joining the square with vertices* $(-1, 0)$, $(1, 0)$, $(1, 2)$, $(-1, 2)$ *and mass density* $\delta_1 = |1-x|$, *and the triangle with vertices* $(1, 0)$, $(1, 2)$, $(3, 0)$ *and mass density* $\delta_2 = y$

(e) *Prove: If a region* $\mathcal{D}$ *is subdivided into two subregions* $\mathcal{D}_1$, *and* $\mathcal{D}_2$ *by an arc* $\mathcal{C}$, *and if the subregions* $\mathcal{D}_i$ *have densities* δ_i, *then*

$$\overline{x} = \frac{\overline{x}_1 M_1 + \overline{x}_2 M_2}{M_1 + M_2}, \quad \text{and} \quad \overline{y} = \frac{\overline{y}_1 M_1 + \overline{y}_2 M_2}{M_1 + M_2}, \tag{5.31}$$

where M_i *is the mass of subregion* $\mathcal{D}_i$ *and* $\overline{x}_i$, *and* $\overline{y}_i$ *are the subregion* x *and* y *centroids.*

Exercise 5.24. *Test formulæ(5.31) using the* MATHEMATICA *function* `twoCentroid`. *If* MATHEMATICA *is unable to do the computation symbolically use* `N` *to evaluate the integrals.*

(a) The upper semicircle of $x^2 + y^2 = a^2$ *with mass density* $\delta_1 = e^x$ *joined to the lower semicircle with mass density* $\delta_2 = e^y$

(b) The circles $x^2 + y^2 = a^2$ *and* $(x - 2a)^2 + y^2 = a^2$, *where one circle has mass density* $\delta_1 = \sqrt{a^2 - x^2 - y^2}$ *and the other* $\delta_2 = \sqrt{a^2 - (x - 2a)^2 - y^2}$

(c) The region made up by joining the upper part of the ellipse $\frac{x^2}{a^2} + \frac{y^2}{b^2} = 1$ *with the lower part of the circle* $x^2 + y^2 = a^2$. *Assume that the mass density of the ellipse is* $\delta_1 = \sqrt{\frac{x^2}{a^2} + \frac{y^2}{b^2}}$ *and that of the circle is* $\delta_2 = \sqrt{x^2 + y^2}$

5.6 Areas of surfaces in $\mathbb{R}^3$

Consider a curved surface in $\mathbb{R}^3$, given by the equation

$$z = z(x, y) . \tag{5.32}$$

with at most one value $z(x, y)$ for any point (x, y) in the plane[5]. It is required to compute the area of a region $\mathcal{S}$ lying on the surface, see Figure 5.13. Let $\mathcal{A}$ denote the projection of the region $\mathcal{S}$ in the xy–plane. The area of $\mathcal{S}$ is computed by integration over the projected region $\mathcal{A}$, which is easier, since $\mathcal{A}$ is in a plane, and $\mathcal{S}$ is curved. We use an internal partition $\mathcal{P}_A$ (see Defini-

[5]Equivalently, parallels to the z–axis cut the surface in at most one point.

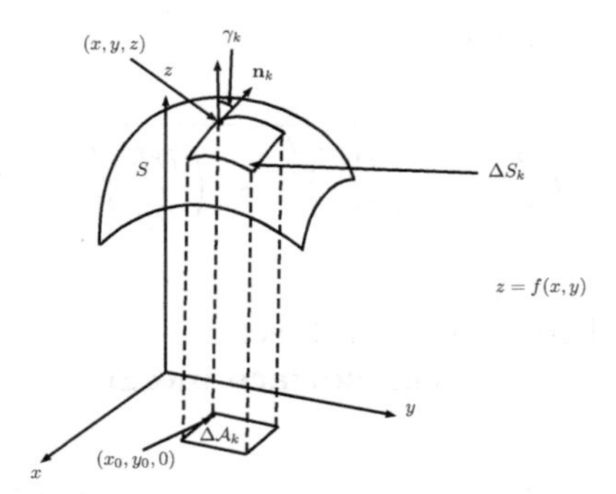

FIGURE 5.13: Computation of surface area.

tion 5.4) of the projected region $\mathcal{A}$, which consists of small rectangles ΔA_k lying entirely within $\mathcal{A}$. These rectangles induce a partition of the region $\mathcal{S}$ into "curvilinear quadrilaterals" ΔS_k , lying over ΔA_k . At some point of ΔS_k we erect the normal vector $\mathbf{n}_k$ to the surface. It makes an acute angle γ_k with the z–axis, see Figure 5.13. The area ΔS_k of the curvilinear quadrilateral is related to the area ΔA_k of its projection (in the xy–plane) as follows

$$\Delta S_k \approx \Delta A_k \sec \gamma_k , \tag{5.33}$$

see Figure 5.13. The area S of the region $\mathcal{S}$ is therefore approximated by the Riemann sum

$$S \approx \sum_k \Delta A_k \sec \gamma_k$$

and in the limit, as $\|\mathcal{P}_A\| \to 0$, the integral

$$S = \iint_{\mathcal{A}} \sec \gamma \, dA = \iint_{\mathcal{A}} \sec \gamma(x,y) \, dx \, dy , \tag{5.34}$$

where $\gamma = \gamma(x,y)$ is the acute angle between the z–axis and the normal $\mathbf{n}$ to the surface at the point $(x, y, z(x,y))$. What makes (5.34) useful is that $\sec \gamma$ can be computed using the normal to the surface

$$\mathbf{n} = -\frac{\partial z}{\partial x}\mathbf{i} - \frac{\partial z}{\partial y}\mathbf{j} + \mathbf{k} . \tag{5.35}$$

Therefore

$$\cos \gamma = \frac{1}{\sqrt{\left(\frac{\partial z}{\partial x}\right)^2 + \left(\frac{\partial z}{\partial x}\right)^2 + 1}} , \tag{5.36}$$

and formula (5.34) becomes

$$S = \iint\limits_{\mathcal{A}} \sqrt{\left(\frac{\partial z}{\partial x}\right)^2 + \left(\frac{\partial f}{\partial y}\right)^2 + 1} \quad dx\, dy \ . \tag{5.37}$$

Sometimes we need to integrate a function $f(x, y, z)$ over a region $\mathcal{S}$ of the curved surface $z = z(x, y)$. This **surface integral** is defined as

$$\iint\limits_{\mathcal{S}} f(x, y, z)\, dS = \iint\limits_{\mathcal{A}} f(x, y, z(x, y)) \sqrt{\left(\frac{\partial z}{\partial x}\right)^2 + \left(\frac{\partial z}{\partial y}\right)^2 + 1} \quad dx\, dy \ , \tag{5.38}$$

where the domain of integration $\mathcal{A}$ is the projection of $\mathcal{S}$ on the xy–plane. The area integral (5.37) is a special case, with the integrand $f \equiv 1$.

Mathematica session 5.9. *The following* MATHEMATICA *procedure computes the surface integrals* ***surfaceIntegral[default_, z_, S_, T_]:=***

Module[{sa}, sa = Simplify[Sqrt[D[z, s]^2 + D[z, t]^2 + 1]];

Integrate[default * sa, {S[[1]], S[[2]], S[[3]]}, {T[[1]], T[[2]], T[[3]]}]]

surfaceIntegral[1, 5 − x^2 − y^2, {x, −2, 2

}, {y, −Sqrt[4 − x^2], Sqrt[4 − x^2]}]

4π

Integrate[x^2 + y^2, {x, −2, 2}, {y, −Sqrt[4 − x^2], Sqrt[4 − x^2]}]

8π

♦

Integration is carried out iteratively, first $T[[1]]$ (which is either x or y) and then $S[[1]]$ (the remaining variable). The area formula (5.37) is obtained for `default1 := 1`. Replacing `default1` by a function $f(x, y, z(x, y))$ gives the surface integral (5.38). Example: integrate $f(x, y) = 2x + 3y$ on that part of the paraboloid $z = 5 - x^2 - y^2$ lying above the plane $z = 1$. We input

```
surfaceIntegral[2 * x + 3 * y, 5 - x^ - y^2, {x, -2, 2}, {y, -Sqrt[1 - x^2],
Sqrt[1 - x^2]}]
```

0

Mathematica session 5.10. *We now turn our attention to finding the centroid of a* **shell**, *i.e. of a thin suface. First we must load the function*

```
surfaceCentroid[default_, z_, S_, T_]:=
Module[{sa, num, denom}, sa = Simplify[Sqrt[D[z, s]^2 + D[z, t]^2 + 1]];
num = Integrate[default * sa * {x, y, z}, {S[[1]], S[[2]], S[[3]]},
{T[[1]], T[[2]], T[[3]]}];
denom = Integrate[default * sa, {S[[1]], S[[2]], S[[3]]}, {T[[1]], T[[2]], T[[3]]}];
num/denom]
surfaceCentroid[1, 5 - x^2 - y^2, {x, -2, 2}, {y, -Sqrt[4 - x^2], Sqrt[4 - x^2]}]
```

$\{0, 0, 3\}$

```
surfaceCentroid[x^2 + y^2, 5 - x^2 - y^2, {x, -2, 2}, {y, -Sqrt[4 - x^2],
Sqrt[4 - x^2]}]
```

$\{0, 0, \frac{7}{3}\}$

```
surfaceCentroid[x^2 + y^2, 1 - x - y, {x, -2, 2}, {y, -Sqrt[4 - x^2],
Sqrt[4 - x^2]}]
```

$\{0, 0, 1\}$

```
surfaceCentroid[Abs[x * y], 5 - x^2 - y^2, {x, -1, 1}, {y, -1, 1}]
```

$\{0, 0, 4\}$

```
surfaceCentroid[Abs[x * y], 1 + x + y, {x, -1, 1}, {y, -1, 1}]
```

{0, 0, 1}

***surfaceCentroid*[*Abs*[*x* + *y*], 1 + *x* + *y*, {*x*, −1, 1}, {*y*, −1, 1}]**

{0, 0, 1}

♦

Consider now the case where the curved surface is given in the implicit form

$$F(x, y, z) = 0.$$

Here the normal $\mathbf{n}$ to the surface, at any point (x, y, z), is given by

$$\mathbf{n} = \frac{\partial F}{\partial x}\mathbf{i} + \frac{\partial F}{\partial y}\mathbf{j} + \frac{\partial F}{\partial z}\mathbf{k}.$$

Consequently, the cosine of γ is given by

$$\cos\gamma = \frac{\frac{\partial F}{\partial z}}{\sqrt{\left(\frac{\partial F}{\partial x}\right)^2 + \left(\frac{\partial F}{\partial y}\right)^2 + \left(\frac{\partial F}{\partial z}\right)^2}}, \tag{5.39}$$

and the area formula (5.34) becomes

$$S = \iint_{\mathcal{A}} \frac{\sqrt{\left(\frac{\partial F}{\partial x}\right)^2 + \left(\frac{\partial F}{\partial y}\right)^2 + \left(\frac{\partial F}{\partial z}\right)^2}}{\frac{\partial F}{\partial z}}\, dx\, dy. \tag{5.40}$$

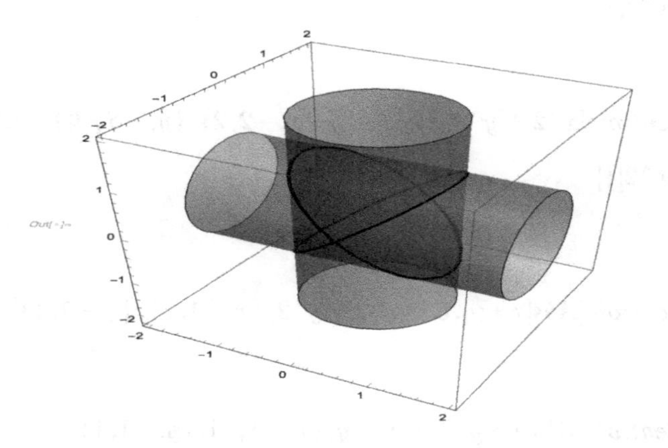

FIGURE 5.14: Illustration of example 5.25.

Example 5.25 (Sphere intersected by a cylinder). Consider the circular cylinder, with radius $\frac{a}{2}$, parallel to the z–axis,

$$\left(x - \frac{a}{2}\right)^2 + y^2 \;=\; \frac{a^2}{4}\;,$$

and the sphere of radius a

$$x^2 + y^2 + z^2 \;=\; a^2\;.$$

It is required to compute the area which the cylinder cuts from the sphere. To simplify our calculations we restrict attention to the upper semi-sphere and the first quadrant, where one fourth of the area (the shaded area $\mathcal{S}$ in Figure 5.14 is situated. The domain of integration is the projection on the xy–plane, which is the shaded semi-disk $\mathcal{A}$ in Figure 5.14. Since the sphere is given implicitly as

$$F(x, y, z) \;:=\; x^2 + y^2 + z^2 - a^2 \;=\; 0\;,$$

we compute the partial derivatives

$$F_x \;=\; 2\,x\;,\quad F_y \;=\; 2\,y\;,\quad \text{and}\quad F_z \;=\; 2z\;,$$

and from the implicit representation (5.39),

$$\sec\gamma \;=\; \frac{2\,\sqrt{x^2 + y^2 + z^2}}{2\,z} \;=\; \frac{a}{z}\;.$$

It is convenient to use here polar coordinates, so we write

$$\sec\gamma \;=\; \frac{a}{\sqrt{a^2 - r^2}}\;,\quad \text{where}\quad r \;=\; \sqrt{x^2 + y^2}\;.$$

The domain of integration $\mathcal{A}$ of Figure 5.25 is

$$\begin{aligned} \mathcal{A} \;&:=\; \{(x, y) :\; \left(x - \frac{a}{2}\right)^2 + y^2 \;\le\; \frac{a^2}{4}\;,\; x \ge 0\;,\; y \ge 0\} \\ &=\; \{(r, \theta) \colon\; 0 \le r \le a\,\cos\theta\;,\quad 0 \le \theta \le \frac{\pi}{2}\},\ \text{in polar coordinates (verify!)}, \end{aligned}$$

and the area is

$$4\,S \;=\; 4\,\iint\limits_{\mathcal{A}} \sec\gamma\, dA \;=\; 4\int_0^{\frac{\pi}{2}} d\theta \int_0^{a\cos\theta} \frac{ar}{\sqrt{a^2 - r^2}}\, dr \;=$$

$$-4\int_0^{\frac{\pi}{2}} d\theta\, a\,\sqrt{a^2 - r^2}\Big|_0^{a\cos\theta} = 4\,a^2 \int_0^{\frac{\pi}{2}} (1 - \sin\theta)\, d\theta \;=\; 2\,a^2\,(\pi - 2)\;.$$

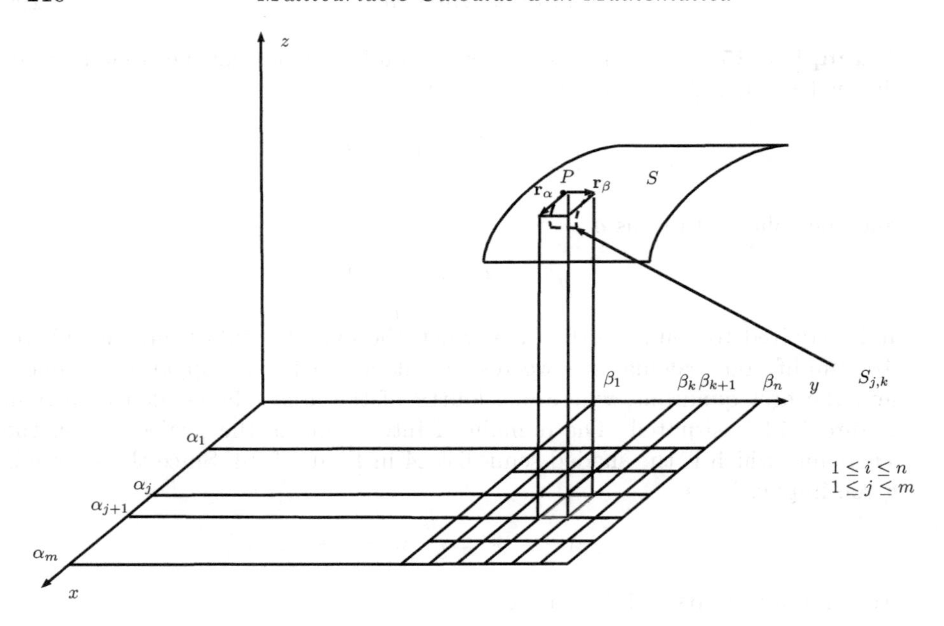

FIGURE 5.15: Parametrizing a surface.

Surfaces were described above in explicit form[6] or in implicit form, i.e. $F(x, y, z) = 0$. Sometimes it is useful to represent a surface in terms of parameters. Since surfaces are two dimensional, we need two parameters to represent a surface. Let α and β denote the parameters, with values in the **parameter domain**, which for convenience is taken as the unit square $[0, 1] \times [0, 1]$, see Figure 5.16. The position vector on the surface $\mathcal{S}$ is represented, in parametrized form, as

$$\mathbf{r}(\alpha, \beta) \;=\; x(\alpha, \beta)\,\mathbf{i} + y(\alpha, \beta)\,\mathbf{j} + z(\alpha, \beta)\,\mathbf{k}\,. \tag{5.41}$$

In what follows we assume that the position vector $\mathbf{r}$ and its first partial derivatives

$$\begin{aligned}
\frac{\partial \mathbf{r}(\alpha, \beta)}{\partial \alpha} &= \frac{\partial x(\alpha, \beta)}{\partial \alpha}\mathbf{i} + \frac{\partial y(\alpha, \beta)}{\partial \alpha}\mathbf{j} + \frac{\partial z(\alpha, \beta)}{\partial \alpha}\mathbf{k} \\
\frac{\partial \mathbf{r}(\alpha, \beta)}{\partial \beta} &= \frac{\partial x(\alpha, \beta)}{\partial \beta}\mathbf{i} + \frac{\partial y(\alpha, \beta)}{\partial \beta}\mathbf{j} + \frac{\partial z(\alpha, \beta)}{\partial \beta}\mathbf{k}
\end{aligned}$$

are continuous in the parameter domain. To define a surface integral, we first partition the unit square into $n \times m$ rectangles

$$\{(\alpha, \beta) \,:\, \alpha_j \;\le\; \alpha \;\le\; \alpha_{j+1}\,, \quad \beta_k \;\le\; \beta \;\le\; \beta_{k+1}\}$$

[6]Sometimes referred to as Monge gauge $z = f(x, y)$

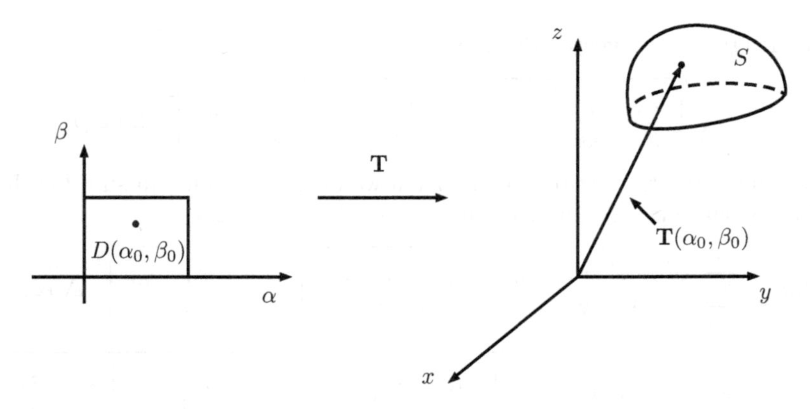

FIGURE 5.16: Projecting elements in the $\alpha - \beta$ plane onto surface elements in $\mathbb{R}^3$.

as illustrated in Figure 5.15. The $(j,k)^{th}$ rectangle is mapped into a curvilinear quadrilateral $\mathcal{S}_{j,k}$ on the surface $\mathcal{S}$, bounded by the curves

$$\mathbf{r}(\alpha_j,\beta)\ ,\ \mathbf{r}(\alpha_{j+1},\beta)\ ,\ \mathbf{r}(\alpha,\beta_k)\ ,\quad \text{and}\quad \mathbf{r}(\alpha,\beta_{k+1})\ .$$

We next approximate the area of the curvilinear quadrilateral $\mathcal{S}_{j,k}$. Its vertex $P\ :=\ \mathbf{r}(\alpha_j,\beta_k)$ is at the intersection of the curves

$$\{\mathbf{r}(\alpha_j,\beta)\ :\ 0\le\beta\le 1\}\quad\text{and}\quad\{\mathbf{r}(\alpha,\beta_k)\ :\ 0\le\alpha\le 1\}\ .$$

The two tangents of these curves at the point P, see Figure 5.15, are partial derivatives of $\mathbf{r}$: • the tangent vector to $\mathbf{r}(\alpha,\beta_k)$ at P is $\frac{\partial\mathbf{r}}{\partial\alpha}(\alpha_j,\beta_k)$, denoted $\mathbf{r}_\alpha(\alpha_j,\beta_k)$, • the tangent vector to $\mathbf{r}(\alpha_j,\beta)$ at P is $\frac{\partial\mathbf{r}}{\partial\beta}(\alpha_j,\beta_k)$, denoted $\mathbf{r}_\beta(\alpha_j,\beta_k)$.

These two vectors determine the tangent plane to $\mathcal{S}$ at $\mathbf{r}(\alpha_j,\beta_k)$. Since the tangent plane is a good approximation of the surface near to the point of contact, the parallelogram with adjacent sides

$$\mathbf{r}_\alpha(\alpha_j,\beta_k)\,\Delta\alpha\quad\text{and}\quad\mathbf{r}_\beta(\alpha_j,\beta_k)\,\Delta\beta$$

is a good approximation to the curvilinear quadrilateral $\mathcal{S}_{j,k}$, provided $\Delta\alpha$ and $\Delta\beta$ are sufficiently small.

We recall that the area of the parallelogram is given by length of the normal vector[7]

$$(\mathbf{r}_\alpha(\alpha_j,\beta_k)\times\mathbf{r}_\beta(\alpha_j,\beta_k))\ \Delta\alpha\,\Delta\beta\ = \tag{5.42}$$

$$\left(\begin{vmatrix} x_\alpha & z_\alpha \\ y_\beta & z_\beta \end{vmatrix}\ \text{ß}\ +\ \begin{vmatrix} z_\alpha & x_\alpha \\ z_\beta & x_\beta \end{vmatrix}\ \text{æ}\ +\ \begin{vmatrix} x_\alpha & y_\alpha \\ x_\beta & y_\beta \end{vmatrix}\ \mathbf{k}\right)\ \Delta\alpha\,\Delta\beta\ . \tag{5.43}$$

[7]We assume that the normal vector is nowhere zero, i.e. at each point (α,β) in the

The length of the vector (5.43) is

$$\sqrt{\begin{vmatrix} x_\alpha & z_\alpha \\ y_\beta & z_\beta \end{vmatrix}^2 + \begin{vmatrix} z_\alpha & x_\alpha \\ z_\beta & x_\beta \end{vmatrix}^2 + \begin{vmatrix} x_\alpha & y_\alpha \\ x_\beta & y_\beta \end{vmatrix}^2} \; \Delta\alpha \, \Delta\beta \, .$$

Summing over the entire $\alpha\beta$–partition we obtain a Riemann sum for the area of the surface $\mathcal{S}$. Taking the limit as $n, m \to \infty$ we get

$$\lim_{n,m\to\infty} \sum_{i=1}^{n} \sum_{j=1}^{m} \sqrt{\begin{vmatrix} x_\alpha & z_\alpha \\ y_\beta & z_\beta \end{vmatrix}^2 + \begin{vmatrix} z_\alpha & x_\alpha \\ z_\beta & x_\beta \end{vmatrix}^2 + \begin{vmatrix} x_\alpha & y_\alpha \\ x_\beta & y_\beta \end{vmatrix}^2} \; \Delta\alpha\Delta\beta =$$

$$= \int_0^1 d\alpha \int_0^1 d\beta \sqrt{\begin{vmatrix} x_\alpha & z_\alpha \\ y_\beta & z_\beta \end{vmatrix}^2 + \begin{vmatrix} z_\alpha & x_\alpha \\ z_\beta & x_\beta \end{vmatrix}^2 + \begin{vmatrix} x_\alpha & y_\alpha \\ x_\beta & y_\beta \end{vmatrix}^2} \, . \tag{5.44}$$

The term

$$dS := \sqrt{\begin{vmatrix} x_\alpha & z_\alpha \\ y_\beta & z_\beta \end{vmatrix}^2 + \begin{vmatrix} z_\alpha & x_\alpha \\ z_\beta & x_\beta \end{vmatrix}^2 + \begin{vmatrix} x_\alpha & y_\alpha \\ x_\beta & y_\beta \end{vmatrix}^2} \; d\alpha \, d\beta \tag{5.45}$$

is called the **surface differential**. The following three examples illustrate the computation of the surface differential for the most popular coordinate systems.

Example 5.26 (Cartesian coordinates)**.** We take the Cartesian coordinates (x, y) as the parameters (α, β) . Here the surface $\mathcal{S}$ is given as

$$z = z(x, y) \, . \tag{5.32}$$

Taking $x = \alpha$ and $y = \beta$ we calculate the surface differential in (5.45)

$$\begin{aligned} dS &= \sqrt{\begin{vmatrix} 0 & z_x \\ 1 & z_y \end{vmatrix}^2 + \begin{vmatrix} z_x & 1 \\ z_y & 0 \end{vmatrix}^2 + \begin{vmatrix} 1 & 0 \\ 0 & 1 \end{vmatrix}^2} \; dx \, dy \\ &= \sqrt{z_x^2 + z_y^2 + 1} \; dx \, dy \, , \end{aligned} \tag{5.46}$$

and the integral for surface area is then, in agreement with (5.37),

$$S = \iint\limits_{\mathcal{A}} \sqrt{\left(\frac{\partial z}{\partial x}\right)^2 + \left(\frac{\partial z}{\partial y}\right)^2 + 1} \;\; dx \, dy \, .$$

Here $\mathcal{A}$ is the projection of the surface $\mathcal{S}$ on the xy–plane.

---

parameter domain, at least one of the 2×2 determinants of the matrix

$$\begin{pmatrix} \dfrac{\partial x(\alpha,\beta)}{\partial \alpha} & \dfrac{\partial y(\alpha,\beta)}{\partial \alpha} & \dfrac{\partial z(\alpha,\beta)}{\partial \alpha} \\ \dfrac{\partial x(\alpha,\beta)}{\partial \beta} & \dfrac{\partial y(\alpha,\beta)}{\partial \beta} & \dfrac{\partial x(\alpha,\beta)}{\partial \beta} \end{pmatrix}$$

is different from zero.

Example 5.27 (Spherical coordinates)**.** Consider a region $\mathcal{S}$ on a sphere, centered at the origin. We compute the surface differential, using spherical coordinates,

$$x := \rho \sin\varphi \cos\theta \ , \quad y := \rho \sin\varphi \sin\theta \ , \quad z := \rho \cos\varphi \ ,$$

with φ , θ as parameters (the coordinate ρ is constant, since $\mathcal{S}$ is on a sphere). Substituting in (5.45) we obtain, after simplification,

$$\sqrt{\begin{vmatrix} y_\varphi & z_\varphi \\ y_\theta & z_\theta \end{vmatrix}^2 + \begin{vmatrix} z_\varphi & x_\varphi \\ z_\theta & x_\theta \end{vmatrix}^2 + \begin{vmatrix} x_\varphi & y_\varphi \\ x_\theta & y_\theta \end{vmatrix}^2}$$

$$= \sqrt{\begin{vmatrix} r\cos\varphi\sin\theta & -\rho\sin\varphi \\ \rho\sin\varphi\cos\theta & 0 \end{vmatrix}^2 + \begin{vmatrix} -\rho\sin\varphi & \rho\cos\varphi\cos\theta \\ 0 & -\rho\sin\varphi\sin\theta \end{vmatrix}^2 + \begin{vmatrix} \rho\cos\varphi\cos\theta & \rho\cos\varphi\sin\theta \\ -\rho\sin\varphi\sin\theta & \rho\sin\varphi\cos\theta \end{vmatrix}^2} =$$

$$\rho^2 \sin\varphi \ .$$

Therefore, the spherical surface differential is

$$dS = \rho^2 \sin\varphi \, d\varphi \, d\theta \ . \tag{5.47}$$

Example 5.28 (Cylindrical coordinates)**.** Consider a surface parametrized by polar coordinates

$$z = z(r, \theta)$$

and recall the formulæ for cylindrical coordinates,

$$x := r \cos\theta \ , \quad y := r \sin\theta \ , \quad z := z \ . \tag{5.71}$$

The cylindrical surface differential is then

$$dS = \sqrt{\begin{vmatrix} y_r & z_r \\ y_\theta & z_\theta \end{vmatrix}^2 + \begin{vmatrix} z_r & x_r \\ z_\theta & x_\theta \end{vmatrix}^2 + \begin{vmatrix} x_r & y_r \\ x_\theta & y_\theta \end{vmatrix}^2} \, dr \, d\theta =$$

$$= \sqrt{\begin{vmatrix} \sin\theta & z_r \\ r\cos\theta & z_\theta \end{vmatrix}^2 + \begin{vmatrix} z_r & \cos\theta \\ z_\theta & -r\sin\theta \end{vmatrix}^2 + \begin{vmatrix} \cos\theta & \sin\theta \\ -r\sin\theta & r\cos\theta \end{vmatrix}^2} \, dr \, d\theta$$

which after squaring the determinants, and quite fortunate cancellations, becomes

$$dS = \sqrt{r^2 + r^2 \left(\frac{\partial z}{\partial r}\right)^2 + \left(\frac{\partial z}{\partial \theta}\right)^2} \, dr \, d\theta \ . \tag{5.48}$$

The surface–area is then given by

$$S = \iint_{\mathcal{A}} \sqrt{r^2 + r^2 \left(\frac{\partial z}{\partial r}\right)^2 + \left(\frac{\partial z}{\partial \theta}\right)^2} \, dr \, d\theta \ . \tag{5.49}$$

Mathematica session 5.11. *We implement the area integral (5.49) by the following*

***polarSurfaceIntegral*[*default*_ , *z*_ , *T*_ , *R*_]:=**

```
Module[{sa}, sa = Sqrt[Simplify[R[[1]]^2 + R[[1]]^2 * D[z, R[[1]]]^2
+D[z, T[[1]]]^2]];
Integrate[default * sa, {T[[1]], T[[2]], T[[3]]}, {R[[1]], R[[2]], R[[3]]} ]]
```

The limits of integration for the variables r and θ are listed in the usual way in this function. [8]

```
polarSurfaceIntegral[1, r^2 * Cos[theta] * Sin[theta], {theta, π/2, π},
{r, 0, -2 * Cos[theta]}]
```

$\frac{1}{18}\left(-3\pi + 2\sqrt{5}\left(12 EllipticE\left[\frac{4}{5}\right] - EllipticK\left[\frac{4}{5}\right]\right)\right)$

```
N[%]
```

2.42917

♦

Example 5.29. Consider the spherical cap on the sphere $x^2 + y^2 + z^2 = 4$, lying above the horizontal plane $z = \sqrt{2}$. The upper hemisphere is expressed, in cylindrical coordinates, as

$$z = \sqrt{4 - r^2}\,.$$

Substituting

$$\frac{\partial z}{\partial r} = \frac{-r}{\sqrt{4-r^2}}\,, \quad \frac{\partial z}{\partial \theta} = 0$$

in (5.49), we get the area of the spherical cap

$$S = \int_0^{\sqrt{2}} \int_0^{2\pi} \frac{2r}{\sqrt{4-r^2}}\, dr\, d\theta = 4\pi \int_0^{\sqrt{2}} \frac{r}{\sqrt{4-r^2}}\, dr = 4\pi\left(2 - \sqrt{2}\right).$$

Since we are dealing here with an area on a sphere, it is more natural to use spherical coordinates (ρ, φ, θ). The sphere is given by $\rho = 2$ and the spherical cap by

$$0 \le \varphi \le \frac{\pi}{4}\,, \quad 0 \le \theta \le 2\pi\,.$$

The surface differential is, by (5.47), $dS = 2^2 \sin\varphi\, d\varphi\, d\theta$. The area of the cap is therefore

$$S = 4 \int_0^{2\pi} \left(\int_0^{\pi/4} d\varphi \, \sin\varphi \right) d\theta = 4\pi\left(2 - \sqrt{2}\right).$$

△

[8]In MATHEMATICA to input a greek letter, say α, rather typing in its name we use ESC + ALPHA ESC .

Consider now the integral

$$\iint_{\mathcal{S}} f(\mathbf{r})\, dS \tag{5.50}$$

of a function f defined on the surface $\mathcal{S}$ in the (x, y, z)–space. We assume that lines parallel to the z–axis intersect $\mathcal{S}$ in at most one point, which we abbreviate by saying that the surface $\mathcal{S}$ is z–**simple**. Let the surface be parametrized by

$$\mathbf{r} \;:=\; \mathbf{r}(\alpha, \beta)\ , \quad \text{with parameters } \alpha_1 \le \alpha \le \alpha_2\ , \quad \beta_1 \le \beta \le \beta_2\ .$$

Then, in analogy with (5.44), the integral (5.50) becomes,

$$\int_{\alpha_1}^{\alpha_2} d\alpha \int_{\beta_1}^{\beta_2} d\beta\, f(\mathbf{r}(\alpha, \beta)) \sqrt{\begin{vmatrix} x_\alpha & z_\alpha \\ y_\beta & z_\beta \end{vmatrix}^2 + \begin{vmatrix} z_\alpha & x_\alpha \\ z_\beta & x_\beta \end{vmatrix}^2 + \begin{vmatrix} x_\alpha & y_\alpha \\ x_\beta & y_\beta \end{vmatrix}^2}\ . \tag{5.51}$$

For example, if the surface $\mathcal{S}$ is parametrized by polar coordinates

$$z \;=\; z(r, \theta)\ , \quad r_1 \le r \le r_2\ ,\ \theta_1 \le \theta \le \theta_2\ ,$$

then, using (5.79), the integral of a function $f(r, \theta, z)$ over the given surface is

$$\iint_{\mathcal{S}} f(r, \theta, z)\, dS = \int_{\theta_1}^{\theta_2} \int_{r_1}^{r_2} f(r, \theta, z(r, \theta)) \sqrt{r^2 \;+\; r^2 \left(\frac{\partial z}{\partial r}\right)^2 \;+\; \left(\frac{\partial z}{\partial \theta}\right)^2}\, dr\, d\theta\ . \tag{5.52}$$

If the surface $\mathcal{S}$ is not z–simple, we can try to cut $\mathcal{S}$ into regions which are z–simple. For example, the unit sphere $\mathcal{S}$ given by $x^2 + y^2 + z^2 \;=\; 1$ is not z–simple, but its upper and lower hemispheres, $\mathcal{S}_+$ and $\mathcal{S}_-$, are z–simple. An integral over $\mathcal{S}$ is therefore broken into integrals over $\mathcal{S}_+$ and $\mathcal{S}_-$,

$$\iint_{\mathcal{S}} f(\mathbf{r})\, dS \;=\; \iint_{\mathcal{S}_+} f(\mathbf{r})\, dS \;+\; \iint_{\mathcal{S}_-} f(\mathbf{r})\, dS\ .$$

Mathematica session 5.12. *We illustrate now a* MATHEMATICA *implementation of the integration formula (5.51).* ***surfaceIntegral[f_, surf_, S_, T_]:=***

```
Module[{X1, X2, R, F, I}, X1 = D[surf, S[[1]]];
X2 = D[surf, T[[1]]]; R = Simplify[X1 × X2];
F = Simplify[f /.{x → surf[[1]], y → surf[[2]], z → surf[[3]]}];
I = Simplify[Sqrt[Dot[R, R]]];
Integrate[I * F, {S[[1]], S[[2]], S[[3]]}, {T[[1]], T[[2]], T[[3]]}]]
```

```
PowerExpand[
surfaceIntegral[1, {rho * Sin[phi] * Cos[theta], rho * Sin[phi] * Sin[theta], rho *
Cos[phi]},
{phi, 0, Pi}, {theta, 0, 2 * Pi}]]
```

$4\pi rho^2$

```
sphere = {rho * Sin[phi] * Cos[theta], rho * Sin[phi] * Sin[theta], rho * Cos[phi]}
```

$\{rhoCos[theta]Sin[phi], rhoSin[phi]Sin[theta], rhoCos[phi]\}$

```
PowerExpand[surfaceIntegral[1, sphere, {phi, 0, Pi}, {theta, 0, 2 * Pi}]]
```

$4\pi rho^2$

```
PowerExpand[surfaceIntegral[x^2 + y^2, sphere, {phi, 0, Pi}, {theta, 0, 2 * Pi}]]
```

$\frac{8\pi rho^4}{3}$

```
PowerExpand[surfaceIntegral[Sqrt[x^2 + y^2], sphere, {phi, 0, Pi}, {theta, 0, 2 *
Pi}]]
```

$\pi^2 rho^3$

```
PowerExpand[surfaceIntegral[Sqrt[x^2 + y^2 + z^2], sphere,
{phi, 0, Pi}, {theta, 0, 2 * Pi}]]
```

$4\pi rho^3$

```
PowerExpand[surfaceIntegral[x * y * z, sphere, {phi, 0, Pi/2}, {theta, 0, Pi/2}]]
```

$\frac{rho^5}{8}$

```
Simplify[D[sphere, theta] × D[sphere, phi]]
```

$\{-rho^2 \, Cos[theta] \, Sin[phi]^2, -rho^2 \, Sin[phi]^2 \, Sin[theta], -rho^2 \, Cos[phi] \, Sin[phi]\}$

```
Simplify[
Dot [{-rho^2 Cos[theta] Sin[phi]^2, -rho^2 Sin[phi]^2 Sin[theta], -rho^2 Cos[phi] Sin[phi]},
{-rho^2 Cos[theta] Sin[phi]^2, -rho^2 Sin[phi]^2 Sin[theta], -rho^2 Cos[phi] Sin[phi]}]]
```

$rho^4 \, Sin[phi]^2$

♦

Example 5.30 (Centroid, and moment of inertia of a lamina)**.** Consider a curved surface in $\mathbb{R}^3$ where each surface element $\Delta\mathcal{S}$ has mass ΔM. The **surface density** at a point $\mathbf{r}$ on $\mathcal{S}$, denoted $\delta(\mathbf{r})$, is defined as the limit

$$\delta(\mathbf{r}) := \lim_{\Delta S \to 0} \frac{\Delta M}{\Delta S}\ , \quad \text{where } \Delta S \text{ is the area of an element } \Delta\,\mathcal{S}, \text{ containing the point } \mathbf{r}.$$

In general, the density δ will be different, from point to point[9]. A connected region on such a curved surface is called a **lamina**. Surface integrals can be used to compute the centroids, and moments of inertia, of laminæ. Let $\mathcal{S}$ be a lamina, with a surface density $\delta(\mathbf{r})$. The mass of a surface element $\Delta\,\mathcal{S}$, of area ΔS, called a **mass element**, is then

$$\Delta M \ \approx\ \delta(\mathbf{r}^*)\,\Delta S\ ,$$

where $\mathbf{r}^*$ is some point on $\Delta\mathcal{S}$. Summing over all mass elements, and taking the limit as $\max \Delta S \to 0$, we express the total mass M of the lamina as the surface integral of the density

$$M \ =\ \iint_{\mathcal{S}} \delta(\mathbf{r})\,dS\ .$$

The moment (w.r.t. origin) of the mass element ΔM, situated at the point $\mathbf{r}^*$, is given by

$$\mathbf{r}^*\,\Delta M \ \approx\ \mathbf{r}^*\,\delta(\mathbf{r}^*)\,\Delta S\ .$$

The **mass centroid** $\bar{\mathbf{r}}$ of the lamina is then

$$\bar{\mathbf{r}} \ =\ \frac{1}{M}\iint_{\mathcal{S}} \mathbf{r}\,\delta(\mathbf{r})\,dS\ .$$

Similarly, the moment of inertia (w.r.t. origin) of the lamina is given by

$$I \ =\ \iint_{\mathcal{S}} \delta(\mathbf{r})\ \|\mathbf{r}\|^2\ dS\ .$$

---

[9]For example, the surface is made of different materials at different points.

Exercises

Exercise 5.31. *Determine the area of the sphere* $x^2 + y^2 + z^2 = 2\,a\,z$ *which is intersected by the cylinder whose projection on the xy–plane is a petal of the rose* $r = a\cos(n\theta)$.

Exercise 5.32. *Compute the area of the surface* $z = 2\,x + y^2$ *lying above the triangle, in the xy–plane, bounded by the lines* $y = 0$, $x = y$, *and and* $x = 1$.

Exercise 5.33. *Consider the cylinder bounded by the planes* $x = \pm 1$, $y = \pm 1$, *the plane* $z = 0$, *and the surface* $z = 2 + x + y + x^2 + y^2$. *Find the total area of the surface.*

Exercise 5.34. *Find the total surface area of the torus, generated by rotating the circle* $(x - 2\,a)^2 + y^2 = a^2$ *about the y–axis.*

Exercise 5.35. *Verify (5.79).*

Exercise 5.36. *A surface* $\mathcal{S}$ *is expressed as* $x = x(y, z)$ *for* $(y, z) \in \mathcal{D}$. *Show that its area iis*

$$\iint_{\mathcal{D}} \sqrt{\left(\frac{\partial x}{\partial y}\right)^2 + \left(\frac{\partial x}{\partial z}\right)^2 + 1}\; dy\, dz\,.$$

Find a similar formula for surfaces represented by the formula $y = y(x, z)$.

Exercise 5.37. *Find the area of the surface cut from the paraboloid* $z = x^2 + y^2$ *by the planes:*

(a) $z = 4$ (b) $z = x + y + 2$ (c) $x = 1$ *and* $x = 5$

Exercise 5.38. *Find the area of the surface cut from the plane* $x + y + z = 1$ *by the cylinder whose lateral sides are given by* $y = x^2$ *and by* $y = 4 - x^2$.

Exercise 5.39. *Find the area of the cap cut from the sphere* $x^2 + y^2 + z^2 = 4$ *by the cone* $z = \sqrt{x^2 + y^2}$. *What is the area of the cap plus the capped cone?*

Exercise 5.40. *Find the area of the ellipse cut from the plane* $x + y + z = 1$ *by the cylinder* $x^2 + y^2 = 1$.

Exercise 5.41. *Integrate the function* $f = xy + z$ *over all sides of the unit cube in the first octant.*

Exercise 5.42. *Integrate* $f = xyz$ *over all sides of the tetrahedron whose sides are given by* $x + y + z = 1$, $x = 0$, $y = 0$, *and* $z = 0$.

Exercise 5.43. *Integrate* $f := x^2 + y^2$ *over the portion of the paraboloid* $z = x^2 + y^2$ *that lies in the first octant.*

Exercise 5.44. *Use the function* `polarSurfaceIntegral` *of* MATHEMATICA–*Session 5.11 to integrate the following functions on the surface* $z = xy$ *which lies over the unit circle.*

(a) $g(r, \theta) := 1$ (b) $g(r, \theta) := e^{-r}$

(c) $g(r, \theta) := r^3 \sin\theta$ (d) $g(r, \theta) := r^3 \sin^2\theta$ *explain!*

Exercise 5.45. *Set up the integrals to calculate the mass, the centroids and the moments of inertia about the x and the y axes for the following objects. Evaluate these using the* MATHEMATICA *function* `SURFACE_INT` *and, if necessary, use* `N`.

(a) *The upper, semi-circular annulus of inner radius a and outer radius b, centered at the origin, whose density is $\delta = \frac{\kappa}{r}$, where $r = \sqrt{x^2 + y^2}$*

(b) *The upper, semi-circular annulus of inner radius a and outer radius b, centered at the origin, whose density is $\delta = \kappa r$, where $r = \sqrt{x^2 + y^2}$*

(c) *The spherical shell given by $\{x^2 + y^2 + z^2 = a^2\} \cap \{z \geq \frac{a}{2}\}$, with a density $\delta = z^2$*

(d) *The spherical shell, given by $\{x^2 + y^2 + z^2 = a^2\} \cap \{z \geq \frac{a}{2}\}$, with a density $\delta = x^2 + y^2$*

(e) *The right cone $z = \sqrt{x^2 + y^2}$, whose density is $\delta = \frac{\kappa}{x^2+y^2}$*

(f) *The right cone $z = \sqrt{x^2 + y^2}$, whose density is $\delta = \frac{\kappa}{\sqrt{x^2+y^2}}$*

(g) *The portion of the sphere $x^2 + y^2 + z^2 = 1$ lying in the first octant*

(h) *The portion of the cylinder $x^2 + y^2 = 1$, $0 \leq z \leq 2$ lying in the second octant*

Exercise 5.46. *Calculate the moments of inertia of the objects in Exercise 5.45 around a line parallel to the x–axis and tangent to an outer boundary.*

Exercise 5.47. *Calculate the moments of inertia of the objects in Exercise 5.45 around a line parallel to the y–axis and tangent to an outer boundary.*

5.7 Triple integrals

Triple integrals are generalizations of double integrals, in the same way that double integrals generalize single integrals. Let $f(x, y, z)$ be a continuous function defined in a domain $\mathcal{D}$ which lies inside some rectangular parallelepiped

$$\mathcal{R} \ := \ [a_1, a_2] \times [b_1, b_2] \times [c_1, c_2] \ .$$

We define a (three-dimensional) partition on R as

$$\mathcal{P}_n \ := \ \mathcal{P}_n^{(1)} \times \mathcal{P}_n^{(2)} \times \mathcal{P}_n^{(3)}$$

where $\mathcal{P}_n^{(1)}$, $\mathcal{P}_n^{(2)}$, $\mathcal{P}_n^{(3)}$ are partitions of $[a_1, a_2]$, $[b_1, b_2]$, $[c_1, c_2]$, respectively, using n subintervals. The partition $\mathcal{P}_n$ consists of n^3 blocks

$$\begin{aligned} \mathcal{R}_{ijk} \ &= \ [x_i, x_{i+1}] \times [y_j, y_{j+1}] \times [z_k, z_{k+1}] \ , \quad i, j, k = 1, \ldots, n \ , \\ \text{with } \textbf{volume} \ &= \ (x_{i+1} - x_i)\,(y_{j+1} - y_j)\,(z_{k+1} - z_k) \ , \\ \text{and } \textbf{diameter} \ &= \ \sqrt{|x_{i+1} - x_i|^2 \ + \ |y_{j+1} - y_j|^2 \ + \ |z_{k+1} - z_k|^2} \ . \end{aligned}$$

The **norm** of the partition $\mathcal{P}_n$, denoted $\|\mathcal{P}_n\|$, is defined as the the largest diameter of a block in the partition. We renumber the blocks which lie inside $\mathcal{D}$ as

$$\mathcal{R}_1 \, , \, \mathcal{R}_2 \, , \, \cdots \, , \, \mathcal{R}_{m_n} \ .$$

The number m_n of blocks lying inside $\mathcal{D}$, depends on the partition $\mathcal{P}_n$ and on the domain $\mathcal{D}$.
A **Riemann sum** over the blocks in $\mathcal{D}$ is any sum of the form

$$\sum_{i=1}^{m_n} f(x_i^*, y_i^*, z_i^*)\, \Delta V_i$$

where ΔV_i is the volume of the ith–block, and (x_i^*, y_i^*, z_i^*) is an arbitrary point in that block. If the limit, as $\|\mathcal{P}_n\| \to 0$, of these Riemann sums exists, it is defined as the (**triple**) **integral** of f over $\mathcal{D}$,

$$\iiint\limits_{\mathcal{D}} f(x, y, z)\, dV \;:=\; \lim_{\|\mathcal{P}_n\| \to 0} \sum_{i=1}^{m_n} f(x_i^*, y_i^*, z_i^*)\, \Delta V_i \;, \tag{5.53}$$

in which case we say that f is **integrable** in $\mathcal{D}$. It can be shown that if $f(x, y, z)$ is continuous and the boundary of $\mathcal{D}$ is sufficiently smooth, then f is integrable in $\mathcal{D}$. However, a proof of this fact is beyond the scope of this course. In the simplest cases we can compute triple integrals iteratively.

Mathematica session 5.13. *In this* MATHEMATICA *session we perform several triple integrations using the module* **tripleInt**.

```
tripleInt[f_, U_, param1_, param2_, param3_]:=
Module[{g, J, j}, g:=f/.{x → U[[1]], y->U[[2]], z->U[[3]]};
J = Outer[D, {U[[1]], U[[2]], U[[3]]}, {param1[[1]], param2[[1]], param3[[1]]}];
j = Abs[Simplify[Det[J]]];
Integrate[j * g, {param1[[1]], param1[[2]], param1[[3]]},
{param2[[1]], param2[[2]], param2[[3]]},
{param3[[1]], param3[[2]], param3[[3]]}]]
```

```
tripleInt[1, {r * Cos[theta] * Sin[phi], r * Sin[theta] * Sin[phi], r * Cos[phi]},
{r, 0, 1}, {theta, 0, 2 * Pi}, {phi, 0, Pi}]
```

$\frac{4\pi}{3}$

1/.{x->r * Cos[theta] * Sin[phi], y->r * Sin[theta] * Sin[phi], z->r * Cos[phi]}

1

Outer[D, {r * Cos[theta] * Sin[phi], r * Sin[theta] * Sin[phi], r * Cos[phi]},

{r, theta, phi}]

{{Cos[theta]Sin[phi], −r Sin[phi] Sin[theta],

r Cos[phi] Cos[theta]}, {Sin[phi] Sin[theta], r Cos[theta] Sin[phi], r Cos[phi]

Sin[theta]}, {Cos[phi], 0, −r Sin[phi]}}

Det[%]

$-r^2$ Cos[phi]^2 Cos[theta]^2 Sin[phi] $-$ r^2 Cos[theta]^2 Sin[phi]^3 $-$ r^2 Cos[phi]^2 Sin[phi]

Sin[theta]^2 $-$ r^2 Sin[phi]^3 Sin[theta]^2

Simplify[%]

$-r^2$ Sin[phi]

Integrate [$-r^2$ Sin[phi], {r, 0, 1}, {theta, 0, 2 * Pi}, {phi, 0, Pi}]

$-\frac{4\pi}{3}$

a = 2; Integrate[r^2, {r, 0, a}]

$\frac{8}{3}$

tripleInt[x * y * z, {r * Cos[theta] * Sin[phi], r * Sin[theta] * Sin[phi], r *

Cos[phi]}, {r, 0, 1}, {theta, 0, Pi/2}, {phi, 0, Pi/2}]

$\frac{1}{48}$

x * y * z/.{x → r * Cos[theta] * Sin[phi], y->r * Sin[theta] * Sin[phi], z->r *

Cos[phi]}

$r^3 \, Cos[phi] \, Cos[theta] \, Sin[phi]^2 \, Sin[theta]$

Integrate [r^3 Cos[phi] Cos[theta] Sin[phi]^2 Sin[theta] * (−r^2 Sin[phi]) , {r, 0, 1}, {theta, 0, Pi/2}, {phi, 0, Pi/2}]

$-\frac{1}{48}$

1/.{x → r * Cos[theta] * Sin[phi], y->r * Sin[theta] * Sin[phi], z->r * Cos[phi]}

1

◆

Example 5.48 (Triple iterated integral). Let $\mathcal{D} := [0,1] \times [-1,1] \times [2,4]$, and let $f(x,y,z) := x\,y + x^2\,z + y\,z^3$. Then

$$\iiint_{\mathcal{D}} f(x,y,z)\,dV = \int_0^1 \int_{-1}^1 \int_2^4 (xy + x^2 z + yz^3)\,dz\,dy\,dx =$$

$$\int_0^1 \int_{-1}^1 \left[x\,y\,z + \tfrac{1}{2}\,x^2\,z^2 + y\,z^4\right]_2^4 dy\,dx =$$

$$\int_0^1 \int_{-1}^1 \left(2\,x\,y + 6\,x^2 + 240\,y\right) dy\,dx =$$

$$\int_0^1 \left[x\,y^2 + 6\,y\,x^2 + 120\,y^2\right]_{-1}^1 dx =$$

$$\int_0^1 (2\,x + 240)\,dx = 1 + 240 = 241 .$$

△

Let the domain D be z–**simple**, i.e. any vertical line cuts the boundary of D twice,

at the **upper surface** $\mathcal{S}_{\mathrm{U}}$ described by $z := h_{\mathrm{U}}(x,y)$ and
the **lower surface** $\mathcal{S}_{\mathrm{L}}$ described by $z := h_{\mathrm{L}}(x,y)$.

Let Ω be the projection of $\mathcal{D}$ on the xy–plane. Then

$$D = \{(x,y,z) : (x,y) \in \Omega , \quad h_{\mathrm{L}}(x,y) \le z \le h_{\mathrm{U}}(x,y)\} .$$

In this case the triple integral can be written as the iterated integral

$$\iiint_{\mathcal{D}} f(x,y,z)\,dV = \iint_{\Omega} \left(\int_{h_{\mathrm{L}}(x,y)}^{h_{\mathrm{U}}(x,y)} f(x,y,z)\,dz\right) dA . \qquad (5.54)$$

Here dA stands for an area differential in the xy–plane, $dA = dx\,dy$. Similar results hold for the case that the domain $\mathcal{D}$ is x–simple or y–simple.

Example 5.49 (Volume by triple integration)**.** We compute the volume of the domain $\mathcal{D}$ bounded by the paraboloid of revolution $z = x^2 + y^2$ and the plane $z = 4$. This domain is z–simple, so we use (5.54) to express the volume as a triple integral

$$V = \iiint\limits_{\mathcal{D}} dV = \int_{-2}^{2} dx \int_{-\sqrt{4-x^2}}^{\sqrt{4-x^2}} dy \int_{x^2+y^2}^{4} dz =$$

$$\int_{-2}^{2} dx \int_{-\sqrt{4-x^2}}^{\sqrt{4-x^2}} \left(4 - x^2 - y^2\right) dy \, .$$

Mathematica session 5.14. *We provide a* MATHEMATICA *procedure for finding the average value of a function defined on a three-dimensional region. Three-dimensional centroids may be computed analogously to the two-dimensional case. To start a new session let us wipe the slate clean with the* MATHEMATICA COMMAND **ClearAll.**

```
ClearAll
```

```
ClearAll
```

```
averageValue[default_, R_, S_, T_]:=
Module[{num, denom},
num = Integrate[default, {R[[1]], R[[2]], R[[3]]}, {S[[1]], S[[2]], S[[3]]},
{T[[1]], T[[2]], T[[3]]}];
denom = Integrate[1, {R[[1]], R[[2]], R[[3]]}, {S[[1]], S[[2]], S[[3]]},
{T[[1]], T[[2]], T[[3]]}]; num/denom]
```

```
averageValue[x * y * z, {z, 0, 2}, {y, 0, 2}, {x, 0, 2}]
```

1

```
averageValue[x * Sin[y], {z, 0, 1}, {y, 0, Pi}, {x, 0, 1}]
```

$\frac{1}{\pi}$

```
threeCentroid[default_, R_, S_, T_]:=
Module[{num, denom},
num = Integrate[default * {x, y, z}, {R[[1]], R[[2]], R[[3]]}, {S[[1]], S[[2]], S[[3]]},
{T[[1]], T[[2]], T[[3]]}];
denom = Integrate[1, {R[[1]], R[[2]], R[[3]]}, {S[[1]], S[[2]], S[[3]]},
{T[[1]], T[[2]], T[[3]]}]; num/denom]
```

```
threeCentroid[1, {x, -2, 2}, {y, -Sqrt[4 - x^2], Sqrt[4 - x^2]},
{z, 0, 4 - x^2 - y^2}]
```

$\{0, 0, \frac{4}{3}\}$

```
threeCentroid[x * y * z, {x, -2, 2}, {y, -Sqrt[4 - x^2], Sqrt[4 - x^2]},
{z, 0, 4 - x^2 - y^2}]
```

$\{0, 0, 0\}$

```
threeCentroid[1 + x^2 + y^2 + z^2, {x, -2, 2}, {y, -Sqrt[4 - x^2],
Sqrt[4 - x^2]}, {z, 0, 4 - x^2 - y^2}]
```

$\{0, 0, \frac{136}{15}\}$

```
Integrate[1, {x, -2, 2}, {y, -3 * Sqrt[1 - x^2/4], 3 * Sqrt[1 - x^2/4]},
{z, 0, 4 * Sqrt[1 - x^2/4 - y^2/9]}]
```

16π

```
twoCentroid[default_, S_, T_]:=
Module[{num, denom},
num = Integrate[default * {x, y}, {S[[1]], S[[2]], S[[3]]}, {T[[1]], T[[2]], T[[3]]}];
denom = Integrate[1, {S[[1]], S[[2]], S[[3]]}, {T[[1]], T[[2]], T[[3]]}]; num/denom]
```

♦

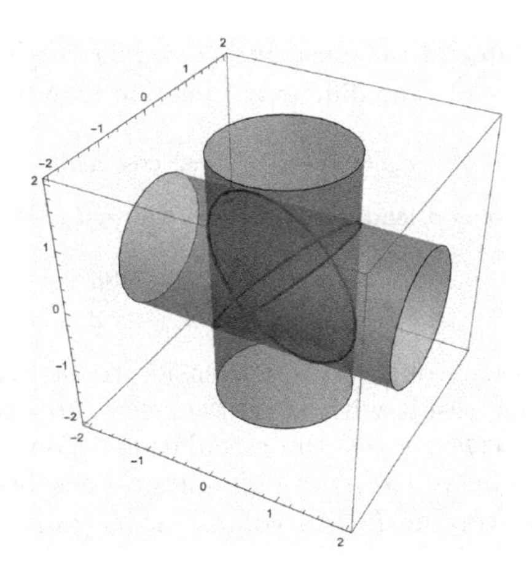

FIGURE 5.17: Intersecting cylinders.

Example 5.50. Consider two circular cylinders of equal radius a , whose axes meet at right angles, see Figure 5.17. It is required to find the volume of the intersection of these cylinders.
Solution: For simplicity, let these cylinders be

$$\begin{aligned} x^2+y^2 &= a^2 \quad , \quad \text{with axis on the } z\text{–axis, and} \\ y^2+z^2 &= a^2 \quad , \quad \text{with axis on the } x\text{–axis .} \end{aligned}$$

The volume may then be set up as the triple integral

$$\int_{-a}^{a} dx \int_{-\sqrt{a^2-x^2}}^{\sqrt{a^2-x^2}} dy \int_{-\sqrt{a^2-y^2}}^{\sqrt{a^2-y^2}} dz \;=\; 8\int_0^a dx \int_0^{\sqrt{a^2-x^2}} \sqrt{a^2-y^2}dy\,.$$

We integrate the y-integral using trigonometric substitution

$$\int_0^{\sqrt{a^2-x^2}} \sqrt{a^2-y^2}dy \;=\; a^2 \int_0^{asin\frac{\sqrt{a^2-x^2}}{a}} \cos^2\theta d\theta \;=$$

$$\frac{a^2}{2}\left(asin\frac{\sqrt{a^2-x^2}}{a} + \frac{1}{2}\sin 2\theta|_0^{asin\frac{\sqrt{a^2-x^2}}{a}} \right)$$

Using that $\sin 2\theta = 2\sin\theta\cos\theta$ this last integral becomes

$$\frac{a^2}{2}\left(\operatorname{asin}\frac{\sqrt{a^2-x^2}}{a} + \frac{x\sqrt{a^2-x^2}}{a^2}\right).$$

We still have another integral to do, namely

$$8\int_0^a \frac{a^2}{2}\left(\operatorname{asin}\frac{\sqrt{a^2-x^2}}{a}+\frac{x\sqrt{a^2-x^2}}{a^2}\right)dx\ .$$

In order to do this integral we substitute $\varphi = \operatorname{asin}\frac{\sqrt{a^2-x^2}}{a}$ or, what is the same thing, $a\sin\varphi = \sqrt{a^2-x^2}$. The differential relation may be calculated from this to be

$$dx = -\frac{a}{x}\sqrt{a^2-x^2}\cos\varphi d\varphi$$

but we also have $x = \cos\varphi$ leading to $dx = -a\sin\varphi d\varphi$. We now write the integral as

$$\frac{8a^3}{2}\int_0^{\frac{\pi}{2}}\varphi\sin\varphi d\varphi \ = \ \frac{8a^3}{2}\ .$$

These integrations would be much easier to do with MATHEMATICA, so let us now check our result with MATHEMATICA. In subsequent problems we would use MATHEMATICA to do the calculations after setting up the iterated integrations. Analogously to the double integral case we may define the mass of an object having volume D and volume density $\delta(x,y,z)$ as

$$M \ := \ \iiint\limits_D \delta(x,y,z)\,dV\ .$$

Likewise, we define the **centroid** $(\overline{x},\overline{y},\overline{z})$ as

$$\overline{x} \ := \ \frac{1}{M}\iiint\limits_D x\,\delta(x,y,z)\,dV\ , \quad \overline{y} \ := \ \frac{1}{M}\iiint\limits_D y\,\delta(x,y,z)\,dV\ ,$$

$$\overline{z} \ := \ \frac{1}{M}\iiint\limits_D z\,\delta(x,y,z)\,dV\ .$$

The moment of inertia about the z-axis is given by

$$I_z \ := \ \iiint\limits_D (x^2+y^2)\,\delta(x,y,z)\,dV.$$

Similarly, the moments of inertia about the x and y axes are given by

$$I_x \ := \ \iiint\limits_D (y^2+z^2)\,\delta(x,y,z)\,dV,\ I_y \ := \ \iiint\limits_D (x^2+z^2)\,\delta(x,y,z)\,dV\ ,$$

respectively. In many integrations it is useful to picture the domain of integration. In Figure 5.18, and Figure 5.19 we illustrate what the region looks like if the upper and and lower bounds may be planes. More complicated regions may also be portrayed this way.

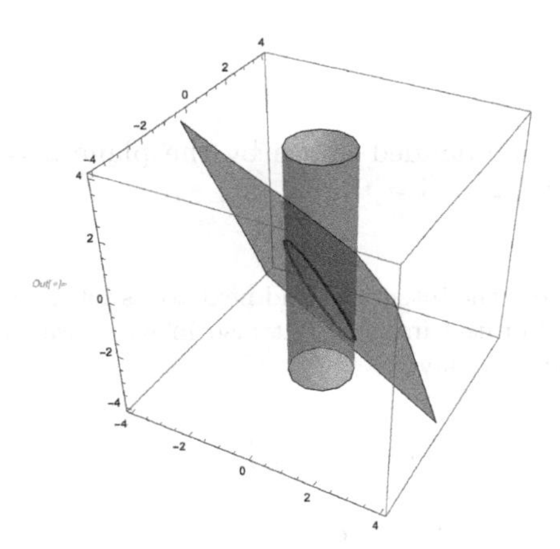

FIGURE 5.18: The cylinder $x^2 + y^2 = 1$, bounded below by the plane $z + x + 2y + 1 = 0$.

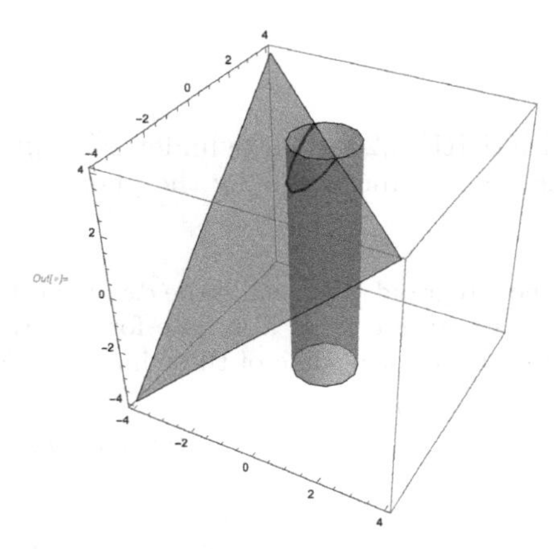

FIGURE 5.19: The cylinder $x^2 + y^2 = 1$, bounded above by the plane $z = x + y + 4$.

Example 5.51 (Moment of inertia of a tetrahedron)**.** Consider a body of constant density which occupies a domain D, bounded by the three coordinate planes and the plane

$$\frac{x}{a} + \frac{y}{b} + \frac{z}{c} = 1 \, .$$

For simplicity assume that $a, b, c > 0$; so the tetrahedron appears as in Figure 5.21. The moment of inertia of such a body is given by the formula

$$I_z := \delta \iiint\limits_D (x^2 + y^2)\, dV = \delta \int_0^a dx \int_0^{b(1-\frac{x}{a})} dy \int_0^{c(1-\frac{x}{a}-\frac{y}{b})} (x^2 + y^2)\, dz \, .$$

In the integral on the right we have indicated that the z integration is done first, then we do the y integration and finally the x integration by placing the dz closest to the integrand, the dy next to its integral to indicate its limits and the next closest to

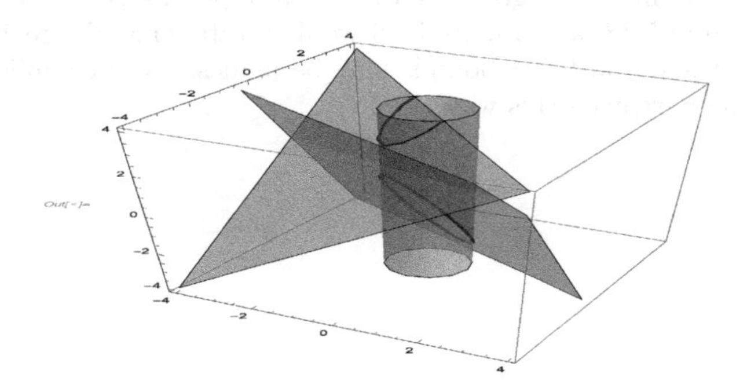

FIGURE 5.20: The cylinder $x^2 + y^2 = 1$ bounded above by the plane $z = x + y + 4$ and below by the plane $z + x + 2y + 1 = 0$.

the integrand, and finally the dx furthest from the integrand and next to its integral. This notation is an alternate form of the iterated integral but also in wide usage. Let us perform some of these integrals. First we have

$$\int_0^{c(1-\frac{x}{a}-\frac{y}{b})} (x^2+y^2)dz = c(x^2+y^2)\left(1-\frac{x}{a}-\frac{y}{b}\right) =$$

$$c\left(x^2(1-\frac{x}{a}) - y\frac{x^2}{b} + (1-\frac{x}{a})y^2 - \frac{y^2}{b}\right)$$

Next we perform the area integral

$$\begin{aligned} &c\int_0^{b(1-\frac{x}{a})}\left(x^2(1-\tfrac{x}{a}) - y\tfrac{x^2}{b} + (1-\tfrac{x}{a})y^2 - \tfrac{y^2}{b}\right)dy \\ &= c\left(\tfrac{x^3}{3} - \tfrac{x^4}{4a}) - y\tfrac{x^3}{3b} + (x - \tfrac{x^2}{2a})y^2 - \tfrac{xy^2}{b}\right)_0^{b(1-\frac{x}{a})} \\ &= \tfrac{bc}{12}\left(6x^2(1-\tfrac{x}{a})^2 + b^2(1-\tfrac{x}{a})^4\right); \end{aligned}$$

hence,

$$\begin{aligned} \int_0^a dx \int_0^{b(1-\frac{x}{a}} dy \int_0^{c(1-\frac{x}{a}-\frac{y}{b})} (x^2+y^2)dz &= \int_0^a \frac{bc}{12}\left(6x^2(1-\frac{x}{a})^2 + b^2(1-\frac{x}{a})^4\right)dx \\ &= \frac{abc}{60}(a^2+b^2)\,. \end{aligned}$$

Example 5.52. We compute the volume of the domain D which is the part of the cylinder

$$x^2 + y^2 \;=\; 1\,,$$

bounded above by the plane $z \;=\; x + y + 4$ and below by $z \;=\; -x - 2\,y - 1$, see

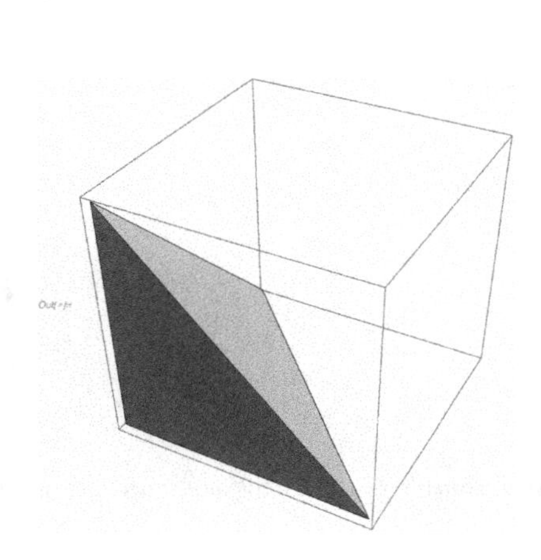

FIGURE 5.21: The tetrahedron $\{x + y + z \leq 1\}$. See exercize (f).

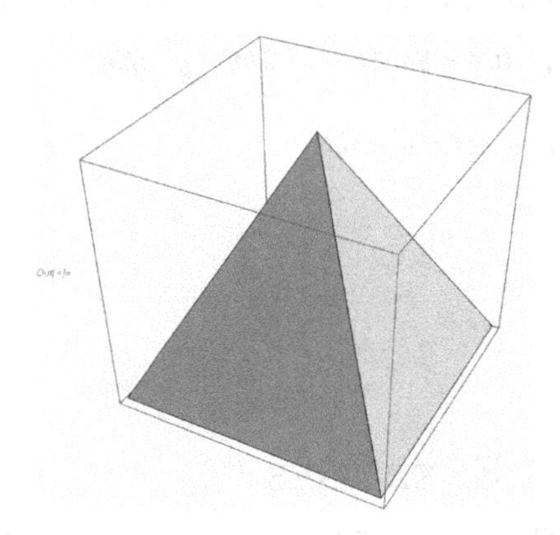

FIGURE 5.22: The pyramid $\{0 \leq x \leq z\} \cap \{0 \leq y \leq z\} \cap \{0 \leq (1 - z) \leq 1\}$ See exersize (g).

Figure 5.20 . The volume is represented by the iterated integral

$$
\begin{aligned}
\iiint dV &= \int_{-1}^{1} dxy - z \int_{-\sqrt{1-x^2}}^{\sqrt{1-x^2}} dy \int_{-x-2y-1}^{x+y+4} dz = \\
&4 \int_{-1}^{1} dx \int_{-\sqrt{1-x^2}}^{\sqrt{1-x^2}} (2x + 3y + 5)\, dy \\
&= \int_{-1}^{1} dx \left(4x\sqrt{1-x^2} + 10\sqrt{1-x^2}\right) \\
&= -\tfrac{4}{3}\left(1 - x^2\right)^{\frac{3}{2}} |_{-1}^{1} + 10 \int_{-1}^{1} \sqrt{1-x^2}dx \\
&= 5\,\pi\ .
\end{aligned}
$$

Example 5.53. Consider the volume of the region bounded by the ellipsoid

$$\frac{x^2}{4} + \frac{y^2}{9} + \frac{z^2}{16} = 1\ ,$$

and the xy–plane. This volume is the iterated integral

$$\iiint_D dV = \left(\int_{-2}^{2} dx \left(\int_{-3\sqrt{1-\frac{x^2}{4}}}^{3\sqrt{1-\frac{x^2}{4}}} dy \left(\int_0^{4\sqrt{1-\frac{x^2}{4}-\frac{y^2}{9}}} dz\right)\right)\right)\ .$$

To compute this integral, use

Mathematica session 5.15.

```
Integrate[1, {x, -2, 2}, {y, -3 * Sqrt[1 - x^2/4], 3 * Sqrt[1 - x^2/4]},
{z, 0, 4 * Sqrt[1 - x^2/4 - y^2/9]}]
```

and get the result

16π

♦

EXERCISES

Exercise 5.54. *Compute the volume V in Example 5.49. Check your answer with* MATHEMATICA.

Exercise 5.55. *Compute the value of the triple integral*

$$\iiint_{\mathcal{D}} f(x, y, z)\, dV$$

for the prescribed f and $\mathcal{D}$ in the problems below:

(a) $f := x + y + z$, *$\mathcal{D}$ bounded by the coordinate planes and $x + y + z = 1$*

(b) $f(x, y, z) := xyz$, *$\mathcal{D}$ as above*

(c) $f := e^{x+y+z}$, *$\mathcal{D}$ is the unit cube lying in the first quadrant.*

(d) $f := 2x + 3y - 4z$, *$\mathcal{D}$ bounded above by $z = 4 - y^2$, and lying over $[0, 1] \times [0, 1]$*

(e) $f := xyz$,*$\mathcal{D}$ bounded above by$z = 4 + x^2$, below by $z = 1 - y^2$over $z = 0$ where $0 \leq x \leq 1, 0 \leq y \leq 2$.*

(f) $f := x$, *$\mathcal{D}$ bounded above by $x + y + z = 2$, over $z = 0$, within $x^2 + y^2 \leq 1$. See Figure 5.21.*

(g) $f := 1$ *where the region is the pyramid whose base is a square, as shown in Figure 5.22.*

Exercise 5.56. *Use triple integration to show that the mass of the tetrahedron, as shown in Figure 5.21, of constant density δ, which is bounded by the coordinate planes and the plane*

$$\frac{x}{a} + \frac{y}{b} + \frac{z}{c} = 0$$

is given by $M = \delta \frac{abc}{6}$. Can you justify this answer geometrically?

Exercise 5.57. *Find the center of gravity of the tetrahedron of the previous problem.*

Exercise 5.58. *The density of a cube is proportional to its distance from the center. Determine the mass of the cube.*

Exercise 5.59. *Find the center of gravity of the symmetric pyramid having the square base of side a and height b.*

Exercise 5.60. *Find the centroid for each of the domains listed below. Assume that each domain has constant density.*

(a) *the cone* $z = \sqrt{x^2 + y^2}$ *bounded above by* $z = 4$

(b) *the paraboloid* $z = x^2 + y^2$ *bounded above by* $z = 4$

(c) *the paraboloid* $z = x^2 + y^2$ *bounded above by* $z = 2x + y + 2$

(d) *the parabolic cylinder* $y = x^2$ *bounded above by* $z = 2y$ *and below by* $z = 0$

(e) *the elliptic cylinder* $\frac{x^2}{4} + \frac{y^2}{1} = 1$ *bounded above by* $z = x + y + 1$ *and below by* $z = 0$

Exercise 5.61. *Find the centroid of the region in the first quadrant interior to the two cylinders* $x^2 + y^2 = 0$ *and* $y^2 + z^2 = 1$.

Exercise 5.62. *Find the volume of the region interior to the elliptic paraboloids* $z = x^2 + 2y^2$ *and* $z = 27 - 2x^2 - y^2$

5.8 Change of variables in multiple integration

Consider a double integral

$$\iint\limits_{\mathcal{A}} f(x, y)\, dx\, dy \tag{5.55}$$

over a domain $\mathcal{A}$ in the xy–plane. Sometimes it is convenient to change variables from x, y to other variables u, v , which are related by a mapping $\mathbf{T}$,

$$(x, y) \;=\; \mathbf{T}(u, v)\ , \quad \text{rewritten as} \quad \begin{array}{rcl} x &=& x(u, v) \\ y &=& y(u, v) \end{array} \tag{5.56}$$

Assuming that $\mathbf{T}$ is one-to-one, its inverse mapping is [10]

$$(u, v) \;=\; \mathbf{T}^{-1}(x, y)\ . \tag{5.57}$$

Then $\mathbf{T}^{-1}(\mathcal{A})$ is the domain in the uv–plane which corresponds to $\mathcal{A}$. Finally, recall the Jacobian of the mapping $\mathbf{T}$,

$$J_{\mathbf{T}} \;:=\; \begin{pmatrix} x_u & x_v \\ y_u & y_v \end{pmatrix} \tag{5.58}$$

[10] See Figure 5.23.

and its determinant

$$\frac{\partial(x,y)}{\partial(u,v)} := \textbf{det}\begin{pmatrix} x_u & x_v \\ y_u & y_v \end{pmatrix} \tag{5.59}$$

The integral (5.55) is evaluated in terms of the variables u, v as follows.

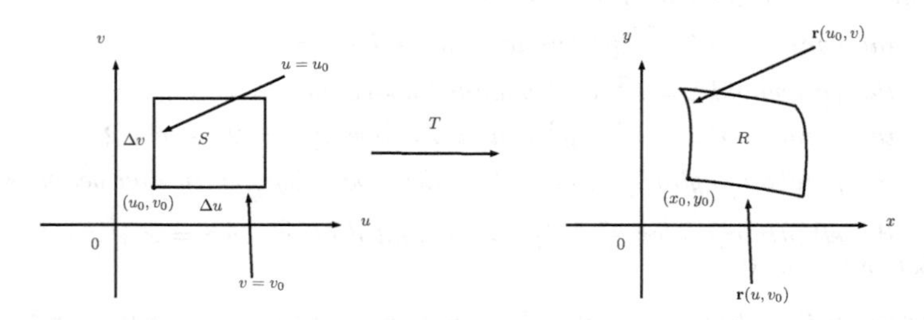

FIGURE 5.23: Mapping from a given coordinate system into another.

Theorem 5.63. Let f, $\mathcal{A}$, $\mathbf{T}$ be as above. If f is integrable in $\mathcal{A}$ then

$$\iint\limits_{\mathcal{A}} f(x,y)\,dx\,dy \;=\; \iint\limits_{\mathbf{T}^{-1}(\mathcal{A})} f(x(u,v),y(u,v)) \left|\frac{\partial(x,y)}{\partial(u,v)}\right| du\,dv. \tag{5.60}$$

Proof:
Follows since the area element $du\,dv$ corresponds to the area element

$$dA = \left|\frac{\partial(x,y)}{\partial(u,v)}\right| du\,dv\ . \tag{5.61}$$

in the xy–plane. □

Example 5.64. For the polar coordinates

$$\begin{aligned} x &= r\cos\theta \\ y &= r\sin\theta \end{aligned} \tag{5.62}$$

the equality (5.60) becomes

$$\iint_{\mathcal{A}} f(x,y)\,dx\,dy\ ,= \iint_{\mathbf{T}^{-1}(\mathcal{A})} f(r\cos\theta\,,\,r\sin\theta) \left|\frac{\partial(x,y)}{\partial(r,\theta)}\right| dr\,d\theta =$$

$$\int\!\!\int\limits_{\mathbf{T}^{-1}(\mathcal{A})} f(r\cos\theta\,,\,r\sin\theta)\,r\,dr\,d\theta\ , \tag{5.63}$$

already seen in (5.8) and (5.22).

Example 5.65 (Confocal parabolas)**.** Consider the system of curvilinear coordinates defined by the families of conformal parabolas [11]

$$\begin{aligned} y^2 &= -2\,u\,x + u^2\,, \\ y^2 &= 2\,v\,x + v^2\,, \end{aligned} \tag{5.64}$$

where u, v are variables. Solving for x from

$$2\,v\,x + v^2 = -2\,u\,x + u^2\,,$$

we get

$$x = \frac{1}{2}(u - v)\,, \tag{5.65a}$$

which substituted in (5.64) gives

$$y = \sqrt{u\,v}\,. \tag{5.65b}$$

The Jacobian of this mapping is

$$\begin{pmatrix} x_u & x_v \\ y_u & y_v \end{pmatrix} = \begin{pmatrix} \frac{1}{2} & -\frac{1}{2} \\ \frac{1}{2}\sqrt{\frac{v}{u}} & \frac{1}{2}\sqrt{\frac{u}{v}} \end{pmatrix}$$

and its determinant is

$$\frac{\partial(x,y)}{\partial(u,v)} = \frac{1}{4}\left(\sqrt{\frac{v}{u}} + \sqrt{\frac{u}{v}}\right)\,.$$

We see that the mapping is one–to–one, if the variables are restricted to $u > 0$ and $v > 0$. For fixed values of u and v the functions (5.64) describe confocal parabolas, see e.g. the parabolas corresponding to $u = v = 1$

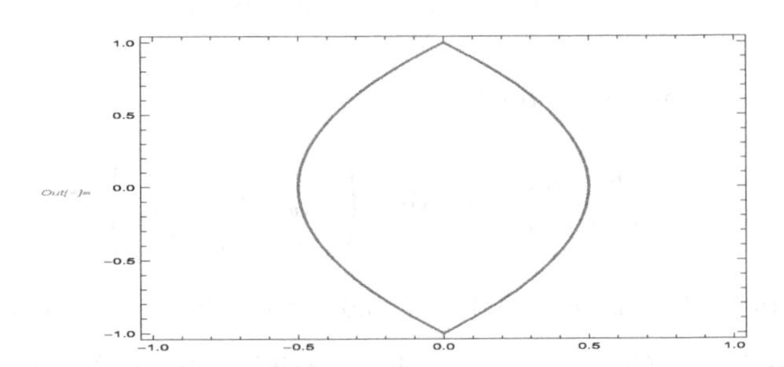

FIGURE 5.24: The two confocal parabolas corresponding to $u = v = 1$.

Consider now the region $\mathcal{A}$ bounded by the curves

$$\begin{aligned} y^2 &= -2\,x + 1\,, &\text{or}\quad u &= 1 \\ y^2 &= -4\,x + 4\,, &\text{or}\quad u &= 2 \\ y^2 &= 6\,x + 9\,, &\text{or}\quad v &= 3 \\ y^2 &= 8\,x + 16\,, &\text{or}\quad v &= 4\,. \end{aligned}$$

[11]See Figure 5.24.

The moment of inertia of $\mathcal{A}$

$$I = \iint_{\mathcal{A}} (x^2 + y^2)\, dx\, dy$$

is computed easily using the uv–variables. First we note that

$$x^2 + y^2 = \frac{1}{4}(u - v)^2 + u\, v = \frac{1}{2}(u + v)^2 .$$

Therefore

$$I = \iint_{\mathcal{A}} (x^2 + y^2)\, dx\, dy = \frac{1}{4} \int_1^2 du \int_3^4 dv\, \frac{1}{2}\,(u + v)^2 \left(\sqrt{\frac{v}{u}} + \sqrt{\frac{u}{v}}\right) .$$

Similarly, a triple integral $\iiint f(x, y, z)\, dx\, dy\, dz$ is sometimes simplified by working with different variables u, v and w,

$$\begin{array}{rcl} x &=& x(u, v, w) \\ y &=& y(u, v, w) \\ z &=& z(u, v, w) \end{array} , \quad \text{or} \quad (x, y, z) = \mathbf{T}\,(u, v, w) . \tag{5.66}$$

The Jacobian of this transformation is

$$J_{\mathbf{T}} = \begin{pmatrix} x_u & x_v & x_w \\ y_u & y_v & y_w \\ z_u & z_v & z_w \end{pmatrix} \tag{5.67}$$

and its determinant is denoted by

$$\frac{\partial(x, y, z)}{\partial(u, v, w)} := \mathbf{det} \begin{pmatrix} x_u & x_v & x_w \\ y_u & y_v & y_w \\ z_u & z_v & z_w \end{pmatrix} . \tag{5.68}$$

The results analogous to (5.61) and (5.60) are collected in the following

Theorem 5.66 (Change of variables in triple integrals)**.** Let $\mathbf{T}$ be a one–to–one mapping from the set $\mathbf{T}^{-1}(\mathcal{D})$ (in the uvw–space) to the set $\mathcal{D}$ in the xyz–space. Then:
(a) The volume element in $\mathcal{D}$ corresponding to $du\, dv\, dw$ is

$$dV = \left| \frac{\partial(x, y, z)}{\partial(u, v, w)} \right| du\, dv\, dw . \tag{5.69}$$

(b) If a function $f(x, y, z)$ is integrable in $\mathcal{D}$, then

$$\iiint_{\mathcal{D}} f(x, y, z)\, dx\, dy\, dz =$$

$$\iiint\limits_{\mathbf{T}^{-1}(\mathcal{D})} f(x(u,v,w), y(u,v,w), z(u,v,w)) \left| \frac{\partial(x,y,z)}{\partial(u,v,w)} \right| du\, dv\, dw \; . \tag{5.70}$$

Proof:
The proof of (5.69) is analogous to the proof of (5.61), page 232. As there, the volume element corresponding to $du\, dv\, dw$ is approximated by the parallelepiped with sides

$$\frac{\partial \mathbf{r}}{\partial u}\, du \; , \quad \frac{\partial \mathbf{r}}{\partial v}\, dv \; , \quad \text{and} \quad \frac{\partial \mathbf{r}}{\partial w}\, dw \; ,$$

whose volume dV is, by Theorem 1.11, given by the absolute value of the triple product

$$\begin{aligned} dV &= \left| \frac{\partial \mathbf{r}}{\partial u} \cdot \frac{\partial \mathbf{r}}{\partial v} \times \frac{\partial \mathbf{r}}{\partial w} \right| du\, dv\, dw \; , \\ &= \left| \mathbf{det} \begin{pmatrix} x_u & x_v & x_w \\ y_u & y_v & y_w \\ z_u & z_v & z_w \end{pmatrix} \right| du\, dv\, dw \; . \end{aligned}$$

The integration formula (5.70) follows from (5.69). □
Theorem 5.66 is illustrated in the following examples.

Example 5.67 (Cylindrical coordinates)**.** For some problems it is convenient to use **cylindrical coordinates**.

$$x := r \cos\theta \; , \quad y := r \sin\theta \; , \quad z := z \; . \tag{5.71}$$

The Jacobian of this mapping is

$$J = \begin{pmatrix} \cos\theta & -r \sin\theta & 0 \\ \sin\theta & r \cos\theta & 0 \\ 0 & 0 & 1 \end{pmatrix}$$

and its determinant is

$$\frac{\partial(x,y,z)}{\partial(r,\theta,z)} = r \; . \tag{5.72}$$

The volume element in the xyz–space corresponding to $dr\, d\theta\, dz$, as illustrated in Figure 5.25 is therefore

$$dV = r\, dr\, d\theta\, dz \; . \tag{5.73}$$

Note that the arc–length corresponding to a radius r and angle $d\theta$ is $r\, d\theta$. The integration formula (5.70) then becomes

$$\iiint\limits_{\mathcal{D}} f(x,y,z)\, dx\, dy\, dz = \iiint\limits_{\mathbf{T}^{-1}(\mathcal{D})} f(r \cos\theta, r \sin\theta, z))\; r\; dr\, d\theta\, dz \; . \tag{5.74}$$

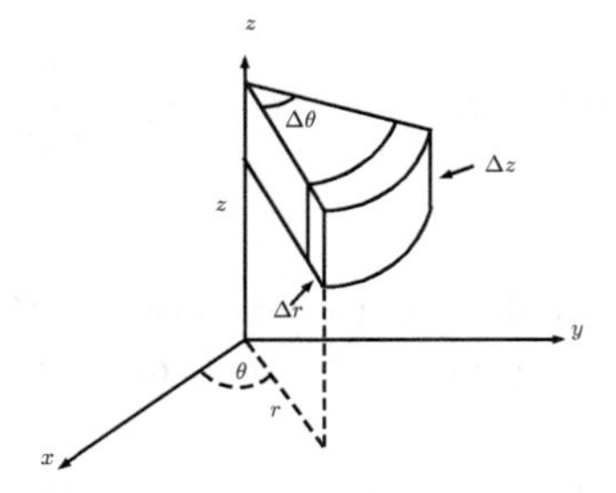

FIGURE 5.25: The volume element in cylindrical coordinates.

Example 5.68 (Spherical coordinates)**.** In problems involving spheres, the coordinates of choice are the **spherical coordinates**

$$x := \rho \sin\varphi \cos\theta \ , \quad y := \rho \sin\varphi \sin\theta \ , \quad z := \rho \cos\varphi \ .$$

Here the Jacobian is

$$J = \begin{pmatrix} \sin\varphi \cos\theta & \rho \cos\varphi \cos\theta & -\rho \sin\varphi \sin\theta \\ \sin\varphi \sin\theta & \rho \cos\varphi \sin\theta & \rho \sin\varphi \cos\theta \\ \cos\varphi & -\rho \sin\varphi & 0 \end{pmatrix}$$

and its determinant is

$$\frac{\partial(x,y,z)}{\partial(\rho,\varphi,\theta)} = \rho^2 \sin\varphi \ . \tag{5.75}$$

Therefore the volume element, in the xyz–space, corresponding to $d\rho\, d\varphi\, d\theta$ is

$$dV = \rho^2 \sin\varphi \, d\rho \, d\varphi \, d\theta \ , \tag{5.76}$$

as illustrated in Figure 5.26, where the volume element is approximated by a box of sides $d\rho$, $\rho\, d\varphi$ and $\rho \sin\varphi \, d\theta$. Finally, the integration formula (5.70) gives

$$\iiint_{\mathcal{D}} f(x,y,z)\, dx\, dy\, dz =$$

$$\iiint_{\mathbf{T}}^{-1} (\mathcal{D}) f(\rho \sin(\varphi) \cos(\theta), \rho \sin(\varphi) \sin(\theta), \rho \cos(\varphi) \ \rho^2 \sin(\varphi) \ d\rho\, d\varphi\, d\theta \ . \tag{5.77}$$

Remark 5.69. For spherical coordinates, the volume element dV, and the spherical surface differential

$$dS = \rho^2 \sin\varphi \, d\varphi \, d\theta \ , \tag{5.78}$$

satisfy

$$dV = d\rho\, dS \ ,$$

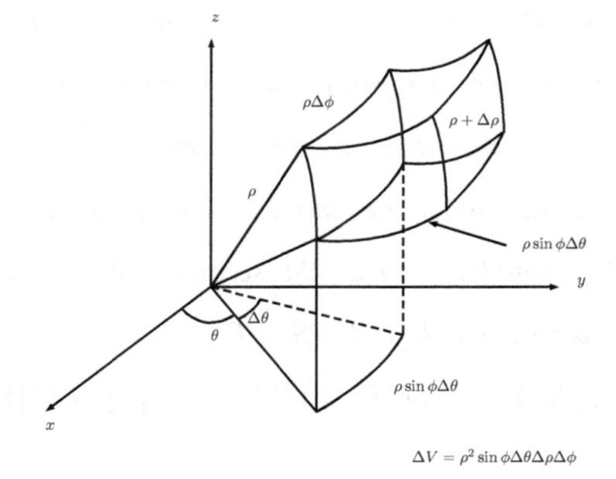

FIGURE 5.26: The volume element in spherical coordinates.

which makes sense (why?). Similarly, in cylindrical coordinates we expect the volume element dV and the surface differential

$$dS = \sqrt{r^2 + r^2 \left(\frac{\partial z}{\partial r}\right)^2 + \left(\frac{\partial z}{\partial \theta}\right)^2}\, dr\, d\theta \,, \tag{5.79}$$

to satisfy

$$dV = dz\, dS \quad \text{(why?)} .$$

This is indeed the case, since r, θ and z are independent and therefore

$$\frac{\partial z}{\partial r} = \frac{\partial z}{\partial \theta} = 0 \,, \quad \text{showing that} \quad dS = r\, dr\, d\theta \,.$$

Example 5.70. Let $\mathcal{D}$ be a uniform hemispherical shell of inner radius ρ_0 and outer radius ρ_1. It is required to find the centroid $(\overline{x}, \overline{y}, \overline{z})$ of $\mathcal{D}$. To simplify the calculations we take the center of the sphere at the origin. Moreover, we assume that the shell is symmetric w.r.t. the x–axis, which guarantees that $\overline{y} = \overline{z} = 0$. The volume V of the shell is the difference of the volume of the outer hemisphere and the volume of the inner hemisphere:

$$V = \frac{2\pi}{3}(\rho_1^3 - \rho_0^3) \,.$$

Then

$$\begin{aligned}
\overline{x} &= \frac{1}{V} \iiint\limits_{\mathcal{D}} x\, dV \\
&= \frac{1}{V} \int_{-\frac{\pi}{2}}^{\frac{\pi}{2}} d\theta \int_0^{\pi} d\varphi \int_{\rho_0}^{\rho_1} (\rho\, \sin\varphi\, \cos\theta)\, \rho^2\, \sin\varphi\, d\rho \,, \quad (\text{since } x = \rho\, \sin\varphi\, \cos\theta) \\
&= \frac{1}{V} \int_{-\frac{\pi}{2}}^{\frac{\pi}{2}} d\theta \int_0^{\pi} d\varphi \int_{\rho_0}^{\rho_1} \rho^3\, \sin^2\varphi\, \cos\theta\, d\rho = \frac{(\pi/4)\, (\rho_1^4 - \rho_0^4)}{(2\pi/3)\, (\rho_1^3 - \rho_0^3)} \,.
\end{aligned}$$

Mathematica session 5.16. *In this session we implement with* MATHEMATICA *the change of integration parameter as the procedure. The coordinate change is indicated as two vectors* $\vec{U} := [u_1, u_2, u_3]$ *and* $[\text{param1}_1, \text{param2}_1, \text{param3}_1]$, *i.e. this is a vector containing as its first component the first parameter, and as the second and third components the parameter*

```
tripleIntegral[default_, R_, S_, T_]:=
Integrate[default, {R[[1]], R[[2]], R[[3]]}, {S[[1]], S[[2]], S[[3]]},
{T[[1]], T[[2]], T[[3]]}]
sphere:={rho * Sin[phi] * Cos[theta], rho * Sin[phi] * Sin, [theta], rho * Cos[phi]}

cylinder = {r * Cos[theta], r * Sin[theta], z}

{r Cos[theta], r Sin[theta], z}

jacobianDeterminant[exprs_, vars_]:=Simplify[Det[Outer[D, exprs, vars]]]

jacobianDeterminant[{rho * Sin[phi] * Cos[theta], rho * Sin[phi] * Sin[theta],
rho * Cos[phi]},
{rho, phi, theta}]
```

$rho^2 Sin[phi]$

```
changeOfVariableTripleIntegral[f_, U_, R_, S_, T_]:=
Module[{j}, j = jacobianDeterminant[U, {R[[1]], S[[1]], T[[1]]}];
tripleIntegral[f * j, R, S, T]]
```

Let us now calculate the volume of a sphere of radius a using the change of integration parameter to spherical coordinates.

```
changeOfVariableTripleIntegral[1,
{rho*Sin[phi]*Cos[theta], rho * Sin[phi] * Sin[theta], rho * Cos[phi]},
{rho, 0, 1}, {phi, 0, Pi}, {theta, 0, 2 * Pi}]
```

$\frac{4\pi}{3}$

(x * y * z)/.{x → rho * Sin[phi] * Cos[theta], y->
rho * Sin[phi] * Sin[theta], z->rho * Cos[phi]}

$rho^3\, Cos[phi]\, Cos[theta]\, Sin[phi]^2\, Sin[theta]$

tripleIntegral [rho³ Cos[phi] Cos[theta] Sin[phi]² Sin[theta] * rho² Sin[phi],
{rho, 0, 1}, {phi, 0, Pi}, {theta, 0, 2 * Pi}]

0

Next let us show that the integral of $f(x, y, z) := xyz$ *over a sphere is zero. This is actually obvious by symmetry arguments.*

changeOfVariableTripleIntegral[
(x * y * z)/.{x → rho * Sin[phi] * Cos[theta], y->rho * Sin[phi] * Sin[theta],
z->rho * Cos[phi]}, {rho * Sin[phi] * Cos[theta],
rho * Sin[phi] * Sin[theta], rho * Cos[phi]},
{rho, 0, 1}, {phi, 0, Pi}, {theta, 0, 2 * Pi}]

0

We next verify the the volume of a cylinder of height h and radius a is $2\pi\, a^{2\,h}$.

changeOfVariableTripleIntegral[1, {r * Cos[theta], z, r * Sin[theta]}, {r, 0, a},
{z, 0, h}, {theta, 0, 2 * Pi}]

$a^2 h\pi$

changeOfVariableTripleIntegral[

```
Simplify[(x^2 + y^2 + z^2)/.{x → r * Cos[theta], y->r * Sin[theta], z → z}],
{r * Cos[theta], z, r * Sin[theta]}, {r, 0, a}, {z, 0, h}, {theta, 0, 2 * Pi}]
```

$\frac{1}{6}a^2h\left(3a^2+2h^2\right)\pi$

```
changeOfVariableTripleIntegral[
Simplify[(x + y + z)/.{x → r * Cos[theta], y->r * Sin[theta], z → z}],
{r * Cos[theta], z, r * Sin[theta]}, {r, 0, a}, {z, 0, h}, {theta, 0, 2 * Pi}]
```

$\frac{1}{2}a^2h^2\pi$

Next we calculate the volume of the sphere which remains after the cone is cut out.

```
changeOfVariableTripleIntegral[1,
{rho * Sin[phi] * Cos[theta], rho * Sin[phi] * Sin[theta], rho * Cos[phi]},
{rho, 0, a}, {phi, Pi/6, Pi}, {theta, 0, 2 * Pi}]
```

$\frac{1}{3}\left(2+\sqrt{3}\right)a^3\pi$

Next we calculate the volume the cone cuts out of the sphere.

```
changeOfVariableTripleIntegral[1,
{rho * Sin[phi] * Cos[theta], rho * Sin[phi] * Sin[theta], rho * Cos[phi]},
{rho, 0, a}, {phi, 0, Pi/6}, {theta, 0, 2 * Pi}]
```

$-\frac{1}{3}\left(-2+\sqrt{3}\right)a^3\pi$

♦

Exercises

Exercise 5.71. *Draw the domain $\mathcal{A}$ in Example 5.65. Compute its:*
(a) *area* , (b) *centroid, and* (c) *moment of inertia* .

Exercise 5.72. *Let the region $\mathcal{A}$ in the xy–plane be bounded by portions of the curves*

$$x\,y \;=\; 1\;, \quad x\,y \;=\; 4\;, \quad y \;=\; \frac{x}{2} \quad \text{and} \quad y \;=\; 2\,x\;.$$

Use the transformation

$$x \;:=\; u\,v\;,\; y \;:=\; \frac{u}{v}\;, \quad \text{with} \quad u>0\;,\; v>0\;,$$

to transform the integral

$$\iint\limits_{\mathcal{A}} \left(\left(\frac{y}{x}\right)^{\frac{3}{2}} + (x\,y)^{\frac{3}{2}}\right)\,dx\,dy\;,$$

to a simpler form, and evaluate it.

Exercise 5.73. *Use the transformation $x \;:=\; a\,u\;,\; y \;:=\; b\,v$ to transform the ellipese*

$$\frac{x^2}{a^2} + \frac{y^2}{b^2} \;=\; 1$$

into the unit circle. Compute the area of the ellipse.

Exercise 5.74. *Use the transformation $x \;:=\; a\,u\;,\; y \;:=\; b\,v$ and $z \;:=\; c\,w$ to compute the volume of the ellipsoid*

$$\frac{x^2}{a^2} + \frac{y^2}{b^2} + \frac{z^2}{c^2} \;=\; 1\;.$$

Exercise 5.75. *Use the transformation $x \;:=\; a\,r\,\cos\theta\;,\; y \;:=\; b\,r\,\sin\theta$ to evaluate the integral*

$$\iint\limits_{\mathcal{A}} (x\,y)\,dx\,dy$$

where $\mathcal{A}$ is the domain bounded by the ellipse

$$\frac{x^2}{a^2} + \frac{y^2}{b^2} \;=\; 1\;.$$

Exercise 5.76. *For each of the following transformations compute the Jacobi matrix and its inverse. Specify where the transformation is one–to–one.*

(a) $x \;:=\; u\;,\; y \;:=\; v\;,\; z \;:=\; w$

(b) $x \;:=\; u\,v\;,\; y \;:=\; u\,\sqrt{1-v^2}\;,\; z \;:=\; w$

(c) $x \;:=\; u\,v\;,\; y \;:=\; v\,w\;,\; z \;:=\; w$

(d) $x \;:=\; \frac{1}{2}(u^2-v^2)\;,\; y \;:=\; u\,v\;,\; z \;:=\; w$

(e) $x \;:=\; u\,v\,w\;,\; y \;:=\; u\,v\,\sqrt{1-w^2}\;,\; z \;:=\; \frac{1}{2}(u^2-v^2)$

Exercise 5.77. *Verify the determinants of the Jacobians:*

(a) (5.72) (b) (5.75)

Hint: Expand each of these determinants in the 3rd row of its Jacobian matrix.

Exercise 5.78. *The density δ of a sphere of radius a is proportional to the inverse of the distance ρ from the origin, say*

$$\delta = k\left(\frac{1}{\rho}\right).$$

Find the mass of the sphere. Note that it is finite, even though the density is infinite at the origin.

Exercise 5.79. *In parts (a)–(f) assume that the solid has uniform density. Find the centroid of:*

(a) *the solid hemisphere* $\{(x, y, z) : x^2 + y^2 + z^2 \leq a^2 \ , \ z \geq 0\}$

(b) *the 1/8 sphere* $\{(x, y, z) : x^2 + y^2 + z^2 \leq a^2 \ , \ x \geq 0 \ , \ y \geq 0 \ , \ z \geq 0\}$

(c) *the solid bounded by* $z = 4(x^2 + y^2)$ *and* $z = 2$

(d) *the solid bounded by* $z = 4(x^2 + y^2)$, $x^2 + y^2 = 16$ *and* $z = 0$

(e) *the portion of the torus* $z^2 + (r - 4)^2 = 1$ *which lies in the second quadrant*

(f) *the solid bounded above by* $z = 4 - x^2 - y^2$ *and below by* $x^2 + y^2 + z^2 = 16$

Exercise 5.80. *The gravitational attraction force between two particles is proportional to the inverse of the square of the distance between them,*

$$\mathbf{F} = \frac{\gamma\, m_1\, m_2}{\|\mathbf{r}_1 - \mathbf{r}_2\|^2},$$

where $\mathbf{r}_i$ is the position vector of particle i, m_i is the mass of particle i, and γ is the gravitational constant. Determine the gravitational force on a particle of unit mass located at the origin, due to:

(a) *the solid hemisphere:* $\{(x, y, z) : x^2 + y^2 + z^2 \leq a^2 \ , \ z \geq 0\}$

(b) *the solid cone:* $\{(x, y, z) : y^2 + z^2 \leq k^2\, x^2 \ , \ 0 \leq x \leq h\}$

(c) *the solid hemispherical shell:* $\{(x, y, z) : \rho_0^2 \leq x^2 + y^2 + z^2 \leq \rho_1^2 \ , \ z \geq 0\}$

Assume the above three solids have uniform density.

Exercise 5.81. *Find the volume of the solid:*

(a) *bounded on the outside by* $x^2 + y^2 + z^2 = 4$ *and inside by* $x^2 + y^2 = 1$

(b) *bounded on the outside by* $x^2 + y^2 + z^2 = 4$ *and inside by* $z^2 = x^2 + y^2$

(c) *bounded by the ellipsoidal cylinder* $\frac{x^2}{4} + y^2 = 1$
and the planes $z = 2\,x + 3\,y + 6$ *and* $z = 0$

Chapter 6

Vector Calculus

6.1 Fields, potentials and line integrals

Let $\mathbf{r}$ denote a **position** $<x, y> = x\,\mathbf{i} + y\,\mathbf{j}$ in $\mathbb{R}^2$, or $<x, y, z> = x\,\mathbf{i} + y\,\mathbf{j} + z\,\mathbf{k}$ in $\mathbb{R}^3$.

A **field** is simply a function of the position $\mathbf{r}$. The **domain** of the field is the set (in $\mathbb{R}^2$ or $\mathbb{R}^3$) where the field is defined. In physics there are two kinds of fields:

1. **Scalar fields** using scalar functions, i.e. functions whose values are single numbers. Examples are **temperature** T and **pressure** p, which depend on the location $\mathbf{r}$.

2. **Vector fields** i.e. functions whose values are vectors. Examples are **force** $\mathbf{F}$, **velocity** $\mathbf{v}$, **electric field** $\mathbf{E}$, **magnetic field** $\mathbf{H}$, **electric displacement** $\mathbf{D}$, **magnetic induction** $\mathbf{B}$ and of course the **position** $\mathbf{r}$ itself.

The above fields may also change with time, in which case we write them as functions of **position** $\mathbf{r}$ and **time** t. For example, the velocity of a fluid measured in time t at point $\mathbf{r}$ is denoted $\mathbf{v}(\mathbf{r}, t)$.

A vector field $\mathbf{F}(\mathbf{r})$ may be 2– or 3–dimensional, independently of $\mathbf{r}$. For example, a 3–dimensional vector field on $\mathbb{R}^3$ is written as

$$\mathbf{F}(x, y, z) := \mathbf{i}\,F_1(x, y, z) + \mathbf{j}\,F_2(x, y, z) + \mathbf{k}\,F_3(x, y, z)\ , \tag{6.1}$$

while a 3-dimensional vector field on $\mathbb{R}^2$ is denoted by

$$\mathbf{F}\,(x, y) := \mathbf{i}\,F_1(x, y) + \mathbf{j}\,F_2(x, y) + \mathbf{k}\,F_3(x, y)\ .$$

Two-dimensional vector fields are denoted similarly,

$$\mathbf{F}\,(x, y, z) := \mathbf{i}\,F_1(x, y, z) + \mathbf{j}\,F_2(x, y, z) \quad (\text{on } \mathbb{R}^3)\ , \tag{6.2}$$

$$\mathbf{F}\,(x, y) := \mathbf{i}\,F_1(x, y) + \mathbf{j}\,F_2(x, y) \quad (\text{on } \mathbb{R}^2)\ . \tag{6.3}$$

In this chapter we concentrate on 3–dimensional vector fields with domains in $\mathbb{R}^3$ (man and his environment are for the most part three-dimensional). Cases where the restriction to two dimensions leads to interesting conclusions or formulations will be treated separately.

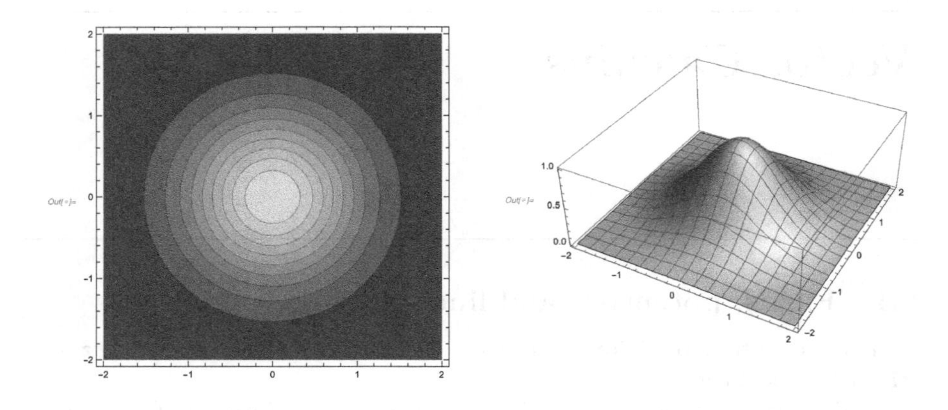

FIGURE 6.1: A contour map of a mountain with gradients.

FIGURE 6.2: The same mountain in three-dimensional perspective.

We recall the definition of the **gradient**,

$$\nabla f(x,y) \;:=\; \frac{\partial f}{\partial x}\,\mathbf{i} + \frac{\partial f}{\partial y}\,\mathbf{j}\,, \quad \text{of a function } f(x,y)\,, \tag{6.4}$$

$$\nabla f(x,y,z) \;:=\; \frac{\partial f}{\partial x}\,\mathbf{i} + \frac{\partial f}{\partial y}\,\mathbf{j} + \frac{\partial f}{\partial z}\,\mathbf{k}\,, \quad \text{of a function } f(x,y,z)\,. \tag{6.5}$$

and define the **gradient operator**, denoted by the Greek symbol ∇, pronounced **del** or **nabla**

$$\nabla \;:=\; \mathbf{i}\,\frac{\partial}{\partial x} + \mathbf{j}\,\frac{\partial}{\partial y} + \mathbf{k}\,\frac{\partial}{\partial z}\,, \tag{6.6}$$

see Exercise 6.14. The following example illustrates several fields with a domain in $\mathbb{R}^2$.

Example 6.1. Consider the mountain of Figure (6.2). At any point (x, y), the **altitude** z is shown in the **contour map**. The altitude is a scalar field, $z(x, y)$. The slope of the mountain at the point (x, y, z) is indicated by the **gradient** of the altitude z

$$\nabla\, z(x,y) \;=\; \frac{\partial z}{\partial x}\,\mathbf{i} + \frac{\partial z}{\partial y}\,\mathbf{j}\,.$$

At any point (x, y) the gradient $\nabla\, z(x, y)$ points in the direction of steepest ascent. The slope in any direction $\mathbf{u}$ is the directional derivative

$$D_{\mathbf{u}}z(x,y) \;=\; \mathbf{u}\,\cdot\,\nabla z(x,y).$$

We place a billiard ball at some point (x, y, z) on the mountain. The ball has **potential energy**,

$$V = m\,g\,z$$

where m is the **mass** of the ball, g is the **gravity acceleration**. The potential energy is a scalar field $V(x, y, z)$ (although, in this problem it depends explicitly only on z). The **gravity force G** on the billiard ball is

$$\mathbf{G} = -m\,g\,\mathbf{k}\,,$$

which points down, hence the negative sign. We can write this as

$$\mathbf{G} = -\nabla V = \nabla\,(-V)\;,$$

a 3–dimensional gradient (as in (6.5)), which is vertical since $\frac{\partial V}{\partial x} = \frac{\partial V}{\partial y} = 0$. The function $-V$ is called the (**scalar**) **potential** of the force **G**.

The ball has a tendency to roll down the mountain, in the direction of $-\nabla\,z(x, y)$, and exchange its potential energy into kinetic energy. To keep it in place, we must apply a **force F** that will prevent it from rolling down. This force is proportional to the slope of the mountain at the position of the ball, i.e.

$$\mathbf{F} = m\,g\,\nabla\,z(x, y)\;,$$

a 2–dimensional gradient, since z is a function of (x, y). The vector field **F** is thus represented by the field of gradients in Figure (6.2).

This example is typical, since many vector fields in physics and engineering are **gradients** of **scalar potentials**.

Mathematica session 6.1. *In this* MATHEMATICA *session we show how to take the curl of vector fields and form the gradients of scalar fields. This will be illustrated for* $\mathbb{R}^n, n = 2, 3$. *Since there is already a three–dimensional curl we convert a vector in* $\mathbb{R}^2$ *to a three dimension vector by adjoining a third coordinate taken to be* 0.

```
twoDimensionalCurl[F_]:=Module[{f}, f = Join[F, {0}];
Dot[Curl[f, {x, y, z}], {0, 0, 1}]]
```

We illustrate this function with several examples below.

```
twoDimensionalCurl[{x/Sqrt[x^2 + y^2], y/Sqrt[x^2 + y^2]}]
```

0

♦

```
twoDimensionalCurl[{y/Sqrt[x^2 + y^2], -x/Sqrt[x^2 + y^2]}]
```

$\frac{x^2}{(x^2+y^2)^{3/2}} + \frac{y^2}{(x^2+y^2)^{3/2}} - \frac{2}{\sqrt{x^2+y^2}}$

twoDimensionalCurl[{y^2, x^2}]

$2x - 2y$

twoDimensionalCurl[{P[x, y], Q[x, y]}]

$-P^{(0,1)}[x, y] + Q^{(1,0)}[x, y]$

♦

Mathematica session 6.2. ***Clear["Global*"]***

We now use MATHEMATICA *to compute gradients and curls in* $\mathbb{R}^3$. *First we introduce the distance to the origin as* R.

R = Sqrt[x^2 + y^2 + z^2]

$\sqrt{x^2 + y^2 + z^2}$

Grad[%, {x, y, z}]

We now take the gradient of the scalar field

$\left\{\frac{x}{\sqrt{x^2+y^2+z^2}}, \frac{y}{\sqrt{x^2+y^2+z^2}}, \frac{z}{\sqrt{x^2+y^2+z^2}}\right\}$

We compute the gradient of the scalar field

$$\frac{xyz}{x^2 + y^2 + z^2}$$

and several other scalar fields.

Grad[(x * y) * z/(x^2 + y^2 + z^2), {x, y, z}]

$\left\{-\frac{2x^2yz}{(x^2+y^2+z^2)^2} + \frac{yz}{x^2+y^2+z^2}, -\frac{2xy^2z}{(x^2+y^2+z^2)^2} + \frac{xz}{x^2+y^2+z^2}, -\frac{2xyz^2}{(x^2+y^2+z^2)^2} + \frac{xy}{x^2+y^2+z^2}\right\}$

Grad[1/Sqrt[x^2 + y^2 + z^2], {x, y, z}]

$$\left\{-\frac{x}{(x^2+y^2+z^2)^{3/2}}, -\frac{y}{(x^2+y^2+z^2)^{3/2}}, -\frac{z}{(x^2+y^2+z^2)^{3/2}}\right\}$$

Grad[1/R^2, {x, y, z}]

$$\left\{-\frac{2x}{(x^2+y^2+z^2)^2}, -\frac{2y}{(x^2+y^2+z^2)^2}, -\frac{2z}{(x^2+y^2+z^2)^2}\right\}$$

v:={x, y, z}

Grad[1/R^2, v]

$$\left\{-\frac{2x}{(x^2+y^2+z^2)^2}, -\frac{2y}{(x^2+y^2+z^2)^2}, -\frac{2z}{(x^2+y^2+z^2)^2}\right\}$$

We may also compute gradients in different coordinate systems, for example cylindrical and spherical coordinates. We first indicate the syntax for cylindrical coordinates.

Grad[f[r, theta, z], {r, theta, z}, "Cylindrical"]

$$\left\{f^{(1,0,0)}[r, theta, z], \frac{f^{(0,1,0)}[r, theta, z]}{r}, f^{(0,0,1)}[r, theta, z]\right\}$$

Grad[r * Cos[theta] * z^2, {r, theta, z}, "Cylindrical"]

$\{z^2\,Cos[theta], -z^2\,Sin[theta], 2rz\,Cos[theta]\}$

We now show the syntax for taking the gradient in spherical coordinates.

w:={r, theta, phi}

Grad[r^2 * Cos[theta] * Sin[phi], w, "Spherical"]

$\{2r\,Cos[theta]\,Sin[phi], -r\,Sin[phi]\,Sin[theta], r\,Cos[phi]\,Cot[theta]\}$

We now shoiw several other examples.

Grad[r * Cos[theta] * z^2, {r, theta, z}, "Cylindrical"]

$\{z^2\,Cos[theta], -z^2\,Sin[theta], 2rz\,Cos[theta]\}$

Grad[r^2 * Cos[theta] * Sin[phi], w, "Spherical"]

$\{2r\,Cos[theta]\,Sin[phi], -r\,Sin[phi]\,Sin[theta], r\,Cos[phi]\,Cot[theta]\}$

♦

Line integrals are integrals of vector fields over curves in the plane, or in 3–dimensional space. The following examples illustrate the importance of this concept.

Example 6.2 (Work). Given a force field

$$\mathbf{F}(\mathbf{r}) = \mathbf{i}\,F_1(\mathbf{r}) + \mathbf{j}\,F_2(\mathbf{r}) + \mathbf{k}\,F_3(\mathbf{r}) ,$$

we compute the work $W_{\mathcal{C}} = W_{\mathcal{C}}(B, E)$ corresponding to the movement of a particle along a curved path $\mathcal{C}$, from a point B (**beginning**) to a point E (**end**). A negative value of $W_{\mathcal{C}}$ means that the movement requires work, a positive value indicates that work is "done" by the movement. Consider first the work ΔW corresponding to an "infinitesimal" movement from the point $\mathbf{r}_0 := <x_0, y_0, z_0>$ to the point $\mathbf{r}_0 + \Delta\mathbf{r} := <x_0 + \Delta x, y_0 + \Delta y, z_0 + \Delta z>$. This work is approximated by the inner product

$$\Delta W \approx F_1\,\Delta x + F_2\,\Delta y + F_3\,\Delta z = \mathbf{F}(\mathbf{r}_0) \cdot \Delta\mathbf{r} \tag{6.7}$$

which is negative if the angle between the vectors $\mathbf{F}(\mathbf{r}_0)$ and $\Delta\mathbf{r}$ is obtuse, i.e. if the movement is "against" the force. If the above angle is acute, the work ΔW is positive, i.e. work is gained if the movement is "with" the force. If the vectors $\mathbf{F}(\mathbf{r}_0)$ and $\Delta\mathbf{r}$ are orthogonal, no work is required or gained. Approximating the path $\mathcal{C}$ as a sum of infinitesimal paths $\Delta\mathbf{r}$, we can approximate $W_{\mathcal{C}}$ as a Riemann sum

$$W_{\mathcal{C}} \approx \sum_{i=0}^{n} \mathbf{F}(\mathbf{r}_i) \cdot \Delta\mathbf{r}_i$$

where $\mathbf{r}_0 = B$, $\mathbf{r}_n = E$. These Riemann sums may converge, in the limit as $\max\limits_{i=1,\dots,n} |\Delta\mathbf{r}_i| \to 0$, to the **line integral** (or **path integral**),

$$W_{\mathcal{C}} = \int_{\mathcal{C}} \mathbf{F}(\mathbf{r}) \cdot d\mathbf{r} \tag{6.8a}$$

$$= \int_{\mathcal{C}} (F_1\,dx + F_2\,dy + F_3\,dz) \tag{6.8b}$$

giving the total work corresponding to the movement along $\mathcal{C}$ from B to E. Let the curve $\mathcal{C}$ be given in parametric form as

$$\mathcal{C} := \{\mathbf{r}(t) : 0 \le t \le T\} \quad \text{where } \mathbf{r}(0) = B\ ,\ \mathbf{r}(T) = E\ .$$

Then the line integral (6.8) becomes

$$W_{\mathcal{C}} = \int_0^T \left(\mathbf{F}(\mathbf{r}(t)) \cdot \frac{d\mathbf{r}(t)}{dt}\right) dt \tag{6.9}$$

where the limits of integration, $t = 0$ and $t = T$, correspond to the points B and E respectively. The line integral (6.9) depends only on the endpoints B, E and the curve $\mathcal{C}$, not on the parametric representation of $\mathcal{C}$ (as long as moving from $t = 0$ to $t = T$ corresponds to moving from B to E along $\mathcal{C}$). In particular, if the parameter is the arclength s, then

$$\mathbf{t}(s) = \frac{d\mathbf{r}}{ds} = \frac{dx}{ds}\mathbf{i} + \frac{dy}{ds}\mathbf{j} + \frac{dz}{ds}\mathbf{k} \tag{6.10}$$

is the **unit tangent** to the curve at the point $\mathbf{r}(s)$, and the line integral (6.9) becomes

$$W_{\mathcal{C}} = \int_0^L (\mathbf{F}(\mathbf{r}(s)) \cdot \mathbf{t}(s))\, ds\,. \tag{6.11}$$

The limits of integration, $s = 0$ and $s = L$, correspond to the points B and E respectively.

Mathematica session 6.3. *We introduce a simple* MATHEMATICA *function to perform line integrals here.*

```
lineIntegral[f_, u_, I_]:=
Module[{r, s, g}, r = D[u, t]; s = Simplify[f/.{x → u[[1]], y → u[[2]], z → u[[3]]}
];
g = Dot[s, r];
Integrate[g, {t, I[[1]], I[[2]]}]]

lineIntegral[{x, y, z}/Sqrt[x^2 + y^2 + z^2], {t + 1, t^2, t^3}, {0, 1}]
```

$-1 + \sqrt{6}$

```
lineIntegral[{x, y, 0}/Sqrt[x^2 + y^2], {Cos[t], Sin[t], 0}, {0, 2 * Pi}]
```

0

```
lineIntegral[{-y, x, 0}/Sqrt[x^2 + y^2], {Cos[t], Sin[t], 0}, {0, 2 * Pi}]
```

2π

```
lineIntegral[{x, y, z}/(x^2 + y^2 + z^2)^(3/2), {t + 1, t^2, t^3}, {0, 1}]
```

$1 - \frac{1}{\sqrt{6}}$

♦

Example 6.3. Consider the force field

$$\mathbf{F} := y\,z\,\mathbf{i} + x\,z\,\mathbf{j} + x\,y\,\mathbf{k} \tag{6.12}$$

and a particle moving from $B := (0,0,0)$ to $E := (1,1,2)$ along a curve $\mathcal{C}_1$ given in parametrized form as

$$\mathcal{C}_1 := \{\mathbf{r}(t) := t\,\mathbf{i} + t^2\,\mathbf{j} + 2\,t\,\mathbf{k} \,:\, 0 \le t \le 1\}$$

The line integral (6.9) is then

$$\begin{aligned} W_{\mathcal{C}_1}(B,E) &= \int_{\mathcal{C}_1} \left(\mathbf{F}(\mathbf{r}(t)) \cdot \frac{d\mathbf{r}(t)}{dt}\right) dt \\ &= \int_0^1 (2\,t^3\,\mathbf{i} + 2\,t^2\,\mathbf{j} + t^3\,\mathbf{k}) \cdot (\mathbf{i} + 2\,t\,\mathbf{j} + 2\,\mathbf{k})\,dt \\ &= \int_0^1 (2\,t^3 + 4\,t^3 + 2\,t^3)\,dt = \int_0^1 (6\,t^3 + 2\,t^3)\,dt = 2\,. \end{aligned}$$

Consider now the straight line $\mathcal{C}_2$ joining the points $(0,0,0)$ and $(1,1,2)$,

$$\mathcal{C}_2 := \{\mathbf{r}(t) := t\,\mathbf{i} + t\,\mathbf{j} + 2\,t\,\mathbf{k} \,:\, 0 \le t \le 1\}\,.$$

Then the line integral (6.9) is

$$\begin{aligned} W_{\mathcal{C}_2}(B,E) &= \int_{\mathcal{C}_2} \left(\mathbf{F}(\mathbf{r}(t)) \cdot \frac{d\mathbf{r}(t)}{dt}\right) dt \\ &= \int_0^1 (2\,t^2\,\mathbf{i} + 2\,t^2\,\mathbf{j} + t^2\,\mathbf{k}) \cdot (\mathbf{i} + \mathbf{j} + 2\,\mathbf{k})\,dt \\ &= \int_0^1 (2\,t^2 + 2\,t^2 + 2\,t^2)\,dt = \int_0^1 (6\,t^2)\,dt = 2\,. \end{aligned}$$

This shows that the work corresponding to movement from $(0,0,0)$ to $(1,1,2)$ along $\mathcal{C}_1$ is the same as the work of movement along $\mathcal{C}_2$.

A force field $\mathbf{F}$ is called **conservative** if, for any two points B and E, the line integrals

$$\int_{\mathcal{C}_1} \mathbf{F} \cdot d\mathbf{r} = \int_{\mathcal{C}_2} \mathbf{F} \cdot d\mathbf{r} \tag{6.13}$$

are equal for all curves $\mathcal{C}_1$, $\mathcal{C}_2$ joining B and E, i.e. the work $W_{\mathcal{C}}(B,E)$ depends

only on the endpoints, not on the path joining them. In this case we call the integral (6.13) **path-independent**.
For example, the force $\mathbf{F}$ of (6.35) is conservative (see Example 6.12), and therefore in Example 6.3, the work along any path from $(0,0,0)$ to $(1,1,2)$ is 2.
If the endpoints B and E coincide, i.e. if $\mathcal{C}$ is a closed curve, the line integral (6.30) is called a **closed line integral** along $\mathcal{C}$, denoted

$$\oint_{\mathcal{C}} \mathbf{F}(\mathbf{r}) \cdot d\mathbf{r} \,, \tag{6.14}$$

here $\oint$ stands for a closed line integral. We assume that $\mathcal{C}$ is **simple**, i.e. it does not intersect itself (except for the endpoints $B = E$). Let P be any point on $\mathcal{C}$ which is different from B. We compute (6.14) by breaking $\mathcal{C}$ into two curves, $\mathcal{C}_1$ from B to P, and $\mathcal{C}_2$ from P to $E = B$,

$$\oint_{\mathcal{C}} \mathbf{F}(\mathbf{r}) \cdot d\mathbf{r} \;=\; \oint_{\mathcal{C}_1} \mathbf{F}(\mathbf{r}) \cdot d\mathbf{r} \;+\; \oint_{\mathcal{C}_2} \mathbf{F}(\mathbf{r}) \cdot d\mathbf{r} \,. \tag{6.15}$$

If the force field $\mathbf{F}$ is conservative, then all closed line integrals are zero,

$$\oint_{\mathcal{C}} \mathbf{F} \cdot d\mathbf{r} \;=\; 0 \tag{6.16}$$

representing the fact that the net work of returning to the same point is zero.

Example 6.4. If a line integral is known to be path-independent, we can sometimes simplify its computation. For example, consider the **inverse–square** force

$$\mathbf{F}(\mathbf{r}) \;=\; \frac{\gamma\,\mathbf{r}}{\|\mathbf{r}\|^3} \;=\; \gamma\,\frac{x\,\mathbf{i} + y\,\mathbf{j} + z\,\mathbf{k}}{(x^2+y^2+z^2)^{\frac{3}{2}}} \,, \tag{6.17}$$

so called because it is inversely proportional to the square of the distance $\|\mathbf{r}\|$ from the origin, indeed

$$\|\mathbf{F}(\mathbf{r})\| \;=\; |\gamma|\,\frac{\|\mathbf{r}\|}{\|\mathbf{r}\|^3} \;=\; \frac{|\gamma|}{\|\mathbf{r}\|^2} \,.$$

The work from $B \;:=\; (1,0,0)$ to $E \;:=\; (2,1,1)$, along the path given parametrically by

$$\mathcal{C} \;:=\; \{\mathbf{r} \;=\; (t+1)\,\mathbf{i} + t^2\,\mathbf{j} + t^3\,\mathbf{k} :\; 0 \le t \le 1\} \,,$$

is the line integral

$$\begin{aligned} W_{\mathcal{C}}(B,E) \;&=\; \gamma \int_0^1 \left(\frac{(t+1)\,\mathbf{i} + t^2\,\mathbf{j} + t^3\,\mathbf{k}}{((t+1)^2 + t^4 + t^i 6)^{\frac{3}{2}}}\right) \left(\frac{\mathbf{i} + 2t\mathbf{j} + 3t^2\,\mathbf{k}}{\sqrt{1+4t^2+9t^4}}\right) \sqrt{1+4t^2+9t^4}dt \\ &=\; \gamma \int_0^1 \frac{1 + t + 2\,t^3 + 3\,t^5}{((1+t)^2 + t^4 + t^6)^{\frac{3}{2}}}\, dt \,, \end{aligned}$$

which is a difficult integral, however not for MATHEMATICA, as may be seen in the next session. However, the force (6.34) is conservative (we show this in Example 6.12), so the above line integral is independent of the path $\mathcal{C}$. We can therefore integrate along a simpler path from B to E. The simplest such path is the straight line

$$\mathcal{C}_1 := \{\mathbf{r} = \mathbf{i}\,(1+t) + \mathbf{j}\,t + \mathbf{k}\,t : \ 0 \le t \le 1\}$$

Then the integral simplifies to

$$\begin{aligned} W_{\mathcal{C}_1}(B,E) &= \gamma \int_0^1 \left(\frac{\mathbf{i}\,(1+t) + \mathbf{j}\,t + \mathbf{k}\,t}{((1+t)^2 + 2t^2)^{\frac{3}{2}}} \right) \cdot (\mathbf{i} + \mathbf{j} + \mathbf{k})\, dt \\ &= \gamma \int_0^1 \frac{1+3t}{1+2t+3t^2} dt = \sqrt{1+2t+3t^2}\,|_0^1 = \gamma \left(\sqrt{6} - 1 \right) . \end{aligned}$$

Example 6.5 (Circulation). Let $\mathbf{v}$ be the velocity field of a fluid, i.e. at any point $\mathbf{r} = <x,y,z>$, the vector $\mathbf{v}(x,y,z)$ is the velocity of the fluid at that point. Let $\mathcal{C}$ be a simple closed curve. The **circulation** (of the fluid) around the curve $\mathcal{C}$ is defined as the closed line integral

$$\begin{aligned} \text{CIRC}\,[\mathbf{v}]_{\mathcal{C}} &:= \oint_{\mathcal{C}} \mathbf{v} \cdot d\mathbf{r} \\ &= \oint_{\mathcal{C}} \left(\mathbf{v} \cdot \frac{d\mathbf{r}}{d\alpha} \right) d\alpha \,, \end{aligned} \tag{6.18}$$

where α corresponds to an arbitrary parameterization of the closed curve $\mathcal{C}$. As a 2–dimensional illustration, consider the circulation of the velocity

$$\mathbf{v} := y\,\mathbf{i} + (x+y)\,\mathbf{j}$$

about the circle parameterized by

$$\mathbf{r}(\alpha) := (\cos\alpha)\,\mathbf{i} + (\sin\alpha)\,\mathbf{j} \,.$$

The velocity, on the circle, takes the form

$$\mathbf{v} := (\sin\alpha)\,\mathbf{i} + (\sin\alpha - \cos\alpha)\,\mathbf{j}$$

and its derivative is $$\frac{d\,\mathbf{r}(\alpha)}{d\alpha} = (-\sin\alpha)\,\mathbf{i} + (\cos\alpha)\,\mathbf{j} \,.$$

Therefore
$$\begin{aligned} \mathbf{v} \cdot \frac{d\mathbf{r}}{d\alpha} &= ((\sin\alpha)\,\mathbf{i} + (\sin\alpha - \cos\alpha)\,\mathbf{j}) \cdot ((-\sin\alpha)\,\mathbf{i} + (\cos\alpha)\,\mathbf{j}) \\ &= -\sin^2\alpha - \cos^2\alpha + \sin\alpha\cos\alpha = -1 + \sin\alpha\cos\alpha, \end{aligned}$$

and $$\text{CIRC}\,[\mathbf{v}]_{\mathcal{C}} := \oint_{\mathcal{C}} \mathbf{v} \cdot \frac{d\mathbf{r}}{d\alpha} d\alpha = \oint_0^{2\pi} (\sin\alpha\cos\alpha - 1)\, d\alpha = -2\,\pi \,.$$

Example 6.6. Consider a simple closed curve $\mathcal{C}$ in the plane, surrounding a planar region $\mathcal{D}$. A fluid flows (in the plane) across $\mathcal{D}$, and it is required to compute the net loss or gain of fluid in the region $\mathcal{D}$. The rate at which the fluid volume V passes over a curve segment of length Δs is

$$\frac{dV}{dt} \approx (\mathbf{v} \cdot \mathbf{n})\, \Delta s$$

where $\mathbf{n}$ is the unit normal to $\mathcal{C}$. Summing along the closed curve $\mathcal{S}$, and taking appropriate limits, we get a closed integral called the **flux** of the fluid across the boundary $\mathcal{C}$,

$$\text{FLUX}\,[\mathbf{v}]_{\mathcal{C}} \ := \ \oint_{\mathcal{C}} (\mathbf{v}\cdot\mathbf{n})\, ds\ . \tag{6.19}$$

The normal vector $\mathbf{n}$ is perpendicular to the **unit tangent vector**

$$\mathbf{t} \ := \ \frac{dx}{ds}\mathbf{i} + \frac{dy}{ds}\mathbf{j}\,,$$

giving two possibilities (according to the $\pm$ sign)

$$\mathbf{n} \ = \ \pm\left(\frac{dy}{ds}\mathbf{i} - \frac{dx}{ds}\mathbf{j}\right)\ . \tag{6.20}$$

The positive sign in (6.20) corresponds to a **positive orientation** of the curve, i.e. $\mathcal{C}$ is traveled in a counter–clockwise manner, so that the region $\mathcal{D}$ is always to the left of the curve, see Figure (6.3)(a). In this case, the normal vector $\mathbf{n}$ points away from $\mathcal{D}$, and is given by

$$\mathbf{n} \ = \ \mathbf{k}\times\mathbf{t}$$

where $\mathbf{k}$ is the unit vector along the z–axis. A choice of minus sign in (6.20) cor-

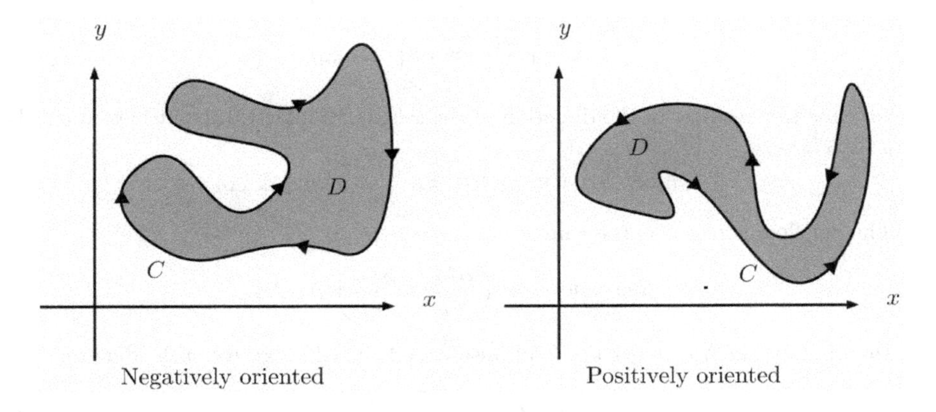

FIGURE 6.3: Negative and positive orientation.

responds to a **negative orientation** of $\mathcal{C}$, see Figure (6.3)(b). Here the curve is traveled clockwise, the region $\mathcal{D}$ is always to the right of $\mathcal{C}$, and the normal vector

$$\mathbf{n} \ = \ -\mathbf{k}\times\mathbf{t}$$

points in a direction inside $\mathcal{D}$. Assuming positive orientation of $\mathcal{C}$, the flux integral (6.19) represents the net outflow rate of the fluid across $\mathcal{C}$, i.e. the rate at which the fluid volume leaves $\mathcal{D}$.

Example 6.7. Consider the infinitesimal rectangle $\mathcal{D}$, of sides Δx and Δy and a vector field

$$\mathbf{F}(x, y) = \mathbf{i}\, F_1(x, y) + \mathbf{j}\, F_2(x, y)\,. \tag{6.21}$$

The flux of $\mathbf{F}$ across the boundary $\mathcal{C}$ is by (6.19),

$$\text{FLUX}\,[\mathbf{F}]_{\mathcal{C}} := \oint_{\mathcal{C}} (\mathbf{F}\cdot\mathbf{n})\, ds\,.$$

We evaluate this integral by summing the flow across the 4 sides of $\mathcal{D}$. The flow through the side BD is

$$\begin{aligned} \text{flow}(BD) &\approx ((F_1\,\mathbf{i} + F_2\,\mathbf{j})\cdot\mathbf{i})\,\Delta y = F_1\,\Delta y \\ &\approx F_1\left(x + \Delta x,\, y + \frac{1}{2}\Delta y\right)\Delta y \end{aligned}$$

and the flow through the opposite side is

$$\begin{aligned} \text{flow}(AC) &\approx (F_1\,\mathbf{i} + F_2\,\mathbf{j})\cdot(-\mathbf{i})\,\Delta y = -F_1\,\Delta y \\ &\approx -F_1\left(x,\, y + \frac{1}{2}\Delta y\right)\Delta y\,, \\ \therefore\ \text{flow}(BD) + \text{flow}(AC) &\approx \left(F_1\left(x + \Delta x, y + \frac{1}{2}\Delta y\right) - F_1\left(x, y + \frac{1}{2}\Delta y\right)\right)\Delta y \\ &\approx \left(\frac{\partial F_1}{\partial x}\left(x,\, y + \frac{1}{2}\Delta x\right)\Delta x\right)\Delta y \\ &\approx \left(\frac{\partial F_1}{\partial x}(x,\, y)\right)\Delta x\,\Delta y, \end{aligned}$$

which is the net flow in the direction of the x–axis. Similarly, the net flow along the y–axis is

$$\text{flow}(CD) + \text{flow}(AB) \approx \frac{\partial F_2}{\partial y}(x,\, y)\,\Delta x\,\Delta y$$

The net flow across $\mathcal{C}$ is the sum

$$\text{FLUX}\,[\mathbf{F}]_{\mathcal{C}} \approx \left(\frac{\partial F_1}{\partial x} + \frac{\partial F_2}{\partial y}\right)\Delta x\,\Delta y.$$

Dividing by $\Delta x\,\Delta y$ we get the **flux**, also called the **divergence** of $\mathbf{F}$, denoted by

$$\text{DIV}\mathbf{F} := \frac{\partial F_1}{\partial x} + \frac{\partial F_2}{\partial y} \tag{6.22}$$

Now we compute the **circulation** of $\mathbf{F}$ along $\mathcal{S}$. The contribution of the side AB is

$$\begin{aligned} \text{circ}(AB) &\approx ((F_1\,\mathbf{i} + F_2\,\mathbf{j})\cdot\mathbf{i})\,\Delta x = F_1\,\Delta x \\ &\approx F_1\left(x + \frac{1}{2}\Delta x,\, y\right)\Delta x \end{aligned}$$

and the contribution of the opposite side is

$$\begin{aligned}
\operatorname{circ}(CD) &\approx (F_1\,\mathbf{i} + F_2\,\mathbf{j}) \cdot (-\mathbf{i})\,\Delta x \;=\; -F_1\,\Delta x \\
&\approx -F_1\left(x + \frac{1}{2}\,\Delta x,\, y + \Delta y\right)\Delta x\,, \\
\therefore\quad \operatorname{circ}(AB) + \operatorname{circ}(CD) &\approx \left(F_1\left(x + \frac{1}{2}\,\Delta x,\, y\right) - F_1\left(x + \frac{1}{2}\,\Delta x,\, y + \Delta y\right)\right)\Delta x \\
&\approx \left(-\frac{\partial F_1}{\partial y}\left(x + \frac{1}{2}\,\Delta x,\, y\right)\Delta y\right)\Delta x \\
&\approx -\frac{\partial F_1}{\partial y}\,(x,\, y)\;\Delta x\,\Delta y.
\end{aligned}$$

Similarly, $$\operatorname{circ}(BD) + \operatorname{circ}(AC) \approx \frac{\partial F_2}{\partial x}\,(x,\, y)\;\Delta x\,\Delta y$$

The circulation along $\mathcal{S}$ is the sum of the above terms

$$\text{CIRC}\,[\mathbf{F}]_{\mathcal{C}} \approx \left(\frac{\partial F_2}{\partial x} - \frac{\partial F_1}{\partial y}\right)\Delta x\,\Delta y.$$

Dividing by $\Delta x\,\Delta y$ we get the **average circulation** $\frac{\partial F_2}{\partial x} - \frac{\partial F_1}{\partial y}$. For reasons to be clear later, we use this expression to define a vertical vector, called the **curl** of **F**, and denoted by

$$\text{CURL}\mathbf{F} := \left(\frac{\partial F_2}{\partial x} - \frac{\partial F_1}{\partial y}\right)\mathbf{k}. \tag{6.23}$$

Mathematica session 6.4. *In this* MATHEMATICA *session we show how to compute the circulation and the flux.*

```
circulation[f_, R_, I_]:=Module[{v, s, g}, v = D[R, t];
F = Simplify[f/.{x → R[[1]], y → R[[2]]}]; g = Dot[F, v];
Integrate[g, {t, I[[1]], I[[2]]}]]
```

```
circulation[{−y, x}/(x^2 + y^2), {Cos[t], Sin[t]}, {0, 2 * Pi}]
```

2π

```
circulation[{−y, x}/(x^2 + y^2), {Cos[t], 2 * Sin[t]}, {0, 2 * Pi}]
```

2π

```
circulation[{−y, x}/(x^2 + y^2), {2 * Cos[t], Sin[t]}, {0, 2 * Pi}]
```

2π

```
circulation[{-y, x}/Sqrt[x^2 + y^2], {Cos[t], Sin[t]}, {0, 2 * Pi}]
```

2π

```
circulation[{-y, x}/Sqrt[x^2 + y^2], {Cos[t], 2 * Sin[t]}, {0, 2 * Pi}]
```

$8EllipticK[-3]$

```
circulation[{-y, x}/Sqrt[x^2 + y^2], {2 * Cos[t], Sin[t]}, {0, 2 * Pi}]
```

$4EllipticK\left[\frac{3}{4}\right]$

```
flux[f_, R_, I_]:=Module[{v, N, F, g}, v = D[R, t]; N = {v[[2]], -v[[1]]};
F = Simplify[f/.{x → R[[1]], y → R[[2]]}]; g = Dot[F, N];
Integrate[g, {t, I[[1]], I[[2]]}]]
```

```
flux[{-y, x}/(x^2 + y^2), {Cos[t], Sin[t]}, {0, 2 * Pi}]
```

0

```
flux[{-y, x}/(x^2 + y^2), {2 * Cos[t], 3 * Sin[t]}, {0, 2 * Pi}]
```

0

```
flux[{-y, x}/(x^2 + y^2), {Abs[a] * Cos[t], Abs[b] * Sin[t]}, {0, 2 * Pi}]
```

0

```
flux[{x/(x^2 + y^2), y/(x^2 + y^2)}, {Cos[t], Sin[t]}, {0, 2 * Pi}]
```

2π

```
circulation[{x/(x^2 + y^2), y/(x^2 + y^2)}, {Cos[t], Sin[t]}, {0, 2 * Pi}]
```

0

♦

Example 6.8 (Electromotive force)**.** The unit of current in the mks system is the [ampere], and the unit of resistance is the [ohm]. $1[\texttt{ampere}] = 1\frac{[\texttt{coulomb}]}{[\texttt{sec}]}$. The [ohm] may be described in terms of work, which is measured in [joule]s and the rate of doing work is measured in [watt]s. More specifically,

$$1\,[\texttt{watt}] = 1\,[\texttt{joule}]\,[\texttt{sec}]^{-1}.$$

These physical quantities are related by

$$1\,[\texttt{ohm}] = 1\,[\texttt{watt}]\,[\texttt{ampere}]^{-2} = 1\,[\texttt{kilogram}]\,[\texttt{meter}]^{2}\,[\texttt{second}]^{-1}\,[\texttt{coulomb}]^{-2};$$

whereas the unit of electric potential is given by

$$1\,[\texttt{volt}] = 1\,[\texttt{watt}]\,[\texttt{ampere}]^{-1} = 1\,[\texttt{kilogram}]\,[\texttt{meter}]^{2}\,[\texttt{second}]^{-2}\,[\texttt{coulomb}]^{-1}.$$

A unit of electric field intensity is obtained from the fact that for most materials the current $\mathbf{J}$ is related to the field intensity linearly, i.e. $\mathbf{J} = \sigma\mathbf{E}$, where σ is a conductivity measured as $[\texttt{ohm}]\,[\texttt{meter}]^{-1}$. As current density is measured in $[\texttt{ampere}]\,[\texttt{meter}]^{2}$ a unit of electric field intensity is one unit of

$$[\texttt{watt}]\,[\texttt{ampere}]\,[\texttt{meter}]^{-1} = [\texttt{volt}]\,[\texttt{meter}]^{-1}.$$

In this example we study forces acted on and by electric charges. Electric charges are measured in [Coulomb]. The charge can have a positive sign, or a negative sign. Consider two electric charges, q and q_1, in positions $\mathbf{r}$ and $\mathbf{r}_1$, respectively. The force $\mathbf{F}_1$ acting on q by q_1, is proportional to the product of the charges, and to the inverse of the square of the distance between them,

$$\mathbf{F}_1 = -\varepsilon \frac{q\,q_1}{\|\mathbf{r}_1 - \mathbf{r}\|^3}\,(\mathbf{r}_1 - \mathbf{r})$$

where $\mathbf{r}_1 - \mathbf{r}$ is the vector pointing from q to q_1. The force $\mathbf{F}_1$ is in the direction of $\mathbf{r}_1 - \mathbf{r}$, i.e. q is attracted to q_1, if they have different signs. If the charges have the same sign (both positive or both negative) then the direction of $\mathbf{F}_1$ is opposite to $\mathbf{r}_1 - \mathbf{r}$, i.e. the charges repulse each other. Consider now a collection of electric charges $q_1, q_2, q_3, \ldots$ in positions $\mathbf{r}_1, \mathbf{r}_2, \mathbf{r}_3, \ldots$, respectively. If we place a charge of q [Coulomb] in position $\mathbf{r}$, it will be subject to a force

$$\mathbf{F}(\mathbf{r}) = -\varepsilon \sum_{k=1,2,3,\ldots} \frac{q\,q_k}{\|\mathbf{r}_k - \mathbf{r}\|^3}\,(\mathbf{r}_k - \mathbf{r})$$

which is the sum of forces due to all the charges. In particular, if $q = 1$, we get

$$\mathbf{E}(\mathbf{r}) = -\varepsilon \sum_{k=1,2,3,\ldots} \frac{q_k}{\|\mathbf{r}_k - \mathbf{r}\|^3}\,(\mathbf{r}_k - \mathbf{r}) \tag{6.24}$$

called the **electric field** at $\mathbf{r}$. It is the force acting on a unit charge in position $\mathbf{r}$. The force acting on a charge of q [Coulomb] in position $\mathbf{r}$ is therefore

$$\mathbf{F}(\mathbf{r}) = q\,\mathbf{E}(\mathbf{r}). \tag{6.25}$$

If a particle of unit charge moves from B to E along a curve $\mathcal{C}$, the work involved is, by (6.30),

$$\int_{\mathcal{C}} \mathbf{E}(\mathbf{r}) \cdot d\mathbf{r} \tag{6.26}$$

called the **electromotive force**, and measured in `[Volts]` (so it is actually **voltage**). For a charge of q `[Coulomb]`, we multiply the electromotive force (6.26) by q to get the actual work

$$W_{\mathcal{C}}(B,E) = q \int_{\mathcal{C}} \mathbf{E}(\mathbf{r}) \cdot d\mathbf{r} \tag{6.27}$$

which has the dimesnion of work, since

$$\texttt{Volt} \times \texttt{Coulomb} = \texttt{Joule}\,.$$

We study next conditions for a line integral to be path–independent.

Theorem 6.9. Let $\mathcal{D} \subset \mathbb{R}^3$ be a simply connected domain, and let the function $\varphi(x,y,z)$ be continuously differentiable in $\mathcal{D}$. If

$$\mathbf{F} = \nabla \varphi \tag{6.28}$$

in $\mathcal{D}$, then for any two points $B, E \in \mathcal{D}$, the line integral from B to E, along a path $\mathcal{C} \subset \mathcal{D}$

$$W_{\mathcal{C}}(B,E) = \int_{\mathcal{C}} \mathbf{F}(\mathbf{r}) \cdot d\mathbf{r} \tag{6.29}$$

is independent of the path $\mathcal{C}$.

Proof:
Let $\mathbf{F} = \nabla \varphi$ be continuous throughout $\mathcal{D}$, let $B := (x_0, y_0, z_0) \in \mathcal{D}$ be an initial point, and $E := (x,y,z)$ be an arbitrary point in $\mathcal{D}$. We connect the points B and E by a curve $\mathcal{C}$ lying in the domain $\mathcal{D}$, and assume that the curve is parametrized by

$$\mathcal{C} := \{\mathbf{r}(t) = x(t)\,\mathbf{i} + y(t)\,\mathbf{j} + z(t)\,\mathbf{k} : 0 \le t \le 1\}$$

where $B = (x(0), y(0), z(0))$, $E = (x(1), y(1), z(1))$ and the functions $x(t), y(t), z(t)$ are continuously differentiable. Then

$$\begin{aligned}
\int_{\mathcal{C}} \mathbf{F}(\mathbf{r}) \cdot d\mathbf{r} &= \int_0^1 \left(\mathbf{F}(\mathbf{r}(t)) \cdot \frac{d\mathbf{r}(t)}{dt} \right) dt\,, \quad \text{by (6.9)}\,, \\
&= \int_0^1 \left(\frac{\partial \varphi}{\partial x}\mathbf{i} + \frac{\partial \varphi}{\partial y}\mathbf{j} + \frac{\partial \varphi}{\partial z}\mathbf{k} \right) \cdot \left(\frac{dx}{dt}\mathbf{i} + \frac{dy}{dt}\mathbf{j} + \frac{dz}{dt}\mathbf{k} \right) dt \\
&= \int_0^1 \left(\frac{\partial \varphi}{\partial x}\frac{dx}{dt} + \frac{\partial \varphi}{\partial y}\frac{dy}{dt} + \frac{\partial \varphi}{\partial z}\frac{dz}{dt} \right) dt \\
&= \int_0^1 \frac{d}{dt}\,\varphi(x(t), y(t), z(t))\,dt \\
&= \varphi(x(1), y(1), z(1)) - \varphi(x(0), y(0), z(0)) = \varphi(E) - \varphi(B)\,,
\end{aligned}$$

showing that the integral depends only on the endpoints B and E, provided the curve $\mathcal{C}$ remains in the domain $\mathcal{D}$. □
We saw in Theorem 6.9 that the condition (6.33) is sufficient for the line integral to be path–independent . The next theorem shows that this condition is also necessary.

Theorem 6.10. Let $\mathcal{D}$ be as above, and let $\mathbf{F}$ be a continuous vector field. If for any two points $B, E \in \mathcal{D}$ the line integral from B to E along paths $\mathcal{C} \subset \mathcal{D}$

$$W_{\mathcal{C}}(B, E) = \int_{\mathcal{C}} \mathbf{F}(\mathbf{r}) \cdot d\mathbf{r} \tag{6.30}$$

is path-independent, then there exists a function $\varphi(x, y, z)$ defined in $\mathcal{D}$ such that

$$\mathbf{F} = \nabla \varphi . \tag{6.31}$$

Proof:
Fix $B := (x_0, y_0, z_0)$ and define

$$\varphi(x, y, z) := \int_{\mathcal{C}} \mathbf{F}(\mathbf{r}) \cdot d\mathbf{r} \tag{6.32}$$

where $\mathcal{C}$ is <u>any</u> path connecting (x_0, y_0, z_0) to (x, y, z). This function is well defined, up to a constant (depending on the choice of the initial point (x_0, y_0, z_0)). Consider now line integrals (6.30) from $B := (x_0, y_0, z_0)$ to $E := (x, y, z)$. Since the integral is path-independent, we take $\mathcal{C}$ as the polygonal path (i.e. a path made of straight lines)

$$\mathcal{C} := \mathcal{C}_x \cup \mathcal{C}_y \cup \mathcal{C}_z$$

where $\mathcal{C}_\alpha$ is a line parallel to the α–axis, $\alpha = x, y, z$. Therefore, $\mathcal{C}_x$, the first leg of the path, is the straight line connecting (x_0, y_0, z_0) to (x, y_0, z_0), $\mathcal{C}_y$ is the straight line connecting (x, y_0, z_0) to (x, y, z_0), and $\mathcal{C}_z$ is the straight line connecting (x, y, z_0) to (x, y, z).
Let the vector field $\mathbf{F}(\mathbf{r})$ be given as

$$\mathbf{F}(x, y, z) := \mathbf{i}\, F_1(x, y, z) + \mathbf{j}\, F_2(x, y, z) + \mathbf{k}\, F_3(x, y, z) .$$

The line integral (6.32) then becomes

$$\begin{aligned}
\varphi(x, y, z) &= \int_{\mathcal{C}_x} (\mathbf{F}(\mathbf{r}) \cdot \mathbf{i})\, dx + \int_{\mathcal{C}_y} (\mathbf{F}(\mathbf{r}) \cdot \mathbf{j})\, dy + \int_{\mathcal{C}_z} (\mathbf{F}(\mathbf{r}) \cdot \mathbf{k})\, dz \\
&= \int_{\mathcal{C}_x} F_1(x, y_0, z_0)\, dx + \int_{\mathcal{C}_y} F_2(x, y, z_0)\, dy + \int_{\mathcal{C}_z} F_3(x, y, z)\, dz
\end{aligned}$$

where (y, z) are fixed in the first integral, (x, z) are fixed in the second, and (x, y) are fixed in the third. We can write this as

$$\varphi(x, y, z) = \int_{x_0}^{x} F_1(t, y_0, z_0)\, dt + \int_{y_0}^{y} F_2(x, t, z_0)\, dt + \int_{z_0}^{z} F_3(x, y, t)\, dt$$

and partial differentiation w.r.t x, y, z then gives

$$\frac{\partial \varphi}{\partial x} = F_1 \,, \quad \frac{\partial \varphi}{\partial y} = F_2 \,, \quad \frac{\partial \varphi}{\partial z} = F_3 \,,$$

proving (6.33). □
We summarize Theorems 6.9–6.10 in the following:

Theorem 6.11. A necessary and sufficient condition for the line integral

$$\varphi(x, y, z) := \int_{C} \mathbf{F}(\mathbf{r}) \cdot d\mathbf{r} = \int_{(x_0, y_0, z_0)}^{(x,y,z)} \mathbf{F}(\mathbf{r}) \cdot d\mathbf{r}$$

to be path-independent is

$$\mathbf{F} = \nabla \varphi \,. \tag{6.33}$$

We recall that a force field $\mathbf{F}$ is **conservative** if its line integrals are path-independent. By Theorem (6.11) this is equivalent to $\mathbf{F} = \nabla \varphi$ for some function φ. The function is not unique (it is determined up to a constant). Any such function φ is called a **scalar potential** of $\mathbf{F}$.

Example 6.12. The inverse square force

$$\mathbf{F}(\mathbf{r}) = \frac{\gamma \mathbf{r}}{\|\mathbf{r}\|^3} = \gamma \frac{x\,\mathbf{i} + y\,\mathbf{j} + z\,\mathbf{k}}{(x^2 + y^2 + z^2)^{\frac{3}{2}}} \,, \tag{6.34}$$

is conservative. A scalar potential for it is

$$\varphi(\mathbf{r}) = -\frac{\gamma}{\|\mathbf{r}\|} = -\frac{\gamma}{(x^2 + y^2 + z^2)^{\frac{1}{2}}} \,.$$

The force

$$\mathbf{F} := y\,z\,\mathbf{i} + x\,z\,\mathbf{j} + x\,y\,\mathbf{k} \tag{6.35}$$

is conservative; a potential is

$$\varphi(x, y, z) = x\,y\,z \,.$$

Example 6.13 (Total energy)**.** If a force field $\mathbf{F}$ is conservative, the work done in moving a particle from point (x_0, y_0, z_0) to point (x, y, z) is independent of the path, and furthermore, this work is

$$\int_{(x_0, y_0, z_0)}^{(x,y,z)} \mathbf{F}(\mathbf{r}) \cdot d\mathbf{r} = \varphi(x, y, z) - \varphi(x_0, y_0, z_0) \tag{6.36}$$

where $\varphi(x, y, z)$ is a potential of $\mathbf{F}$. It is usually called the **potential energy**, so the work is equal to the potential energy difference.

In particular, a closed line integral of a conservative force has zero work

$$\oint \mathbf{F}(\mathbf{r}) \cdot d\mathbf{r} = \varphi(x_0, y_0, z_0) - \varphi(x_0, y_0, z_0) = 0 \,,$$

so the potential energy does not change along a closed path. The force acting on a particle of **mass** m is, by Newton's second law,

$$\mathbf{F} = m\,\frac{d^2\mathbf{r}}{dt^2}$$

where $\frac{d^2\mathbf{r}}{dt^2}$ is the **acceleration**. The work integral is therefore

$$\varphi(\mathbf{r}) - \varphi(\mathbf{r}_0) = \int_{\mathbf{r}_0}^{\mathbf{r}} \mathbf{F}\cdot d\mathbf{r} = \int_0^t \mathbf{F}\cdot\frac{d\mathbf{r}}{dt}\,dt = \int_0^t m\,\frac{d^2\mathbf{r}}{dt^2}\cdot\frac{d\mathbf{r}}{dt}\,dt$$

$$= \frac{m}{2}\int_0^t \frac{d}{dt}\left(\frac{d\mathbf{r}}{dt}\cdot\frac{d\mathbf{r}}{dt}\right)dt = \frac{m}{2}\int_0^t \frac{d}{dt}\|\mathbf{v}(t)\|^2 dt = \frac{m}{2}\|\mathbf{v}\|^2 - \frac{m}{2}\|\mathbf{v}_0\|^2 .$$

The term $\frac{m}{2}\|\mathbf{v}\|^2$ is called the **kinetic energy** of the particle. We conclude, by transposing terms,

$$\varphi(\mathbf{r}) + \frac{m}{2}\|\mathbf{v}\|^2 = \varphi(\mathbf{r}_0) + \frac{m}{2}\|\mathbf{v}_0\|^2,$$

showing that the sum of the potential and kinetic energies is conserved along any path (not just closed path). This sum is called the **total energy** of the particle

$$E = \varphi(\mathbf{r}) + \frac{m}{2}\|\mathbf{v}\|^2 .$$

Since E does not change along any path, and $\varphi(\mathbf{r})$ does not change along closed paths, we conclude that the kinetic energy also does not change along closed paths.

Exercises

Exercise 6.14. *Let f and g be continuously differentiable functions defined in some domain in $\mathbb{R}^3$, and let a and b be constants. Prove that the gradient operator (6.6) satisfies the following rules*

(a) $\nabla(a\,f + b\,g) = a\nabla f + b\nabla g$ (b) $\nabla(f\,g) = f\nabla g + g\nabla f$

These follow directly from the linearity of partial differentiation and the definition (6.6). We summarize these results by saying that ∇ is a linear operator.

Exercise 6.15. *Evaluate the following line integrals for the curve C given parametrically as $x = t$, $y = t^3$ where $t \in [0,1]$*

(a) $\int_C e^x y^2\,dx + e^x x^2\,dy$ (b) $\int_C x^3 y^2\,dx + xy^2\,dy$

Exercise 6.16. *Evaluate the following line integrals over the curve C given parametrically as $x = \cos t$, $y = \sin t$ where $t \in [0,\pi]$*

(a) $\int_C y\,dx + x\,dy$ (b) $\int_C x\,dx - y\,dy$

(c) $\int\limits_{C} y\,dx - x\,dy$ (d) $\int\limits_{C} (y-x)\,dx + xy\,dy$

Exercise 6.17. *Evaluate* $\int\limits_{C} y\,dx - x\,dy$ *over the following positively oriented , closed curves:*

(a) $x^2 + y^2 = 1$ (b) $x^2 + y^2 = a^2$

(c) *the square with vertices* $(\pm 1, \pm 1)$

(d) *the triangle with vertices* $(0,0)$, $(1,0)$, $(0,1)$

Exercise 6.18. *Find the work done in moving a particle against a force*

$$\mathbf{F} := \frac{x}{\|\mathbf{r}\|}\mathbf{i} + \frac{y}{\|\mathbf{r}\|}\mathbf{j}$$

along the curves shown in Figure (6.4).

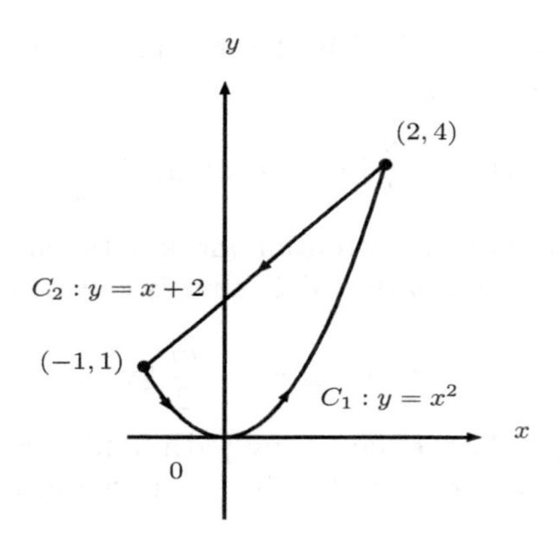

FIGURE 6.4: Ilustration of $\mathcal{C} = \mathcal{C}_1 \cup \mathcal{C}_1$.

Exercise 6.19. *Compute the work done in moving a particle along a curve given parametrically by* $\mathbf{r}(t) = \sin t\mathbf{i} + \cos t\mathbf{j} + t\mathbf{k}$ *for* $t \in [0, 4\pi]$, *when the resisting force is given by*

(a) $\mathbf{F} = x\mathbf{i} + y\mathbf{j} + z\mathbf{k}$ (b) $\mathbf{F} = xy\mathbf{i} - z\mathbf{j} + y\mathbf{k}$

(c) $\mathbf{F} = xz\mathbf{i} + yz\mathbf{j} + \frac{1}{2}(x^2+y^2)\mathbf{k}$ (d) $\mathbf{F} = e^x\mathbf{i} + e^y\mathbf{j} + \sin z\mathbf{k}$

Exercise 6.20. *Use* MATHEMATICA *to check whether the following force-fields have vector potentials, and if they do find that potential. Check your results by doing the hand calculations too.*

(a) $\mathbf{F} = < y^2z^3, 2xyz^3, 3xy^2z^2 >$ (b) $\mathbf{F} = < y^2z^3, 3xz^3, 3xy >$

(c) $\mathbf{F} = < \log\sqrt{x^2+y^2}, \tan\frac{y}{x} >$ (d) $\mathbf{F} = < \cos y \sinh x, \sin y \cosh x >$

(e) $\mathbf{F} = < \cos y \cosh x, \sin y \sinh x >$ (f) $\mathbf{F} = < e^x \sin x, e^x \cos x >$

Exercise 6.21. *A particle of mass m moves along the curve $\mathcal{C}$. At time t its coordinates are given by $\mathbf{r} = \mathbf{r}(t) =< x(t), y(t), z(t) >$, and it is acted on by a force field $\mathbf{F} = k\mathbf{r}$. Show that the force field is conservative. If $\mathcal{C}$ is the closed loop $x = \cos 2t$, $y = \sin 2t$, $z = \sin 4t$ where $t \in [0, \pi]$, compute the work done as the particle traverses a complete cycle of the loop.*

Exercise 6.22. *A particle of mass m moves along the curve $\mathcal{C}$. At time t its coordinates are given by $\mathbf{r} = \mathbf{r}(t) =< x(t), y(t), z(t) >$, and it is acted on by a force field $\mathbf{F} = k\mathbf{r} + \lambda\frac{d\mathbf{r}}{dt}$. Find an expression for the work done as the point traverses a segment of the curve $\mathcal{C}$. Show that this force field is not conservative.*

Exercise 6.23. *Compute the* **circulation**, *and* **flux** *of the velocities:*

(a) $\mathbf{v} = \frac{x}{r^2}\mathbf{i} + \frac{y}{r^2}\mathbf{j}$ (b) $\mathbf{v} = \frac{x-y}{r^2}\mathbf{i} + \frac{x}{r^2}\mathbf{j}$

over the closed curve $x^2 + y^2 = a^2$. Here $r^2 = x^2 + y^2$.

6.2 Green's theorem

The main result in this section is that double integrals over a plane region bounded by a closed curve $\mathcal{C}$ reduce, under certain conditions, to line integrals along $\mathcal{C}$. This gives an alternative method for computing double integrals, which is preferred since line integrals are easier. We recall the definition of **positive orientation** of a closed planar curve $\mathcal{C}$, see Figure (6.3)(a). It corresponds to $\mathbf{r}(t)$ moving around $\mathcal{C}$ in a counter-clockwise direction. Movement in the clockwise direction indicates negative orientation.
It should be noted that the orientation is a **property of the parametrization** of the curve. Indeed if

$$\mathcal{C} = \{\mathbf{r}(t) = x(t)\,\mathbf{i} + y(t)\,\mathbf{j} + z(t)\,\mathbf{k} : 0 \le t \le T\}$$

is a curve with positive orientation, then

$$\mathcal{C} = \{\mathbf{r}(\tau) = x(T-\tau)\,\mathbf{i} + y(T-\tau)\,\mathbf{j} + z(T-\tau)\,\mathbf{k} : 0 \le \tau \le T\}$$

is the same physical curve, but with a negative orientation.
We use the notation

$$\oint_{\mathcal{C}+} (P\,dx + Q\,dy) = \oint_{\mathcal{C}+} (P(x,y)\,dx + Q(x,y)\,dy) \tag{6.37}$$

for a closed line integral, where the curve $\mathcal{C}$ has positive orientation. The closed line integral with negative orientation is denoted

$$\oint_{\mathcal{C}-} (P\,dx + Q\,dy)$$

Theorem 6.24 (Green's Theorem). Let $\mathcal{C}$ be a piece-wise smooth simple curve in the plane, bounding a region $\mathcal{D}$. If the functions $P(x, y)$ and $Q(x, y)$ are continuously differentiable in $\mathcal{D}$, then

$$\oint_{\mathcal{C}+} (P\,dx + Q\,dy) = \iint_{\mathcal{D}} \left(\frac{\partial Q}{\partial x} - \frac{\partial P}{\partial y}\right) dx\,dy\,. \tag{6.38}$$

Proof:
Consider regions $\mathcal{D}$ which are both horizontally simple and vertically simple[1]. We first evaluate the double integral

$$\iint_{\mathcal{D}} \left(\frac{\partial P}{\partial y}\right) dx\,dy\,.$$

Let A and B be the points on the curve $\mathcal{C}$ where the x–coordinate has its minimum and maximum, respectively, and let a and b be the abscisas of these points. Since $\mathcal{D}$ is vertically simple, it is bounded between two curves, $y_1(x)$ and $y_2(x)$, $a \le x \le b$. The double integral may now be written as an iterated integral

$$\iint_{\mathcal{D}} \left(\frac{\partial P}{\partial y}\right) dx\,dy = \int_a^b dx \int_{y_1(x)}^{y_2(x)} \frac{\partial P}{\partial y}\,dy$$

$$= \int_a^b (P(x, y_2(x)) - P(x, y_1(x)))\,dx = -\oint_{\mathcal{C}+} P\,dx\,. \tag{6.39}$$

Similarly, we compute the double integral,

$$\iint_{\mathcal{D}} \left(\frac{\partial Q}{\partial x}\right) dx\,dy = \int_c^d dy \int_{x_1(y)}^{x_2(y)} \left(\frac{\partial Q}{\partial x}\right) dx =$$

$$\int_c^d (Q(x_2(y), y) - Q(x_1(y), y))\,dy = \oint_{\mathcal{C}+} Q\,dy\,, \tag{6.40}$$

and (6.79) follows from (6.39)–(6.40). □
Green's Theorem has numerous applications. Here are some examples.

[1]A more general region may sometimes be broken into regions which are both horizontally simple and vertically simple, and the proof holds with few changes.

Example 6.25 (Area)**.** Let $\mathcal{D}$ be a region bounded by a piecewise smooth curve $\mathcal{C}$, and let A be the area of $\mathcal{D}$. Consider Green's Theorem with $P(x, y) = y$, $Q(x, y) = 0$. Then, using (6.79),

$$-\oint_{\mathcal{C}+} y\,dx = \iint_{\mathcal{D}} 1\,dx\,dy = A ,$$

Similarly, for $P(x, y) = 0$, $Q(x, y) = x$ we get from (6.79),

$$\oint_{\mathcal{C}+} x\,dy = \iint_{\mathcal{D}} 1\,dx\,dy = A .$$

Adding these results gives the area as a line integral over $\mathcal{C}$,

$$A = \frac{1}{2}\oint_{\mathcal{C}+} (x\,dy - y\,dx) . \tag{6.41}$$

Example 6.26 (Centroid)**.** Consider the (area) centroid of the region $\mathcal{D}$, with area A. The x–component of the centroid is

$$\overline{x} := \frac{1}{A}\iint_{\mathcal{D}} x\,dx\,dy.$$

Using (6.79) with

$$\frac{\partial Q}{\partial x} = x , \quad \frac{\partial P}{\partial y} = 0$$

we get an alternative expression

$$\overline{x} = \frac{1}{2A}\oint_{\mathcal{C}_+} x^2\,dy \tag{6.42}$$

Similarly, the y–coordinate of the centroid

$$\overline{y} := \frac{1}{A}\iint_{\mathcal{D}} y\,dx\,dy \tag{6.43}$$

is computed by Green's Theorem, using

$$-\frac{\partial P}{\partial y} = y , \quad \frac{\partial Q}{\partial x} = 0 ,$$

as

$$\overline{y} = -\frac{1}{2A}\oint_{\mathcal{C}_+} y^2\,dx . \tag{6.44}$$

Example 6.27. We apply (6.42)–(6.44) to calculate the centroid of the unit semicircle lying in the upper half plane. The upper boundary of this region can be parametrized by

$$\mathcal{C}_1 := \{<x,y>: \ x := \cos\theta \ , \ y := \sin\theta \ , \quad 0 \le \theta \le \pi\} \ .$$

The bottom boundary (on the x–axis) is parametrized by

$$\mathcal{C}_2 := \{<x,y>: \ x := t \ , \ y := 0 \ , \quad -1 \le t \le 1\} \ .$$

By symmetry, $\overline{x} = 0$, but let us calculate it anyway, using (6.42). Since $dy = 0$ on the x–axis, there is no contribution from the line integral on $\mathcal{C}_2$. Therefore

$$\overline{x} = \frac{1}{\pi} \int_0^\pi (1 - \sin^2(\theta) \cos(\theta))\, d\theta = (\sin(\theta) - \frac{1}{3} sin^3(\theta)|_0^\pi = 0$$

We compute next $\overline{y}$ using (6.44). Since $y = 0$ on the x–axis, there is no contribution from $\mathcal{C}_2$. Therefore

$$\overline{y} = -\frac{1}{\pi} \int_0^\pi \sin^2\theta\, (-\sin\theta)\, d\theta = \frac{1}{\pi} \left(-\cos\theta + \frac{1}{3} \cos^3\theta \right)\Big|_0^\pi = \frac{4}{3\pi} \ .$$

We can apply Green's Theorem to more complicated regions by subdividing them into subregions which are both x–simple and y–simple. For this we need the following

Definition 6.28. A **simply–connected** domain $\mathcal{D}$ is a domain having the property that every simple, closed curve may be continuously contracted to a point in $\mathcal{D}$ without leaving $\mathcal{D}$. A domain without this property is **multiply–connected**.

Mathematica session 6.5. *In this section we use* MATHEMATICA *to present two forms for illustrating the Green's Theorem. One is the the circulation form and the other the flux form.*

```
greenCirculationForm[F_, U_, V_]:=Module[{G}, G:=Curl[{F[[1]], F[[2]]},
{x, y}];
Integrate[G, {U[[1]], U[[2]], U[[3]]}, {V[[1]], V[[2]], V[[3]]}]]
```

```
greenCirculationForm[{x^2, y^2}, {x, -1, 1}, {y, 0, Sqrt[1 - x^2]}]
```

0

```
greenCirculationForm[{-y/(x^2 + y^2), x/(x^2 + y^2)}, {y, 0, Sqrt[1-
x^2]}, {x, -1, 1}]
```

0

```
Curl[{-y/(x^2 + y^2), x/(x^2 + y^2)}, {x, y}]
```

$-\frac{2x^2}{(x^2+y^2)^2} - \frac{2y^2}{(x^2+y^2)^2} + \frac{2}{x^2+y^2}$

```
Integrate[%, {x, -1, 1}, {y, 0, Sqrt[1 - x^2]}]
```

0

```
Curl[{P[x, y, z], Q[x, y, z], R[x, y, z]}, {x, y, z}]
```

$\{-Q^{(0,0,1)}[x,y,z] + R^{(0,1,0)}[x,y,z], P^{(0,0,1)}[x,y,z]- R^{(1,0,0)}[x,y,z], -P^{(0,1,0)}[x,y,z] + Q^{(1,0,0)}[x,y,z]\}$

```
Dot[%, {0, 0, 1}]
```

$-P^{(0,1,0)}[x,y,z] + Q^{(1,0,0)}[x,y,z]$

```
greenCirculationForm2[F_, U_, V_]:=Module[{G, P, Q, R}, P = F[[1]];
Q = F[[2]]; R = F[[3]];
G:=Dot[Curl[{P, Q, R}, {x, y, z}], {0, 0, 1}];
Integrate[G, {U[[1]], U[[2]], U[[3]]}, {V[[1]], V[[2]], V[[3]]}]]
```

```
greenCirculationForm2[{-y/(x^2 + y^2), x/(x^2 + y^2), 0}, {y, 0, Sqrt[1-
x^2]}, {x, -1, 1}]
```

0

```
greenFluxForm[F_, U_, V_]:=Module[{G}, G = Div[F, {x, y}];
Integrate[G, {U[[1]], U[[2]], U[[3]]}, {V[[1]], V[[2]], V[[3]]}]]
```

greenFluxForm$[\{x-y,x\},\{x,-1,1\},\ \{y,0,$ ***Sqrt***$[1-x\char`\^2]\}]$

$\frac{\pi}{2}$

greenCirculationForm$[\{x-y,x\},\{x,-1,1\},\ \{y,0,$ ***Sqrt***$[1-x\char`\^2]\}]$

π

♦

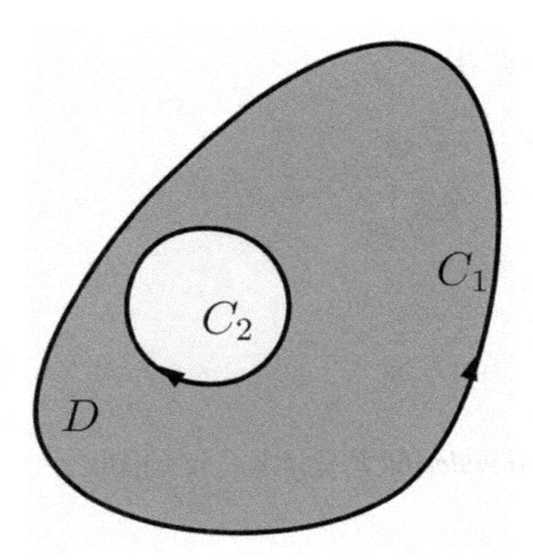

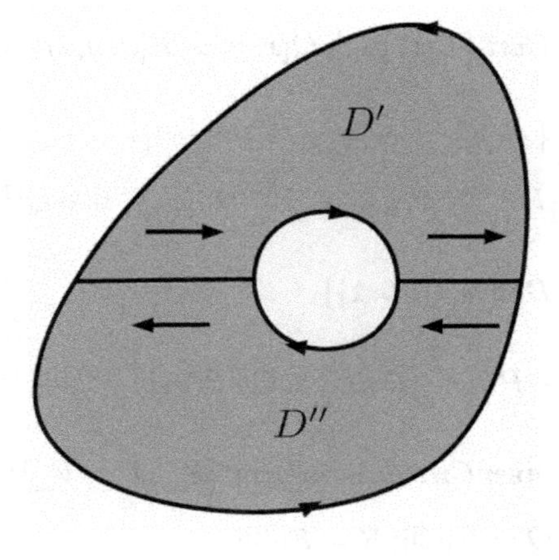

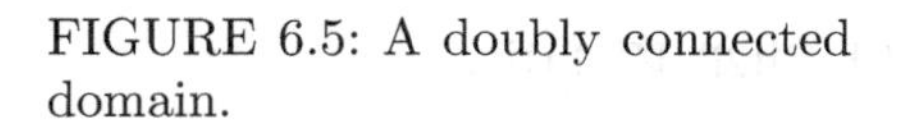

FIGURE 6.5: A doubly connected domain.

FIGURE 6.6: A doubly connected domain made simply connected using cuts.

In the plane, simply–connected regions do not have holes; multiply–connected regions have holes. Another way of expressing this is to say that the boundary of a multiply -connected region in the plane consists of several **components**. For example, the disk in simply–connected, whereas the annulus is not. In three–dimensions the sphere is simply-connected; the torus is not. Simply–connected and multiply–connected domains are illustrated in Figure (6.5), (6.6). For example, in the case of the annulus there are two components. These are the concentric circles $\{(x,y) : \sqrt{x^2+y^2} = a\}$ and $\{(x,y) : \sqrt{x^2+y^2} = b\}$ forming the boundary. The annulus, however, may be converted to a simply-connected region by performing the **cuts** shown in the Figure (6.6). This procedure may be be used in more general multiply-connected regions too. Having converted a multiply-connected region to a singly connected region we may apply Green's Theorem. Note, however, that in performing the integral the cut is traversed twice. Moreover, as these integrals are taken in opposite directions this portion of the integral cancels as the calculation below shows.

$$\iint_{\mathcal{D}} \left(\frac{\partial Q}{\partial x} - \frac{\partial P}{\partial y}\right) dx\, dy =$$

$$\oint_{\mathcal{C}_a^+} Pdx + Qdy + \oint_{\mathcal{C}_b^-} Pdx + Qdy + \int_{\mathcal{L}} Pdx + Qdy + \int_{-\mathcal{L}} Pdx + Qdy =$$

$$\oint_{\mathcal{C}^+} Pdx + Qdy,$$

where $\mathcal{C} = \mathcal{C}_b^+ \cup \mathcal{C}_a^-$ and $\mathcal{L}$ is the cut. $-\mathcal{L}$ is the cut with the opposite orientation.

Theorem 6.29. A two-dimensional vector field is the gradient of a potential if and only if

$$\frac{\partial Q}{\partial x} = \frac{\partial P}{\partial y} . \tag{6.45}$$

Proof:
Recall from Green's Theorem that for a simply-connected domain $\mathcal{D}$

$$\oint_{\mathcal{C}+} P\, dx + Q\, dy = \iint_{\mathcal{D}} \left(\frac{\partial Q}{\partial x} - \frac{\partial P}{\partial y}\right) dx .$$

If (6.45) holds then

$$\oint_{\mathcal{C}+} P\, dx + Q\, dy = 0$$

for any closed curve $\mathcal{C}^+ \subset \mathcal{D}$. Dividing such a curve into two pieces

$$\mathcal{C} = \mathcal{C}_1 \cup \mathcal{C}_2 ,$$

we conclude that

$$\oint_{\mathcal{C}_1} P\, dx + Q\, dy = \oint_{-\mathcal{C}_2} P\, dx + Q\, dy$$

as is shown in Figure (6.7). The condition (6.45) holding in $\mathcal{D}$ is sufficient for the integral to be path-independent. Using Green's Theorem it is easy to show that this condition is also necessary.□

Remark 6.30. We recall, see (6.74), that the vector

$$\left(\frac{\partial Q}{\partial x} - \frac{\partial P}{\partial y}\right) \mathbf{k}$$

is the **two–dimensional curl** of the vector field $P\,(x,y)\,\mathbf{i} + Q(x,y)\,\mathbf{j}$. Here $\mathbf{k}$ is the unit vector perpendicular to the xy–plane having positive orientation. A necessary and sufficient criterion for a planar line integral to be path-independent is for the two-dimensional curl to vanish.

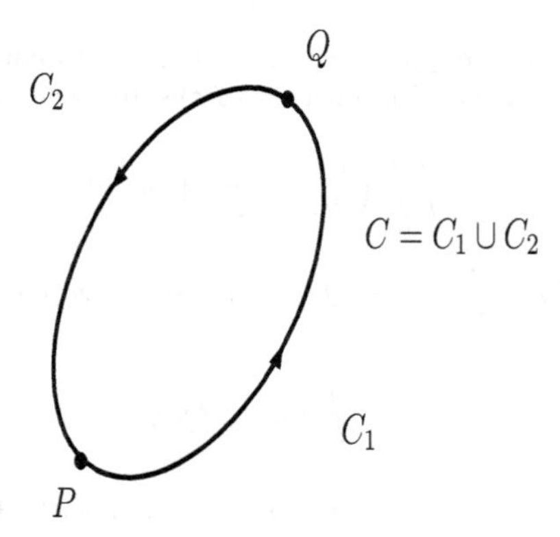

FIGURE 6.7: Illustration of Theorem (6.29) .

We illustrate Green's Theorem for the integral

$$\oint_{C+} \frac{-y\,dx + x\,dy}{x^2+y^2} . \tag{6.46}$$

It turns out that the answer depends on whether the curve $\mathcal{C}$ surrounds the origin or not. Writing

$$P(x,y) = \frac{-y}{x^2+y^2}, \quad Q(x,y) = \frac{x}{x^2+y^2},$$

we first verify the path independency condition (6.45) holds.

$$P_y = \frac{-1}{x^2+y^2} + \frac{2y^2}{(x^2+y^2)^2} = \frac{y^2-x^2}{x^2+y^2},$$

$$Q_x = \frac{1}{x^2+y^2} - \frac{2x^2}{(x^2+y^2)^2} = \frac{y^2-x^2}{x^2+y^2},$$

so

$$\frac{\partial Q}{\partial x} = \frac{\partial P}{\partial y} \quad \text{provided } (x,y) \neq (0,0) .$$

At the origin $(0,0)$, both Q_x and P_y are not defined, so the origin is not part of a domain $\mathcal{D}$ where Green's Theorem is valid.

We now use MATHEMATICA to check this result. We introduce a module called **greenCirculation**, where we begin our module with a lower case letter as all MATHEMATICA commands begin with a capital letter. The difference between these two answers is a multiple of $\frac{\pi}{4}$, which remains constant as long as y does not change sign.

Mathematica session 6.6. *greenCirculationForm2[F_, U_, V_]:=*

```
Module[{G, P, Q, R}, P = F[[1]];
Q = F[[2]];
G:=Dot[Curl[{P, Q}, {x, y}]];
Integrate[G, {U[[1]], U[[2]], U[[3]]}, {V[[1]], V[[2]], V[[3]]}]]
```

```
greenCirculationForm2[{-y/(x^2 + y^2), x/(x^2 + y^2), 0},
{y, Sqrt[1/4 - x^2], Sqrt[1 - x^2]}, {x, -1, 1}]
```

0

```
greenCirculationForm2[{-y/(x^2 + y^2), x/(x^2 + y^2)}, {x, -1, 1},
{y, -Sqrt[1/4 - x^2], Sqrt[1/4 - x^2]}]
```

0

```
greenFluxForm[F_, U_, V_]:=Module[{G}, G = Div[F, {x, y}];
Integrate[G, {U[[1]], U[[2]], U[[3]]}, {V[[1]], V[[2]], V[[3]]}]]
```

```
greenFluxForm[{-y/(x^2 + y^2), x/(x^2 + y^2)}, {x, -1, 1},
{y, -Sqrt[1/4 - x^2], Sqrt[1/4 - x^2]}]
```

0

```
greenCirculationForm2[{-y/(x^2 + y^2), x/(x^2 + y^2)}, {x, -1, 1},
{y, Sqrt[1/4 - x^2], Sqrt[1 - x^2]}]
```

0

```
lineIntegral2d[f_, u_, I_]:=
Module[{r, s, g}, r = D[u, t]; s = Simplify[f/.{x → u[[1]], y → u[[2]]}];
```

```
g = Dot[s, r];
Integrate[g, {t, I[[1]], I[[2]]}]]
```

```
lineIntegral2d[{-y/(x^2 + y^2), x/(x^2 + y^2)}, {Cos[2 * π * t],
Sin[2 * π * t]}, {0, 1}]-
lineIntegral2d[{-y/(x^2 + y^2), x/(x^2 + y^2)},
{1/1000 * Cos[2 * π * t], 1/1000 * Sin[2 * π * t]}, {0, 1}]
```

0

♦

Consider now a curve $\mathcal{C}$ which lies entirely on the upper–half plane, i.e. $y > 0$. Going around $\mathcal{C}$ does not change the potential $\varphi(x, y)$, which we denote by

$$\Delta\varphi(x, y)|_{\mathcal{C}} \;=\; 0\;.$$

The same result holds for a curve $\mathcal{C}$ which lies entirely on the lower–half plane. On the other hand if $\mathcal{C}$ is a circle concentric to the origin then, using the polar coordinates $(r\,,\,\theta)$, we have

$$\varphi(r\,,\,\theta) \;=\; \frac{\pi}{2}\,sign\,\cos\theta \;+\; \theta$$

so the change along $\mathcal{C}$ is

$$\Delta\varphi(r\,,\,\theta)|_{\mathcal{C}} \;=\; 2\,\pi\;.$$

Mathematica session 6.7. *In this session we consider how to determine whether a vector field is generated by a potential.*

```
D[-y/(x^2 + y^2), y] - D[x/(x^2 + y^2), x]
```

$$\frac{2x^2}{(x^2+y^2)^2} + \frac{2y^2}{(x^2+y^2)^2} - \frac{2}{x^2+y^2}$$

```
Simplify[%]
```

0

```
myPotential[f_, P_]:=
```

```
Module[{g}, g = Simplify[(-D[f[[2]], x] + D[f[[1]], y])/.{x → P[[1]], y →
P[[2]]}];
If[Abs[g] > 0, "there is no potential", Integrate[f[[1]], x] + Integrate[f[[2]],
y]]]
```

```
myPotential[{y/(x^2 + y^2), x/(x^2 + y^2)}, {1, 1}]
```

$ArcTan\left[\frac{x}{y}\right] + ArcTan\left[\frac{y}{x}\right]$

```
myPotential[{exp(x) + 2 * x * y, exp(y) + x^2}, {1, 1}]
```

$x^2y + \frac{\exp y^2}{2} + \frac{1}{2}x^2(\exp + 2y)$

```
myPotential[{Exp[x] + 2 * x * y, Exp[y] + x^2}, {0, 0}]
```

$e^x + e^y + 2x^2y$

```
myPotential[{x - y, x^2}, {1, 1}]
```

there is no potential

```
myPotential[{Exp[x] + x^2, y + Exp[y] - 2}, {1, 1}]
```

$e^x + e^y + \frac{x^3}{3} - 2y + \frac{y^2}{2}$

It is possible to show that the change in $\varphi(x, y)$ going around C is the same for all curves which encircle the origin once, see Exercise 6.40. See also Exercise 6.36, where it is required to compute analytically the line integral (6.46) along the unit circle.

♦

Exercises

Exercise 6.31. *Use Green's Theorem to evaluate the line integrals*

$$\oint_C P\,dx + Q\,dy$$

where P, Q, are specified and C is the boundary of the square having vertices at

$(\pm 1, \pm 1)$.

(a) $P = xy,\ Q = x^2$ (b) $P = -2xy,\ Q = x^2 + y^2$

(c) $P = e^{\tan x} + 2xy,\ Q = \cosh y + x^2$ (d) $P = \sin x + y^2,\ Q = \sin y + xy$

Exercise 6.32. *Do the same integrals of the previous problems where C is replaced by the circle $x^2 + y^2 = 1$.*

Exercise 6.33. *Again the same integrals are to be evaluated for the boundary of the region contained between the parabola $y = x^2$ and the line $y = x$.*

Exercise 6.34. *Write a* MATHEMATICA *function which checks (using nested* `IF` *statements) if a vector* **F** *can be represented as a scalar potential, and generates this potential, and if it does not have a representation as a scalar potential reports* `"Not representable as a scalar potential"`. *Use this function to write another one which evaluates the work integral in terms of the potential. Apply this function to the work integrals*

$$W(x, y; C) \ := \int_C \mathbf{F} \cdot \mathbf{t}\, ds,$$

where C joins a fixed point (x_0, y_0) to a variable point (x, y).

(a) $\mathbf{F} = (\sin x + \frac{y^2}{x})\,\mathbf{i} + (y^2 \ln x + y)\,\mathbf{j}$ (b) $\mathbf{F} = y e^{xy}\,\mathbf{i} + x e^{xy}\,\mathbf{k}$

(c) $\mathbf{F} = x \sec^2 y\,\mathbf{i} + \frac{x^2}{2} \sec^2 y\,\mathbf{j}$ (d) $\mathbf{F} = e^x \cosh y\,\mathbf{i} + e^x \sinh y\,\mathbf{j}$

(e) $\mathbf{F} = \left(\ln(\cosh y) + x^2\right)\,\mathbf{i} + x \tanh y\,\mathbf{j}$

(f) $\mathbf{F} = (\sin y + x \cos y)\,\mathbf{i} + (x \cos - \frac{x^2}{2} \sin y)\,\mathbf{j}$

Exercise 6.35. *Compute by hand a potential satisfying*

$$\frac{\partial \varphi}{\partial x} = \frac{-y}{x^2 + y^2}, \quad \text{and} \quad \frac{\partial \varphi}{\partial y} = \frac{x}{x^2 + y^2}.$$

Exercise 6.36. *Use polar coordinates to evaluate the integral*

$$\oint_{C+} \frac{-y\,dx + x\,dy}{x^2 + y^2}$$

about the unit circle.

Exercise 6.37. *Consider the function $u(x, y)$ which is defined by the integral*

$$u(x, y) \ := \int_{(a,0)}^{(x,y)} \frac{y\,dx - x\,dy}{x^2 + y^2},$$

where $a > 0$ and the integration path lies on the circle of radius a. Show that the integral has the value θ where θ is the angle which the line drawn from the origin to the point (x, y) makes with the positive x-axis. Hence, the function $u(x, y) = \theta$ is an infinitely multi-valued function.

Exercise 6.38. *Let* $\mathbf{F} = \mathbf{i}P + \mathbf{j}Q$ *be a conservative vector field in the plane, i.e.* $P_y = Q_x$ *and* $\mathcal{D}$ *be a ring-shaped region bounded by the two smooth curves* $\mathcal{C}_1$ *and* $\mathcal{C}_2$*. Prove:*

$$\int_{\mathcal{C}_1} P\,dx + Q\,dy \;=\; \int_{\mathcal{C}_2} P\,dx + Q\,dy\,.$$

Exercise 6.39. *Let* Σ *be the region exterior to the circles* $(x-2)^2 + y^2 = 1$ *and* $(x+2)^2 + y^2 = 1$*. Discuss the function* $u(x,y)$ *defined by the integral given below where the integration path is to remain in* Σ*. Use* MATHEMATICA *to check whether the integral is path-independent and to evaluate the integral.*

$$\int_{(0,0)}^{(x,y)} \left(\frac{y}{(x-2)^2+y^2} + \frac{y}{(x+2)^2+y^2}\right) dx - \left(\frac{x-2}{(x-2)^2+y^2} + \frac{x+2}{(x+2)^2+y^2}\right) dy.$$

Mathematica session 6.8. *We consider the vector field* $\mathbf{F} = \frac{-\mathbf{yi}+\mathbf{xj}}{\mathbf{X^2+y^2}}$ *to see if it is the gradient of a scalar function* φ*. To this end, we take the two-dimensional curl of this vector and then* **Simplify** *the result.*

```
D[-y/(x^2 + y^2), y] - D[x/(x^2 + y^2), x]
```

$$\frac{2x^2}{(x^2+y^2)^2} + \frac{2y^2}{(x^2+y^2)^2} - \frac{2}{x^2+y^2}$$

```
Simplify[%]
```

0

Knowing that the curl vanishes we may compute the potential using a line integral. We seek a program, **myPotential**, *to to do this.*

```
myPotential[f_]:=Module[{g}, g = Simplify[-D[f[[2]], x] + D[f[[1]], y]];
If[PossibleZeroQ[g], Integrate[f[[1]], x] + Integrate[f[[2]], y],
"there is no potential"]]
```

```
myPotential[{-y/(x^2 + y^2), x/(x^2 + y^2)}]
```

$$-ArcTan\left[\frac{x}{y}\right] + ArcTan\left[\frac{y}{x}\right]$$

myPotential[{y/(x^2 + y^2), x/(x^2 + y^2)}]

there is no potential

myPotential[{Exp[x] + 2 * x * y, Exp[y] + x^2}]

$e^x + e^y + 2x^2y$

myPotential[{x − y, x^2}]

there is no potential

myPotential[{E[x] + x^2, y + Exp[y] − 2}]

$e^y - 2y + \frac{y^2}{2} + \int \left(x^2 + \mathrm{e}[x]\right)\, dx$

Can you explain this by doing some hand calculations?

Exercises

Exercise 6.40. *Construct an argument which shows that the integral*

$$\oint_{C+} \frac{-ydx + xdy}{x^2 + y^2}$$

where C^+ is any nonintersecting curve which winds once around the origin has the same value if C^+ is replaced by the unit circle $x^2 + y^2 = 1$.
Hint: Cut the region between the unit circle and C^+ so that this region is simply connected. Show that the integral around the boundary of this simply connected region is zero, and hence, the integrals over the unit circle and C^+ are the same.

Exercise 6.41. *Let $\mathcal{D}$ be a region in the plane bounded by the smooth curve C, and let $u(x, y)$ and $v(x, y)$ be functions with first order continuous derivatives in $\mathcal{D}$. Prove the identities:*

$$\text{(a)} \qquad \iint_{\mathcal{D}} \frac{\partial(u, v)}{\partial(x, y)}\, dx\, dy = \int_{C} u\, dv\,,$$

$$\text{(b)} \qquad \iint_{\mathcal{D}} u \left(\frac{\partial^2 v}{\partial x^2} + \frac{\partial^2 v}{\partial y^2}\right) dx\, dy + \iint_{\mathcal{D}} u \left(\frac{\partial v}{\partial x}\frac{\partial u}{\partial x} + \frac{\partial v}{\partial y}\frac{\partial u}{\partial y}\right) dxdy = -\int_{C} u \frac{\partial u}{\partial n} ds$$

where ds is the arc-length differential and $\frac{\partial u}{\partial n}$ is the partial derivative of u in the **n** *direction.*
Hint: In (a) use $P = u\frac{\partial v}{\partial x}$, $Q = u\frac{\partial v}{\partial y}$. In (b) use $P = u\frac{\partial v}{\partial y}$, $Q = -u\frac{\partial v}{\partial x}$.

An important operation on a scalar field $\varphi(x, y)$ is the **Laplacian operator**

$$\Delta\varphi := \frac{\partial^2\varphi}{\partial x^2} + \frac{\partial^2\varphi}{\partial y^2} \tag{6.47}$$

It is used in the following exercises.

Exercise 6.42. *If $\varphi(x, y)$ has two continuous derivatives in a region $\mathcal{D}$ which is simply-connected, and bounded by a piecewise smooth curve $\mathcal{C}$, show by Green's Theorem that*

(a) $\displaystyle\oint_{\mathcal{C}} \frac{\partial\varphi}{\partial x}\, dy - \frac{\partial\varphi}{\partial y}\, dx = \iint_{\mathcal{D}} \Delta\varphi\, dx\, dy$ (b) $\displaystyle\oint_{\mathcal{C}} \frac{\partial\varphi}{\partial x}\, dy - \frac{\partial\varphi}{\partial y}\, dx = \oint_{\mathcal{C}} \frac{\partial\varphi}{\partial n}\, ds,$

where n indicates differentiation in the normal direction to the curve $\mathcal{C}$ and s is arc-length.

Exercise 6.43. *Let $\mathcal{D}$, $\mathcal{C}$, $u(x, y)$ and $v(x, y)$ be as in Exercise 6.41. Then prove*

$$\iint_{\mathcal{D}} (u\Delta v + \nabla\cdot\nabla v) = -\int_{\mathcal{C}} u\frac{\partial v}{\partial n} ds \text{ (\textbf{Green's first identity})}$$

Exercise 6.44. *Let $\mathcal{D}$, $\mathcal{C}$, $u(x, y)$ and $v(x, y)$ be as in Exercise 6.41. Prove:*

(a) $\displaystyle\iint_{\mathcal{D}} (u\Delta v - v\Delta u)\, dxdy = -\int_{\mathcal{C}} \left(u\frac{\partial v}{\partial n} - v\frac{\partial u}{\partial n}\right) ds$

The identity above is referred to as the **Green's second identity**.

(b) $\displaystyle\iint_{\mathcal{D}} \Delta u\, dx\, dy = -\int_{\mathcal{C}} \frac{\partial u}{\partial n}\, ds\, .$

*If in addition, u satisfies $\Delta u = 0$ (***Laplace equation***)*

(c) $\displaystyle\iint_{\mathcal{D}} \left(\left(\frac{\partial u}{\partial x}\right)^2 + \left(\frac{\partial u}{\partial y}\right)^2\right) dx\, dy = -\int_{\mathcal{C}} u\frac{\partial u}{\partial n}\, ds$

(d) $\displaystyle\int_{\mathcal{C}} \frac{\partial u}{\partial n} ds = 0\,,$ *and, moreover, if $u \not\equiv 0$,* (e) $\displaystyle\int_{\mathcal{C}} u\frac{\partial u}{\partial n} ds < 0$

6.3 Gauss' divergence theorem

In two dimensions, the flux was shown to be,

$$\text{DIV}\mathbf{F} := \frac{\partial F_1}{\partial x} + \frac{\partial F_2}{\partial y} \tag{6.48}$$

The analogous concept in three dimensions is the **divergence**, defined as follows.

Definition 6.45 (Divergence). Let $\mathbf{F}$ be a vector field whose components have continuous partial derivatives

$$\mathbf{F}(x, y, z) := \mathbf{i}\, F_1(x, y, z) + \mathbf{j}\, F_2(x, y, z) + \mathbf{k}\, F_3(x, y, z) \,.$$

The **divergence** of $\mathbf{F}(x, y, z)$ is defined as

$$\operatorname{DIV} \mathbf{F}(x, y, z) := \frac{\partial F_1}{\partial x}(x, y, z) + \frac{\partial F_2}{\partial y}(x, y, z) + \frac{\partial F_3}{\partial z}(x, y, z) \,. \tag{6.49}$$

A convenient notation for $\operatorname{DIV} \mathbf{F}$ is

$$\nabla \cdot \mathbf{F} = \left(\mathbf{i}\,\frac{\partial}{\partial x} + \mathbf{j}\,\frac{\partial}{\partial y} + \mathbf{k}\,\frac{\partial}{\partial z}\right) \cdot (\mathbf{i}\, F_1 + \mathbf{j}\, F_2 + \mathbf{k}\, F_3) = \frac{\partial F_1}{\partial x} + \frac{\partial F_2}{\partial y} + \frac{\partial F_3}{\partial z}$$

considered as a formal inner product of the gradient operator

$$\nabla = \mathbf{i}\,\frac{\partial}{\partial x} + \mathbf{j}\,\frac{\partial}{\partial y} + \mathbf{k}\,\frac{\partial}{\partial z}$$

and the vector $\mathbf{F}$. We will use both $\operatorname{DIV} \mathbf{F}$ and $\nabla \cdot \mathbf{F}$ to denote the divergence of $\mathbf{F}$.

Example 6.46.
(a) The divergence of the position vector $\mathbf{r}$,

$$\nabla \cdot \mathbf{r} = \nabla \cdot (x\,\mathbf{i} + y\,\mathbf{j} + z\,\mathbf{k}) = \frac{\partial x}{\partial x} + \frac{\partial y}{\partial y} + \frac{\partial z}{\partial z} = 3 \,. \tag{6.50}$$

(b) An important operation on a scalar field φ is the **Laplacian operator** $\Delta\varphi$, defined as the divergence of a gradient,

$$\Delta\varphi := \nabla \cdot \nabla\varphi = \frac{\partial^2 \varphi}{\partial x^2} + \frac{\partial^2 \varphi}{\partial y^2} + \frac{\partial^2 \varphi}{\partial z^2} \,, \tag{6.51}$$

see the two–dimensional version in (6.47). The Laplacian is used in Exercises 6.58–6.61.

The following theorem is the main result of this section, giving a relation between a triple integral over a region $\Omega \subset \mathbb{R}^3$ and a double integral over the surface $\mathcal{S}$ of Ω.

Theorem 6.47 (Gauss' Divergence Theorem). Let $\mathcal{S}$ be a piecewise smooth surface bounding a region $\Omega \subset \mathbb{R}^3$, and let

$$\mathbf{F}(x, y, z) := \mathbf{i}\, F_1(x, y, z) + \mathbf{j}\, F_2(x, y, z) + \mathbf{k}\, F_3(x, y, z)$$

be a vector field with continuous derivatives on Ω. Then

$$\iiint_{\Omega} \nabla \cdot \mathbf{F}\, dV = \iint_{\mathcal{S}} \mathbf{F} \cdot \mathbf{n}\, dS \tag{6.52}$$

where $\mathbf{n}$ is the outer (unit) normal to the surface $\mathcal{S}$.

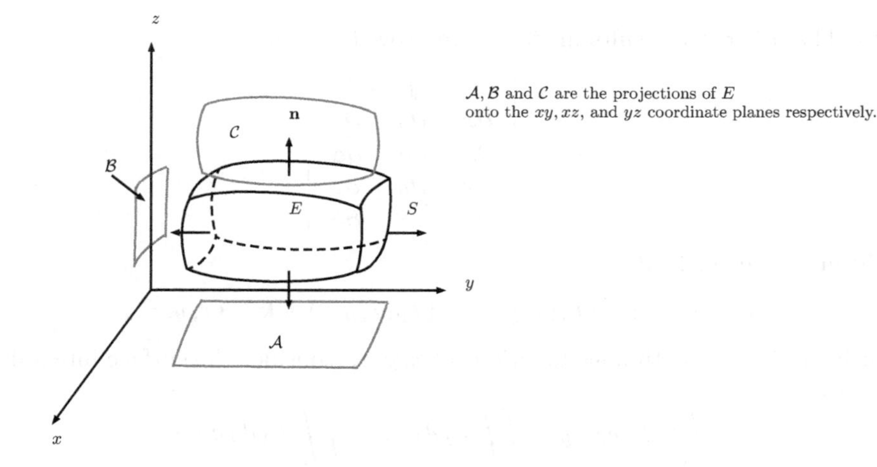

FIGURE 6.8: A region Ω, its surface $\mathcal{S}$ and the projections $\mathcal{A}$, $\mathcal{B}$ and $\mathcal{C}$, which are useful for visualizing the proof of the Gauss' Theorem.

Proof:
To prove the theorem we consider double integrals over the planar sets $\mathcal{A}$, $\mathcal{B}$ and $\mathcal{C}$, the projections of the region Ω on the xy–plane, xz–plane and yz–plane, respectively, see Figure (6.8). For any integrable function $f(x, y, z)$ we have:

$$\iint_{\mathcal{A}} f(x,y,z)\,dx\,dy = \iint_{\mathcal{S}} f(x,y,z)\,\cos\gamma\,dS \tag{6.53a}$$

$$\iint_{\mathcal{B}} f(x,y,z)\,dx\,dz = \iint_{\mathcal{S}} f(x,y,z)\,\cos\beta\,dS \tag{6.53b}$$

$$\iint_{\mathcal{C}} f(x,y,z)\,dy\,dz = \iint_{\mathcal{S}} f(x,y,z)\,\cos\alpha\,dS \tag{6.53c}$$

where $\cos\alpha$, $\cos\beta$, $\cos\gamma$ are the **direction cosines** of the unit normal $\mathbf{n}$

$$\mathbf{n} = \mathbf{i}\cos\alpha + \mathbf{j}\cos\beta + \mathbf{k}\cos\gamma\,, \tag{6.54}$$

in other words,

$\alpha :=$ the angle between $\mathbf{n}$ and $\mathbf{i}$,
$\beta :=$ the angle between $\mathbf{n}$ and $\mathbf{j}$, and
$\gamma :=$ the angle between $\mathbf{n}$ and $\mathbf{k}$.

For example, (6.53a) follows from

$$dx\,dy = dS\,\cos\gamma$$

relating the area element $dx\,dy$ (in $\mathcal{A}$) and the corresponding surface element

dS. The other two results in (6.53) are proved similarly.

$$\mathbf{N} = \begin{vmatrix} \mathbf{i} & \mathbf{j} & \mathbf{k} \\ \dfrac{\partial x}{\partial \alpha} & \dfrac{\partial y}{\partial \alpha} & \dfrac{\partial z}{\partial \alpha} \\ \dfrac{\partial x}{\partial \beta} & \dfrac{\partial y}{\partial \beta} & \dfrac{\partial z}{\partial \beta} \end{vmatrix} .$$

Given the vector field

$$\mathbf{F}(x, y, z) := \mathbf{i}\, F_1(x, y, z) + \mathbf{j}\, F_2(x, y, z) + \mathbf{k}\, F_3(x, y, z) ,$$

with F_1, F_2, F_3 continuous functions of x, y, z, consider the surface integrals, see (6.53),

$$\iint_{\mathcal{A}} F_1\, dx\, dy + \iint_{\mathcal{B}} F_2\, dx\, dz + \iint_{\mathcal{C}} F_3\, dy\, dz =$$

$$\iint_{\mathcal{S}} (F_1 \cos\gamma + F_2 \cos\beta + F_3 \cos\alpha)\, dS = \iint_{\mathcal{S}} \mathbf{F} \cdot \mathbf{n}\, dS , \tag{6.55}$$

by (6.54). In the special instance where the surface is parameterized by $z = l(x, y)$ the integral takes on a special form since x, and y may now be used as the integration parameters. We compute readily that $\frac{\partial(y,z)}{\partial(x,y)} = -\frac{\partial l}{\partial x}$, $\frac{\partial(z,x)}{\partial(x,y)} = -\frac{\partial l}{\partial y}$, and that $\frac{\partial(y,z)}{\partial(x,y)} = 1$. This leads to the special form

$$\int_{\mathcal{S}}\int \mathbf{F} \cdot \mathbf{n} dS = \int_{D}\int \left(-P\frac{\partial l}{\partial x} - Q\frac{\partial l}{\partial y} + R\right) dxdy,$$

where D is meant to be the projection of $\mathcal{S}$ onto the xy plane. To relate these surface integrals to the volume integral in the left side of (6.61), we compute

$$\iiint_{\Omega} \frac{\partial F_3}{\partial z}\, dV = \iint_{\mathcal{A}} dx\, dy \int_{z_1(x,y)}^{z_2(x,y)} \frac{\partial F_3}{\partial z}\, dz$$

$$= \iint_{\mathcal{A}} dx\, dy\, (F_3(x, y, z_2(x, y)) - F_3(x, y, z_1(x, y)))$$

$$= \iint_{\mathcal{A}} F_3(x, y, z_2(x, y))\, dx\, dy - \iint_{\mathcal{A}} F_3(x, y, z_1(x, y))\, dx\, dy \tag{6.56}$$

where $z = z_1(x, y)$ is the graph of the bottom $\mathcal{S}_1$ of the surface $\mathcal{S}$, and $z = z_2(x, y)$ is the graph of the top $\mathcal{S}_2$, see Figure (6.8). The first integral in (6.56) is

$$\iint_{\mathcal{A}} F_3(x, y, z_2(x, y))\, dx\, dy = \iint_{\mathcal{S}_2} F_3(x, y, z_2(x, y)) \cos\gamma_2\, dS , \text{ by (6.53a).}$$

Note that this is an integral over the top surface $\mathcal{S}_2$. Similarly,

$$\iint\limits_{\mathcal{A}} F_3(x,y,z_1(x,y))\,dx\,dy \;=\; -\iint\limits_{\mathcal{S}_1} F_3(x,y,z_1(x,y))\,\cos\gamma_1\,dS$$

an integral over the bottom surface $\mathcal{S}_1$. The minus sign is explained since the angle γ_1 is obtuse, see Figure (6.8). Therefore

$$\begin{aligned}\iiint_\Omega \frac{\partial F_3}{\partial z}\,dV &= \iint_{\mathcal{S}_2} F_3(x,y,z_2(x,y))\,\cos\gamma_2\,dS \\ &\quad + \iint_{\mathcal{S}_1} F_3(x,y,z_1(x,y))\,\cos\gamma_1\,dS \\ &= \iint_{\mathcal{S}} F_3(x,y,z)\,\cos\gamma\,dS,\end{aligned}$$

an integral over the whole surface $\mathcal{S}$. Similarly,

$$\iiint\limits_{\Omega} \frac{\partial F_1}{\partial x}\,dV \;=\; \iint\limits_{\mathcal{S}} F_1(x,y,z)\,\cos\alpha\,dS$$

$$\iiint\limits_{\Omega} \frac{\partial F_2}{\partial y}\,dV \;=\; \iint\limits_{\mathcal{S}} F_2(x,y,z)\,\cos\beta\,dS\,.$$

Adding the last three integrals, and using (6.57), we get (6.61).
This proof clearly holds for regions Ω which can be subdivided into subregions that are simultaneously x–simple, y–simple and z–simple. □
The **flux** of a vector field $\mathbf{F}$ across a closed surface $\mathcal{S}$ is defined as

$$\iint\limits_{\mathcal{S}} \mathbf{F}\cdot\mathbf{n}\,dS\,. \tag{6.57}$$

Gauss' Divergence Theorem can be used to compute this flux as the volume integral of $\nabla\cdot\mathbf{F}$.

Remark 6.48. Green's Theorem 6.24 can be regarded as a special, two–dimensional, case of Gauss' Theorem 6.47. Consider the vector field in $\mathbb{R}^2$,

$$\mathbf{F}(x,y) \;=\; \mathbf{i}\,F_1(x,y)+\mathbf{j}\,F_2(x,y) \tag{6.58}$$

and a closed curve $\mathcal{C} := \{x(s)\,\mathbf{i}+y(s)\,\mathbf{j} \,:\, 0\le s\le L\}$ surrounding a domain $\mathcal{D}$. The normal to $\mathcal{C}$ is

$$\mathbf{n} \;=\; \frac{dy}{ds}\,\mathbf{i}-\frac{dx}{ds}\,\mathbf{j}$$

and the **flux** of $\mathbf{F}$ across $\mathcal{C}$ is

$$\begin{aligned}
\oint_{\mathcal{C}+} \mathbf{F}\cdot\mathbf{n}\,ds &= \oint_{\mathcal{C}+} \left(F_1\,\frac{dy}{ds} - F_2\,\frac{dx}{ds}\right) ds \\
&= \iint\limits_{\mathcal{D}} \left(\frac{\partial F_1}{\partial x} - \frac{\partial(-F_2)}{\partial y}\right) dx\,dy\ , \quad \text{by Green's Theorem} \\
&= \iint\limits_{\mathcal{D}} \left(\frac{\partial F_1}{\partial x} + \frac{\partial F_2}{\partial y}\right) dx\,dy \;=\; \iint\limits_{\mathcal{D}} (\nabla\cdot\mathbf{F})\,dx\,dy\ .
\end{aligned}$$

This is a two–dimensional version of the Gauss Theorem

$$\iint\limits_{\mathcal{S}} \mathbf{F}\cdot\mathbf{n}\,dS = \iiint\limits_{\Omega} \nabla\cdot\mathbf{F}\,dV\ . \tag{6.59}$$

Mathematica session 6.9. *We use* MATHEMATICA *to evaluate the surface integrals in (6.61), where the integration surfaces are either spherical or cylindrical.* ***Clear["Global*"]***

```
divergenceIntegral[F_, X_, Y_, S_, T_]:=
Module[{function, cp, ncp, main},
function = F/.{X[[1]] → Y[[1]], X[[2]] → Y[[2]], X[[3]] → Y[[3]]};
cp = D[Y, S[[1]]] × D[Y, T[[1]]];
ncp = Simplify[Sqrt[Dot[cp, cp]]];
main = Simplify[function * ncp]];
Integrate[main, {S[[1]], S[[2]], S[[3]]}], {T[[1]], T[[2]], T[[3]]}
```

```
PowerExpand[divergenceIntegral[Sqrt[x^2 + y^2 + z^2], {x, y, z},
{r * Sin[phi] * Cos[theta], r * Sin[phi] * Sin[theta],
r * Cos[phi]}, {theta, 0, 2 * Pi},
{phi, 0, Pi}]]
```

$r^3\,Sin[phi]$

```
divergenceIntegral[x * y * z, {x, y, z},
{r Cos[theta] Sin[phi], r Sin[phi] Sin[theta], r Cos[phi]}, {phi, 0, Pi},
{theta, 0, 2 * Pi}]
```

$r^5\, Cos[phi]\, Cos[theta]\, Sin[phi]^3\, Sin[theta]$

```
PowerExpand[TransformedField["Cartesian" → "Spherical", Sqrt[x^2+
y^2 + z^2], {x, y, z} → {r, theta, phi}]]
```

$r\sqrt{Cos[theta]^2 + Cos[phi]^2 Sin[theta]^2 + Sin[phi]^2 Sin[theta]^2}$

```
Simplify[Sqrt[x^2 + y^2 + z^2]/.
{x->r * Sin[phi] * Cos[theta], y->r * Sin[phi] * Sin[theta], z->r * Cos[phi]}]
```

$\sqrt{r^2}$

```
Integrate[x * y * z/.{x->r * Sin[phi] * Cos[theta], y → Sin[phi] * Sin[theta],
z → Cos[phi]}, {phi, 0, Pi}, {theta, 0, 2 * Pi}]
```

0

```
D[{r * Sin[phi] * Cos[theta], r * Sin[phi] * Sin[theta], r * Cos[phi]}, phi]×
D[{r * Sin[phi] * Cos[theta], r * Sin[phi] * Sin[theta], r * Cos[phi]}, theta]
```

$\{r^2 Cos[theta] Sin[phi]^2, r^2 Sin[phi]^2 Sin[theta], r^2 Cos[phi] Cos[theta]^2 Sin[phi]+$
$r^2 Cos[phi] Sin[phi] Sin[theta]^2\}$

```
Simplify[Sqrt[Dot[%, %]]]
```

$\sqrt{r^4 Sin[phi]^2}$

```
Integrate[r^2 * Sin[phi], {phi, 0, Pi}, {theta, 0, 2 * Pi}]
```

$4\pi r^2$

```
Y = {r * Sin[phi] * Cos[theta], r * Sin[phi] * Sin[theta], r * Cos[phi]};
S = {phi, 0, Pi};
T = {theta, 0, 2 * Pi}
```

$\{theta, 0, 2\pi\}$

```
D[Y, S[[1]]]
```

$\{r\,Cos[phi]\,Cos[theta], r\,Cos[phi]\,Sin[theta], -r\,Sin[phi]\}$

```
D[Y, T[[1]]]
```

$\{-r\,Sin[phi]\,Sin[theta], r\,Cos[theta]\,Sin[phi], 0\}$

```
{r Cos[phi] Cos[theta], r Cos[phi] Sin[theta], -r Sin[phi]} ×
{-r Sin[phi] Sin[theta], r Cos[theta] Sin[phi], 0}
```

$\{r^2\,Cos[theta]\,Sin[phi]^2, r^2\,Sin[phi]^2\,Sin[theta], r^2\,Cos[phi]\,Cos[theta]^2\,Sin[phi]$
$+r^2\,Cos[phi]\,Sin[phi]\,Sin[theta]^2\}$

```
Simplify[Sqrt[Dot[%, %]]]
```

$\sqrt{r^4\,Sin[phi]^2}$

```
Integrate[Sqrt[r^4 Sin[phi]^2], {phi, 0, Pi}, {theta, 0, 2 * Pi}]
```

$4\pi\sqrt{r^4}$

```
PowerExpand[Sqrt[r^4 Sin[phi]^2]]
```

$r^2\,Sin[phi]$

```
Integrate[%, {phi, 0, Pi}]
```

$2r^2$

```
Clear[All["Global*"]]

divergenceIntegral[F_, X_, Y_, S_, T_]:=
Module[{function, cp, ncp, main},
function = F/.{X[[1]] → Y[[1]], X[[2]] → Y[[2]], X[[3]] → Y[[3]]};
cp = D[Y, S[[1]]] × D[Y, T[[1]]];
ncp = Simplify[Sqrt[Dot[cp, cp]]];
main = PowerExpand[Simplify[function * ncp]];
Integrate[main, {S[[1]], S[[2]], S[[3]]}, {T[[1]], T[[2]], T[[3]]}]]

PowerExpand[divergenceIntegral[Sqrt[x^2 + y^2 + z^2], {x, y, z},
{r * Sin[φ] * Cos[θ], r * Sin[φ] * Sin[θ], r * Cos[φ]},
{θ, 0, 2 * Pi}, {φ, 0, Pi}]]
```

$4\pi r^3$

```
integrand[F_, X_, Y_, S_, T_]:=
Module[{function, cp, ncp, main},
function = F/.{X[[1]] → Y[[1]], X[[2]] → Y[[2]], X[[3]] → Y[[3]]};
cp = D[Y, S[[1]]] × D[Y, T[[1]]];
ncp = Simplify[Sqrt[Dot[cp, cp]]];
main = PowerExpand[Simplify[function * ncp]]]

integrand[Sqrt[x^2 + y^2 + z^2], {x, y, z},
{r * Sin[φ] * Cos[θ], r * Sin[φ] * Sin[θ], r * Cos[φ]},
{θ, 0, 2 * Pi}, {φ, 0, Pi}]
```

$r^3 Sin[\varphi]$

```
Integrate[Sqrt[r^2] Sqrt[r^4 Sin[φ]^2], {theta, 0, 2 * Pi}, {phi, 0, Pi}]

PowerExpand[Sqrt[r^2] Sqrt[r^4 Sin[φ]^2]]
```

$r^3 Sin[\varphi]$

Recall that the surface differential can be approximated by

$$dS \approx \left| \frac{\partial \mathbf{r}}{\partial \alpha} \times \frac{\partial \mathbf{r}}{\partial \beta} \right| d\alpha d\beta,$$

and therefore a surface integral may be represented by

$$\iint_{S} f(\mathbf{r})\, dS = \iint_{\square} f(\mathbf{r}(\alpha,\beta)) \left| \frac{\partial \mathbf{r}}{\partial \alpha} \times \frac{\partial \mathbf{r}}{\partial \beta} \right| d\alpha\, d\beta\,. \tag{6.60}$$

We next use MATHEMATICA *to compute the* **directed** *surface differential*

```
surfaceIntegral[default_, z_, S_, T_]:=
Module[{sa}, sa = Simplify[Sqrt[D[z, s]^2 + D[z, t]^2 + 1]];
Integrate[default * sa, {S[[1]], S[[2]], S[[3]]}, {T[[1]], T[[2]], T[[3]]}]]


surfaceIntegral[1, 5 − x^2 − y^2, {x, −2, 2
}, {y, −Sqrt[4 − x^2], Sqrt[4 − x^2]}]
```

4π

```
Simplify[Sqrt[D[5 − x^2 − y^2, x]^2 + D[5 − x^2 − y^2, y]^2]]
```

$2\sqrt{x^2 + y^2}$

```
Integrate[1, {x, −2, 2}, {y, −Sqrt[4 − x^2], Sqrt[4 − x^2]}]
```

4π

```
Integrate[2 * x + 3 * y, {x, −2, 2}, {y, −Sqrt[4 − x^2], Sqrt[4 − x^2]}]
```

0

```
Integrate[x^2 + y^2, {x, -2, 2}, {y, -Sqrt[4 - x^2], Sqrt[4 - x^2]}]
```

8π

```
Clear["Global*"]
```

In order to verify the Divergence Theorem, we construct a procedure for evaluating the volume integral version of the Divergence Theorem

```
divergenceIntegral[F_, X_, Y_, S_, T_]:=
Module[{function, cp, ncp, main},
function = F/.{X[[1]] → Y[[1]], X[[2]] → Y[[2]], X[[3]] → Y[[3]]};
cp = D[Y, S[[1]]] × D[Y, T[[1]]];
ncp = Simplify[Sqrt[Dot[cp, cp]]];
main = Simplify[function * ncp]];
Integrate[main, {S[[1]], S[[2]], S[[3]]}], {T[[1]], T[[2]], T[[3]]}
```

```
PowerExpand[divergenceIntegral[Sqrt[x^2 + y^2 + z^2], {x, y, z},
{r * Sin[phi] * Cos[theta], r * Sin[phi] * Sin[theta], r * Cos[phi]}, {theta, 0, 2 *
Pi}, {phi, 0, Pi}]]
```

$r^3 Sin[phi]$

```
divergenceIntegral[x * y * z, {x, y, z},
{r Cos[theta] Sin[phi], r Sin[phi] Sin[theta], r Cos[phi]}, {phi, 0, Pi},
{theta, 0, 2 * Pi}]
```

$r^5 Cos[phi] Cos[theta] Sin[phi]^3 Sin[theta]$

```
PowerExpand[TransformedField["Cartesian" → "Spherical", Sqrt[x^2 + y^2+
z^2], {x, y, z} → {r, theta, phi}]]
```

$r\sqrt{Cos[theta]^2 + Cos[phi]^2 Sin[theta]^2 + Sin[phi]^2 Sin[theta]^2}$

Simplify[Sqrt[x^2 + y^2 + z^2]/.

{x->r * Sin[phi] * Cos[theta], y->r * Sin[phi] * Sin[theta], z->r * Cos[phi]}]

$\sqrt{r^2}$

Integrate[x * y * z/.{x->r * Sin[phi] * Cos[theta], y → Sin[phi] * Sin[theta],

z → Cos[phi]}, {phi, 0, Pi}, {theta, 0, 2 * Pi}]

0

D[{r * Sin[phi] * Cos[theta], r * Sin[phi] * Sin[theta], r * Cos[phi]}, phi] ×

D[{r * Sin[phi] * Cos[theta], r * Sin[phi] * Sin[theta], r * Cos[phi]}, theta]

$\{r^2 Cos[theta] Sin[phi]^2, r^2 Sin[phi]^2 Sin[theta], r^2 Cos[phi] Cos[theta]^2 Sin[phi]$

$+r^2 Cos[phi] Sin[phi] Sin[theta]^2\}$

Simplify[Sqrt[Dot[%, %]]]

$\sqrt{r^4 Sin[phi]^2}$

Integrate[r^2 * Sin[phi], {phi, 0, Pi}, {theta, 0, 2 * Pi}]

$4\pi r^2$

Y = {r * Sin[phi] * Cos[theta], r * Sin[phi] * Sin[theta], r * Cos[phi]};

S = {phi, 0, Pi};

T = {theta, 0, 2 * Pi}

$\{theta, 0, 2\pi\}$

D[Y, S[[1]]]

$\{r Cos[phi] Cos[theta], r Cos[phi] Sin[theta], -r Sin[phi]\}$

= **D[Y, T[[1]]]**

$\{-r\,Sin[phi]\,Sin[theta], r\,Cos[theta]\,Sin[phi], 0\}$

{r Cos[phi] Cos[theta], r Cos[phi] Sin[theta], −r Sin[phi]}×
{−r Sin[phi] Sin[theta], r Cos[theta] Sin[phi], 0}

$\{r^2\,Cos[theta]\,Sin[phi]^2, r^2\,Sin[phi]^2\,Sin[theta], r^2\,Cos[phi]\,Cos[theta]^2\,Sin[phi]+$
$r^2\,Cos[phi]\,Sin[phi]\,Sin[theta]^2\}$

Simplify[Sqrt[Dot[%, %]]]

$\sqrt{r^4\,Sin[phi]^2}$

Integrate $\left[\sqrt{r^4\,Sin[phi]^2}, \{phi, 0, Pi\}, \{theta, 0, 2 * Pi\}\right]$

$4\pi\sqrt{r^4}$

PowerExpand $\left[\sqrt{r^4\,Sin[phi]^2}\right]$

$r^2\,Sin[phi]$

Integrate[%, {phi, 0, Pi}]

$2r^2$

♦

Example 6.49. In Example 6.7 we computed the flux of a two–dimensional vector field across an infinitesimal rectangle. Here we do the same for a three–dimensional vector field

$$\mathbf{F}(x,y,z) := \mathbf{i}\,F_1(x,y,z) + \mathbf{j}\,F_2(x,y,z) + \mathbf{k}\,F_3(x,y,z)\ ,$$

and an infinitesimal cube of sides Δx, Δy and Δz, see Figure (6.9). The surface $\mathcal{S}$ of the cube is made of six sides, *ABCD*, *EFGH*, *BDFH*, etc. We use (6.57) to compute the flux (net flow) of $\mathbf{F}$ through the surface $\mathcal{S}$. The normal to the side *BDFH* is $\mathbf{j}$, and the flow through *BDFH* is approximately,

$$\begin{aligned} \text{flow}(BDFH) &\approx ((F_1\,\mathbf{i} + F_2\,\mathbf{j} + F_3\,\mathbf{k})\cdot\mathbf{j})\ \Delta x\,\Delta z \\ &= F_2\,\Delta x\,\Delta z \\ &\approx F_2\left(x + \frac{1}{2}\,\Delta x,\ y + \Delta y,\ z + \frac{1}{2}\,\Delta z\right)\ \Delta x\,\Delta z\ . \end{aligned}$$

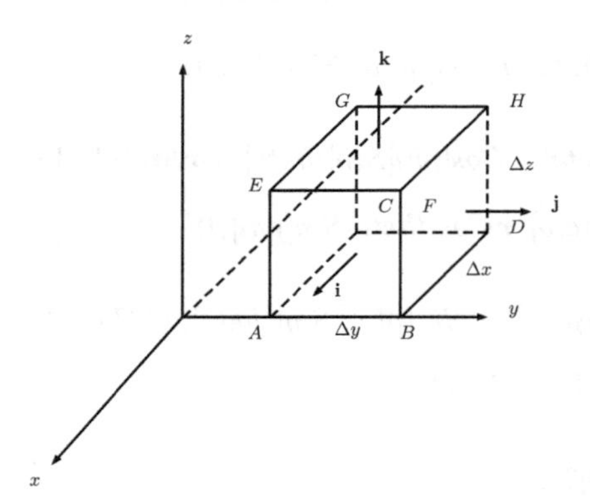

FIGURE 6.9: Illustration of Example(6.49).

The flow through the opposite side $ACEG$ (with normal $-\mathbf{j}$) is approximately

$$\begin{aligned}\text{flow}(ACEG) &\approx ((F_1\,\mathbf{i} + F_2\,\mathbf{j} + F_3\,\mathbf{k}) \cdot (-\mathbf{j}))\,\Delta x\,\Delta z \\ &= -F_2\,\Delta x\,\Delta z \\ &\approx -F_2\left(x + \frac{1}{2}\,\Delta x,\, y, z + \frac{1}{2}\,\Delta z\right)\,\Delta x\,\Delta z\,,\end{aligned}$$

so the net flow in the direction of the y–axis is approximately,

$$\text{flow}(BDFH) + \text{flow}(ACEG) \approx$$

$$\left(F_2\left(x + \frac{1}{2}\,\Delta x,\, y + \Delta y,\, z + \frac{1}{2}\,\Delta z\right) - F_2\left(x + \frac{1}{2}\,\Delta x,\, y,\, z + \frac{1}{2}\,\Delta z\right)\right)\Delta x\,\Delta z$$

$$\approx \left(\frac{\partial F_2}{\partial y}\left(x + \frac{1}{2}\,\Delta x,\, y,\, z + \frac{1}{2}\,\Delta z\right)\,\Delta y\right)\,\Delta x\,\Delta z$$

$$\approx \frac{\partial F_2}{\partial y}(x,\, y,\, z)\;dV.$$

Similarly, the net flow in the x–direction,

$$\text{flow}(ABEF) + \text{flow}(CDGH) \approx \frac{\partial F_1}{\partial x}(x,\, y,\, z)\;dV$$

and in the z–direction,

$$\text{flow}(EFGH) + \text{flow}(ABCD) \approx \frac{\partial F_3}{\partial z}(x,\, y,\, z)\;dV$$

Summing these terms we get the net flow through the surface of the cube

$$\text{FLUX}\,[\mathbf{F}]_{\mathcal{S}} \approx \left(\frac{\partial F_1}{\partial x}(x,\, y,\, z) + \frac{\partial F_2}{\partial y}(x,\, y,\, z) + \frac{\partial F_3}{\partial z}(x,\, y,\, z)\right)\,dV = (\nabla \cdot \mathbf{F})\;dV.$$

giving an interpretation of the divergence as the rate of flux. The Gauss Divergence Theorem

$$\iiint_{\Omega} \nabla \cdot \mathbf{F}\,dV = \iint_{\mathcal{S}} \mathbf{F} \cdot \mathbf{n}\,dS \tag{6.61}$$

can be thought of as summing the differentials $(\nabla \cdot \mathbf{F})\ dV$ to obtain the flux across $\mathcal{S}$.

Example 6.50. Let $\mathcal{S}$ be a closed surface in a region filled with fluid. At any point $\mathbf{r}$, let $\delta(\mathbf{r})$ denote the **density** of the fluid, and $\mathbf{v}(\mathbf{r})$ its **velocity**. Consider the vector field

$$\mathbf{F}(\mathbf{r}) \ := \ \delta(\mathbf{r})\,\mathbf{v}(\mathbf{r})\ , \tag{6.62}$$

in the direction of the fluid flow $\mathbf{v}$. The magnitude of $\mathbf{F}$, $\|\mathbf{F}(\mathbf{r})\| \ = \ \delta(\mathbf{r})\,\|\mathbf{v}(\mathbf{r})\|$, is the rate of flow of fluid mass at the point $\mathbf{r}$. The flux of the fluid across $\mathcal{S}$ is the surface integral $\displaystyle\iint_{\mathcal{S}} \mathbf{F}\cdot\mathbf{n}\,dS$. If instead of fluid we have electric charges, then the flux $\mathbf{F}$ is **electric current**, and δ is the **charge density**. Let Δm be the fluid mass passing through a surface element ΔS in time Δt. The rate of flow is approximately

$$\frac{\Delta m}{\Delta t} \ \approx \ \delta\ (\mathbf{v}\cdot\mathbf{n})\ \Delta S$$

and in the limit, the rate of change of mass across the surface $\mathcal{S}$,

$$\begin{aligned}\frac{dm}{dt} \ &= \ \iint_{\mathcal{S}} \delta\ (\mathbf{v}\cdot\mathbf{n})\ dS \ = \ \iint_{\mathcal{S}} (\mathbf{F}\cdot\mathbf{n})\ dS\ , \quad \text{by (6.62)} \\ &= \ \iiint_{\Omega} (\nabla\cdot\mathbf{F})\ dV\ , \quad \text{by Gauss' Theorem}\end{aligned} \tag{6.63}$$

where Ω is the region enclosed by $\mathcal{S}$. Let Ω be small, and let V be its volume. Then the outflow across $\mathcal{S}$ is by (6.63),

$$\frac{\Delta m}{\Delta t} \ \approx \ (\nabla\cdot\mathbf{F})\ V.$$

Also

$$\frac{\Delta m}{\Delta t} \ \approx \ -\left(\frac{\Delta\delta}{\Delta t}\right)\ V$$

since a net outflow of mass Δm decreases the mass inside Ω by the amount $\Delta\delta\, V$, approximately. Comparing the last two results we conclude

$$\nabla\cdot\mathbf{F} \ \approx \ -\frac{\Delta\delta}{\Delta t}$$

and in the limit, as the volume V goes to zero, we get the equation

$$\nabla\cdot\mathbf{F} \ + \ \frac{\partial\delta}{\partial t} \ = \ 0\ , \tag{6.64}$$

called the **continuity equation** of the fluid. Equation (6.64) is also called the **mass conservation equation**, since it states that the mass of the fluid is unchanged.

Example 6.51. A body $\mathcal{D}$ with a closed surface $\mathcal{S}$ is submerged in fluid. The force on a surface element dS is

$$d\mathbf{F} = -(p\,dS)\,\mathbf{n}\,,$$

where $\mathbf{n}$ is the **unit normal** to the surface element, and p is the **hydrostatic pressure**.The **net force** on the body is the surface integral

$$\mathbf{F} = -\iint_{\mathcal{S}} p\,\mathbf{n}\,dS \tag{6.65}$$

written by its components as

$$\mathbf{F}(x,y,z) := \mathbf{i}\,F_1(x,y,z) + \mathbf{j}\,F_2(x,y,z) + \mathbf{k}\,F_3(x,y,z)\,. \tag{6.66}$$

The component F_3 is

$$F_3 = \mathbf{k}\cdot\mathbf{F} = -\iint_{\mathcal{S}} (p\,\mathbf{k})\cdot\mathbf{n}\,dS = -\iiint_{\mathcal{D}} \nabla\cdot(p\,\mathbf{k})\,dV \tag{6.67}$$

by Theorem 6.47. We use Cartesian coordinates with the fluid boundary as the xy–plane, and the z–axis pointing down. The pressure at the point (x,y,z) in the fluid is then

$$p(x,y,z) = \delta\,z\,, \tag{6.68}$$

where δ is the fluid **density**. Note that the pressure does not depend on x and y. Therefore

$$\nabla\cdot(p\,\mathbf{k}) = \frac{\partial p}{\partial z} = \delta$$

and, by (6.67),

$$F_3 = -\iiint_{\mathcal{D}} \delta\,dV \tag{6.69}$$

which is **Archimedes' Law**: The bouyancy force on a body in fluid is equal to the weight of the fluid which the body displaces. The other two components of $\mathbf{F}$ are zero (see Exercise 6.54)

$$F_1 = F_2 = 0\,,$$

so the body does not move sideways.

Example: We compute the force of buoyancy on a spherical buoy, 4 `[feet]` in diameter, which is half submerged in water. The pressure is, by (6.68),

$$p(x,y,z) = 62.4\,z\;[\mathtt{lb\ ft}^{-2}]\,,$$

and the bouyancy force is, by (6.67),

$$F_3 = 62.4\,\frac{1}{2}\,\frac{4\,\pi\,2^3}{3}\;[\mathtt{lb}] = 41.6\,\pi\;[\mathtt{lb}]$$

Exercises

Exercise 6.52. *The divergence operator satisfies, for any (differentiable) vector fields* $\mathbf{F}$, $\mathbf{G}$, *constants* a, b *and a (differentiable) scalar field* f

(a) $\nabla \cdot (a\mathbf{F} + b\mathbf{G}) = a\nabla \cdot \mathbf{F} + b\nabla \cdot \mathbf{G}$ (b) $\nabla \cdot (f\mathbf{F}) = f\nabla \cdot \mathbf{F} + (\nabla f) \cdot \mathbf{F}$

These follow from the linearity of differentiation and the definition of the divergence. Prove (a) and (b).

Exercise 6.53. *Illustrate Theorem 6.47 by computing the surface integral and the volume integral for the following vector fields and surfaces.*

(a) $\mathbf{F} = \mathbf{r} := x\mathbf{i} + y\mathbf{j} + z\mathbf{k}$, *and* $\mathcal{S}$ *is the unit sphere.*

(b) $\mathbf{F} = \mathbf{r}\,\|\mathbf{r}\|^n$, *and* $\mathcal{S}$ *is the unit sphere.*

(c) $\mathbf{F} = x^n\mathbf{i} + y^n\mathbf{j} + z^n\mathbf{k}$, *and* $\mathcal{S}$ *is the cube with vertices.*$(\pm 1, \pm 1, \pm 1)$

(d) $\mathbf{F} = x^{2m+1}\mathbf{i} + y^{2m+1}\mathbf{j} + z^{2m+1}\mathbf{k}$, *and* $\mathcal{S}$ *is the unit tube of length 4.*

Exercise 6.54. *Prove the conclusion* $F_1 = F_2 = 0$ *in Example 6.51.*

Exercise 6.55. *Using nested if statements write a function which evaluates the surface integral*

$$\int\!\!\int_{S} \mathbf{F} \cdot \mathbf{n}\, dS$$

first by attempting to do the integral directly, and if you are unable to do this, use the Divergence Theorem to evaluate it. Test your function on the examples of the previous problem.

Exercise 6.56. *Prove the following representations for the volume* V *of a region* $\mathcal{D}$ *having a piecewise smooth boundary* S

$$V = \iint_{\mathcal{S}} z\, dx\, dy = \iint_{\mathcal{S}} x\, dy\, dz = \iint_{\mathcal{S}} y\, dz\, dx\,. \tag{6.70}$$

Exercise 6.57. *Use the Divergence Theorem to derive the following alternate forms for the centroid of a region* $\mathcal{D}$, *with volume* V, *having a piecewise smooth boundary* S

$$\begin{aligned} \overline{x} &= \tfrac{1}{2\,V} \int_{S} x^2\, dy\, dz \\ \overline{y} &= \tfrac{1}{2\,V} \int_{S} x^2\, dz\, dx \\ \overline{z} &= \tfrac{1}{2\,V} \int_{S} x^2\, dx\, dy \end{aligned} \tag{6.71}$$

Exercise 6.58. *Let* $u(x, y, z)$ *have continuous derivatives up to first order, and let* $v(x, y, z)$ *have continuous derivatives up to second order, in a region* $\Omega \subset \mathbb{R}^3$ *bounded by a smooth surface* $\mathcal{S}$. *Prove the integral identity*

$$\iiint_{\Omega} u\, \Delta v\, dV \; +$$

$$\iiint\limits_{\Omega} u \left(\frac{\partial v}{\partial x} \frac{\partial u}{\partial x} + \frac{\partial v}{\partial y} \frac{\partial u}{\partial y} + \frac{\partial v}{\partial z} \frac{\partial u}{\partial z} \right) dV = - \iint\limits_{\mathcal{S}} u \frac{\partial v}{\partial n} dS . \tag{6.72}$$

Compare this identity with Exercise 6.44(a).
Hint: Use Green's Theorem with $F_1 = u \frac{\partial v}{\partial x}$, $F_2 = u \frac{\partial v}{\partial y}$, $F_3 = u \frac{\partial v}{\partial z}$.

Exercise 6.59. *Let u be a twice continuously differentiable solution of Laplace's equation, such that*

$$\Delta u = 0 \text{ in } \Omega , \tag{6.73a}$$

$$u \equiv 0 \text{ on the boundary } \mathcal{S} \text{ of } \Omega . \tag{6.73b}$$

(a) Use (6.72) to show

$$\iiint\limits_{\Omega} \|\nabla u\|^2 \, dV = 0 .$$

Hint: Use $v \equiv u$ *in (6.72).*
(b) Conclude that u is constant in Ω*. Since it is zero on the boundary of* Ω*, conclude that* $u \equiv 0$ *in* Ω*.*
(c) Consider the **Dirichlet problem***: Find a solution u of the equations*

$$\Delta u = 0 \text{ in } \Omega ,$$

$$u \equiv f \text{ on the boundary } \mathcal{S} \text{ of } \Omega ,$$

where f is a given continuous function. Use (b) to show that a solution u, if it exists, must be unique.
Hint: If there are two solutions, u_1 *and* u_2*, then their difference* $u_1 - u_2$ *satisfies (6.73a)–(6.73b).*

Exercise 6.60. *Consider the* **Neumann problem***: Find a solution u of the equations*

$$\Delta u = 0 \text{ in } \Omega ,$$

$$\frac{\partial u}{\partial n} \equiv f \text{ on the boundary } \mathcal{S} \text{ of } \Omega .$$

where f is a given continuous function, and $\frac{\partial u}{\partial n}$ *is the inward normal derivative of u on the boundary. Show that a solution u, if it exists, must be unique.*
Hint: Follow the steps of Exercise 6.59. Note that from a zero normal derivative on the boundary, you may only conclude that u is a constant. How do you show that constant is zero?

Exercise 6.61. *Let* Ω *and* $\mathcal{S}$ *be as in Theorem 6.47, and let* $u(x, y, z)$ *and* $v(x, y, z)$ *have continuous derivatives up to order 2 in* Ω*. Prove:*

(a) $$\iiint\limits_{\Omega} (u \, \Delta v - v \, \Delta u) \, dV = - \iint\limits_{\mathcal{S}} \left(u \frac{\partial v}{\partial n} - v \frac{\partial u}{\partial n} \right) dS$$

Hint: Interchange u and v in (6.72).

(b) $$\iiint\limits_{\Omega} \Delta v \, dV \;=\; -\iint\limits_{\mathcal{S}} \frac{\partial v}{\partial n}\, dS\,, \quad \textit{(a special case of (a))}$$

*Moreover, if u satisfies $\Delta u = 0$ (**Laplace equation**),*

(c) $$\iint\limits_{\mathcal{S}} \frac{\partial u}{\partial n}\, dS \;=\; 0$$

(d) $$\iint\limits_{\mathcal{S}} u\,\frac{\partial u}{\partial n}\, dS \;\le\; 0$$

Exercise 6.62. *Let the vector $\mathbf{F}$ be continuously differentiable in a region of $\mathbb{R}^3$, containing the set $\mathcal{S}$. Show that statements (a) and (b) below are equivalent.*

(a) $$\iint\limits_{\mathcal{S}} \mathbf{F}\cdot\mathbf{n}\, dS \;=\; 0$$

(b) $$\nabla\cdot\mathbf{F} \;=\; 0$$

6.4 The curl

The **circulation** of a two–dimensional vector field is given by the vector

$$\text{CURL}\,\mathbf{F} \;:=\; \left(\frac{\partial F_2}{\partial x} - \frac{\partial F_1}{\partial y}\right)\mathbf{k} \tag{6.74}$$

The analogous result in three–dimensions is defined next.

Definition 6.63. The **curl**[2] of the continuously differentiable vector field

$$\mathbf{F}(x,y,z) \;:=\; \mathbf{i}\,F_1(x,y,z) + \mathbf{j}\,F_2(x,y,z) + \mathbf{k}\,F_3(x,y,z) \tag{6.75}$$

is the vector

$$\text{CURL}\,\mathbf{F} \;:=\; \nabla \times \mathbf{F} \;=\; \mathbf{det}\begin{pmatrix} \mathbf{i} & \mathbf{j} & \mathbf{k} \\ \frac{\partial}{\partial x} & \frac{\partial}{\partial y} & \frac{\partial}{\partial z} \\ F_1 & F_2 & F_3 \end{pmatrix}\,. \tag{6.76}$$

This is not an ordinary determinant, since its matrix has vectors $(\mathbf{i}\,,\mathbf{j}\,,\mathbf{k})$ in the first row, operations $(\frac{\partial}{\partial x}\,,\frac{\partial}{\partial y}\,,\frac{\partial}{\partial z})$ in the second row, and functions $(F_1(x,y,z)\,, F_2(x,y,z)\,, F_3(x,y,z))$ in the third row. Treating these as numbers, we expand the determinant along the first row to give

$$\text{CURL}\,\mathbf{F} \;=\; \mathbf{i}\left(\frac{\partial F_3}{\partial y} - \frac{\partial F_2}{\partial z}\right) + \mathbf{j}\left(\frac{\partial F_1}{\partial z} - \frac{\partial F_3}{\partial x}\right) + \mathbf{k}\left(\frac{\partial F_2}{\partial x} - \frac{\partial F_1}{\partial y}\right)\,. \tag{6.77}$$

If $\mathbf{F}$ is a two–dimensional vector field

$$\mathbf{F}(x,y) \;=\; \mathbf{i}\,F_1(x,y) + \mathbf{j}\,F_2(x,y) \tag{6.78}$$

[2]Also **rotation** (abbreviated **rot**). Both terms are in common usage, **curl** is more popular in the U.S., and **rot** is used more commonly in Europe.

then (6.76) reduces to the two–dimensional curl (6.74).
The rules for taking the curl of sums and products are given in Exercise 6.65.

Mathematica session 6.10. ***Clear["Global*"]***

R = Sqrt[x^2 + y^2 + z^2]

$\sqrt{x^2+y^2+z^2}$

Grad[%, {x, y, z}]

$\left\{\frac{x}{\sqrt{x^2+y^2+z^2}}, \frac{y}{\sqrt{x^2+y^2+z^2}}, \frac{z}{\sqrt{x^2+y^2+z^2}}\right\}$

Grad[(x * y) * z/(x^2 + y^2 + z^2), {x, y, z}]

$\left\{-\frac{2x^2yz}{(x^2+y^2+z^2)^2} + \frac{yz}{x^2+y^2+z^2}, -\frac{2xy^2z}{(x^2+y^2+z^2)^2} + \frac{xz}{x^2+y^2+z^2}, -\frac{2xyz^2}{(x^2+y^2+z^2)^2} + \frac{xy}{x^2+y^2+z^2}\right\}$

Grad[1/Sqrt[x^2 + y^2 + z^2], {x, y, z}]

$\left\{-\frac{x}{(x^2+y^2+z^2)^{3/2}}, -\frac{y}{(x^2+y^2+z^2)^{3/2}}, -\frac{z}{(x^2+y^2+z^2)^{3/2}}\right\}$

Grad[1/R^2, {x, y, z}]

$\left\{-\frac{2x}{(x^2+y^2+z^2)^2}, -\frac{2y}{(x^2+y^2+z^2)^2}, -\frac{2z}{(x^2+y^2+z^2)^2}\right\}$

v:={x, y, z}

Grad[1/R^2, v]

$\left\{-\frac{2x}{(x^2+y^2+z^2)^2}, -\frac{2y}{(x^2+y^2+z^2)^2}, -\frac{2z}{(x^2+y^2+z^2)^2}\right\}$

Grad[f[r, theta, z], {r, theta, z}, "Cylindrical"]

$\left\{f^{(1,0,0)}[r, theta, z], \frac{f^{(0,1,0)}[r, theta, z]}{r}, f^{(0,0,1)}[r, theta, z]\right\}$

Grad[r * Cos[theta] * z^2, {r, theta, z}, "Cylindrical"]

$\{z^2 Cos[theta], -z^2 Sin[theta], 2rz\,Cos[theta]\}$

w:={r, theta, phi}

Grad[r^2 * Cos[theta] * Sin[phi], w, "Spherical"]

$\{2r\,Cos[theta]Sin[phi], -r\,Sin[phi]Sin[theta], r\,Cos[phi]\,Cot[theta]\}$

Grad[r * Cos[theta] * z^2, {r, theta, z}, "Cylindrical"]

$\{z^2 Cos[theta], -z^2 Sin[theta], 2rz\,Cos[theta]\}$

Grad[r^2 * Cos[theta] * Sin[phi], w, "Spherical"]

$\{2r\,Cos[theta]Sin[phi], -r\,Sin[phi]Sin[theta], r\,Cos[phi]\,Cot[theta]\}$

The MATHEMATICA *function* `curl` *computes the* **curl** *of a vector field* $\mathbf{F}(x, y, z)$. ***Curl[{P[x, y, z], Q[x, y, z], R[x, y, z]}, {x, y, z}]***

$$\{-Q^{(0,0,1)}[x, y, z] + R^{(0,1,0)}[x, y, z], P^{(0,0,1)}[x, y, z] - R^{(1,0,0)}[x, y, z], -P^{(0,1,0)}[x, y, z] + Q^{(1,0,0)}[x, y, z]\}$$

Curl[{P[r, θ, z], Q[r, θ, z], R[r, θ, z]}, {r, θ, z}, "Cylindrical"]

$$\left\{-Q^{(0,0,1)}[r, \theta, z] + \frac{R^{(0,1,0)}[r,\theta,z]}{r}, P^{(0,0,1)}[r, \theta, z] - R^{(1,0,0)}[r, \theta, z], -\frac{-Q[r,\theta,z]+P^{(0,1,0)}[r,\theta,z]}{r} + Q^{(1,0,0)}[r, \theta, z]\right\}$$

Curl[{P[r, θ, φ], Q[r, θ, φ], R[r, θ, φ]}, {r, θ, φ}, "Spherical"]

$$\left\{-\frac{Csc[\theta]\left(-Cos[\theta]R[r,\theta,\varphi]+Q^{(0,0,1)}[r,\theta,\varphi]\right)}{r} + \frac{R^{(0,1,0)}[r,\theta,\varphi]}{r}, \frac{Csc[\theta]}{r}\left(-R[r, \theta, \varphi]Sin[\theta] + P^{(0,0,1)}[r, \theta, \varphi]\right) - R^{(1,0,0)}[r, \theta, \varphi], -\frac{-Q[r,\theta,\varphi]+P^{(0,1,0)}[r,\theta,\varphi]}{r} + Q^{(1,0,0)}[r, \theta, \varphi]\right\}$$

We do this by entering the gradient of an arbitrary function. Notice the syntax requires two arguments. The first is an expression and the second prescribes the variable. Hence we see that the curl of a gradient of an arbitrary function with two continuous derivatives is zero. Next we introduce the divergence function, and test it on some vector fields. Now define the distance from the origin in $\mathbb{R}^3$ *as* `R(x,y,z):=SQRT(x^2+y^2+z^2)` *and denote the vector from the origin to the point* (x, y, z) *as* `[x, y, z]`. *We may now use* MATHEMATICA *to obtain several simple identities:*

Curl[{P[x, y, z], Q[x, y, z], R[x, y, z]}, {x, y, z}]

$$\{-Q^{(0,0,1)}[x,y,z] + R^{(0,1,0)}[x,y,z], P^{(0,0,1)}[x,y,z] - R^{(1,0,0)}[x,y,z], -P^{(0,1,0)}[x,y,z] + Q^{(1,0,0)}[x,y,z]\}$$

Curl[{P[r, θ, z], Q[r, θ, z], R[r, θ, z]}, {r, θ, z}, "Cylindrical"]

$$\left\{-Q^{(0,0,1)}[r,\theta,z] + \frac{R^{(0,1,0)}[r,\theta,z]}{r}, P^{(0,0,1)}[r,\theta,z] - R^{(1,0,0)}[r,\theta,z], -\frac{-Q[r,\theta,z]+P^{(0,1,0)}[r,\theta,z]}{r} + Q^{(1,0,0)}[r,\theta,z]\right\}$$

Curl[{P[r, θ, φ], Q[r, θ, φ], R[r, θ, φ]}, {r, θ, φ}, "Spherical"]

$$\left\{-\frac{Csc[\theta]\left(-Cos[\theta]R[r,\theta,\varphi]+Q^{(0,0,1)}[r,\theta,\varphi]\right)}{r} + \frac{R^{(0,1,0)}[r,\theta,\varphi]}{r}, \frac{Csc[\theta]}{r}(-R[r,\theta,\varphi]Sin[\theta]+ P^{(0,0,1)}[r,\theta,\varphi]) - R^{(1,0,0)}[r,\theta,\varphi], -\frac{-Q[r,\theta,\varphi]+P^{(0,1,0)}[r,\theta,\varphi]}{r} + Q^{(1,0,0)}[r,\theta,\varphi]\right\}$$

$$\boldsymbol{Div\left[\{-Q^{(0,0,1)}[x,y,z] + R^{(0,1,0)}[x,y,z], P^{(0,0,1)}[x,y,z] - R^{(1,0,0)}[x,y,z], -P^{(0,1,0)}[x,y,z] + Q^{(1,0,0)}[x,y,z]\}, \{x,y,z\}\right]}$$

0

Simplify[

$$\boldsymbol{Div\left[\left\{-Q^{(0,0,1)}[r,\theta,z] + \frac{R^{(0,1,0)}[r,\theta,z]}{r}, P^{(0,0,1)}[r,\theta,z] - R^{(1,0,0)}[r,\theta,z], -\frac{-Q[r,\theta,z]+P^{(0,1,0)}[r,\theta,z]}{r} + Q^{(1,0,0)}[r,\theta,z]\right\}, \{r,\theta,z\}, \text{"Cylindrical"}\right]\Big]}$$

0

Simplify[

$$\boldsymbol{Div\Bigg[\Bigg\{-\frac{Csc[\theta]\left(-Cos[\theta]R[r,\theta,\varphi]+Q^{(0,0,1)}[r,\theta,\varphi]\right)}{r} + \frac{R^{(0,1,0)}[r,\theta,\varphi]}{r},}$$

$$\frac{Csc[\theta]\left(-R[r,\theta,\varphi]Sin[\theta]+P^{(0,0,1)}[r,\theta,\varphi]\right)}{r}-R^{(1,0,0)}[r,\theta,\varphi],$$
$$-\frac{-Q[r,\theta,\varphi]+P^{(0,1,0)}[r,\theta,\varphi]}{r}+Q^{(1,0,0)}[r,\theta,\varphi]\Big\},\{r,\theta,\varphi\},\text{"Spherical"}\Big]\Big]$$

♦

Exercises

Exercise 6.64. *Calculate by hand the curl and divergence of each of the following vector fields. Check your results with* MATHEMATICA.

(a) $\mathbf{F}(x,y) := x\mathbf{i} + y\mathbf{j}$ (b) $\mathbf{F}(x,y) := y\mathbf{i} - x\mathbf{j}$

(c) $\mathbf{F}(x,y) := 2xy\mathbf{i} + y^2\mathbf{j}$ (d) $\mathbf{F}(x,y) := \sin(xy)\mathbf{i} + \cos(xy)\mathbf{j}$

(e) $\mathbf{F}(x,y,z) := x\mathbf{i} + y\mathbf{j} + z\mathbf{k}$ (f) $\mathbf{F}(x,y,z) := xz\mathbf{i} + yz\mathbf{j} + xy\mathbf{k}$

(g) $\mathbf{F}(x,y,z) := x^2\mathbf{i} + y^2\mathbf{j} + z^2\mathbf{k}$

(h) $\mathbf{F}(x,y,z) := (y^2+z^2)\mathbf{i} + (x^2+z^2)\mathbf{j} + (x^2+y^2)\mathbf{k}$

(i) $\mathbf{F}(x,y,z) := e^z\cos x\mathbf{i} + e^z\cos y\mathbf{j} + e^z\mathbf{k}$

(j) $\mathbf{F}(x,y,z) := \cos x\sin y\mathbf{i} + \sin x\sin y\mathbf{j} + \sin x\mathbf{k}$

Exercise 6.65. *Let* $\mathbf{F}$, $\mathbf{G}$ *be vector fields,* f *and* φ *scalar fields, and let* a *and* b *be constants. Then:*

(a) $\nabla\times(a\mathbf{F}+b\mathbf{G}) = a\nabla\times\mathbf{F} + b\nabla\times\mathbf{G}$

(b) $\nabla\times(f\mathbf{F}) = f(\nabla\times\mathbf{F}) + (\nabla f)\times\mathbf{F}$

(c) $\nabla\cdot(\mathbf{F}\times\mathbf{G}) = \mathbf{G}\cdot(\nabla\times\mathbf{F}) + \mathbf{F}\cdot(\nabla\times\mathbf{G})$

(d) $\nabla\cdot(\nabla\times\mathbf{F}) = 0$ (e) $\nabla\times(\nabla\varphi) = 0$

Exercise 6.66. *Let* $\mathbf{v} := v_1\mathbf{i}+v_2\mathbf{j}$ *be the velocity field of a fluid. If the divergence vanishes*

$$\nabla\cdot\mathbf{v} = \frac{\partial v_1}{\partial x} + \frac{\partial v_2}{\partial y} = 0$$

the flow is called **source free** *(i.e. it has no sources or sinks). If the curl of the velocity field also vanishes*

$$\nabla\times\mathbf{v} = \frac{\partial v_1}{\partial y} - \frac{\partial v_2}{\partial x} = \mathbf{0},$$

the flow is called **irrotational**. *Show that the components* v_1, v_2 *of a source–free, irrotational,* $\mathbf{v}$ *satisfy the* **Laplace equation**

$$\Delta v := \frac{\partial^2 v}{\partial x^2} + \frac{\partial^2 v}{\partial x^2} = 0.$$

Exercise 6.67. *Repeat the previous problem for three-dimensions and show that the components* v_1, v_2, v_3 *of a source–free, irrotational,* $\mathbf{v}$ *satisfy the (three-dimensional)* **Laplace equation**

$$\Delta v := \frac{\partial^2 v}{\partial x^2} + \frac{\partial^2 v}{\partial x^2} + \frac{\partial^2 v}{\partial z^2} = 0.$$

Note: As in the two–dimensional case, a **source-free** *flow satisfies* $\nabla\cdot\mathbf{v} = 0$, *and an irrotational field satisfies* $\nabla\times\mathbf{v} = \mathbf{0}$.

6.5 Stokes' theorem

Let

$$\mathbf{F}(x,y) \;=\; \mathbf{i}\,F_1(x,y)+\mathbf{j}\,F_2(x,y)$$

be a two–dimesnional vector field. Its **circulation** around the boundary $\mathcal{C}$ of a domain $\mathcal{D}$ is

$$\begin{aligned}
\text{CIRC}[\mathbf{F}]_{\mathcal{C}} \;&:=\; \oint_{\mathcal{C}} \mathbf{F}\cdot d\mathbf{r} = \oint_{\mathcal{C}} (F_1\,\mathbf{i}+F_2\,\mathbf{j})\cdot\left(\frac{dx}{ds}\,\mathbf{i}+\frac{dy}{ds}\,\mathbf{j}\right)ds = \oint_{\mathcal{C}} (F_1\,dx+F_2\,dy)\\
&=\; \iint_{\mathcal{D}} \left(\frac{\partial F_2}{\partial x}-\frac{\partial F_1}{\partial y}\right)dx\,dy\;, \quad \text{by Green's Theorem}\;,\\
&=\; \iint_{\mathcal{D}} \left(\left(\frac{\partial F_2}{\partial x}-\frac{\partial F_1}{\partial y}\right)\mathbf{k}\right)\cdot\mathbf{k}\,dx\,dy \;=\; \iint_{\mathcal{D}} (\text{CURL}\mathbf{F})\cdot\mathbf{k}\,dx\,dy.
\end{aligned}$$

See (6.74) for the above.

The analogous result in three dimensions is called **Stokes' Theorem**. It gives the circulation of a vector field along a curve $\mathcal{C}$ as the surface integral of CURL $\mathbf{F}$. Stokes' Theorem will be proved using the two-dimensional case, i.e. Green's Theorem,

$$\oint_{\mathcal{C}+} (P\,dx+Q\,dy) \;=\; \iint_{\mathcal{D}} \left(\frac{\partial Q}{\partial x}-\frac{\partial P}{\partial y}\right)dx\,dy \tag{6.79}$$

where the **orientation** of the curve $\mathcal{C}$ plays an important role. We have a similar consideration in three dimensions. A surface $\mathcal{S}$ in $\mathbb{R}^3$ has two sides. We define the **orientation** of the surface in terms of these two sides.

Definition 6.68. A surface $\mathcal{S}$ is orientable if its two sides are distinguishable, i.e. if no point can belong to both sides of $\mathcal{S}$.

This last sentence may sound strange: how can a point belong to both sides of a surface? However, strange things can happen in $\mathbb{R}^3$.

Example 6.69. The **Möbius strip**, shown in Figure (6.10), is non–orientable. The sphere, ellipsoid and torus shown in Figure (6.11),(6.12), and (6.13) are orientable surfaces. Indeed, most of the surfaces you will meet in the applications are orientable.

If a surface $\mathcal{S}$ is orientable, we can call one of its sides the **inside**, so the other side is the **outside**[3]. Then at any point on $\mathcal{S}$ we can distinguish between an **outer normal** (pointing out) and an **inner normal**. Figures (6.11),(6.12), and (6.13) shows orientable surfaces and several (outer) normals.

[3]**Q**: On which side does a chicken have more feathers? **A**: On the outside.

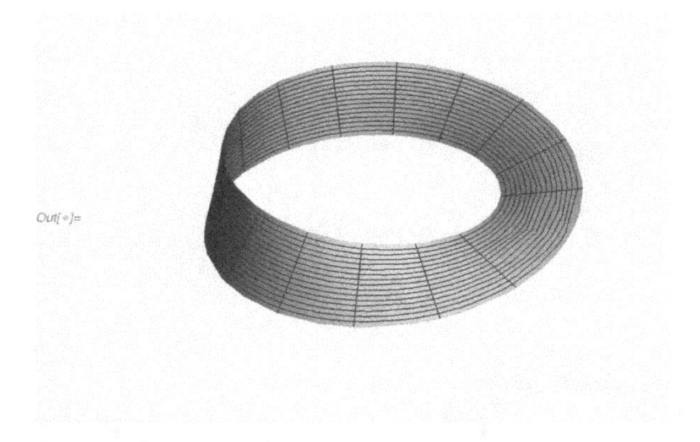

FIGURE 6.10: Making a Möbius strip in two easy steps.

Theorem 6.70 (Stokes). Let $\mathcal{S}$ be a bounded, orientable surface, bounded by a closed curve $\mathcal{C}$, and let

$$\mathbf{F}(x, y, z) := \mathbf{i}\, F_1(x, y, z) + \mathbf{j}\, F_2(x, y, z) + \mathbf{k}\, F_3(x, y, z) \tag{6.80}$$

be a vector field with continuously differentiable components in a domain $\mathcal{D}$ which contains $\mathcal{S}$. Then

$$\oint_{\mathcal{C}} \mathbf{F} \cdot \mathbf{t}\, ds = \iint_{\mathcal{S}} (\nabla \times \mathbf{F}) \cdot \mathbf{n}\, dS\,, \tag{6.81}$$

where the orientation of the surface $\mathcal{S}$ is chosen to match the orientation of the curve $\mathcal{C}$.

Proof:
We denote the projections on the xy–plane of the surface $\mathcal{S}$ and the curve $\mathcal{C}$ by $\mathcal{A}$ and Γ respectively and rewrite (6.81) in terms of components of vectors as follows

$$\oint_{\mathcal{C}} F_1\, dx + F_2\, dy + F_3\, dz =$$

$$\iint_{\mathcal{S}} \left\{ \left(\frac{\partial F_3}{\partial y} - \frac{\partial F_2}{\partial z} \right) \cos\alpha + \left(\frac{\partial F_1}{\partial z} - \frac{\partial F_3}{\partial x} \right) \cos\beta + \left(\frac{\partial F_2}{\partial x} - \frac{\partial F_1}{\partial y} \right) \cos\gamma \right\} dS. \tag{6.82}$$

We prove this first for a special case of a surface $\mathcal{S}$ represented by $z = z(x, y)$, where the function $z(x, y)$, and its partial derivatives, are continuous in the planar region $\mathcal{A}$.

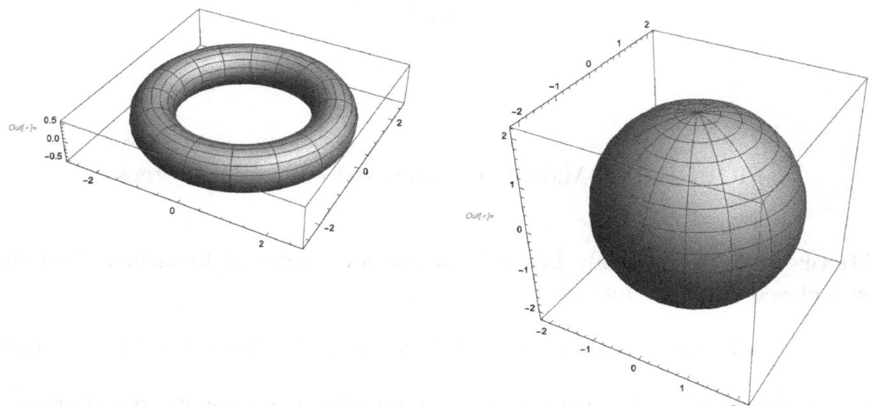

FIGURE 6.11: A torus is orientable.

FIGURE 6.12: A sphere is orientable.

Using $dz = \frac{\partial z}{\partial x}\,dx + \frac{\partial z}{\partial y}\,dy$ we rewrite the line integral along $\mathcal{C}$ as a line integral along Γ,

$$\begin{aligned}\oint\limits_{\mathcal{C}} F_1\,dx + F_2\,dy + F_3\,dz &= \oint\limits_{\Gamma} \left(F_1 + F_3\,\frac{\partial z}{\partial x}\right)\,dx + \left(F_2 + F_3\,\frac{\partial z}{\partial y}\right)\,dy \\ &= \oint\limits_{\Gamma} P(x,y)\,dx + Q(x,y)\,dy \qquad (6.83)\end{aligned}$$

where $P := F_1 + F_3\,\frac{\partial z}{\partial x}$, $Q := F_2 + F_3\,\frac{\partial z}{\partial y}$. Using Green's Theorem

$$\oint\limits_{\Gamma} P\,dx + Q\,dy = \iint\limits_{\mathcal{A}} \left(\frac{\partial Q}{\partial x} - \frac{\partial P}{\partial y}\right)\,dx\,dy$$

where the integrand is, in detail,

$$\frac{\partial Q}{\partial x} - \frac{\partial P}{\partial y} = \frac{\partial F_2}{\partial x} - \frac{\partial F_1}{\partial y} + \left(\frac{\partial F_3}{\partial x} - \frac{\partial F_1}{\partial z}\right)\frac{\partial z}{\partial y} + \left(\frac{\partial F_2}{\partial z} - \frac{\partial F_3}{\partial y}\right)\frac{\partial z}{\partial x}$$

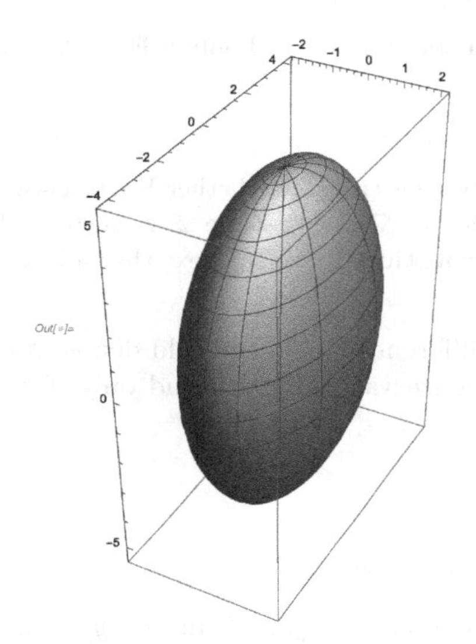

FIGURE 6.13: A prolate ellipsoid is orientable

The normal to $\mathcal{S}$ is

$$\mathbf{n} = -\frac{\partial z}{\partial x}\mathbf{i} - \frac{\partial z}{\partial y}\mathbf{j} + \mathbf{k}\,, \tag{6.84}$$

and its direction angles are

$$\cos\alpha = \frac{-z_x}{\sqrt{1+z_x^2+z_y^2}}, \quad \cos\beta = \frac{-z_y}{\sqrt{1+z_x^2+z_y^2}}, \quad \cos\gamma = \frac{1}{\sqrt{1+z_x^2+z_y^2}},$$

where the direction angle γ is acute (assuming a suitable orientation of the surface $\mathcal{S}$). Therefore

$$\begin{aligned}
&\iint_{\mathcal{A}} \left(\frac{\partial Q}{\partial x} - \frac{\partial P}{\partial y}\right) dx\,dy \\
= &\iint_{\mathcal{A}} \left\{\left(\frac{\partial F_2}{\partial x} - \frac{\partial F_1}{\partial y}\right)\cos\gamma + \left(\frac{\partial F_3}{\partial y} - \frac{\partial F_2}{\partial z}\right)\cos\alpha + \left(\frac{\partial F_1}{\partial z} - \frac{\partial F_3}{\partial x}\right)\right. \\
&\quad \left.\cos\beta\right\} \sqrt{1+z_x^2+z_y^2}\,dx\,dy \\
= &\iint_{\mathcal{S}} \left\{\left(\frac{\partial F_3}{\partial y} - \frac{\partial F_2}{\partial z}\right)\cos\alpha + \left(\frac{\partial F_1}{\partial z} - \frac{\partial F_3}{\partial x}\right)\cos\beta + \left(\frac{\partial F_2}{\partial x} - \frac{\partial F_1}{\partial y}\right)\right. \\
&\quad \left.\cos\gamma\right\} dS\,, \quad \text{by (5.46)}\,.
\end{aligned} \tag{6.85}$$

This proves the theorem, in the case, where the surface $\mathcal{S}$ is represented by $z = z(x, y)$. It can be proved similarly for the case that $\mathcal{S}$ is represented as $x = x(y, z)$ or $y = y(x, z)$. Therefore, the proof holds also for surfaces $\mathcal{S}$ which can be broken into parts represented as above. □

In Theorem 6.11 we saw that a field **F** is **conservative** if and only if $\mathbf{F} = \nabla \varphi$ for some scalar potential φ , in which case,

$$\nabla \times \mathbf{F} = \nabla \times \nabla \varphi = 0 ,$$

see Exercise 6.65(e). Therefore $\nabla \times \mathbf{F} = 0$ provides a check whether **F** has a scalar potential, i.e. a **necessary condition** for $\mathbf{F} = \nabla \varphi$ with some φ. A vector field **F** whose curl is zero everywhere is called **irrotational**. We see next that a field is conservative if and only if it is irrotational.

Theorem 6.71. If **F** is a continuously differentiable vector field defined in a simply–connected domain $\mathcal{D}$, then **F** is conservative in $\mathcal{D}$ if and only if it is irrotational in $\mathcal{D}$.

Proof:
We need only prove that

$$\mathbf{F} \text{ irrotational} \quad \Longrightarrow \quad \mathbf{F} \text{ conservative} ,$$

i.e. $\nabla \times \mathbf{F} = 0$ implies that, for any two points (x_0, y_0, z_0) and (x, y, z), and any two curves $\mathcal{C}_1$, $\mathcal{C}_2$ joining them,

$$\int_{\mathcal{C}_1} \mathbf{F} \cdot d\mathbf{r} = \int_{\mathcal{C}_2} \mathbf{F} \cdot d\mathbf{r} . \tag{6.86}$$

Let $\mathcal{C}_1$, $\mathcal{C}_2$ be two such curves, joining arbitrary points (x_0, y_0, z_0) and (x, y, z) in $\mathcal{D}$, and define the closed curve

$$\mathcal{C} := \mathcal{C}_1 \cup \{-\mathcal{C}_2\} .$$

Let $\mathcal{S}$ be a surface bounded by $\mathcal{C}$ and lying in the domain $\mathcal{D}$. If **F** is irrotational, then by Stokes' Theorem,

$$\oint_{\mathcal{C}} \mathbf{F} \cdot \mathbf{t}\, ds = \iint_{\mathcal{S}} (\nabla \times \mathbf{F}) \cdot \mathbf{n}\, dS = 0 ,$$

proving (6.86). □

If a vector field satisfies (6.86), then the line integrals along $\mathcal{C}$ from a fixed initial point (x_0, y_0, z_0) to a variable terminal point (x, y, z) depend only on the endpoint, and define a <u>potential</u>

$$\varphi(x, y, z) := \oint_{(x_0,y_0,z_0)}^{(x,y,z)} \mathbf{F} \cdot \mathbf{t}\, ds \tag{6.87}$$

where

$$\nabla\varphi(x,y,z) \;=\; \mathbf{F}(x,y,z)\,.$$

Example 6.72. Determine if there is a potential $\varphi(x,y,z)$ such that

$$\nabla\varphi(x,y,z) \;=\; (x+z)\,\mathbf{i} - (y+z)\,\mathbf{j} + (x-y)\,\mathbf{k}\,,$$

in which case find such a potential.
The above vector field is irrotational

$$\nabla \times ((x+z)\,\mathbf{i} - (y+z)\,\mathbf{j} + (x-y)\,\mathbf{k}) \;=\; \mathbf{0} \qquad \text{(verify!)}$$

Therefore φ exists, and satisfies:

$$\begin{aligned} \frac{\partial\varphi}{\partial x} &= x+z \\ \frac{\partial\varphi}{\partial y} &= -y-z \\ \frac{\partial\varphi}{\partial z} &= x-y \end{aligned}$$

We integrate the first equation to obtain

$$\varphi \;=\; \frac{x^2}{2} + x\,z + f(y,z)\,, \quad \text{where } f \text{ is some function of } (y,z)\,.$$

The second equation then gives

$$\begin{aligned} \frac{\partial\varphi}{\partial y} &= -y-z \;=\; \frac{\partial f}{\partial y} \\ \therefore\quad f(y,z) &= -\frac{y^2}{2} - y\,z + g(z)\,, \quad \text{where } g \text{ is some function of } z\,, \end{aligned}$$

and the third equations lead to

$$\begin{aligned} \frac{\partial\varphi}{\partial z} &= x-y \;=\; \frac{\partial f}{\partial z} + x \\ \therefore\quad f(y,z) &= -y\,z + h(y)\,, \quad \text{where } h \text{ is some function of } y\,. \end{aligned}$$

The above conclusions hold by choosing

$$g(z) \;\equiv\; 0 \quad \text{and} \quad h(y) \;=\; -\frac{y^2}{2}\,,$$

in which case a suitable potential is

$$\varphi(x,y,z) \;=\; \frac{1}{2}\left(x^2-y^2\right) \;+\; z\,(x-y) \;+\; C\,,$$

where C is an arbitrary constant.

Remark 6.73. Our MATHEMATICA function **myThreePotential** may be used to solve for a scalar potential much in the same way as we did for **my-Potential** as may be seen below.

Mathematica session 6.11. ***myThreePotential*[*f*_, *P*_]:=**

***Module*[{*g*}, *g* = /.{*x* → *P*[[1]], *y* → *P*[[2]], *z* → *P*[[3]]}];**

***If*[*Norm*[*g*] > 0, "*there is no potential*",**

***Integrate*[*f*[[1]], *x*] + *Integrate*[*f*[[2]], *y*] + *Integrate*[*f*[[3]], *z*]]]**

***myThreePotential*[{*x* + *z*, −*y* − *z*, *x* − *y*}, {1, 1, 1}]**

$\frac{x^2}{2} - \frac{y^2}{2} + xz + (x - y)z - yz$

***myThreePotential*[{*x*, *y*, *z*}/*Sqrt*[*x*^2 + *y*^2 + *z*^2], {1, 1, 1}]**

$3\sqrt{x^2 + y^2 + z^2}$

***myThreePotential*[{*x*, *y*, *z*}/(*x*^2 + *y*^2 + *z*^2), {1, 1, 1}]**

$\frac{3}{2}Log\left[x^2 + y^2 + z^2\right]$

***myThreePotential*[{*x* + 2*y*, *x*^2 + *z*, *y*^3 + *z*^2}, {1, 1, 1}]**

there is no potential

♦

Mathematica session 6.12. *The line integral and the surface integrals occurring in the statement of* STOKES *theorem is*

***stokesLineIntegral*[*f*_, *u*_, *I*_]:=**

***Module*[{*r*, *s*, *g*}, *r* = *D*[*u*, *I*[[1]]]; *s* =**

***Simplify*[*f*/.{*x* → *u*[[1]], *y* → *u*[[2]], *z* → *u*[[3]]}];**

***g* = *Dot*[*s*, *r*];**

***Integrate*[*g*, {*I*[[1]], *I*[[2]], *I*[[3]]}]]**

***ball* = {*r* ∗ *Sin*[*φ*] ∗ *Cos*[*θ*], *r* ∗ *Sin*[*φ*] ∗ *Sin*[*θ*], *r* ∗ *Cos*[*φ*]}**

$\{r\,Cos[\theta]\,Sin[\varphi], r\,Sin[\theta]\,Sin[\varphi], r\,Cos[\varphi]\}$

```
circle = {r * Cos[θ], r * Sin[θ], 0}
```

$\{r\,Cos[\theta], r\,Sin[\theta], 0\}$

```
stokesSurfaceIntegral[f_, U_, S_, T_]:=Module[{g, N, h}, g = Curl[f,
{x, y, z}]; N = D[U, S[[1]]] × D[U, T[[1]]];
h = Dot[g, N]/.{x → U[[1]], y → U[[2]], z → U[[3]]};
Integrate[h, {S[[1]], S[[2]], S[[3]]}, {T[[1]], T[[2]], T[[3]]}]]
```

```
stokesSurfaceIntegral[{x^2, y^2, z^2}, ball, {φ, 0, π/2}, {θ, 0, 2 * π}]
```

0

```
stokesLineIntegral[{x^2, y^2, z^2}, circle, {θ, 0, 2π}]
```

0

```
stokesSurfaceIntegral[{x + y, y^2, z − x}, ball, {φ, 0, π/2}, {θ, 0, 2 * π}]
```

$-\pi r^2$

```
stokesLineIntegral[{x + y, y^2, z − x}, circle, {θ, 0, 2π}]
```

$-\pi r^2$

```
parabola = {x, y, x^2 + y^2}
```

$\{x, y, x^2 + y^2\}$

```
stokesSurfaceIntegral[{x^2, y^2, z^2}, parabola, {y, −Sqrt[1 − x^2],
Sqrt[1 − x^2]}, {x, −1, 1}]
```

0

```
stokesLineIntegral[{x^2, y^2, 1}, {Cos[θ], Sin[θ], 1}, {θ, 0, 2 * π}]
```

0

♦

Example 6.74 (Water circulating in a tank). Because the tank has a circular cross section it is convenient to describe the rotating fluid using cylindrical coordinates, with the z axis of the coordinate system aligned along the axis of the tank. In cylindrical coordinates the velocity vector of a particle of fluid is written as

$$\mathbf{v} = \mathbf{r}\, v_r + \theta\, v_\theta + \mathbf{k}\, v_z ,$$

where $\mathbf{r}$, θ, and $\mathbf{k}$ are unit vectors in the radial, angular, and z–directions respectively. In order to describe a rotating fluid whose angular velocity ω is uniform, we prescribe the components of $\mathbf{v}$ as

$$v_r = 0 , \quad v_\theta = \omega r , \quad v_z = 0 .$$

Let us check what the curl of such a vector is. The curl is written, in cylindrical coordinates, as

$$\nabla \times \mathbf{A} = \begin{vmatrix} \frac{1}{r}\mathbf{r} & \theta & \frac{1}{r}\mathbf{k} \\ \frac{\partial}{\partial r} & \frac{\partial}{\partial \theta} & \frac{\partial}{\partial z} \\ A_r & r\, A_\theta & A_z \end{vmatrix} .$$

Using this we compute

$$\nabla \times \mathbf{v} = 2\,\omega\,\mathbf{k} ,$$

hence, the velocity $\mathbf{v}$ is not irrotational and therefore cannot be written as the gradient of a scalar potential. We show now that $\mathbf{v}$ is the CURL of some vector field

$$\mathbf{v} = \nabla \times \mathbf{A} \quad \Leftrightarrow ,\nabla \cdot \mathbf{v} = 0. \tag{6.88}$$

where $\mathbf{A}$ is called a **vector potential** of $\mathbf{v}$. To visualize $\mathbf{A}$, suppose we have placed little paddle-wheels in the fluid, whose axes are aligned with the lines of $\mathbf{v}$, in order to measure the circulation by turning with the fluid. These meters can be thought of as measuring the curl of $\mathbf{A}$. Conversely, we might think of these little paddle-wheels as turning at the proper speed as to drive the fluid at a velocity equivalent to that imparted by the vector potential $\mathbf{A}$. This construction with the paddle-wheels suggests a possible solution of our problem would be to choose the vector potential $\mathbf{A}$ so that it is parallel to $\mathbf{k}$.[4] Since $\mathbf{v}$ vanishes at the origin we seek an A_z which also vanishes at the origin and varies with the distance from the center of the tank. Because the rotation of fluid is uniform the strength of the vector potential will not vary with θ or z. With these considerations in mind we seek a vector potential $\mathbf{A}$ whose components are of the form

$$A_r = 0 , \quad A_\theta = 0 , \quad A_z = A(r) . \tag{6.89}$$

In order that $\mathbf{v} = \nabla \times \mathbf{A}$ we need the components of both sides of this equation to

[4]To avoid ∇A having a component in the $\mathbf{k}$ direction we set the r and θ components equal to zero. This insures that the curl must lie in the r-θ plane.

match, that is

$$\frac{1}{r}\left(\frac{\partial A_z}{\partial \theta} - \frac{\partial (r\,A_\theta)}{\partial z}\right) = 0 \tag{6.90a}$$

$$\frac{\partial A_r}{\partial z} - \frac{\partial A_z}{\partial r} = \omega\, r \tag{6.90b}$$

$$\frac{1}{r}\left(\frac{\partial (r\,A_\theta)}{\partial r} - \frac{\partial A_r}{\partial \theta}\right) = 0\,. \tag{6.90c}$$

We check now if the guess (6.89) is right, by substituting these values in (6.90). The second equation (6.90b) becomes

$$\frac{\partial A_z}{\partial r} = -\omega\, r\,,$$

which we integrate to obtain

$$A_z = -\frac{\omega\, r^2}{2} + C\,,$$

where C is a constant of integration. So a suitable vector potential $\mathbf{A}$ is given by

$$\mathbf{A} := \left(-\frac{\omega\, r^2}{2} + C\right)\mathbf{k}\,.$$

This vector potential has zero divergence

$$\nabla \cdot \mathbf{A} = 0\,. \tag{6.91}$$

It is easy to see that the vector potential is not unique. Indeed, let $\mathbf{A}_1$ be any gradient of a scalar field

$$\mathbf{A}_1 := \nabla\,\Phi\,.$$

Then $\nabla \times \mathbf{A}_1 = \nabla \times \nabla\,\Phi \equiv 0$, and therefore,

$$\nabla \times (\mathbf{A} + \mathbf{A}_1) = \mathbf{v}\,,$$

showing that $\mathbf{A} + \mathbf{A}_1$ also satisfies (6.88). However, it can be shown that there is only one vector potential with zero divergence. In other words, the conditions (6.88) and (6.91) determine a unique vector potential

$$A_r = 0\,, \quad A_\theta = 0\,, \quad A_z = -\frac{\omega\, r^2}{2} + C\,.$$

Exercises

Exercise 6.75. *Suppose we define the scalar valued function φ as $\varphi(\mathbf{x}) := |\mathbf{x}|$ find $\nabla\,\varphi$.*

Exercise 6.76. *If $\lambda(\mathbf{x}) := |\mathbf{x}|^n$, find $\nabla\,\lambda$.*

Exercise 6.77. *Evaluate the expression*

$$\nabla \cdot (\mathbf{x} \times \mathbf{a})\,,$$

where $\mathbf{a}$ is a constant vector.

Exercise 6.78. *Find the potentials φ for the following relations*

(a) $\nabla \varphi = < \sin y - y \sin x + z \,,\, \cos x + x \cos y + z \,,\, x + y >$

(b) $\nabla \varphi = < 2\,x\,e^{y} + z^2 \,,\, x^2\,e^{y} + z^2 - 2\,z \,,\, 2\,(z-1)\,y + 2\,x\,z >$

Exercise 6.79. *Sketch a proof for the fact that a three-dimensional vector field* **F** *is the gradient of a potential if and only if*

$$\nabla \times \mathbf{F} = 0 \,.$$

Hint: Generalize the ideas in the proofs of Theorems 6.10 and 6.29 to three-dimensions.

Exercise 6.80. *Recall that it is possible to compute the flow of a velocity* **V** *along a space curve $\mathcal{C}$. If C is the helix* $\mathbf{r}(t) := \cos t\mathbf{i} + \sin t\mathbf{j} + t\mathbf{k}$ *compute the flows for the following velocities.*

(a) $\mathbf{v} := x\,\mathbf{i} + y\,\mathbf{j} + z\,\mathbf{k}$ (b) $\mathbf{v} := y\,\mathbf{i} + x\,\mathbf{j} + 1\mathbf{k}$

(c) $\mathbf{v} := (x-y)\,\mathbf{i} + y\,\mathbf{j} + (y-z)\,\mathbf{k}$

Exercise 6.81. *Use* MATHEMATICA *to check whether there exists a scalar potential whose gradient equals the vector* **F**.

(a) $\mathbf{F} :=< 2xy\sin z, x^2\sin z, x^2y\cos z >$ (b) $\mathbf{F} :=< 2xze^{y}, x^2ze^{y} + y^3, x^2e^{y} >$

(c) $\mathbf{F} :=< x^3y + xyz + \sin y, \frac{1}{4}x^4 + \frac{1}{2}x^2z + x\cos y, \frac{1}{2}x^2y >$

(d) $\mathbf{F} :=< e^{x}\tan y\cos z, e^{x}\sec^2 y\cos z, -e^{x}\tan y\sin z >$

Exercise 6.82. *Use* MATHEMATICA *to compute* $\nabla \, \|\mathbf{r}\|^n \, \mathbf{a} \times \mathbf{r}$*, where* **a** *is a constant vector, and* $\mathbf{r} := < x \,,\, y \,,\, z >$*. Show that you can decompose the curl of* $\|\mathbf{r}\|^n \, \mathbf{a} \times \mathbf{r}$ *into a linear combination of* **a** *and* **r***, with scalar coefficients.*

Exercise 6.83. *Use* MATHEMATICA *to compute the following*

(a) $\nabla \left(\dfrac{\mathbf{a} \cdot \mathbf{r}}{\|\mathbf{r}\|^3} \right)$ (b) $\nabla \times \left(\dfrac{\mathbf{a} \times \mathbf{r}}{|\mathbf{r}|^3} \right)$

Exercise 6.84. *Suppose*

$$\nabla \times \mathbf{F} = < a \,,\, b \,,\, c > \,,$$

and

$$\nabla \cdot \mathbf{F} = p\,x + q\,y + r\,z \,,$$

where a, b, c, p, q, and r are constants. Show that

$$\nabla F_1 = p \,, \quad \nabla F_2 = q \,, \quad \nabla F_3 = r \,.$$

Exercise 6.85. *Write a* MATHEMATICA *function which tests whether a vector field can be written as the gradient of a potential, and if so produces that potential. Hint: First solve for the potential in the form*

$$\varphi(x,y,z) = \int^{x} F_1(x,y,z)\,dx + h(y,z),$$

then use the equations

$$\frac{\partial \varphi}{\partial y} = F_2, \text{ and } \frac{\partial \varphi}{\partial z} = F_3$$

to obtain a formula for $h(y,z)$.

Mathematica session 6.13. ***Clear["Global*"]***

```
Grad[f[x, y, z], {x, y, z}]
```

$\{f^{(1,0,0)}[x,y,z], f^{(0,1,0)}[x,y,z], f^{(0,0,1)}[x,y,z]\}$

```
Curl[{f^(1,0,0)[x, y, z], f^(0,1,0)[x, y, z], f^(0,0,1)[x, y, z]}, {x, y, z}]
```

$\{0,0,0\}$

```
R = Sqrt[x^2 + y^2 + z^2]
```

$\sqrt{x^2+y^2+z^2}$

```
Div[{x/R, y/R, z/R}, {x, y, z}]//Simplify
```

$\frac{2}{\sqrt{x^2+y^2+z^2}}$

```
Grad[%, {x, y, z}]
```

$\left\{-\frac{2x}{(x^2+y^2+z^2)^{3/2}}, -\frac{2y}{(x^2+y^2+z^2)^{3/2}}, -\frac{2z}{(x^2+y^2+z^2)^{3/2}}\right\}$

```
Curl[{x/R, y/R, z/R}, {x, y, z}]//Simplify
```

$\{0,0,0\}$

```
Div[{x/R^2, y/R^2, z/R^2}, {x, y, z}]//Simplify
```

$\frac{1}{x^2+y^2+z^2}$

```
Div[{x/R^3, y/R^3, z/R^3}, {x, y, z}]//Simplify
```

0

```
Curl[{x/R^3, y/R^3, z/R^3}, {x, y, z}]//Simplify
```

$\{0,0,0\}$

♦

6.6 Applications of Gauss' theorem and Stokes' theorem

Example 6.86 (Electrostatics). The **electric field strength E** is determined by the force **F** on a charged **test particle** that is placed within the electric field using the following relation between force and field strength

$$\mathbf{F} = q\,\mathbf{E}, \tag{6.92}$$

where q is the charge strength. It is well known that if a charged particle traverses any closed curve $\mathcal{C}$ in an electrostatic field and returns to the point at it which it began that no work is done either on or by the particle, i.e.

$$\oint_{\mathcal{C}} \mathbf{F}\cdot\mathbf{t}\,ds = 0\,.$$

Using (6.92) for the relation between the force on the charged particle and the electrostatic field we have

$$q\oint_{\mathcal{C}} \mathbf{E}\cdot\mathbf{t}\,ds = 0\,, \tag{6.93}$$

and from Stokes' Theorem this becomes

$$\oint_{\mathcal{C}} \mathbf{E}\cdot\mathbf{t}\,ds = \iint_{\mathcal{S}} (\nabla\times\mathbf{E}\cdot\mathbf{n})\,dS = 0\,, \tag{6.94}$$

where $\mathcal{S}$ is any (smooth) surface having $\mathcal{C}$ as its boundary. Since this is true of all closed paths we have that $\nabla\times\mathbf{E} = 0$ for all surfaces that lie in the region where **E** is defined. It follows that

$$\nabla\times\mathbf{E} = 0 \tag{6.95}$$

everywhere and that **E** is said to be **irrotational**[5] and a scalar potential V exists, such that

$$\mathbf{E} = -\nabla V\,, \tag{6.96}$$

where we have put a minus sign before the gradient as is conventional in physics. Suppose we have an arbitrary closed surface $\mathcal{S}$ bounding a region Ω where an electric field exists. We further assume that the surface does not pass through any other material that might be a boundary for where the electric field exists. Suppose we now determine the field strength at each point of the surface using the test particle and the procedure mentioned above we conclude that the integral

$$\iint_{\mathcal{S}} (\mathbf{E}\cdot\mathbf{n})\,dS$$

[5] Some authors refer to such fields as being lamellar.

is proportional to the total amount of charge within Ω. Indeed, if we use as the constant of proportionality the **dielectric constant** ε then [6]

$$\iint\limits_{\mathcal{S}} \varepsilon_0 \left(\mathbf{E}\cdot\mathbf{n}\right) dS \;=\; Q\,. \tag{6.97}$$

We refer to $\varepsilon\,\mathbf{E}$ as the **electrostatic flux**. Gauss' Theorem then leads to the following result concerning the electrostatic flux

$$\iint\limits_{\mathcal{S}} \mathbf{D}\cdot\mathbf{n}\,dS \;=\; \iiint\limits_{\Omega} \left(\nabla\cdot\mathbf{D}\right) dV \;=\; Q\,.$$

If we now introduce ρ as the charge density within the region Ω, we then have equality of the two volume integrals

$$\iiint\limits_{\Omega} \left(\nabla\cdot\mathbf{D}\right) dV \;=\; \iiint\limits_{\Omega} \rho\,dV\,. \tag{6.98}$$

Since we postulated that the surface $\mathcal{S}$ is arbitrary the integrands in (6.98) must be equal, that is we have the

$$\nabla\cdot\mathbf{D} \;=\; \rho\,. \tag{6.99}$$

For a **homogeneous material** that is one where the dielectric constant ε does not change from point to point we may rewrite (6.99) as

$$\nabla\cdot\mathbf{E} \;=\; \frac{\rho}{\varepsilon}\,.$$

Recalling that the electric potential V is related to $\mathbf{E}$ by (6.92) this equation becomes

$$\nabla\cdot(\nabla V) \;=\; -\frac{\rho}{\varepsilon}\,, \tag{6.100}$$

called the **Poisson equation**. Notice that if the charge density is everywhere zero this is the **Laplace equation**.

Green's first identity If we consider the case where u and v are both continuously differentiable in a region $\lceil$, and let $\mathcal{D}\subset\mathcal{V}$ then

$$\int\limits_{\mathcal{D}} \left[\nabla u\cdot\nabla v \,+\, u\Delta v\right] dx\,dy \,+\, \int \partial\mathcal{D} u\frac{\partial v}{\partial n}\,dS \;=\; 0. \tag{6.101}$$

We provide another a physical interpretation of Gauss' Divergence Theorem.

---

[6] If we use the [mks] units, with $\|\mathbf{E}\|$ measured in [volts] $[\text{meter}]^{-1}$, length in [meters], and charge in [coulombs], then the value of the dielectric constant in a vacuum, $\varepsilon_0$is about 8.86×10^{-12}. In a material the dielectric constant is given by $\varepsilon \;=\; \kappa\varepsilon_0$, where κ is a number between 0 and 1 and is a measure of how much weaker the electrostatic forces are in a material.

The flux also plays a role in the theory of heat transfer. Let $\mathbf{q}$ be the **heat flux vector** on the orientable surface $\mathcal{S}$; then the **total heat flux** across $\mathcal{S}$ is represented by the integral

$$\iint_{\mathcal{S}} \mathbf{q} \cdot \mathbf{n}\, dS .$$

Moreover, if $\mathcal{S}$ is a closed surface Gauss' Theorem permits this integral to be written as the volume integral

$$\iint_{\mathcal{S}} \mathbf{q} \cdot \mathbf{n}\, dS = \iiint_{\mathcal{D}} \nabla \cdot \mathbf{q}\, dV . \tag{6.102}$$

If we divide both sides of (6.102) by the volume $|D|$ and use the triple integral Mean Value Theorem we get

$$\frac{1}{|D|} \iint_{\mathcal{S}} \mathbf{q} \cdot \mathbf{n}\, dS = \frac{1}{|D|} \iiint_{\mathcal{D}} \nabla \cdot \mathbf{q}\, dV = \nabla \cdot \mathbf{q}\,(x_0, y_0, z_0) ,$$

where (x_0, y_0, z_0) is some point interior to $\mathcal{D}$. Now let us take the limit as $|D| \to 0$ to get

$$\nabla \cdot \mathbf{q}\,(x_0, y_0, z_0) = \lim_{|D| \to 0} \frac{1}{|D|} \iint_{\mathcal{S}} \mathbf{q} \cdot \mathbf{n}\, dS \tag{6.103}$$

The quantity $\nabla \cdot \mathbf{q}\,(x_0, y_0, z_0) := \rho(x_0, y_0, z_0)$ is known as the source density of the heat. Consider now an application of heat transfer due to heat **conduction**. According to **Fourier's Law** heat flows from points of higher temperature to points of lower temperature, and this flow is proportional to the gradient of the temperature. Let us denote the temperature at time t at a point (x, y, z) of a conducting solid by $T(x, y, z, t)$. Fourier's Law states that the flow of heat across the surface $\mathcal{S}$ is given by

$$q_S = - \iint_{\mathcal{S}} K \frac{\partial T}{\partial n}\, dS , \tag{6.104}$$

where q_S is measured in `[cal] [sec]`$^{-1}$. Note in the above statement we are dealing with the **total heat flux** across the surface $\mathcal{S}$ and this is a scalar and not a vector. In the case of nonsteady flow of heat it may be more useful to consider Q_S, the total heat which has flowed across the surface since the beginning of the process. Q_S is related to q_S by $q_S =: \frac{\partial Q_S}{\partial t}$, and equation (6.104) is replaced by

$$\frac{\partial Q_S}{\partial t} = - \iint\limits_{\mathcal{S}} K \frac{\partial T}{\partial n} dS . \tag{6.105}$$

Remark 6.87. The axiom of heat accumulation states if the temperature $T(x, y, z, t)$ and its time derivative $\frac{\partial T}{\partial t}$ are a continuous functions of x, y, z, and t for $(x, y, z) \in \mathcal{D}$, a bounded domain and $0 \leq t \leq t_0$ for t_0 a fixed but bounded positive number, then the rate at which heat accumulates in $\mathcal{V} \subset \mathcal{D}$ is given by

$$\frac{\partial Q}{\partial t} = \iiint\limits_{\mathcal{V}} C \frac{\partial u}{\partial t} dV . \tag{6.106}$$

Here C is known as the specific heat coefficient and may be either a constant or a function of x, y, z.

We now derive the equation of heat conduction assuming that the temperature together with its partial derivatives of the first two orders are continuous in the variables x, y, z, and t. We let $\mathcal{V}$ represent an arbitrary subregion in the region of flow $\mathcal{D}$. Since we assume there are no heat sources or sinks in $\mathcal{V}$ the rate at which heat accumulates in $\mathcal{V}$ must satisfy the **heat balance equation**, namely

$$\iiint\limits_{\mathcal{V}} C \frac{\partial T}{\partial t} dV = - \iiint\limits_{\mathcal{V}} K \frac{\partial T}{\partial n} dS .$$

Here n refers to the inner normal to the surface ∂V. The conductivity is usually taken to be constant. In this case by using

$$\int\limits_{V}\int \Delta u dV = - \int\limits_{\partial V} \frac{\partial T}{\partial n} dS, \tag{6.107}$$

we obtain

$$\int\limits_{V}\int \left(C \frac{\partial T}{\partial t} - K \Delta u \right) dV = 0. \tag{6.108}$$

Since the integrand of equation (6.108) is continuous at each point of the flow, the only way this integral can vanish for each $V \subset D$ is for the integrand to be identically zero. We show this as follows. Suppose for example that the integrand is greater than 0 at some point $\mathbf{P} \in D$; then because of continuity the integrand must remain positive in a neighborhood of the point **P**. This contradicts equation (6.108) which states that the integrand must be zero over such a neighborhood; hence, we are led to the condition that

$$\frac{\partial T}{\partial t} = \kappa \Delta T, \tag{6.109}$$

where $\kappa = \frac{K}{C}$. This is known as the **heat conduction equation.**

Definition 6.88. A **steady state** heat flow is one in which the temperature at any given point is independent of time. In this instance, the temperature obeys **Laplace's Equation**

$$\frac{\partial^2 T}{\partial x^2} + \frac{\partial^2 T}{\partial y^2} + \frac{\partial^2 T}{\partial z^2} = 0. \tag{6.110}$$

The Laplace equation turns up in many applications of physics and engineering. In addition to steady state heat flow, there are steady state fluid flow problems, electrostatics, and steady state elastic deformations. In many of these problems the geometry of the physical objects involved suggest that a special coordinate system be used. For example, the study of spherical elastic shells and the cooling of a sphere suggest that spherical polar coordinates might be helpful in exploiting the geometry; whereas, heat flow in a cylinder suggests the use of cylindrical coordinates.

Example 6.89 (The Charged Sphere). When an electrostatic field acts upon a conducting material, or conductor, electric charges can flow. Because charges can and do flow within a conductor, in the electrostatic case there can be no electrostatic field within the conductor. If there were an electrostatic field it would rearrange the charges in the conductor until the field vanished. Hence, if $\mathbf{E} = 0$ in a conductor, then V is a constant. Moreover, since $\mathbf{E} = 0$ it follows that $\nabla \cdot \mathbf{E} = \rho = 0$. There is no charge at any point within a conductor. The surface of the conductor is another matter. It is, however, possible for the surface of a conductor to possess a charge, and for that matter for there to exist a nonvanishing electric field on the surface of a conductor.
Let us consider a conductor which is a spherical shell $\mathcal{S}$ of radius r_0 and has infinitesimal thickness. Suppose further that the conductor has a charge Q distributed on it. From our above comments the conductor must be at a constant electric potential V_0. Suppose we wish to find the electric field about the spherical shell, assuming that it is **isolated** in $\mathbb{R}^3$. To do this we are required to solve the Laplace equation in the region $\mathbb{R}^3/\mathcal{S}$. Because the surface of the sphere is an **equipotential** it is convenient for us to consider this problem in spherical coordinates. From the MATHEMATICA Session (6.2) the Laplace equation in spherical coordinates is

$$\Delta V := \frac{\partial^2 V}{\partial r^2} + \frac{2}{r}\frac{\partial V}{\partial r} + \frac{1}{r^2}\frac{\partial^2 V}{\partial \theta^2} + \frac{\cot\theta}{r^2}\frac{\partial V}{\partial \theta} + \frac{1}{r^2 \sin^2\theta}\frac{\partial^2 V}{\partial \varphi^2} = 0\,. \tag{6.111}$$

Because the charge is uniformly distributed on the shell and the shell is a sphere it is clear that the solution must possess some symmetry properties. In particular, the solution must be **spherically symmetric**, which means the solution should not depend on either the angles θ or φ. Consequently we are led to solve the spherically symmetric Laplace equation

$$\Delta V \;:=\; \frac{\partial^2 V}{\partial r^2} + \frac{2}{r}\frac{\partial V}{\partial r} \;=\; 0\,. \tag{6.112}$$

It is easy to demonstrate that a solution to the **differential equation** (6.112) is given by

$$V \;=\; \frac{C_1}{r} + C_2,$$

where C_1 and C_2 are arbitrary constants. These constants, however, must be chosen to meet certain physical requirements. For example, very far away from the sphere the electric potential caused by the charge distribution on the sphere is insignificant. This is expressed mathematically by

$$\lim_{r\to 0} V \;=\; 0;$$

consequently, to have this occur we set $C_2 = 0$. To determine the constant C_1 we use Gauss's Theorem. Notice that the charge density on the spherical shell is given by

$$q = \frac{Q}{4\pi r_0^2}.$$

From Gauss' Divergence Theorem we have

$$\oint_S \mathbf{D}\cdot\mathbf{n}da \;=\; \oint_S \frac{Q}{4\pi r_0^2}\, da,$$

from which we conclude that on the shell the normal component of the electric field is given by

$$\mathbf{E}\cdot\mathbf{n} \;=\; \frac{Q}{4\pi\varepsilon r_0^2}\,.$$

Since the potential does not vary with θ or φ the equipotentials must be concentric spheres; hence, only the radial component of the gradient will appear in the calculation of the electic field. From the MATHEMATICA session the gradient in spherical coordinates is given by

$$\nabla V \;=\; \frac{\partial V}{\partial r}\mathbf{r} + \frac{1}{r}\frac{\partial V}{\partial \theta}\theta \;+\; \frac{\partial V}{\partial r}\varphi, \tag{6.113}$$

where $\mathbf{r}$, θ, and φ are unit vectors in the r, θ, and φ directions respectively. Moreover, since the potential does not vary with θ or φ the gradient of the potential will be radial, that is

$$\mathbf{E} \;=\; -\frac{\partial V}{\partial r}\mathbf{r} \;=\; \frac{C_1}{r^2}\mathbf{r},$$

which is true everywhere, and in particular, on the surface $\mathcal{S}$. We conclude that the potential and the electric field due to a charged sphere are given by

$$V = \frac{Q}{4\pi\varepsilon r} \quad \text{and } \mathbf{E} = \frac{Q}{4\pi r^2}\mathbf{r} \tag{6.114}$$

respectively.

Example 6.90 (Magnetic Field)**.** If a wire loop is put into a magnetic field and the wire loop is carrying a current it is well known that depending on the orientation of the loop there will be a **magnetic force** exhibited on the wire loop. Indeed, this magnetic force on a straight segment of wire may be expressed as

$$\mathbf{F} = I\,\mathbf{l}\times\mathbf{B}, \tag{6.115}$$

where I is the current, $\mathbf{B}$ the **magnetic flux density**[7] , and $\mathbf{l}$ is a vector in the direction of the wire and having its length as magnitude. If I is measured in `[amperes]` and $|\mathbf{l}|$ in `[meters]`, and $\mathbf{F}$ in `[newtons]` then $\mathbf{B}$ is measured in `[webers] [meters]`$^{-2}$. Magnetic flux Φ, like electric flux, is defined in terms of an integral over an area, namely

$$\Phi = \int_{\mathcal{S}} \mathbf{B}\cdot\mathbf{n}\,da. \tag{6.116}$$

Faraday's Law of Induction[8] states, in its differential form, that the **electromotive force** induced in a loop of wire by a magnetic field is given in terms of the magnetic flux as

$$\oint_{\mathcal{C}} \mathbf{E}\cdot\mathbf{t}ds = -\frac{\partial\Phi}{\partial t}. \tag{6.117}$$

By using Stokes' Theorem to rewrite the left hand side of (6.117) and (6.116) to replace the right hand side we get

$$\int_{\mathcal{S}} (\nabla\mathbf{E})\cdot\mathbf{n}da = -\int_{\mathcal{S}} \frac{\partial\mathbf{B}}{\partial t}\cdot\mathbf{n}da. \tag{6.118}$$

Since this equation holds for all loops $\mathcal{C}$ and all surfaces $\mathcal{S}$ bounded by such a loop we conclude that the two integrands must be equal, that is

$$\nabla\mathbf{E} = -\frac{\partial\mathbf{B}}{\partial t}. \tag{6.119}$$

Equation (6.119) indicates that the curl of $\mathbf{E}$ does not vanish if the magnetic

[7] $\mathbf{B}$ is also known as the **magnetic induction**.

[8] In 1831 the British scientist Michael Faraday discovered that an electric field could be produced magnetically. This was also discovered independently a few months later by the American scientist Joseph Henry.

flux density varies with time.
Let us now consider a **magnetostatic field**. It is known that the integral of the normal component of **B** over all closed surfaces

$$\oint_{\mathcal{S}} \mathbf{B} \cdot \mathbf{n} da$$

vanishes, i.e. there are no magnetic monopoles! Consequently, using Gauss' Divergence Theorem the magnetic field **B** has no divergence

$$\nabla \cdot \mathbf{B} = 0 . \tag{6.120}$$

Finally, if we consider the tangential component of the magnetic field along a closed loop we find

$$\oint_{\mathcal{C}} \frac{1}{\mu} \mathbf{B} \cdot \mathbf{t} ds = I,$$

where I is the induced current, and μ is the permeability of the material. If we introduce the **magnetic intensity H** defined as $\mathbf{B} = \mu \mathbf{H}$, then this becomes

$$\oint_{\mathcal{C}} \mathbf{H} \cdot \mathbf{t} ds = I. \tag{6.121}$$

If we can replace the current I with a current density over a surface $\mathcal{S}$ bounded by $\mathcal{C}$, then

$$\oint_{\mathcal{C}} \mathbf{H} \cdot \mathbf{t}\, ds = \int_{\mathcal{S}} \mathbf{J} \cdot \mathbf{n}\, da , \tag{6.122}$$

and by using Stokes' Theorem once more we establish that

$$\nabla \times \mathbf{H} = \mathbf{J} . \tag{6.123}$$

In any region where $\mathbf{J} \equiv 0$ the curl of **H** is zero and there will be a scalar magnetic potential; however, where $\nabla \times \mathbf{H} \neq 0$ this is not the case. On the other hand, as the magnetic field always has vanishing divergence, we can always construct a vector potential for **H**. Let us call this potential **A** and then

$$\mathbf{H} = \nabla \times \mathbf{A} \tag{6.124}$$

Mathematica session 6.14. *Clear["Global*"]*

divergenceIntegral[F_, X_, Y_, S_, T_]:=

Module[{function, cp, ncp, main},

function = F/.{X[[1]] → Y[[1]], X[[2]] → Y[[2]], X[[3]] → Y[[3]]};

```
cp = D[Y, S[[1]]] × D[Y, T[[1]]];
ncp = Simplify[Sqrt[Dot[cp, cp]]];
main = Simplify[function * ncp]];
Integrate[main, {S[[1]], S[[2]], S[[3]]}], {T[[1]], T[[2]], T[[3]]}
```

```
PowerExpand[divergenceIntegral[Sqrt[x^2 + y^2 + z^2], {x, y, z},
{r * Sin[phi] * Cos[theta], r * Sin[phi] * Sin[theta], r * Cos[phi]},
{theta, 0, 2 * Pi},
{phi, 0, Pi}]]
```

$r^3 Sin[phi]$

```
divergenceIntegral[x * y * z, {x, y, z},
{r Cos[theta] Sin[phi], r Sin[phi] Sin[theta], r Cos[phi]}, {phi, 0, Pi},
{theta, 0, 2 * Pi}]
```

$r^5 Cos[phi] Cos[theta] Sin[phi]^3 Sin[theta]$

```
PowerExpand[TransformedField["Cartesian" → "Spherical", Sqrt[x^2 + y^2
+z^2], {x, y, z} → {r, theta, phi}]]
```

$r\sqrt{Cos[theta]^2 + Cos[phi]^2 Sin[theta]^2 + Sin[phi]^2 Sin[theta]^2}$

```
Simplify[Sqrt[x^2 + y^2 + z^2]/.
{x->r * Sin[phi] * Cos[theta], y->r * Sin[phi] * Sin[theta], z->r * Cos[phi]}]
```

$\sqrt{r^2}$

```
Integrate[x * y * z/.{x->r * Sin[phi] * Cos[theta], y → Sin[phi] * Sin[theta],
z → Cos[phi]}, {phi, 0, Pi}, {theta, 0, 2 * Pi}]
```

0

D[{r * Sin[phi] * Cos[theta], r * Sin[phi] * Sin[theta], r * Cos[phi]}, phi] ×
D[{r * Sin[phi] * Cos[theta], r * Sin[phi] * Sin[theta], r * Cos[phi]}, theta]

$\{r^2 Cos[theta] Sin[phi]^2, r^2 Sin[phi]^2 Sin[theta], r^2 Cos[phi] Cos[theta]^2 Sin[phi] + r^2 Cos[phi] Sin[phi] Sin[theta]^2\}$

Simplify[Sqrt[Dot[%, %]]]

$\sqrt{r^4 Sin[phi]^2}$

Integrate[r^2 * Sin[phi], {phi, 0, Pi}, {theta, 0, 2 * Pi}]

$4\pi r^2$

Y = {r * Sin[phi] * Cos[theta], r * Sin[phi] * Sin[theta], r * Cos[phi]};
S = {phi, 0, Pi};
T = {theta, 0, 2 * Pi}

$\{theta, 0, 2\pi\}$

D[Y, S[[1]]]

$\{r Cos[phi] Cos[theta], r Cos[phi] Sin[theta], -r Sin[phi]\}$

D[Y, T[[1]]]

$\{-r Sin[phi] Sin[theta], r Cos[theta] Sin[phi], 0\}$

{r Cos[phi] Cos[theta], r Cos[phi] Sin[theta], −r Sin[phi]} ×
{−r Sin[phi] Sin[theta], r Cos[theta] Sin[phi], 0}

$\{r^2 Cos[theta] Sin[phi]^2, r^2 Sin[phi]^2 Sin[theta], r^2 Cos[phi] Cos[theta]^2 Sin[phi]+$
$r^2 Cos[phi] Sin[phi] Sin[theta]^2\}$

Simplify[Sqrt[Dot[%, %]]]

$\sqrt{r^4 Sin[phi]^2}$

Integrate $\left[\sqrt{r^4 Sin[phi]^2}, \{phi, 0, Pi\}, \{theta, 0, 2 * Pi\}\right]$

$4\pi\sqrt{r^4}$

PowerExpand $\left[\sqrt{r^4 Sin[phi]^2}\right]$

$r^2 Sin[phi]$

Integrate[%, {phi, 0, Pi}]

$2r^2$

♦

Example 6.91 (Electromagnetics)**.** Light waves, radio waves, television waves, etc. are all examples of electromagnetic waves. Moreover, all electromagnetic fields are generated by distributions of electric charge [9] and current.Information about electric and magnetic fields existed in the middle of the nineteeth century. However, most scientists still believed in **action at a distance**. Indeed, Faraday appears to have been the first to actually believe in a **field theory**. The law relating current to the magnetic circulation,

$$\nabla \times \mathbf{B} = \mu_0 \mathbf{J}$$

is called **Ampère's Law**[10]. Ampère noticed that two wires carrying a current attracted one another; however, there was no intermediate agency for applying

[9]For present discussions let us assume that these distributions are continuous rather than discrete, that is all point charges and all current filaments are smoothed out.
[10]André Marie Ampère [1775–1836].

the force on the other wire. Hence, the assumption of action at a distance was used to explain this phenomenon. This was similar to **Newton's Law of Gravitation** which also assumed action at a distance. The physical idea of the electric and magnetic fields was conceived by Faraday[11] on the basis of experiments in electric and magnetic fields. Farady discovered the law

$$\nabla \times \mathbf{B} = -\frac{\partial}{\partial t}\mathbf{B}\,. \tag{6.125}$$

The mathematical theory of electric and magnetic fields was developed by Maxwell [12]in his *Treatise on Electricity and Magnetism.* Maxwell realized that the existing theory was incomplete, and he completed it to give an electromagnetic description of light. This was an accomplishment of great creativity. Maxwell showed mathematically that energy would be transmitted as electromagnetic waves which travelled at the speed of light. This was a death blow to the action at a distance theory which predicted that electric action occurred instantly at all points no matter how remote they were. Indeed, the speed of the wave motion could be computed and measured.
According to Maxwell a magnetic field $\mathbf{H}$ is produced by the sum of the **conductive current** and the **displacement current**. In the following equation the term $\frac{\partial \mathbf{D}}{\partial t}$ corresponds to the **displacement current density**, and $\mathbf{J}$ is the **vector conductive current density**.

$$\nabla \times \mathbf{H} = \frac{\partial \mathbf{D}}{\partial t} + \mathbf{J}\,. \tag{6.126}$$

We have already seen the next relation as (6.119)

$$\nabla \times \mathbf{E} + \frac{\partial \mathbf{B}}{\partial t} = 0\,, \tag{6.127}$$

If ρ is the **electric charge density** then $\mathbf{J}$ is related to ρ through the equation

$$\nabla \cdot \mathbf{J} + \frac{\partial \rho}{\partial t} = 0\,. \tag{6.128}$$

We have already seen the other two fundamental field equations, namely (6.120)

$$\nabla \cdot \mathbf{B} = 0\,, \tag{6.129}$$

and (6.99)

$$\nabla \cdot \mathbf{D} = \rho \tag{6.130}$$

The four equations (6.126), (6.127), (6.129), and (6.130) are known as **Maxwell's Equations**. The electric flux [13] is a function of the electric field $\mathbf{E}$, and the magnetic field $\mathbf{H}$ is a function of the magnetic induction $\mathbf{B}$. This

[11] Michael Faraday [1791–1867].
[12] James Clerk Maxwell [1831–1879].
[13] Some authors call $\mathbf{D}$ the **electric displacement**

functional relation depends on the physical properties of the material. In a vacuum the vectors **D** and **E** differ only by a constant, as is also the case for **H** and **B**, namely

$$\mathbf{D} = \varepsilon_0 \mathbf{E}\,, \quad \mathbf{H} = \frac{1}{\mu_0}\mathbf{B}\,. \tag{6.131}$$

At each point of an **isotropic** material **D** is parallel to **E**, and **H** is parallel to **B**, so for an isotropic, linear material[14] we have

$$\mathbf{D} = \varepsilon(\mathbf{r})\,\mathbf{E}\,, \quad \mathbf{H} = \frac{1}{\mu(\mathbf{r})}\mathbf{B}\,, \tag{6.132}$$

where $\varepsilon(\mathbf{r})$ and $\mu(\mathbf{r})$ are functions of the position vector $\mathbf{r}$. Consider Maxwell's equations in a vacuum. Then $\varepsilon = \varepsilon_0$ and $\mu = \mu_0$ are constant and this will simplify the analysis. We shall show how in this case it is easy to reduce the solution of the Maxwell equations to the solution of the **wave equation**. First we compute the curl of both sides of equation (6.127) to obtain

$$\nabla \times (\nabla \times \mathbf{E}) = -\mu_0\, \nabla \times \frac{\partial \mathbf{H}}{\partial t} = -\mu_0 \frac{\partial\, \nabla\, \mathbf{H}}{\partial t}. \tag{6.133}$$

In a vacuum (6.126) becomes

$$\nabla \times \mathbf{H} = \varepsilon_0 \frac{\partial \mathbf{E}}{\partial t}\,, \tag{6.134}$$

which we can now use to replace **H** in equation

$$\nabla \times (\nabla \times \mathbf{E}) = -\mu_0\, \varepsilon_0 \frac{\partial^2 \mathbf{E}}{\partial t^2}\,. \tag{6.135}$$

At this point we use the vector identity (6.139) (see Exercise 6.92), which gives

$$\nabla \times (\nabla \times \mathbf{E}) = -\nabla^2 \mathbf{E}\,, \quad \text{since } \nabla \cdot \mathbf{E} \equiv 0 \text{ in a vacuum}\,.$$

Substituting this in (6.133) yields the **wave equation** for the electric field

$$\nabla^2 \mathbf{E} = \mu_0\, \varepsilon_0 \frac{\partial^2 \mathbf{E}}{\partial t^2}\,, \tag{6.136}$$

written componentwise as

$$\begin{aligned} \nabla^2 E_x &= \mu_0\, \varepsilon_0 \frac{\partial^2 E_x}{\partial t^2} \\ \nabla^2 E_y &= \mu_0\, \varepsilon_0 \frac{\partial^2 E_y}{\partial t^2} \\ \nabla^2 E_z &= \mu_0\, \varepsilon_0 \frac{\partial^2 E_z}{\partial t^2}\,. \end{aligned} \tag{6.137}$$

[14]Linear means here that the constitutive equations relating **D** to **E** and **H** to **B** are linear.

Mathematica session 6.15. MATHEMATICA *has the facility of making coordinate changes and computing the curl, divergence, and gradient in a variety of orthogonal coordinate systems. These calculations necessitate the use of* `linalg`.

Grad[f[r, θ, z], {r, θ, z}, "Cylindrical"]

$$\left\{f^{(1,0,0)}[r,\theta,z], \frac{f^{(0,1,0)}[r,\theta,z]}{r}, f^{(0,0,1)}[r,\theta,z]\right\}$$

Grad[Cos[x^2 + y^2], {x, y}]

$$\{-2xSin\left[x^2+y^2\right], -2ySin\left[x^2+y^2\right]\}$$

$\nabla_{\{x,y\}} f[x,y]$

$$\{f^{(1,0)}[x,y], f^{(0,1)}[x,y]\}$$

Grad[f[r, θ, z], {r, θ, z}, "Spherical"]

$$\left\{f^{(1,0,0)}[r,\theta,z], \frac{f^{(0,1,0)}[r,\theta,z]}{r}, \frac{Csc[\theta]f^{(0,0,1)}[r,\theta,z]}{r}\right\}$$

Div[{P[x, y, z], Q[x, y, z], R[x, y, z]}, {x, y, z}]

$$R^{(0,0,1)}[x,y,z] + Q^{(0,1,0)}[x,y,z] + P^{(1,0,0)}[x,y,z]$$

Div[{P[r, θ, z], Q[r, θ, z], R[r, θ, z]}, {r, θ, z}, "Cylindrical"]

$$R^{(0,0,1)}[r,\theta,z] + \frac{P[r,\theta,z]+Q^{(0,1,0)}[r,\theta,z]}{r} + P^{(1,0,0)}[r,\theta,z]$$

Div[{P[r, θ, φ], Q[r, θ, φ], R[r, θ, φ]}, {r, θ, φ}, "Spherical"]

$$\frac{Csc[\theta]\left(Cos[\theta]Q[r,\theta,\varphi]+P[r,\theta,\varphi]Sin[\theta]+R^{(0,0,1)}[r,\theta,\varphi]\right)}{r} + \frac{P[r,\theta,\varphi]+Q^{(0,1,0)}[r,\theta,\varphi]}{r}$$
$$+ P^{(1,0,0)}[r,\theta,\varphi]$$

Curl[{P[x, y, z], Q[x, y, z], R[x, y, z]}, {x, y, z}]

$\{-Q^{(0,0,1)}[x,y,z]+R^{(0,1,0)}[x,y,z],P^{(0,0,1)}[x,y,z]$
$-R^{(1,0,0)}[x,y,z],-P^{(0,1,0)}[x,y,z]+Q^{(1,0,0)}[x,y,z]\}$

Curl[{P[r, θ, z], Q[r, θ, z], R[r, θ, z]}, {r, θ, z}, "Cylindrical"]

$\Big\{-Q^{(0,0,1)}[r,\theta,z]+\frac{R^{(0,1,0)}[r,\theta,z]}{r},P^{(0,0,1)}[r,\theta,z]$
$-R^{(1,0,0)}[r,\theta,z],-\frac{-Q[r,\theta,z]+P^{(0,1,0)}[r,\theta,z]}{r}+Q^{(1,0,0)}[r,\theta,z]\Big\}$

Curl[{P[r, θ, φ], Q[r, θ, φ], R[r, θ, φ]}, {r, θ, φ}, "Spherical"]

$\Big\{-\frac{Csc[\theta]\left(-Cos[\theta]R[r,\theta,\varphi]+Q^{(0,0,1)}[r,\theta,\varphi]\right)}{r}+\frac{R^{(0,1,0)}[r,\theta,\varphi]}{r},\frac{Csc[\theta]}{r}(-R[r,\theta,\varphi]Sin[\theta]$
$+P^{(0,0,1)}[r,\theta,\varphi])-R^{(1,0,0)}[r,\theta,\varphi],-\frac{-Q[r,\theta,\varphi]+P^{(0,1,0)}[r,\theta,\varphi]}{r}+Q^{(1,0,0)}[r,\theta,\varphi]\Big\}$

We recall that in Cartesian coordinates that

Laplacian[f[x, y, z], {x, y, z}]

$f^{(0,0,2)}[x,y,z]+f^{(0,2,0)}[x,y,z]+f^{(2,0,0)}[x,y,z]$

Div[Grad[f[x, y, z], {x, y, z}], {x, y, z}]

$f^{(0,0,2)}[x,y,z]+f^{(0,2,0)}[x,y,z]+f^{(2,0,0)}[x,y,z]$

Div[Grad[f[r, θ, z], {r, θ, z}, "Cylindrical"], {r, θ, z}, "Cylindrical"]

$f^{(0,0,2)}[r,\theta,z]+\frac{\frac{f^{(0,2,0)}[r,\theta,z]}{r}+f^{(1,0,0)}[r,\theta,z]}{r}+f^{(2,0,0)}[r,\theta,z]$

Laplacian[f[r, θ, z], {r, θ, z}, "Cylindrical"]

$f^{(0,0,2)}[r,\theta,z]+\frac{\frac{f^{(0,2,0)}[r,\theta,z]}{r}+f^{(1,0,0)}[r,\theta,z]}{r}+f^{(2,0,0)}[r,\theta,z]$

Div[Grad[f[r, θ, φ], {r, θ, φ}, "Spherical"], {r, θ, φ}, "Spherical"]

$\frac{Csc[\theta]\left(\frac{Csc[\theta]f^{(0,0,2)}[r,\theta,\varphi]}{r}+\frac{Cos[\theta]f^{(0,1,0)}[r,\theta,\varphi]}{r}+Sin[\theta]f^{(1,0,0)}[r,\theta,\varphi]\right)}{r}+f^{(2,0,0)}[r,\theta,\varphi]+$
$\frac{\frac{f^{(0,2,0)}[r,\theta,\varphi]}{r}+f^{(1,0,0)}[r,\theta,\varphi]}{r}$

Laplacian[f[r, θ, φ], {r, θ, φ}, "Spherical"]

$$\frac{Csc[\theta]\left(\frac{Csc[\theta]f^{(0,0,2)}[r,\theta,\varphi]}{r}+\frac{Cos[\theta]f^{(0,1,0)}[r,\theta,\varphi]}{r}+Sin[\theta]f^{(1,0,0)}[r,\theta,\varphi]\right)}{r} + f^{(2,0,0)}[r,\theta,\varphi] + \frac{\frac{f^{(0,2,0)}[r,\theta,\varphi]}{r}+f^{(1,0,0)}[r,\theta,\varphi]}{r}$$

♦

Mathematica session 6.16. *There is another way to compute the gradient, divergence, and curl for orthogonal coordinates and this is by introducing the* **scales** *associated with that coordinate system. For a particular coordinate system its scales are simply the set of factors* $[h_1, h_2, h_3]$ *which are needed to express the square of the infintesimal distance in that geometry as*

$$ds^2 = (h_1\, dx_1)^2 + (h_2\, dx_2)^2 + (h_3\, dx_3)^2.$$

For cylindrical coordinates the scales are $[1, r, 1]$ *and for spherical coordinates the scales are* $[1, \rho, \rho\sin(\theta)]$. *For the next example we compute the gradient in* **prolate spheroidal coordinates.** *In* MATHEMATICA *see what coordinates are available.* ***SetCoordinates[ProlateSpherpidal]***

SetCoordinates[ProlateSpherpidal]

Grad[f[ξ, η, φ], {ξ, η, φ}, "ProlateSpheroidal"]

$$\left\{\frac{\sqrt{2}f^{(1,0,0)}[\xi,\eta,\varphi]}{\sqrt{-Cos[2\eta]+Cosh[2\xi]}}, \frac{\sqrt{2}f^{(0,1,0)}[\xi,\eta,\varphi]}{\sqrt{-Cos[2\eta]+Cosh[2\xi]}}, Csc[\eta]\,Csch[\xi]\,f^{(0,0,1)}[\xi,\eta,\varphi]\right\}$$

Div[{f[ξ, η, φ], g[ξ, η, φ], h[ξ, η, φ]}, {ξ, η, φ}, "ProlateSpheroidal"]

$$Csc[\eta]\,Csch[\xi]\left(\frac{\sqrt{2}Cosh[\xi]f[\xi,\eta,\varphi]Sin[\eta]}{\sqrt{-Cos[2\eta]+Cosh[2\xi]}}+\frac{\sqrt{2}Cos[\eta]g[\xi,\eta,\varphi]Sinh[\xi]}{\sqrt{-Cos[2\eta]+Cosh[2\xi]}}+h^{(0,0,1)}[\xi,\eta,\varphi]\right) +$$

$$\frac{\sqrt{2}\left(g[\xi,\eta,\varphi]\left(-\frac{Sin[2\eta]}{Cos[2\eta]-Cosh[2\xi]}-\frac{Sin[2\eta]}{-Cos[2\eta]+Cosh[2\xi]}\right)-\frac{f[\xi,\eta,\varphi]Sinh[2\xi]}{Cos[2\eta]-Cosh[2\xi]}+g^{(0,1,0)}[\xi,\eta,\varphi]\right)}{\sqrt{-Cos[2\eta]+Cosh[2\xi]}}$$

$$+\frac{\sqrt{2}\left(-\frac{g[\xi,\eta,\varphi]Sin[2\eta]}{Cos[2\eta]-Cosh[2\xi]}+f[\xi,\eta,\varphi]\left(-\frac{Sinh[2\xi]}{Cos[2\eta]-Cosh[2\xi]}-\frac{Sinh[2\xi]}{-Cos[2\eta]+Cosh[2\xi]}\right)+f^{(1,0,0)}[\xi,\eta,\varphi]\right)}{\sqrt{-Cos[2\eta]+Cosh[2\xi]}}$$

There are many other coordinate systems which are useful for solving the Laplace equation. We shall meet some of these in the Exercises.

Exercises

Exercise 6.92. *Let*

$$\mathbf{F}(x,y,z) \;:=\; \mathbf{i}\,F_1(x,y,z) \;+\; \mathbf{j}\,F_2(x,y,z) \;+\; \mathbf{k}\,F_3(x,y,z) \tag{6.138}$$

be a vector field with two times continuously differentiable components. Use MATHEMATICA *to establish the identity*

$$\nabla \times (\nabla \times \mathbf{F}) \;=\; \nabla(\nabla \cdot \mathbf{F}) \;-\; \nabla^2 \mathbf{F}\,, \tag{6.139}$$

where $\nabla^2\mathbf{F}$ is shorthand for

$$\begin{aligned} \nabla^2\mathbf{F} \;&:=\; \mathbf{i}\,\Delta F_1 \;+\; \mathbf{j}\,\Delta F_2 \;+\; \mathbf{k}\,\Delta F_3 \\ &=\; \mathbf{i}\left(\frac{\partial^2}{\partial x^2}+\frac{\partial^2}{\partial y^2}+\frac{\partial^2}{\partial z^2}\right)F_1 \;+\; \mathbf{j}\left(\frac{\partial^2}{\partial x^2}+\frac{\partial^2}{\partial y^2}+\frac{\partial^2}{\partial z^2}\right)F_2 \\ &\quad +\mathbf{k}\left(\frac{\partial^2}{\partial x^2}+\frac{\partial^2}{\partial y^2}+\frac{\partial^2}{\partial z^2}\right)F_3\,. \end{aligned}$$

Hint: Expand both sides and identify terms.

Exercise 6.93. *Find the electric field due to two charged, concentric spheres. Suppose that one sphere carries a total positive charge Q, whereas the other sphere carries an equal but negative charge.*

Exercise 6.94. *Find the electric field due to two charged, coaxial cylinders. Suppose that the inner cylinder carries a total positive charge Q; whereas the outer cylinder carries an equal but negative charge.*

Exercise 6.95. *The two concentric spheres and the two coaxial spheres mentioned in the previous two problems constitute what are called* **capacitors**. *More generally any two charged surfaces $\mathcal{S}_1$, $\mathcal{S}_2$ can form capacitors. For a given charge on one (or both) of the surfaces $\mathcal{S}_1$, $\mathcal{S}_2$ there is a voltage difference between the two surfaces. This voltage difference may be found by solving for the potental between the two surfaces. We say that the two surfaces have a* **capacitance**, *and this capacitance C is defined to be*

$$C \;=\; \frac{Q}{V_{12}}' \tag{6.140}$$

where V_{12} is the voltage difference between the two surfaces of the capacitor. Find the capacitance of the spherical capacitor. Repeat the problem for the cylindrical capacitor.

Exercise 6.96. *Suppose that the y and z components of the electric field $\mathbf{E}$ vanish. Moreover, assume that E_x is only a function of z. Using equation (6.137) shows that E_x satisfies the partial differential equation*

$$\frac{\partial^2 E_x}{\partial z^2} = \mu_0\varepsilon_0\frac{\partial^2 E_x}{\partial t^2}.$$

Show further that solutions exist of the form

$$E_x(z) := f\,(z \;-\; ct)\,,$$

where $c = (\mu_0 \varepsilon_0)^{-\frac{1}{2}}$ *and* f *is any function of one variable with two continuous derivatives. Use* MATHEMATICA *to plot the particular solution*

$$E_x(z) := \sin(z - ct),$$

for various values of ct*. Conclude that the* **disturbance** $E_x(z) := \sin z$ *travels with the speed* c*.*

Exercise 6.97. *Using the Maxwell equations in a vacuum*

$$\nabla \times \mathbf{H} = \varepsilon_0 \frac{\partial \mathbf{E}}{\partial t},$$

and

$$\nabla \times \mathbf{E} + \mu_0 \frac{\partial \mathbf{H}}{\partial t} = 0,$$

eliminate $\mathbf{E}$ *to obtain the wave equation for* $\mathbf{H}$

$$\nabla^2 \mathbf{H} = \mu_0 \varepsilon_0 \frac{\partial^2 \mathbf{H}}{\partial t^2}. \tag{6.141}$$

Exercise 6.98. *Show that the induced current density* $\mathbf{J}$ *due to a magnetic field can be written in terms of a magnetic vector potential as*

$$\mathbf{J} = \nabla \times \nabla \times \mathbf{A}.$$

Exercise 6.99. *Show that a necessary and sufficient condition for a steady heat flow is : if* V *is an arbitrary subregion in the region of flow* D*, then* $\frac{\partial Q}{\partial t} = 0$ *for this region.*

Exercise 6.100. *Suppose* K *and its first derivatives with regard to* $x, y,$ *and* z *are continuous. Derive the following form for the heat equation*

$$\frac{\partial u}{\partial t} = \frac{1}{C} \left\{ \frac{\partial}{\partial x} \left(K \frac{\partial u}{\partial x} \right) + \frac{\partial}{\partial y} \left(K \frac{\partial u}{\partial y} \right) + \frac{\partial}{\partial z} \left(K \frac{\partial u}{\partial z} \right) \right\}.$$

Exercise 6.101. *From the four relations describing the transition of the electromagnetic field from medium 1 to medium 2*

$$\mathbf{n} \cdot (\mathbf{B}_2 - \mathbf{B}_1) = 0, \quad \mathbf{n} \times (\mathbf{H}_2 - \mathbf{H}_1) = I$$

$$\mathbf{n} \times (\mathbf{E}_2 - \mathbf{E}_1) = 0, \quad \mathbf{n} \cdot (\mathbf{D}_2 - \mathbf{D}_1) = \omega \tag{6.142}$$

derive the transition conditions for the normal components of $\mathbf{E}$ *and* $\mathbf{H}$*, namely*

$$\mathbf{n} \cdot \left(\mathbf{H}_2 - \frac{\mu_1}{\mu_2} \mathbf{H}_1 \right) = 0, \quad \mathbf{n} \cdot \left(\mathbf{E}_2 - \frac{\varepsilon_1}{\varepsilon_2} \mathbf{E}_1 \right) = \frac{\omega}{\varepsilon_2}. \tag{6.143}$$

Exercise 6.102. *Derive the following transition equations for the tangential components of* $\mathbf{B}$ *and* $\mathbf{D}$*,*

$$\mathbf{n} \times \left(\mathbf{B}_2 - \frac{\mu_2}{\mu_1} \mathbf{B}_1 \right) = \mu_2 \mathbf{I}, \quad \mathbf{n} \cdot \left(\mathbf{D}_2 - \frac{\varepsilon_2}{\varepsilon_1} \mathbf{D}_1 \right) = 0.$$

Exercise 6.103. *Let* **A** *and* **B** *be two vector fields which satisfy the following field equations in all space*

$$\nabla \times \mathbf{A} = 0, \quad \nabla \cdot \mathbf{B} = 0. \tag{6.144}$$

Assume, moreover, that **A** *and* **B** *are continuous everywhere in* $\mathbb{R}^3$ *except for a smooth surface* $\mathcal{S}$. *Show that the tangential components of* **A** *and the normal components of* **B** *are continuous, that is show that*

$$\mathbf{n} \times [\mathbf{A}]_{\mathcal{S}} = 0, \quad \mathbf{n} \cdot [\mathbf{B}]_{\mathcal{S}} = 0. \tag{6.145}$$

Exercise 6.104. *In the following problems* **orthogonal curvilinear coordinate** *systems are given in the form*

$$\xi_1 = \xi_1(x, y, z),\ \xi_2 = \xi_2(x, y, z),\ \xi_3 = \xi_3(x, y, z),$$

and the unit vectors associated with these coordinates are be be denoted by $\mathbf{a}_1$, $\mathbf{a}_2$, *and* $\mathbf{a}_3$. *In the curvilinear coordinates the differential gradient differential operator is given by*

$$\nabla\varphi = \mathbf{a}_1 \frac{1}{h_1}\frac{\partial\varphi}{\partial\xi_1} + \mathbf{a}_2 \frac{1}{h_2}\frac{\partial\varphi}{\partial\xi_2} + \mathbf{a}_3 \frac{1}{h_3}\frac{\partial\varphi}{\partial\xi_3}, \tag{6.146}$$

where the **scale factors** h_n *are given by*

$$h_n^2 := \left[\left(\frac{\partial\xi_n}{\partial x}\right)^2 + \left(\frac{\partial\xi_n}{\partial y}\right)^2 + \left(\frac{\partial\xi_n}{\partial z}\right)^2\right]^{-1} = \left[\left(\frac{\partial x}{\partial\xi_n}\right)^2 + \left(\frac{\partial y}{\partial\xi_n}\right)^2 + \left(\frac{\partial z}{\partial\xi_n}\right)^2\right]. \tag{6.147}$$

In the problems below the curvilinear coordinates are specified as well as the scale factors. Write a MATHEMATICA *function which computes the gradient differential operator.*

(a) $x := \xi_1$, $y := \xi_2$, $x := \xi_3$, $h_1 = h_2 = h_3 = 1$

(b) $x := \xi_1\,\xi_2$, $y := \xi_1\sqrt{1-\xi_2^2}$, $z := \xi_3$, $h_1 = h_3 = 1$, $h_2 = \frac{\xi_1}{\sqrt{1-\xi_2^2}}$

(c) $x := \xi_1\,\xi_2$, $y := \sqrt{(\xi_1^2 - d^2)(1-\xi_2^2)}$,

$$h_1 = \sqrt{(\xi_1^2 - d^2\,\xi_2^2)/(\xi_2^2 - d^2)}\ ,\ h_2 = \sqrt{(\xi_1^2 - d^2\,\xi_2^2)(1-\xi_2^2)}\ ,\ h_3 = 1$$

(d) $x := \frac{1}{2}(\xi_1^2 - \xi_2^2)$, $y := \xi_1\,\xi_2$, $z := \xi_3$, $h_1 = h_2 = \sqrt{\xi_1^2 + \xi_2^2}$ $h_3 = 1$.

(e) $x := \xi_1\,\xi_2\,\xi_3$, $y := \xi_1\,\xi_2\sqrt{1-\xi_3^2}$, $z := \frac{1}{2}(\xi_1^2 - \xi_2^2)$

$$h_1 = h_2 = \sqrt{\xi_1^2 + \xi_2^2}\ ,\ h_3 = \frac{\xi_1\,\xi_2}{\sqrt{1-\xi_3^2}}\ .$$

Exercise 6.105. *The divergence is given in terms of the orthogonal, curvilinear coordinates as*

$$\nabla \cdot \mathbf{F} = \frac{1}{h_1\,h_2\,h_3}\sum_{k=1}^{3}\frac{\partial}{\partial\xi_k}\left(h_1\,h_2\,h_3\,\frac{F_k}{h_k}\right). \tag{6.148}$$

Write a MATHEMATICA *function which computes the divergence for the orthogonal curvilinear coordinate systems. Illustrate the use of your* MATHEMATICA *function by calculating the divergence for the systems given in the problem.*

Exercise 6.106. *In an orthogonal, curvilinear coordinate system the curl is given by*

$$\nabla \times \mathbf{F} = \frac{1}{h_1\, h_2\, h_3} \sum_{i,j,k} h_i\, \mathbf{a}_i \left[\frac{\partial}{\partial \xi_j} (h_k\, F_k) - \frac{\partial}{\partial \xi_k} (h_j\, F_j) \right] , \tag{6.149}$$

where $\{i,j,k\} = \{1,2,3\}$ *or* $\{i,j,k\} = \{2,3,1\}$ *or* $\{i,j,k\} = \{3,1,2\}$*. Write a* MATHEMATICA *function which computes the curl and use it to compute the curl for the orthogonal coordinate systems of Exercise 6.104.*

Exercise 6.107. *The Laplacian in curvilinear coordinates takes the form*

$$\nabla \cdot \nabla \varphi = \frac{1}{h_1\, h_2\, h_3} \sum_{k=1}^{3} \frac{\partial}{\partial \xi_k} \left(\frac{h_1\, h_2\, h_3}{h_k^2} \frac{\partial \varphi}{\partial \xi_k} \right) . \tag{6.150}$$

Write a MATHEMATICA *function which computes the Laplacian for the curvilinear coordinates of Exercise 6.104.*

Chapter 7

Elements of Tensor Analysis

7.1 Introduction to tensor calculus

This section is optional. It can be skipped at first reading. The material covered here is not used elsewhere in this book.

In this section we generalize the idea of $\mathbb{R}^3$, Cartesian space to $\mathbb{R}^n$ where $n = 3, 4, \ldots$ and to a curved space known as Riemannian space. Much of what we have studied concerning curves in space etc. will be found to have analogues here. Let us begin once more with $\mathbb{R}^3$ and use as Cartesian coordinates ξ^1, ξ^2, ξ^3, where we have replaced subscripts by superscripts to indicate the particular coordinate. The reason for the change of notation in this chapter will become evident later. In $\mathbb{R}^3$ the square of the distance differential is written in the form

$$ds^2 = d\xi^1\, d\xi^1 + d\xi^2\, d\xi^2 + d\xi^3\, d\xi^3$$

If the coordinates, $\xi(\mathbf{x})_j$, $j = 1, 2, 3$ are functions of x then the differential would be written as

$$ds^2 = \sum_{m=1,n=1}^{3,3} g_{mn} dx^m\, dx^n$$

In the future we will use the simplified notation

$$ds^2 = \varepsilon g_{mn} dx^m\, dx^n, \tag{7.1}$$

where repeated indices indicate summation over these indices. [1] Note that since the coefficient of $dx^m\, dx^n$ is $g_{mn} + g_{nm}$ we may assume that these terms may be taken to be equal to one another so that the matrix (g_{mn}) is symmetric, i.e. $g_{mn} = g_{nm}$. Moreover, $\varepsilon = \pm 1$, is the indicator which is introduced so that $ds > 0$.

Suppose we represent the coordinates on a surface $\mathcal{S} \subset \mathbb{R}^3$ by $\mathbf{y} = (y^1, y^2, y^3)$. Then we may describe a patch of this surface using a mapping from a subset of $\mathbb{R}^2$ to $\mathcal{S}$ using

$$y^1 = f^1(x^1, x^2),\ y^2 = f^2(x^1, x^2),\ y^3 = f^3(x^1, x^2),$$

[1] This notation is due to A. Einstein.

Then it is easily seen that

$$g_{mn} = \frac{\partial y^1}{\partial x^m}\frac{\partial y^1}{\partial x^n} + \frac{\partial y^2}{\partial x^m}\frac{\partial y^2}{\partial x^n} + \frac{\partial y^3}{\partial x^m}\frac{\partial y^3}{\partial x^n}. \tag{7.2}$$

The differential expression (7.1) is known as the **Fundamental Form** of the space we are describing.

7.1.1 Covariant and contravariant tensors

It is helpful at this time to introduce the concept of a tensor. We do this by noting how the tensor changes under a coordinate transformation. Suppose a differential displacement in the coordinate system x^r is expressed as dx^r and that in terms of another coordinate system x'^r the same differential as dx'^r. We are speaking about the same displacement here! If there is a differentiable coordinate transform from one system into the other the representation, namely $\mathbf{x}' = \mathbf{x}'(\mathbf{x})$, then

$$dx'^r = \frac{\partial x'^r}{\partial x^s}\, dx^s. \tag{7.3}$$

If the original point from which the differential is held fixed then the coefficients $\frac{\partial x'^r}{\partial x^s}$ remain fixed and define the g_{mn}. This then represents an affine transformation between the differentials, namely

$$dx'^r = g_{rs}\, dx^s.$$

We say that these differentials transform from one coordinate system to another **contravariently**. Indeed, any objects defined at the fixed point P which transform according to the rule

$$T'^r = T^s \frac{\partial x'^r}{\partial x^s} \tag{7.4}$$

are said to transform contravariantly. Furthermore, a set of objects T^{mn} is said to transform contravariantly from the coordinates $\mathbf{x}$ to $\mathbf{x}'$ if they transform according to the rule

$$T'^{rs} = T^{mn} \frac{\partial x'^r}{\partial x^m}\frac{\partial x'^s}{\partial x^n} \tag{7.5}$$

The object T^{mn} is called a contravariant tensor of second order. The vector T^s is a contravariant vector of order one.
Let us now consider how partial derivatives transform under a coordinate transformation. For example from the chain rule we have that

$$\frac{\partial F}{\partial x'^r} = \frac{\partial F}{\partial x^s}\frac{\partial x^s}{\partial x'^r}$$

If we compare the above with the transformation (7.3) we notice that the coefficients are the reciprocals of those in (7.3). We say that the partial derivatives transform **covariantly** and we follow the convention of using subscripts for covariant tensors. Moreover, any tensor that transforms according to this rule

$$T'_r = T_s \frac{\partial x^s}{\partial x'^r}$$

is covariant. Hence, a second order covariant tensor transforms as

$$T'_{rs} = T_{mn} \frac{\partial x^m}{\partial x'^s} \frac{\partial x^n}{\partial x'^s}$$

A **mixed tensor** may be both contravariant and covariant and such a third order tensor transforms according to a rule such as

$$T'^r_{st} = T^m_{np} \frac{\partial x'^r}{\partial x^m} \frac{\partial x^n}{\partial x'^s} \frac{\partial x^p}{\partial x'^t}$$

A covariant tensor T_{rs} is said to be symmetric if it follows the rule $T_{rs} = T_{sr}$ and antisymmetric if it follows the rule $T_{rs} = -T_{sr}$. A similar definition applies to contravariant tensors. Symmetry and antisymmetry are preserved under coordinate transforms, since

$$T^{rs} - T^{sr} \quad \text{and} \quad T^{rs} + T^{sr},$$

because of the transformation rules, will vanish in all coordinate systems if they vanish in one system. Moreover, it is obvious that any tensor may be written in terms of its symmetric and antisymmetric parts as

$$T^{rs} = \frac{1}{2}(T^{rs} - T^{sr}) + \frac{1}{2}(T^{rs} + T^{sr}).$$

If A^{rs} is symmetric and B^{rs} is antisymmetric then from

$$A^{rs}B^{rs} = -A^{sr}B^{sr} = -A^{rs}B^{rs},$$

we conclude that $A^{rs}B^{rs} = 0$. Hence, the metric $g_{ij}x^ix^j$ is not changed by replacing the coefficients g_{ij} by their symmetric part, i.e.

$$g_{ij}x^ix^j = \frac{1}{2}(g_{ij} + g_{ji})\, x^ix^j$$

We define a tensor multiplication of the mixed tensors S^{ij}_k, T^{pq}_i as a contraction of the tensors

$$R^{jq}_p = S^{ij}T^q_{ip}.$$

Clearly several such contractions are possible.

Definition 7.1. A space $\mathbb{V}$ is said to be Riemannian if there exists in $\mathbb{V}$ a metric or fundamental covariant tensor of second order which is symmetric, but not necessarily positive-definite. The distance between adjacent points is given by

$$ds^2 = g_{mn}dx^m dx^n.$$

It is clear that in a Riemannian space the distance between two points might be zero .

Starting with a covariant metric tensor g_{mn}, let Δ^{mn} be its cofactor, then from the algebra of matrices we have

$$g_{mr}\Delta^{ms} = g_{rm}\Delta^{sm} = \delta^s_r\, a,$$

where $a = \det(\{g_{ij}\}^n_{i,j=1})$ and δ is the Kronecher Delta. If we define $g^{mn} := \frac{\Delta^{mn}}{a}$ we shall assume, until otherwise stated, that the matrix g is positive definite [2] it follows that

$$g_{mr}\, g^{ms} = g_{rm}\, g^{sm} = \delta^s_r \tag{7.6}$$

This shows that $g_{rm}\, g^{ms} = \delta^s_r$, i.e. the matrices $\{g_{ij}\}^n_{i,j=1}$ and $\{g^{ij}\}^n_{i,j=1}$ are the inverse of each other.

This shows that g^{ij} is symmetric if and only if g_{ij} is symmetric. (Why?) We can show that g^{ij} is a contravariant tensor of second order and is said to be the conjugate to the tensor g_{ij}.

7.1.2 Raising and lowering indices.

For reasons to be made explicit later we shall follow the notation of not indicating subscripts and superscripts on the same vertical line, i.e. we shall write $T^{ij}_{\cdot\cdot k\ell}$ rather than $T^{ij}_{k\ell}$. We shall use the covariant fundamental tensor $\{g_{ij}\}$ and its contravariant tensor $\{g^{ij}\}$ for the purpose of raising and lowering indices.

Beginning with the mixed tensor $T^{j}_{\cdot k\ell}$ we construct the tensor

$$T_{ik\ell} := g_{ij}\, T^{j}_{\cdot k\ell},$$

which we say is the lowered tensor because one superscript becomes a subscript, i.e. the type of tensor is changed after the operation. Alternately, we may raise the index by contraction multiplication with $\{g^{ij}\}$

$$T^{j}_{\cdot k\ell} = g^{ij}\, T_{ik\ell}$$

and obtain the original tensor. This is not surprising because the matrix $\{g^{ij}\}$ is the inverse of $\{g_{ij}\}$.

In a Riemannian space with a metric the magnitude of a contravariant vector x^r is defined by

$$\|x\|^2 := \varepsilon g_{mn} x^m x^n,$$

where ε is the indicator of x^r and takes one of the two values ± 1. In the case of covariant vector we use

$$\|x\|^2 := \varepsilon g^{mn} x_m x_n.$$

We remark that in the case where the metric form is indefinite there will be zero distances for distinct points.

[2]In General Relativity this will not hold in general.

7.1.3 Geodesics

It is possible to introduce the concept of the angle between two intersecting curves in an N dimensional Riemannian space $\mathbb{V}$, with a positive definite fundamental form. The idea follows from the argument using the law of cosines for $\mathbb{R}^N$. Suppose the two curves are parameterized as $\xi^m(t)$ and $\eta^n(t)$ and intersect at $t = 0$. We indicate the length of the "secants" to these points from $t = 0$, in terms of their components as

$$d\xi(t) := \sqrt{g_{mn}(0)\,(\xi^m(t) - \xi^m(0))\,(\xi^n(t) - \xi^n(0))},$$

and

$$d\eta(t) := \sqrt{g_{mn}(0)\,(\eta^m(t) - \eta^m(0))\,(\eta^n(t) - \eta^n(0))};$$

moreover, the cosine of the angle between the two secants is seen to be the limit

$$\cos(\theta) \;:= \lim_{t\to 0} \frac{g_{mn}(0)(\xi^m(t) - \xi^m(0))(\eta^n(t) - \eta^n(0))}{d\xi(t)\,d\eta(t)}. \tag{7.7}$$

We next show that the angle θ is uniquely determined when the fundamental form is assumed positive definite, i.e.

$$g_{mn}\,(\xi^m(t) + s\eta^m)\,(\xi^n(t) + s\eta^n) \,\geq\, 0,$$

where ξ^m and η^m are unit vectors and $s \in \mathbb{R}^1$. Expanding the above we have

$$s^2 \,+\, 2s\,g_{mn}\xi^n\eta^n + 1 \geq 0.$$

Completing the square leads to a positive quantity

$$(s + g_{mn}\xi^m\eta^n)^2 \,+\, \left(1 \,-\, (g_{mn}\xi^m\eta^n)^2\right) \,\geq\, 0.$$

Choosing s to make the first bracketed term vanish we obtain the inequality

$$|g_{mn}\xi^m\eta^n| \,\leq\, 1,$$

which shows that equation (7.7) is uniquely solvable. We shall adopt the definition of perpendicularity for indefinite Riemann spaces as

$$g_{mn}\xi^m\eta^n \,=\, 0. \tag{7.8}$$

In contrast to an N-dimensional Euclidian space $\mathbb{R}^N$, there are no straight lines; however, there are curves which play a similar role; these are called geodesics . These curves are defined as the shortest curve between two points. Using the ideas of the variational calculus we seek the curve $\mathcal{C}$, connecting the points **P** and **Q**, whose length has a stationary value with respect to infinitesimal variations of the curve, i.e. we seek the curve whose integration of the arc-length differential between the points **P** and **Q** is minimal; see

Figure 7.1. We consider the family of smooth [3] $\mathcal{C}^v$ parameterized by t and distinguished by v, namely $x^m = x^m(t,v)$. Define the arc-length as

$$L := \varepsilon \int_{t_2}^{t_1} \left(g_{mn} \frac{\partial x^m}{\partial t} \frac{\partial x^n}{\partial t} \right)^{\frac{1}{2}} dt$$

where $u^r(t_1, v) = P^r$ and $u^r(t_2, v) = Q^r$. Note that L is a smooth function of v. Using $p^m = \frac{\partial x^m}{\partial t}$, we write this variational integral as

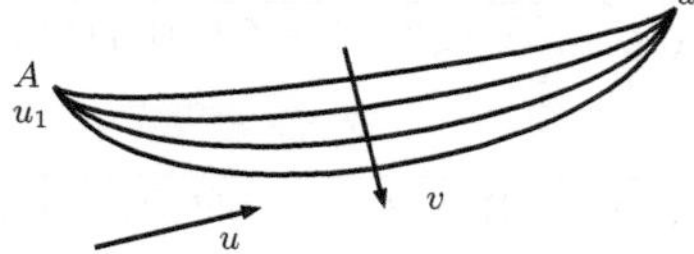

FIGURE 7.1: A family of smooth curves joining the points P to Q.

$$\frac{dL}{dv} = \int_{t_1}^{t_2} \frac{\varepsilon \partial \left(g_{mn} p^m p^n \right)^{\frac{1}{2}}}{\partial v} dt. \tag{7.9}$$

To simplify the differentiations we replace the integrand by setting $w := g_{mn} p^m p^n$. Then

$$\frac{\partial(\varepsilon w)^{\frac{1}{2}}}{\partial v} = \frac{\partial(\varepsilon w)^{\frac{1}{2}}}{\partial x^r} \frac{\partial x^r}{\partial v} + \frac{\partial(\varepsilon w)^{\frac{1}{2}}}{\partial p^r} \frac{\partial p^r}{\partial v}.$$

which is then used substituted into the variational equation for L, namely (7.9). Integration by parts, using $\delta L := \frac{dL}{dv} \delta v, \quad \delta x^r := \frac{\partial x^r}{\partial v} \delta v$, leads to

$$\delta L = \left[\frac{\partial(\varepsilon w)^{\frac{1}{2}}}{\partial p^r} \delta x^r \right]_{t_1}^{t_2} - \int_{t_1}^{t_2} \left(\frac{\partial}{\partial t} \frac{\partial(\varepsilon w)^{\frac{1}{2}}}{\partial p^r} - \frac{\partial(\varepsilon w)^{\frac{1}{2}}}{\partial x^r} \right) \delta x^r \, dt.$$

The reader should check the integration by parts above. Since the end points $u^r(t_1, v) = P^r$ and $u^r(t_2, v) = Q^r$ have been assumed fixed, δx^r is fixed there; hence the end point contribution at t_1 and t_2 vanishes and we have

$$\delta L = - \int_{t_1}^{t_2} \left(\frac{d}{dt} \frac{\partial(\varepsilon w)^{\frac{1}{2}}}{\partial p^r} - \frac{\partial(\varepsilon w)^{\frac{1}{2}}}{\partial x^r} \right) \frac{\partial x^r}{\partial v} dt \tag{7.10}$$

If the curve is a geodesic, it vanishes for all arbitrary variations δx^r. Using the fundamental theorem of variational calculus the integrand must vanish [2, 8]

$$\frac{d}{dt} \frac{\partial(\varepsilon w)^{\frac{1}{2}}}{\partial p^r} - \frac{\partial(\varepsilon w)^{\frac{1}{2}}}{\partial x^r} = 0.$$

[3] Smooth curves are ontinuously differential curves.

By differentiating and collecting terms this can easily be put in the form

$$\frac{d}{dt}\frac{\partial w}{\partial p^r} - \frac{\partial w}{\partial x^r} = \frac{1}{2w}\frac{\partial w}{\partial t}\frac{\partial w}{\partial p^r}.$$

Hereafter, we replace the partial derived $\frac{\partial}{\partial t}$ with $\frac{d}{dt}$ because we are talking about following a curve that satisfies the condition $\delta L = 0$ and there should be no chance of confusion. Instead of using arbitrary t to parameterize the curve let us use the arc-length s with $p^r := \frac{dx^r}{ds}$. Then $w = g_{mn}p^n p^m = 1$ and $\frac{dw}{dt} = 0$; hence the geodesic equations take a simple form known as the Euler-Legrange equations

$$\frac{d}{ds}\frac{\partial w}{\partial p^r} - \frac{\partial w}{\partial x^r} = 0. \tag{7.11}$$

Substituting for w the term $g_{mn}p^n p^m$ in (7.11) leads to

$$g_{rm}\frac{dp^m}{ds} + \frac{\partial g_{rm}}{\partial x^m} p^m p^n - \frac{1}{2}\frac{\partial g_{rm}}{\partial x^r} p^m p^n = 0. \tag{7.12}$$

It is easily seen that by rearranging the summation one obtains

$$\frac{\partial g_{rm}}{\partial x^n} p^m p^n = \frac{1}{2}\left(\frac{\partial g_{rm}}{\partial x^n} + \frac{\partial g_{rn}}{\partial x^m}\right) p^m p^n. \tag{7.13}$$

It is useful at this point to introduce some notation, namely the Christoffel symbols of the first kind [4]

$$\Gamma_{rmn} := \frac{1}{2}\left(\frac{\partial g_{rm}}{\partial x^n} + \frac{\partial g_{rn}}{\partial x^m} - \frac{\partial g_{mn}}{\partial x^r}\right);$$

whereas the Christoffel symbol of the second kind is given by

$$\Gamma^r_{\cdot mn} = g^{rq}\Gamma_{qmn}.$$

Multiplying the equation (7.13) by g^{rs}, followed by renaming the dummy variable from q to r, yields

$$\frac{dp^r}{ds} + \Gamma^r_{\cdot mn} p^m p^n = 0, \tag{7.14}$$

or by using $p^r = \frac{dx^r}{ds}$

$$\frac{d^2 x^r}{ds^2} + \Gamma^r_{\cdot mn}\frac{dx^m}{ds}\frac{dx^n}{ds} = 0. \tag{7.15}$$

Replacing the arc-length s in equation (7.11) by the arbitrary parameter v we have

$$\frac{d^2x^r}{dv^2} + \Gamma^r_{\cdot mn}\frac{dx^m}{dv}\frac{dx^m}{dv} = -\frac{\frac{d^2v}{ds^2}}{(\frac{dv}{ds})^2}\frac{dx^r}{dv} = \lambda\frac{dx^r}{dv} \tag{7.16}$$

[4] Older notation is related to the Γ notations by $\Gamma_{cab} = [ab, c] = [ba, c] := g_{cd}\Gamma^d_{\cdot ab}$.

$$\frac{d}{d\,u}\frac{\partial w}{\partial p^r} - \frac{\partial w}{\partial x^r} = 0.$$

We note that by a suitable choice of the parameter v the function λ of (7.16) may be made into any preassigned function of v we wish. Hence, it is sufficient for a curve to be a geodesic for

$$\frac{d^2 x^r}{dv^2} + \Gamma^r_{\cdot mn}\frac{dx^m}{dv}\frac{dx^m}{dv}$$

to be proportional to $\frac{dx^r}{dv}$ and for

$$a_{mn}\frac{dx^m}{dv}\frac{dx^n}{dv} = 0.$$

If we multiply the above by p^r and integrate by parts we have

$$\frac{d}{d\,u}\left(p^r\frac{\partial w}{\partial\,p^r}\right) - \frac{d\,p^r}{d\,u}\frac{\partial\,w}{\partial\,p^r} - \frac{d\,x^r}{ds\,u}\frac{\partial\,w}{\partial\,x^r} = 0,$$

which simplifies to

$$\frac{d}{d\,u}\left(p^r\frac{\partial w}{\partial\,p^r} - w\right) = 0,$$

and

$$a_{mn}\frac{\partial x^m}{\partial u}\frac{\partial x^n}{\partial u}$$

Hence w is a first integral of the differential equation as inserting w into the differential equation reduces it to a constant; i.e.

$$p^r\frac{\partial w}{\partial\,p^r} - w = \text{constant}.$$

This implies $w = g_{mn}p^m\,p^n =$ constant and if u is replaced by the arc-length the constant is $\varepsilon \pm 1$.

Mathematica session 7.1. *In this section we show how to calculate the metric tensor for several coordinate systems. First we consider the spherical coordinates, which we have seen before in Chapter 6. Using the equation (7.2) we calculate the metric tensor for the spherical coordinates, which are parametrically represented in* $\mathbb{R}^3$ *as:*

$$x = \rho\sin(\varphi)\cos(\theta), \quad y = \rho\sin(\varphi)\sin(\theta), \quad z = \rho\cos(\varphi)$$

```
Table[Sum[D[{ρ * Cos[θ] * Sin[φ], ρ * Sin[θ] * Sin[φ], ρ * Cos[φ]}[[c]], {ρ, θ, φ}[[a]]]*
D[{ρ * Cos[θ] * Sin[φ], ρ * Sin[θ] * Sin[φ], ρ * Cos[φ]}[[c]], {ρ, θ, φ}[[b]]], {c, 1, 3}],
{a, 1, 3}, {b, 1, 3}];
```

where we have placed a ";" at the end of the command line to suppress the output. We use **MatrixForm** *on the saved output to get*

MatrixForm[Simplify[%]]

$$\begin{pmatrix} 1 & 0 & 0 \\ 0 & \rho^2 Sin[\varphi]^2 & 0 \\ 0 & 0 & \rho^2 \end{pmatrix}$$

The above suggests we write a MATHEMATICA **Function** *called* **metricTensor** *to calculate the metric tensors for different coordinate systems. We do this as*

metricTensor[Y_, X_]:=Simplify[MatrixForm[Table[Sum[D[Y[[c]], X[[a]]]* D[Y[[c]], X[[b]]], {c, 1, 3}], {a, 1, 3}, {b, 1, 3}]]]

metricTensor[{ρ * Cos[θ] * Sin[φ], ρ * Sin[θ] * Sin[φ], ρ * Cos[φ]}, {ρ, θ, φ}]

$$\begin{pmatrix} 1 & 0 & 0 \\ 0 & \rho^2 Sin[\varphi]^2 & 0 \\ 0 & 0 & \rho^2 \end{pmatrix}$$

The array component [2, 1] *is printed by using the* MATHEMATICA **Part** *command*

$$\begin{pmatrix} \mathbf{1} & \mathbf{0} & \mathbf{0} \\ \mathbf{0} & \boldsymbol{r^2 Sin[\varphi]^2} & \mathbf{0} \\ \mathbf{0} & \mathbf{0} & \boldsymbol{r^2} \end{pmatrix} \mathbf{[[2, 1]]}$$

0

The [2, 2] *term is obtained again by using* **Part***.*

$$\begin{pmatrix} \mathbf{1} & \mathbf{0} & \mathbf{0} \\ \mathbf{0} & \boldsymbol{r^2 Sin[\varphi]^2} & \mathbf{0} \\ \mathbf{0} & \mathbf{0} & \boldsymbol{r^2} \end{pmatrix}[[2,2]]$$

$r^2 Sin[\varphi]^2$

Next try **metricTensor** *on some other coordinate systems. The first is for the* **cylindrical coordinates** *which have the parametric representation in* $\mathbb{R}^3$ *as* $r\cos(\theta), rsin(\theta), z$

metricTensor[{r * Cos[θ], r * Sin[θ], z}, {r, θ, z}]

$$\begin{pmatrix} 1 & 0 & 0 \\ 0 & r^2 & 0 \\ 0 & 0 & 1 \end{pmatrix}$$

The **Prolate Spheroidal Coordinates** *which may be specified parametrically as*

$$x = a\sinh(u)\sin(v)\cos(\varphi), y = a\sinh(u)\sin(v)\sin(\varphi), z = a\cosh(u)\cos(v)$$

metricTensor[{Sinh[u] * Sin[v] * Cos[φ], Sinh[u] * Sin[v] * Sin[φ], Cosh[u] Cos[v]}, {u, v, φ}].

$$\begin{pmatrix} \frac{1}{2}(-Cos[2v] + Cosh[2u]) & 0 & 0 \\ 0 & \frac{1}{2}(-Cos[2v] + Cosh[2u]) & 0 \\ 0 & 0 & Sin[v]^2 Sinh[u]^2 \end{pmatrix}$$

Next we consider the **Oblate Spherical Coordinates**. *Note that when ellipses are rotated about their minor axis the surfaces, which are generated are referred to as oblate spheroids.*

metricTensor[{Cosh[u] * Sin[v] * Cos[φ], Coshh[u] * Sin[v] * Sin[φ], Sinh[u] Cos[v]}, {u, v, φ}];

The output is too large to print in a textbook; hence we have suppressed it with a ;t. The reader is encouraged to run the the program and observe some of the symmetries in the output.
Next we consider the **Parabolic Coordinates**, *which occur from the families of parabolas*

$$x^2 = 2\xi^2(z + \xi^2/2), \quad x^2 = -2\eta^2(z + \eta^2/2).$$

These coordinate surfaces are paraboloids of revolution extending in the direction of the positive z axis and parabolas of revolution extending towards the negative z axis.

metricTensor[{ξ ∗ η ∗ Cos[φ], ξ ∗ η ∗ Sin[φ], (η^2 − ξ^2/2)}, {ξ, η, φ}]

$$\begin{pmatrix} \eta^2 + \xi^2 & -\eta\xi & 0 \\ -\eta\xi & 4\eta^2 + \xi^2 & 0 \\ 0 & 0 & \eta^2\xi^2 \end{pmatrix}$$

♦

7.1.4 Derivatives of tensors

The derivative of a tensor is usually not a tensor; however, by adding some additional terms we can make it a tensor. This is a very useful idea in differential geometry and mechanics. We have seen that f^r which is defined by

$$f^r = \frac{d^2x^r}{d^2u} + \Gamma^r_{\cdot mn}\frac{dx^m}{du}\frac{dx^n}{du} \tag{7.17}$$

is a contravariant tensor (of order 1). Hence, f^r transforms as

$$f'^r = f^r\frac{\partial x'^r}{\partial x^s}.$$

The Christoffel symbols, of the second kind, transform according to the formulae as

Theorem 7.2.

$$\Gamma^{r\prime}_{\cdot pq} = \Gamma^s_{\cdot pq}\frac{\partial x'^r}{\partial x^s}\frac{\partial x^p}{\partial x'^m}\frac{\partial x^s}{\partial x'^n} + \frac{\partial x'^r}{\partial x^s}\frac{\partial^2 x^s}{\partial x'^m \partial x'^n} \tag{7.18}$$

Proof. To see how $\Gamma^{r}_{.mn}$ transforms with coordinate changes, we start with the prime system. Recall that the Christoffel symbols of the second kind are given in terms of the metric tensor g'^{rq}

$$\Gamma'^{r}_{.mn} = g'^{rq}\Gamma'_{qmn}. \tag{7.19}$$

with

$$\Gamma'_{qmn} := \frac{1}{2}\left(\frac{\partial g'_{qm}}{\partial x'^n} + \frac{\partial g'_{qn}}{\partial x'^m} - \frac{\partial g'_{mn}}{\partial x'^q}\right)$$

Note that the metric tensor g'^{rq} transforms as follows

$$g'^{rq} = g^{\rho\kappa}\frac{\partial x'^r}{\partial x^\rho}\frac{\partial x'^q}{\partial x^\kappa}$$

and the first term in (7.1.4) can expressed as

$$\frac{\partial g'_{qm}}{\partial x'^n} = \frac{\partial g_{\kappa\mu}}{\partial x^\nu}\frac{\partial x^\kappa}{\partial x'^q}\frac{\partial x^\mu}{\partial x'^m}\frac{\partial x^\nu}{\partial x'^n} + g_{\kappa\mu}\left(\frac{\partial^2 x^\kappa}{\partial x'^n \partial x'^q}\frac{\partial x^\mu}{\partial x'^m} + \frac{\partial^2 x^\mu}{\partial x'^n \partial x'^m}\frac{\partial x^\kappa}{\partial x'^q}\right)$$

Applying a similar operation to the second term and the third term, followed by taking into account the symmetry of γ_{mn} and rearranging the indices, we have

$$\frac{\partial g'_{qn}}{\partial x'^m} = \frac{\partial g_{\kappa\nu}}{\partial x^\mu}\frac{\partial x^\kappa}{\partial x'^q}\frac{\partial x^\mu}{\partial x'^m}\frac{\partial x^\nu}{\partial x'^n} + g_{\kappa\mu}\left(\frac{\partial^2 x^\kappa}{\partial x'^m \partial x'^q}\frac{\partial x^\mu}{\partial x'^n} + \frac{\partial^2 x^\mu}{\partial x'^m \partial x'^n}\frac{\partial x^\kappa}{\partial x'^q}\right)$$

$$\frac{\partial g'_{mn}}{\partial x'^q} = \frac{\partial g_{\mu\nu}}{\partial x^\kappa}\frac{\partial x^\kappa}{\partial x'^q}\frac{\partial x^\mu}{\partial x'^m}\frac{\partial x^\nu}{\partial x'^n} + g_{\kappa\mu}\left(\frac{\partial^2 x^\kappa}{\partial x'^m \partial x'^q}\frac{\partial x^\mu}{\partial x'^n} + \frac{\partial^2 x^\kappa}{\partial x'^q \partial x'^n}\frac{\partial x^\mu}{\partial x'^m}\right)$$

Substituting these three expressions into (7.1.4), we see four terms cancel out. Finally, (7.1.4) leads to

$$\Gamma'^{r}_{.mn} = \frac{\partial x'^r}{\partial x^\rho}\frac{\partial x^\mu}{\partial x'^m}\frac{\partial x^\nu}{\partial x'^n}\Gamma^{\rho}_{\cdot\mu\nu} + \delta^\rho_\mu\frac{\partial x'^r}{\partial x^\rho}\frac{\partial^2 x^\mu}{\partial x'^m \partial x'^n} \tag{7.20}$$

□

Let $x^r := x^r(u)$ define a curve $\mathcal{C}$, then we define the absolute derivative of T^r along $\mathcal{C}$ as

$$\frac{\delta T^r}{\delta u} = \frac{dT^r}{du} + \Gamma^{\cdot r}_{mn}T^m\frac{dx^n}{du}. \tag{7.21}$$

Theorem 7.3. *The absolute derivative of a contravariant (covariant) tensor is a contravariant (covariant) tensor.*

Proof. Multiply the equation 7.29 by $\frac{\partial x'^q}{\partial x^r}$ to get

$$\frac{\partial^2 x'^q}{\partial x^p \partial x^s} + \frac{\partial^2 x^r}{\partial x'^m \partial x'^s}\frac{\partial x'^m}{\partial x^p}\frac{\partial x'^n}{\partial x^s}\frac{\partial x'^q}{\partial x^r} = 0.$$

Now using the fact that the contravariant vectors transform according to the rule

$$T'^r = T^s \frac{\partial x'^s}{\partial x^s},$$

we have

$$\frac{\delta T'^r}{\delta u} = \frac{\delta T^s}{\delta u}\frac{\partial x'^r}{\partial x^s} = T^m \frac{dx^m}{du}\left(\frac{\partial^2 x'^r}{\partial x^m \partial x^n} + \frac{\partial^2 x^s}{\partial x'^p \partial x'^q}\frac{\partial x'^r}{\partial x^s}\frac{\partial x'^p}{\partial x^m}\frac{\partial x'^q}{\partial x^n}\right),$$

which vanishes because multiplying equation (7.29) by $\frac{\partial x'^q}{\partial x^r}$ generates this term. This concludes the proof. □

Definition 7.4. *If the vector T^r satisfies the differential equation*

$$\frac{\delta T^r}{d\delta} := \frac{dT^r}{du} + \Gamma^r_{\cdot mn} T^m \frac{dx^n}{du} \tag{7.22}$$

along the curve $\mathcal{C} := x^r(u)$. We say the vector is said to be propagated parallelly along the curve.

It is possible to find out how covariant vectors propagate parallelly along a curve using the same type of argument we used above; however, a simple trick suffices. Suppose S^r parallel propagates along $\mathcal{C}$. As the term $T^r S_r$ is a tensor of zero order it must be an invariant tensor . Moreover, $\frac{dT^r S_r}{du}$ is also an invariant. Hence, its transformation property is

$$\frac{d}{du}(T_r S^r) = \frac{dT_r}{du}S^r + T_r\frac{dS^r}{du} = \left(\frac{dT_r}{du} - \Gamma^r_{\cdot rn} T^m \frac{dx^n}{du}\right) = 0.$$

As this equation holds for all parallel propagated S^r we conclude that

$$\frac{\delta T_r}{\delta u} := \left(\frac{dT_r}{du} - \Gamma^r_{\cdot rn} T^m \frac{dx^n}{du}\right)$$

is a covariant vector.

Definition 7.5.

$$\frac{\delta T_r}{\delta u} := \frac{dT_r}{du} - \Gamma^r_{\cdot rn} T^m \frac{dx^n}{du} \tag{7.23}$$

is called the absolute derivative of T_r.

If a tensor is defined in a domain Ω then it is possible to define its absolute derivative in all of Ω since for any curve $\mathcal{C} \subset \Omega$ we have

$$\frac{\delta T_r}{\delta u} := \left(\frac{dT_r}{dx^n} - \Gamma^r_{\cdot rn} T^m \frac{dx^n}{du}\right)\frac{dx^n}{du}.$$

Since $\frac{dx^n}{du}$ is a contravariant tensor we conclude that $\frac{\delta T_r}{\delta u}$ is a mixed tensor. We shall use the notation $T^r_{|n}$ for the covariant derivative of T_r which is given as

$$\frac{\delta T_r}{\delta u} := \frac{dT_r}{dx^n} - \Gamma^r_{\cdot rn} T^m \frac{dx^n}{du}. \tag{7.24}$$

Mathematica session 7.2. *In this section we show how to calculate the Chistoffel symbols.*

We introduce a MATHEMATICA *miniprogram to calculate the Christoffel symbol of the first kind called* **christoffelOne**. *We illustrate the construction of this program for the spherical coordinate fundamental form.*

```
christoffelOne[a_, x_, r_, m_, n_]:=
1/2 * (D[a[[r, m]], x[[n]]] + D[a[[r, n]], x[[m]]] − D[a[[m, n]], x[[r]]]);
```

We have suppressed the output as it is too long. The reader is urged to run this on their computer.

```
christoffelTwo[a_, x_, r_, m_, n_]:=
Sum[Inverse[a][[r, s]] * christoffelOne[a, x, s, m, n], {s, 1, 2}]
```

For the Christoffel of the second kind we use the MATHEMATICA *command* **christoffelTwo**. *The command makes use of the inverse of the matrix expression for the fundamental form. We again do the calculation for the spherical coordinates.*

$$Table\left[christoffelTwo\left[\begin{pmatrix} 1 & 0 & 0 \\ 0 & \rho^2 Sin[\varphi]^2 & 0 \\ 0 & 0 & \rho^2 \end{pmatrix}, \{\rho, \theta, \varphi\}, r, m, n\right], \{r, 1, 3\}, \{m, 1, 3\}, \{n, 1, 3\}\right];$$

Where we have suppressed the output because of length.

♦

Exercises

Exercise 7.6. *Use this approach to define covariant derivatives of a covariant tensor and a mixed tensor. In particular prove the following:*

$$T_{r|n}; = \frac{\partial T_r}{\partial x^n} - \Gamma^{m}_{\cdot rn} T_m$$

$$T^{rs}_{|n} = \frac{\partial T^{rs}}{\partial x^n} + \Gamma^{r}_{\cdot mn} T^{ms} + \Gamma^{s}_{\cdot mn} T^{rm},$$

$$T_{rs|n} = \frac{\partial T^{r}_{\cdot s}}{\partial x^n} - \Gamma^{m}_{rn} T_{ms} - \Gamma^{m}_{\cdot sn} T_{rm},$$

$$T^{r}_{\cdot s|n} := \frac{\partial T^{r}_{\cdot s}}{\partial x^n} + \Gamma^{r}_{\cdot mn} T^{m}_{\cdot s} - \Gamma^{m}_{\cdot sn} T^{r}_{\cdot m}.$$

We now compute the absolute derivative of the metric tensor

$$g_{rs|t} = \frac{\partial g_{ms}}{\partial x^s} - \Gamma^m_{\cdot rt} g_{ms} - \Gamma^m_{\cdot st} g_{rm} = \frac{\partial g_{ms}}{\partial x^s} - \Gamma_{srt} - \Gamma_{rst}$$

$$= \frac{\partial g_{ms}}{\partial x^s} - \frac{1}{2}\left(g_{sr,t} + g_{st,r} - g_{rt,s}\right) - \frac{1}{2}\left(g_{rs,t} + g_{rt,s} - g_{st,r}\right) = 0. \quad (7.25)$$

[5]

Example 7.7. *We consider the case of a sphere $\mathcal{S} \subset \mathbb{R}^3$. That is we consider the sphere to be imbedded into three-dimensional Euclidean space. The spherical coordinates are $(x^1 = r, x^2 = \theta, x^3 = \varphi)$ and as these coordinates are mutually orthogonal we have just the terms g_{11}, g_{22}, g_{33} occurring in the metric tensor,i.e.*

$$ds^2 = (dx^1)^2 + (x^1 dx^2)^2 + (x^1 \sin(x^2) dx^3)^2;$$

i.e.

$$g_{11} = 1, \quad g_{22} = r^2, \quad g_{33} = r^2 \sin^2(\theta).$$

Using the notation $p^j = \frac{dx^j}{du}$

$$w = \left(p^1\right)^2 + \left(x^1 p^2\right)^2 + \left(x^1 \sin(x^2) p^3\right)^2.$$

If we use the definition (7.17) for general f^r, then we can formally compute these terms for the case at hand

$$f_r = \frac{1}{2}\left[\frac{d}{du}\left(\frac{\partial w}{\partial p^r}\right) - \frac{\partial w}{\partial x^r}\right] \quad (7.26)$$

then

$$\left\{ \begin{array}{ll} f_1 = & \frac{d}{du}\left(p^1\right) - x^1\left(p^2\right)^2 - x^1\left(\sin(x^2)p^3\right)^2 \\ f_2 = & \frac{d}{du}\left((x^1)^2 p^2\right) - \sin(x^2)^2\cos(x^2)(x^1p^3)^2 \\ f_3 = & \frac{d}{du}\left((x^1\sin(x^2)^2 p^3\right) \end{array} \right. . \quad (7.27)$$

Example 7.8. *We generalize the above problem to the $n-1$ dimensional hypersphere imbedded in $\mathbb{R}^n$. The hyperspherical coordinates are given for $n \geq 3$ as $x^1 = r$, $x^2 = \theta_1$, $\ldots x^{n-1} = \theta_{n-2}$, $x^n = \varphi$,given for $n \geq 3$ as*

$$\left\{ \begin{array}{ll} y^1 = & x^1\cos(x^2) \\ y^2 = & x^1\sin(x^2)\cos(x^3) \\ y^3 = & x^1\sin(x^2)\sin(x^3)\cos(x^4) \\ \cdots & \cdots\cdots \\ y^{n-2} = & x^1\sin(x^2)\sin(x^3)\cdots\sin(x^{n-3})\cos(x^{n-2}) \\ y^{n-1} = & x^1\sin(x^2)\sin(x^3)\cdots\sin(x^{n-2})\cos(x^{n-1}) \\ y^n = & x^1\sin(x^2)\sin(x^3)\cdots\sin(x^{ni3})\sin(x^{n-2}) \end{array} \right.$$

[5] In the above equations we have used a comma for partial differentiation and a pipe for covariant differentiation.

where $0 \leq \pi, j = 2, 3, \cdots n-1$ $, 0 \leq x^n \leq 2\pi$. In these coordinates the n dimensional volume element is given by

$$dV = (x^1)^{n-1}(\sin(x^2))^{n-2}(\sin(x^3))^{n-3}\cdots(\sin(x^{n-2}))dx^1\, dx^2 \cdots dx^{n-2}dx^{n-1}.$$

Let us consider first a hypersphere of three dimensions imbedded in $\mathbb{R}^4$. Here we have the coordinates

$$\begin{cases} y^1 &= x^1\cos(x^2), & x^2 &= x^1\sin(x^2)\cos(x^3) \\ y^3 &= x^1\sin(x^2)\sin(x^3)\cos(x^4), & y^4 &= x^1\sin(x^2)\sin(x^3)\sin(x^4). \end{cases}$$

As the hyperspherical coordinates are orthogonal to one another we will not have mixed terms in the metric tensor. Moreover, by hand calculation it is easy to demonstrate that the metric tensor is diagonal and these terms are given as

$$g_{11} = 1, \quad g_{22} = (x^1)^2, \quad g_{33} = (x^1)^2\sin^2(x^2), \quad g_{44} = (x^1)^2\sin^2(x^2)\sin^2(x^3).$$

We see the pattern; hence, in the general case the hyperspherical metric has the components

$$g_{11} = 1, \quad g_{mm} = \left(x^1\right)^2 \prod_{j=1}^{m-2} \left(\sin(x^j)\right)^2, \quad m \leq n-2$$

and the fundamental form is

$$\left(x^1\right)^2 (dx^1)^2 + \sum_{k=2}^{n-2} \left(x^1\right) \prod_{j=1}^{k-1} dx^j dx^j \left(\sin(x^k)\right)^2. \tag{7.28}$$

Note that this metric tensor is restricted to the surface of the hypersphere $\mathcal{S}_n$.

Exercises

Exercise 7.9. *Use*

$$f_r = \frac{1}{2}\left[\frac{d}{du}\left(\frac{\partial w}{\partial p^r}\right) - \frac{\partial w}{\partial x^r}\right],$$

to compute the Christoffel symbols for $n = 4$.

Exercise 7.10. *Derive the equations (7.12) and (7.13).*

Exercise 7.11. *Starting with the identity*

$$\frac{\partial x^r}{\partial x'^n}\frac{\partial x'^n}{\partial x^s} = \delta^r_s,$$

differentiate with respect to x^p *to obtain*

$$\frac{\partial^2 x^r}{\partial x'^m \partial x'^n}\frac{\partial x'^m}{\partial x^p}\frac{\partial'^n}{\partial x^s} + \frac{\partial x^r}{\partial x'^s}\frac{\partial^2 x'^n}{\partial x^p \partial x^s} = 0. \tag{7.29}$$

Exercise 7.12. *Prove that*

$$T^{n}_{\cdot rm|t} = g_{rs}T^{sn}_{\cdot\cdot m|t}.$$

Exercise 7.13. *Show that* $\Gamma_{rmn} + \Gamma_{rnm} = \frac{\partial g_{mn}}{\partial x^r}$.

Exercise 7.14. *Show that in the case of Euclidean space* $\mathbb{R}^3$ *the Christoffel symbols vanish.*

Exercise 7.15. *Using the method of parallel propagation show that*

$$\frac{\delta T_{rs}}{\delta u} := \frac{dT_{rs}}{du} - \Gamma^{m}_{\cdot rn}T_{ms}\frac{dx^n}{du} - \Gamma^{m}_{\cdot sn}T_{rm}\frac{dx^n}{du} \tag{7.30}$$

is a second order contravariant tensor.

Exercise 7.16. *Derive the expressions for the components* f^r *in the expression (7.27) and conclude*

$$\Gamma^{1}_{\cdot 22} = -x^1, \qquad \Gamma^{1}_{\cdot 33 = -x^1\left(\sin(x^2)\right)}$$
$$\Gamma^{2}_{\cdot 33} = -(x^1)^2\sin(x^2)\cos(x^2), \quad \Gamma^{3}_{\cdot 23} = (x^1)^2\sin(x^2)\cos(x^2), \quad \Gamma^{3}_{\cdot 31} = x^1\left(\sin(x^2)^2\right).$$

7.1.5 Frenet formulas

As we have seen in Chapter 3 for any curve in $\mathbb{R}^3$ there is an orthogonal set of 3 unit vectors, known as the tangent, normal and binormal. In this section we will show that a similar result holds for an n-dimensional Riemannian surface. To this end, we introduce a curve parameterized by the arc length s. Then the tangent vector to the curve is given by $T = \frac{d\mathbf{X}}{ds}$ and

$$g_{mn}T^mT^m = \varepsilon.$$

Moreover,

$$\frac{d\left(g_{mn}T^m\,T^n\right)}{du} = 2g_{mn}T^m\frac{\delta T^n}{\delta s} = 2T_m\frac{\delta T^n}{\delta s} = 0.^6$$

Hence, the unit vector, $N_{(1)^r}$, which is co-directional with $\frac{\delta T^r}{\delta s}$ is perpendicular to T^r. This vector is called the first unit-normal to the curve, and the magnitude of this vector, $\|\frac{\delta T^r}{\delta s}\|$, is the first curvature $\kappa_{(1)}$. Next we define a unit vector $N^r_{(2)}$ and a coefficient $\kappa_{(2)}$ using the equations

$$\frac{\delta N_{(1)}}{\delta s} = \kappa_{(2)}N_{(2)})^r - \varepsilon_{(1)}\kappa(1)T^r, \quad \varepsilon_{(2)}N_{(2)\mu}N^{\mu}_{(2)} = 1, \tag{7.31}$$

where $\varepsilon_{(2)}$ is the indicator of $N^r_{(2)}$. By multiplying 7.31 by T_r and $N_{(1)r}$ we see that $N_{(1)r}$ are is perpendicular to $N^r_{(2)}$. We may repeat the above procedure with the equation

$$\frac{\delta N_{(2)}}{\delta s} = \kappa_{(3)}N^r_{3)} - \varepsilon_{(1)}\varepsilon_{(2)}N^r_{(1)},\, \varepsilon_{(3)}N_{(3)n}N$$

[6]Where $\frac{\delta T^n}{\delta s}$ denotes that the absolute derivative is meant.

and continue this process to the case where we seek $N_{(n-1)}$ as we have now created n mutually orthogonal vectors $\left(T^r, N_{(1)}, N_{(2)}, \cdots N_{(n-1)}\right)$ and there are no more than n such vectors in an n-dimensional space.

7.1.6 Curvature

In Chapter 3 the concept of curvature was introduced for Euclidian space $\mathbb{R}^3$. Extending these ideas to the higher dimensional Euclidean space, $\mathbb{R}^n$ is formal. Extending these ideas to a curved space is conceptually different. We say a space is **flat** when it is possible to choose a coordinate system such that the metric takes a diagonal form

$$\varepsilon_1(dx^1)^2 + \varepsilon_2(dx^2)^2 + \cdots + \varepsilon_1(dx^n)^2,$$

where the ε are ± 1. To determine whether a fundamental form can be put into diagonal form, we start with the formula for taking the covariant derivative of a covariant tensor, namely

$$T_{r|m} = \frac{\partial T_r}{\partial x^m} - \Gamma^p_{\cdot rm} T_p.$$

Taking the covariant derivative one more time leads to

$$T_{r|mn} = \frac{\partial T_{r|m}}{\partial x^n} - \Gamma^s_{rn} T_{s|m} - \Gamma^s_{\cdot mn} T_{r|s}.$$

If we now interchange m and n and subtract this term from the former we have

$$T_{r|mn} - T_{r|nm} =$$

$$\frac{\partial}{\partial x^n}\left(\frac{\partial T_r}{\partial x^m} - \Gamma^s_{\cdot rm} T_s\right) - \Gamma^s_{\cdot rn}\left(\frac{\partial T_s}{\partial x^m} - \Gamma^s_{\cdot rm} T_s\right)$$

$$\frac{\partial}{\partial x^m}\left(\frac{\partial T_r}{\partial x^n} - \Gamma^s_{\cdot rn} T_s\right) - \Gamma^s_{\cdot rm}\left(\frac{\partial T_s}{\partial x^n} - \Gamma^s_{\cdot rm} T_s\right)$$

Due to cancellations this simplifies to

$$T_{r|mn} - T_{r|nm} = R^s_{\cdot rmn} T_s.$$

$R^s_{\cdot rmn}$, the coefficient of T_s, is known as the mixed curvature tensor and has the representation

$$R^s_{\cdot rmn} = \frac{\partial}{\partial x^m}\Gamma^s_{\cdot rn} - \frac{\partial}{\partial x^n}\Gamma^s_{\cdot rm} + \Gamma^p_{\cdot rn}\Gamma^s_{\cdot pm} - \Gamma^p_{\cdot rm}\Gamma^s_{\cdot pm}. \tag{7.32}$$

From inspection of 7.32 it is easy to show that the curvature tensor is anti-symmetric in the m, n subscripts, i.e.

$$R^s_{\cdot rm} = -R^s_{\cdot mr}.$$

By hand calculation or by using MATHEMATICA it is easy to show the cyclic symmetry

$$R^{s}_{\cdot rmn} + R^{s}_{\cdot nmr} + R^{s}_{\cdot nrm} = 0. \tag{7.33}$$

As usual we may lower the superscript of the mixed curvature tensor by

$$R_{rsmn} = g_{r\sigma} R^{\sigma}_{\cdot rmn}.$$

It is easily checked with MATHEMATICA that

$$R_{rsmn} = \frac{\partial}{\partial x^m}\Gamma_{rsn} - \frac{\partial}{\partial x^n}\Gamma_{rsm} + \Gamma^{p}_{rn}\Gamma_{pm,s} - \Gamma^{p}_{rm}\Gamma_{pn,s}; \tag{7.34}$$

moreover,

$$R_{rsmn} = \frac{1}{2}\left(\frac{\partial^2 g_{rn}}{\partial x^s \partial x^m} + \frac{\partial^2 g_{sm}}{\partial x^r \partial x^n} - \frac{\partial^2 g_{rm}}{\partial x^s \partial x^n} - \frac{\partial^2 g_{sn}}{\partial x^r \partial x^m}\right)$$

$$+ g^{pq}\left(\Gamma_{prn}\Gamma_{qsm} - \Gamma_{prm}\Gamma_{qsn}\right). \tag{7.35}$$

The covariant form of the stress tensor has additional symmetries compared to the mixed version, namely

$$R_{rsmn} = -R_{srmn}, \quad R_{rsmn} = -R_{srnm}, \quad R_{rsmn} = -R_{mnrs} \tag{7.36}$$

$$R_{rsmn} + R_{rmns} + R_{rnsm} = 0. \tag{7.37}$$

It is important to note that at the beginning of a Riemannian coordinate system that the Christoffel symbols of both kind vanish and the first order partial derivatives of the metric tensor vanish. To this end, we consider the geodesics drawn through the origin of the Riemannian coordinates. Through the origin each geodesic satisfies the differential equation

$$\frac{d^2 x'^r}{ds^2} + (\Gamma^{r}_{\cdot mn})' \frac{dx'^m}{ds}\frac{dx'^n}{ds} = 0,$$

where the $\frac{d'^m}{ds}$ have been chosen arbitrarily; hence we conclude [8] that

$$(\Gamma^{r}_{\cdot mn})' = 0$$

at the origin of the Riemannian coordinates. Because of this the covariant derivative of the curvature tensor, evaluated at the origin just consists of the partial differentials and is the same expression as the

$$R_{rsmn|t} = \frac{1}{2}\left(\frac{\partial^3 g_{rn}}{\partial x^s \partial x^m \partial x^t} + \frac{\partial^3 g_{sm}}{\partial x^r \partial x^n \partial x^t} - \frac{\partial^3 g_{rm}}{\partial x^s \partial x^n \partial x^t} - \frac{\partial^3 g_{sn}}{\partial x^r \partial x^m \partial x^t}\right). \tag{7.38}$$

By permuting the last three subscripts one obtains the **Bianchi identity**

$$R_{rsmn|t} + R_{rsnt|m} + R_{rstm|n} = 0. \tag{7.39}$$

As this is a tensor equation it holds in general. If we contract the tensor $R^{n}_{\cdot rst}$ as follows we obtain the **Ricci** tensor

$$R_{rs} = R^{n}_{\cdot rsn}, \qquad \text{or as} \quad R_{rs} = g^{tn} R_{trsn}. \tag{7.40}$$

By using (7.38) successively, it is easy to show that the Ricci Tensor is symmetric. From the expression for the mixed curvature tensor we have the identity

$$R_{rm} = \frac{\partial}{\partial x^m}\Gamma^{n}_{\cdot rn} - \frac{\partial}{\partial x^n}\Gamma^{n}_{\cdot rm} + \Gamma^{p}_{\cdot rn}\Gamma^{n}_{\cdot pm} - \Gamma^{p}_{\cdot rm}\Gamma^{n}_{\cdot pn}. \tag{7.41}$$

Finally there is a **curvature invariant**

$$R := g^{mn} R_{mn} = R^{n}_{\cdot n}. \tag{7.42}$$

Mathematica session 7.3. *In this* MATHEMATICA *session we illustrate some things about the mixed curvature tensor; in particular, we show for Euclidean space* $\mathbb{R}^n$ *that the curvature tensor vanishes and illustrate that this is a necessary condition for flatness. A proof of this fact will be given shortly. First we provide a short program for calculating the* **riemannTensor**. *We then test this program for the spherical coordinates. We compute the metric tensor first, namely*

$$\begin{pmatrix} 1 & 0 & 0 \\ 0 & \rho^2 Sin[\varphi]^2 & 0 \\ 0 & 0 & \rho^2 \end{pmatrix}$$

Next we show how to pick out the various components of the metric tensor using **Part**. [[2, 1]] *picks out the second row and first column.*

$$\begin{pmatrix} \mathbf{1} & \mathbf{0} & \mathbf{0} \\ \mathbf{0} & \boldsymbol{r^2 Sin[\varphi]^2} & \mathbf{0} \\ \mathbf{0} & \mathbf{0} & \boldsymbol{r^2} \end{pmatrix}\mathbf{[[2,1]]}$$

0

; whereas [[2, 2]] *picks out the second row and second column. Experiment with*

$$\mathbf{Part.}\begin{pmatrix} 1 & 0 & 0 \\ 0 & \boldsymbol{r^2 Sin[\varphi]^2} & 0 \\ 0 & 0 & \boldsymbol{r^2} \end{pmatrix}\mathbf{[[2,2]]}$$

$r^2 Sin[\varphi]^2$

We design a function to compute the Christoffel symbol of the first kind, called **christoffelOne**.

```
christoffelOne[a_, x_, r_, m_, n_]:=
1/2 * (D[a[[r, m]], x[[n]]] + D[a[[r, n]], x[[m]]] - D[a[[m, n]], x[[r]]])
```

Next we write a function for the Christoffel symbol of the second kind, **christoffelTwo**.

```
christoffelTwo[a_, x_, r_, m_, n_]:=
Sum[Inverse[a][[r, s]] * christoffelOne[a, x, s, m, n], {s, 1, 3}]
```

Using the above two functions it is easy to write a function for computing the **riemann curvature tensor**.

```
riemannTensor[a_, x_, s_, r_, m_, n_]:=
D[christoffelOne[a, x, s, n, r], x[[m]]] - D[christoffelOne[a, x, s, m, r], x[[n]]]+
Sum[christoffelTwo[a, x, s, m, p] * christoffelOne[a, x, r, n, p], {p, 1, 3}]-
Sum[christoffelTwo[a, x, s, m, p] * christoffelOne[a, x, r, m, p], {p, 1, 3}]
```

The Ricci tensor, which is a contracted Riemann curvature tensor is formed as follows:

```
ricci[a_, x_, r_, m_]:=Sum[riemannTensor[a, x, n, r, m, n], {n, 1, 3}];
```

We have suppressed the output here, as it is too long, with a ";"

We suggest the student remove the "; " and actually view the entire output on your computer screen.

$$Table\left[Sum\left[D\left[christoffelOne\left[\begin{pmatrix} 1 & 0 & 0 \\ 0 & \rho^2 Sin[\varphi]^2 & 0 \\ 0 & 0 & \rho^2 \end{pmatrix}, \{\rho, \theta, \varphi\}, r, m, n\right], \{\rho, \theta, \varphi\}[[s]]\right], \{s, 1, 3\}\right], \{r, 1, 3\}, \{m, 1, 3\}, \{n, 1, 3\}];$$

$$Table\left[Sum\left[D\left[christoffelTwo\left[\begin{pmatrix}1&0&0\\0&\rho^2 Sin[\varphi]^2&0\\0&0&\rho^2\end{pmatrix},\{\rho,\theta,\varphi\},r,m,n\right],\{\rho,\theta,\varphi\}[[s]]\right],\{s,1,3\}],\{r,1,3\},\{m,1,3\},\{n,1,3\}];$$

We have sduppressed the output here. The reader is urged to run the input line

$$Table\left[riemannTensor\left[\begin{pmatrix}1&0&0\\0&\rho^2 Sin[\varphi]^2&0\\0&0&\rho^2\end{pmatrix},\{\rho,\theta,\varphi\},s,r,m,n\right],\right.$$
$$\{s,1,3\},\{r,1,3\},\{m,1,3\},\{n,1,3\}]\ ;$$

$$Table\left[ricci\left[\begin{pmatrix}1&0&0\\0&\rho^2 Sin[\varphi]^2&0\\0&0&\rho^2\end{pmatrix},\{\rho,\theta,\varphi\},r,m\right],\{r,1,3\},\{m,1,3\}\right];$$

We have suppressed the output, which is still too long. We suggest when you are reviewing this session on your computer let the output show.

R[a_, x_]:=Sum[ricci[a, x, r, r], {r, 1, 3}]

We evaluate the **Riemann invariant** *using the function* **ricci** *and then perform the contraction.*

$$Sum\left[ricci\left[\begin{pmatrix}1&0&0\\0&\rho^2 Sin[\varphi]^2&0\\0&0&\rho^2\end{pmatrix},\{\rho,\theta,\varphi\},r,r\right],\{r,1,3\}\right]$$

$$1-\rho^2+\rho^2 Cos[\varphi]^2+Sin[\varphi]^2-2\rho^2 Sin[\varphi]^2-\rho^2 Sin[\varphi]^4.$$

Finally we use the function **R** *directly, and notice that this produces, as expected, the same result* $R\left[\begin{pmatrix}1&0&0\\0&\rho^2 Sin[\varphi]^2&0\\0&0&\rho^2\end{pmatrix},\{\rho,\theta,\varphi\}\right]$

$$1-\rho^2+\rho^2 Cos[\varphi]^2+Sin[\varphi]^2-2\rho^2 Sin[\varphi]^2-\rho^2 Sin[\varphi]^4$$

♦

We consider the rather complicated appearing, orientable, surface from Chapter 6, shown again below

$$x = \cos(\theta), \quad y = \sin(\theta) + \cos(\varphi), \quad z = sin(\varphi).$$

Suppose that $\mathbf{V}_N$ is an N dimensional Riemannian space. Spain [9] presents a quick proof that necessary and sufficient conditions for $\mathbf{V}_N$ to be flat, are namely that the Riemannian curvature tensor vanishes. [7] To this end consider the invariant

$$R_{rjnp}X^r X^n Y^j Y^p.$$

By replacing X^i and Y^i by the linear combinations

$$X^i = \alpha A^i + \beta B^i, \quad Y^i = \delta X^I + \gamma Y^i$$

one obtains

$$R_{rjnp}X^r X^n Y^j Y^p = (\alpha\gamma - \beta\delta) R_{rjnp}A^r A^n B^j B^p.$$

Next, consider the invariant

$$(g_{rn}g_{jp} - g_{rp}g_{jn}) X^r X^n Y^j Y^p = (\alpha\gamma - \beta\delta) (g_{rn}g_{jp} - g_{rp}g_{jn}) 2A^r A^n B^j B^p$$

It is clear that

$$K = \frac{R_{rjnp}A^r A^n B^j B^p}{(g_{rn}g_{jp} - g_{rp}g_{jn})}, \tag{7.43}$$

known as the Riemannian curvature of the space $\mathbb{V}_N$ is an invariant, whose value is not changed when the two vectors A^r, BV^r are replaced by any linear combination, i.e. K is invariant under linear transformations.
A space, $\mathbb{V}_N$, is called flat if K vanishes at every point of $\mathbb{V}_N$. From (7.43) it follows the necessary and sufficient condition is

$$R_{rjnp}A^r A^n B^j B^p = 0,$$

for all vectors A^r and B^r. Using Equations (7.36) we easily establish

$$R_{rjnp} + R_{njrp} + R_{nprj} + R_{rpnj} = 0 \tag{7.44}$$

by first noticing that the above equation is equivalent to

$$R_{rpnj} + R_{rjnp} = 0.$$

Equation (7.36) allows us to write the second term to get

$$R_{rjnp} = R_{rpjn}$$

[7]B. Spain credits the idea to L. Lovitch .

Now change j, n, p cyclically in the above equation to get

$$R_{rjnp} = R_{rpjn} = R_{rnpi},$$

which on substituting into (7.44) implies

$$R_{rjnp} = 0. \tag{7.45}$$

On the other hand, if these terms are all zero then $K = 0$. This proves that necessary and sufficient conditions for $\mathbb{V}_N$ to be flat is that the Riemann curvature tensor be identically zero.

Mathematica session 7.4. *Let us see if we may find some flat surfaces. We begin by listing of some of our own* MATHEMATICA *functions which will be useful. Recall that unless you are working in the same session* MATHEMATICA *may have forgotten our self-constructed functions.*

```
metricTensor[Y_, X_]:=Simplify[MatrixForm[Table[Sum[D[Y[[c]], X[[a]]]*D[Y[[c]], X[[b]]], {c, 1, 3}], {a, 1, 2}, {b, 1, 2}]]]

christoffelOne[a_, x_, r_, m_, n_]:=
1/2 * (D[a[[r, m]], x[[n]]] + D[a[[r, n]], x[[m]]] - D[a[[m, n]], x[[r]]])

christoffelTwo[a_, x_, r_, m_, n_]:=
Sum[Inverse[a][[r, p]] * christoffelOne[a, x, p, m, n], {p, 1, 2}]

reimannTensor[a_, x_, s_, r_, m_, n_]:=
D[christoffelOne[a, x, s, n, r], x[[m]]] - D[christoffelOne[a, x, s, m, r], x[[n]]]+
Sum[christoffelTwo[a, x, s, m, p] * christoffelOne[a, x, r, n, p], {p, 1, 2}]-
Sum[christoffelTwo[a, x, s, m, p] * christoffelOne[a, x, r, m, p], {p, 1, 2}]
```

We now consider the surface of an infinite cylinder of radius a and check whether it is **flat**. *First we compute the metric tensor and notice it is a matrix of constants.*

```
metricTensor[{a * Cos[θ], a * Sin[θ], z}, {θ, z}]
```

$$\begin{pmatrix} 1 & 0 \\ 0 & 1 \end{pmatrix}$$

Next we compute the Christoffel symbols of the first and second kind.

$$\textbf{Table}\left[\textbf{christoffelOne}\left[\begin{pmatrix} 1 & 0 \\ 0 & 1 \end{pmatrix}, \{\theta, z\}, r, m, n\right], \{r, 1, 2\}, \{m, 1, 2\}, \{n, 1, 2\}\right]$$

{{{0,0},{0,0}},{{0,0},{0,0}}}

$$\textbf{Table}\left[\textbf{christoffelTwo}\left[\begin{pmatrix} 1 & 0 \\ 0 & 1 \end{pmatrix}, \{\theta, z\}, r, m, n\right], \{r, 1, 2\}, \{m, 1, 2\}, \{n, 1, 2\}\right]$$

{{{0,0},{0,0}},{{0,0},{0,0}}}

As the Riemann tensor has the form given next, it is easy to see that the cylinder must be flat.

```
reimannTensor[a_, x_, s_, r_, m_, n_]:=
D[christoffelOne[a, x, s, n, r], x[[m]]] − D[christoffelOne[a, x, s, m, r], x[[n]]]+
Sum[christoffelTwo[a, x, s, m, p] ∗ christoffelOne[a, x, r, n, p], {p, 1, 2}]−
Sum[christoffelTwo[a, x, s, m, p] ∗ christoffelOne[a, x, r, m, p], {p, 1, 2}]
```

And performing this computation we see indeed, that the cylinder is flat.

$$\textbf{Table}\left[\textbf{reimannTensor}\left[\begin{pmatrix} 1 & 0 \\ 0 & 1 \end{pmatrix}, \{\theta, z\}, s, r, m, n\right], \{s, 1, 2\}, \{r, 1, 2\}, \{m, 1, 2\}, \{n, 1, 2\}\right]$$

{{{{0,0},{0,0}},{{0,0},{0,0}}},{{{0,0},{0,0}},{{0,0},{0,0}}}}

♦

Let us try a more difficult object now which is illustrated in Figure 7.2.

♦

Mathematica session 7.5. ***metricTensor*[{*Cos*[θ], *Sin*[θ] + *Cos*[φ], *Sin*[φ]}, {θ, φ}]**

$$\begin{pmatrix} 1 & -Cos[\theta]Sin[\varphi] \\ -Cos[\theta]Sin[\varphi] & 1 \end{pmatrix}$$

We apply the function **Riemann** *to this orientable surface and obtain*

riemannTensor[a_, x_, s_, r_, m_, n_]:=

D[christoffelOne[a, x, s, n, r], x[[m]]] − D[christoffelOne[a, x, s, m, r], x[[n]]]+

Sum[christoffelTwo[a, x, s, m, p] * christoffelOne[a, x, r, n, p], {p, 1, 2}]−

Sum[christoffelTwo[a, x, s, m, p] * christoffelOne[a, x, r, m, p], {p, 1, 2}]

$$\boldsymbol{riemannTensor\left[\begin{pmatrix} 1 & -Cos[\theta]Sin[\varphi] \\ -Cos[\theta]Sin[\varphi] & 1 \end{pmatrix}, \{\theta, \varphi\}, 2, 1, 2, 1\right]}$$

$$Cos[\varphi]Sin[\theta] - \frac{Cos[\theta]^3 Cos[\varphi]^2 Sin[\varphi]}{1 - Cos[\theta]^2 Sin[\varphi]^2}$$

Using the MATHEMATICA *function* **Table** *allows us to visualize*

$$\boldsymbol{Table\left[riemannTensor\left[\begin{pmatrix} 1 & -Cos[\theta]Sin[\varphi] \\ -Cos[\theta]Sin[\varphi] & 1 \end{pmatrix}, \{\theta, \varphi\}, s, r, m, n\right],\right.}$$
$$\boldsymbol{\{s, 1, 2\}, \{r, 1, 2\}, \{m, 1, 2\}, \{n, 1, 2\}]}$$

We suppress the MATHEMATICA *output here as it is very large. If the reader will perform this computation, it will be clear that the orientable surface above is definitely non-flat. We consider next the surface of a sphere expressed in spherical coordinates as*

metricTensor[{2 * Cos[θ] * Cos[φ], 2 * Sin[θ] * Cos[φ], 2 * Sin[φ]}, {θ, φ}]

$$\begin{pmatrix} 4Cos[\varphi]^2 & 0 \\ 0 & 4 \end{pmatrix}$$

$$\boldsymbol{Table\left[riemannTensor\left[\begin{pmatrix} 4Cos[\varphi]^2 & 0 \\ 0 & 4 \end{pmatrix}, \{\theta, \varphi\}, s, r, m, n\right], \{s, 1, 2\},\right.}$$
$$\boldsymbol{\{r, 1, 2\}, \{m, 1, 2\}, \{n, 1, 2\}]}$$

We suppress the MATHEMATICA *output here also as it is very large. The reader should let it run to see the result.*

ParametricPlot3D[{Cos[θ], Sin[θ] + Cos[φ], Sin[φ]}, {θ, 0, 2 * π}, {φ, −π, π}]

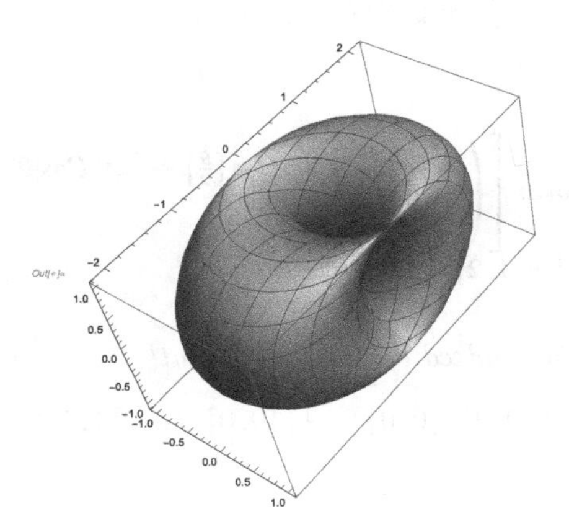

FIGURE 7.2: The orientable surface
$\{\cos(\theta), \sin(\theta) + \cos(\varphi), \sin(\varphi)\}, \{\theta, 0, 2\pi\}, \{\varphi, -\pi/2, \pi/2\}$.

♦

Mathematica session 7.6. *We investigate next the Klein bottle and compute its metric tensor. We input a parametric representation of the Klein bottle as a mapping from $\mathbb{R}^4$. There are other representations but this is easy to implement.*

$$\boldsymbol{Table\left[riemannTensor\left[\begin{pmatrix} 1 & 0 \\ 0 & \frac{1}{4}\left(4+3s^2-2s^2 Cos[t]+8sSin\left[\frac{t}{2}\right]\right) \end{pmatrix}, \{s,t\}, p, r, m, n\right],}$$
$$\boldsymbol{\{p,1,2\},\{r,1,2\},\{m,1,2\},\{n,1,2\}]}$$

$$\boldsymbol{metricTensor[\{(2+(\varphi/2)Cos[\theta/2])Cos[\theta], (2+(\varphi/2)Cos[\theta/2])Sin[\theta], (\varphi/2)Sin[\theta/2]\},}$$
$$\boldsymbol{\{\theta,\varphi\}]}$$

Let us investigate the Mobius strip. We will check out if it is a flat space.

$$\boldsymbol{metricTensor[\{(2+(\varphi/2)Cos[\theta/2])Cos[\theta], (2+(\varphi/2)Cos[\theta/2])Sin[\theta], (\varphi/2)Sin[\theta/2]\},}$$
$$\boldsymbol{\{\theta,\varphi\}]}$$

$$\begin{pmatrix} 4+\frac{3\varphi^2}{16}+2\varphi Cos\left[\frac{\theta}{2}\right]+\frac{1}{8}\varphi^2 Cos[\theta] & 0 \\ 0 & \frac{1}{4} \end{pmatrix}$$

$$\textbf{Table}\left[\textbf{reimannTensor}\left[\begin{pmatrix} 4+\frac{3\varphi^2}{16}+2\varphi Cos\left[\frac{\theta}{2}\right]+\frac{1}{8}\varphi^2 Cos[\theta] & 0 \\ 0 & \frac{1}{4} \end{pmatrix}, \{s,t\}, p, r, m, n\right]\right.$$
$$\left., \{p,1,2\}, \{r,1,2\}, \{m,1,2\}, \{n,1,2\}\right]$$

We find out that indeed the Mobius is flat!

$\{\{\{\{0,0\},\{0,0\}\},\{\{0,0\},\{0,0\}\}\},\{\{\{0,0\},\{0,0\}\},\{\{0,0\},\{0,0\}\}\}\}$

♦

Mathematica session 7.7. *We now test whether the Klein bottle is flat, namely whether are all the components of its Riemann tensor are zero. Graphs of the Klein bottle are shown in Figures 7.3, and 7.4.*

Simplify[

metricTensorFour $\left[\left\{(2+Cos[v]) * Cos[u], (2+Cos[v]) * Sin[u], Cos\left[\frac{u}{2}\right] Sin[v], Sin\left[\frac{u}{2}\right] Sin[v]\right\}, \{u,v\}\right]\right]$

$$\begin{pmatrix} \frac{37}{8}+4Cos[v]+\frac{3}{8}Cos[2v] & 0 \\ 0 & 1 \end{pmatrix}$$

christoffelOne[a_, x_, r_, m_, n_]:=

1/2 * (D[a[[r, m]], x[[n]]] + D[a[[r, n]], x[[m]]] − D[a[[m, n]], x[[r]]])

$$\textbf{Table}\left[\textbf{christoffelOne}\left[\begin{pmatrix} \frac{37}{8}+4Cos[v]+\frac{3}{8}Cos[2v] & 0 \\ 0 & 1 \end{pmatrix}, \{\theta, v\}, r, m, n\right],\right.$$
$$\left.\{r,1,2\}, \{m,1,2\}, \{n,1,2\}\right]$$

$$\left\{\left\{\left\{0, \frac{1}{2}\left(-4Sin[v]-\frac{3}{4}Sin[2v]\right)\right\}, \left\{\frac{1}{2}\left(-4Sin[v]-\frac{3}{4}Sin[2v]\right), 0\right\}\right\}, \left\{\left\{\frac{1}{2}\left(4Sin[v]+\frac{3}{4}Sin[2v]\right), 0\right\}, \{0,0\}\right\}\right\}$$

$$\textit{Table}\left[\textit{christoffelTwo}\left[\begin{pmatrix} \frac{37}{8}+4\textit{Cos}[v]+\frac{3}{8}\textit{Cos}[2v] & 0 \\ 0 & 1 \end{pmatrix}, \{\theta,\varphi\}, r, m, n\right],\right.$$
$$\left.\{r,1,2\},\{m,1,2\},\{n,1,2\}\right]$$

$\{\{\{0,0\},\{0,0\}\},\{\{0,0\},\{0,0\}\}\}$

reimannTensor[a_, x_, s_, r_, m_, n_]:=

D[christoffelOne[a, x, s, n, r], x[[m]]] − D[christoffelOne[a, x, s, m, r], x[[n]]]+

Sum[christoffelTwo[a, x, s, m, p] * christoffelOne[a, x, r, n, p], {p, 1, 2}]−

Sum[christoffelTwo[a, x, s, m, p] * christoffelOne[a, x, r, m, p], {p, 1, 2}]

$$\textit{Table}\left[\textit{reimannTensor}\left[\begin{pmatrix} \frac{37}{8}+4\textit{Cos}[v]+\frac{3}{8}\textit{Cos}[2v] & 0 \\ 0 & 1 \end{pmatrix}, \{\theta,\varphi\}, s, r, m, n\right],\right.$$
$$\left.\{s,1,2\},\{r,1,2\},\{m,1,2\},\{n,1,2\}\right]$$

$\{\{\{\{0,0\},\{0,0\}\},\{\{0,0\},\{0,0\}\}\},\{\{\{0,0\},\{0,0\}\},\{\{0,0\},\{0,0\}\}\}\}$

{(a + bCos[v]) * Cos[u], (a + bCos[v]) * Sin[u], b * Sin[v] * Cos[u/2], b * Sin[v]Sin[u/2]};

We use a slight variant of **metricTensor** *and call it* **metricTensorTwo**.

metricTensorTwo[Y_, X_]:=Simplify[MatrixForm[Table[Sum[D[Y[[c]],

X[[a]]] * D[Y[[c]], X[[b]]], {c, 1, 4}], {a, 1, 2}, {b, 1, 2}]]]

metricTensorTwo[{(2 + Cos[v])Cos[u], (2 + Cos[v])Sin[u],

$$\mathbf{2}\textit{\textbf{Cos}}\left[\tfrac{u}{2}\right]\textit{\textbf{Sin}}[v], \mathbf{2}\textit{\textbf{Sin}}\left[\tfrac{u}{2}\right]\textit{\textbf{Sin}}[v]\}, \{u, v\}]$$

$$\begin{pmatrix} 5+4\textit{Cos}[v] & 0 \\ 0 & \frac{1}{2}(5+3\textit{Cos}[2v]) \end{pmatrix},$$

where we have set the parameters $a = 2$, $b = 1$.

{a * (1 + Sin[u]) * Cos[u] − c * (1 − Cos[u]/2) * Cos[v], b * Sin[u], (1 − Cos[u]/2) * Sin[v]}

$$\left\{-c\left(1-\frac{\textit{Cos}[u]}{2}\right)\textit{Cos}[v]+a\textit{Cos}[u](1+\textit{Sin}[u]), b\textit{Sin}[u], \left(1-\frac{\textit{Cos}[u]}{2}\right)\textit{Sin}[v]\right\}$$

```
metricTensorThree[Y_, X_]:=Simplify[MatrixForm[Table[Sum[D[Y[[c]],
X[[a]]] * D[Y[[c]], X[[b]]], {c, 1, 3}], {a, 1, 2}, {b, 1, 2}]]]

metricTensorThree[{3 * (1 + Sin[u]) * Cos[u] − 2(1 − Cos[u]/2) * Cos[v], 4 *
Sin[u], 2 * (1 − Cos[u]/2) * Sin[v]}, {uy, v}]
```

$$\begin{pmatrix} 0 & 0 \\ 0 & (-2 + Cos[u])^2 \end{pmatrix}$$

```
metricTensorThree[{3 * (1 + Sin[u]) * Cos[u] + 2(1 − Cos[u]/2) * Cos[v], 4 *
Sin[u], 3 * (1 − Cos[u]/2) * Sin[v]}, {uy, v}]
```

$$\begin{pmatrix} 0 & 0 \\ 0 & \frac{1}{8}(-2 + Cos[u])^2(13 + 5Cos[2v]) \end{pmatrix}$$

```
ParametricPlot3D[{3 * (1 + Sin[u]) * Cos[u] + 2 * (1 − Cos[u]/2) * Cos[v], 2 *
Sin[u], 3 * (1 − Cos[u]/2) * Sin[v]}, {u, 0, 2π}, {v, 0, 2 * π}]
```

♦

```
ParametricPlot3D[{3 * (1 + Sin[u]) * Cos[u] − 2 * (1 − Cos[u]/2) * Cos[v], 2 *
Sin[u], 3 * (1 − Cos[u]/2) * Sin[v]}, {u, 0, 2π}, {v, 0, 2 * π}]
```

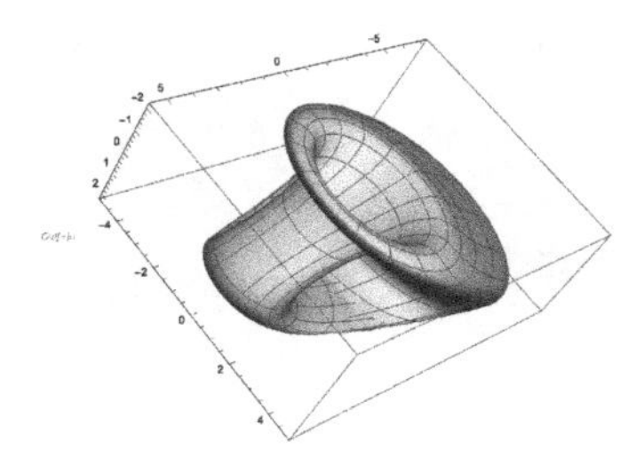

FIGURE 7.3: A view of the Klein Bottle

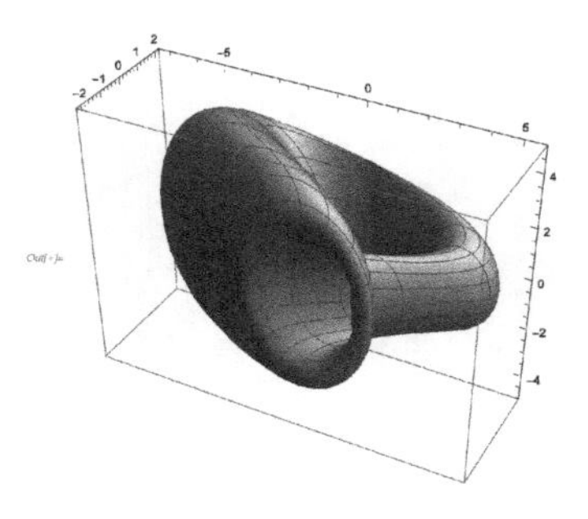

FIGURE 7.4: Another view of the Klein Bottle

Chapter 8

Partial Differential Equations

In this chapter we solve several, elementary, partial differential equations generated by mathematical physics. We start with first-order transport equations, first using the built in MATHEMATICA command **DSolve** and then show how we may combine this with the method of characteristics to solve more general first-order equations. In later sections we consider the heat, potential and wave equations using the method of separation of variables. The potential equation, also known as Laplace's equation is solved in two and three space variables. We make frequent use of the MATHEMATICA command **DSolve** to find the general form of a solution. For second order equations we mostly use a separation of variables technique; however, there is a built in MATHEMATICA command for solving Laplace's equation. Whereas, we want to show the possibilities to use MATHEMATICA directly, the purpose of this book is to study mathematics and to use the mathematics to help you write your own programs. The use of writing programs is meant to reinforce the mathematical ideas.

8.1 First order partial differential equations

8.1.1 Linear, equations

We first consider first-order, linear, transport equations which describe a disturbance traveling in **one direction** with velocity $c(x,t)$. Such solutions are frequently referred to as traveling waves

$$\frac{\partial u}{\partial x} - \frac{1}{a(x,t)}\frac{\partial u}{\partial t} = 0.$$

For the case of a constant coefficient $a(x,t)$ and several variable cases, **DSolve** will solve the transport equation. For more complicated equations a numerical approach using the method of characteristics will be presented.

Mathematica session 8.1. $\boldsymbol{pde1 = D[u[x,t],t] + a * D[u[x,t],x] == 0}$

$u^{(0,1)}[x,t] + au^{(1,0)}[x,t] == 0$

$\boldsymbol{soln1 = DSolve[pde1, u[x,t], \{x,t\}]}$

$C[1]\left[\frac{a-tx}{a}\right]$ *means the* **arbitrary** *function* $C_1\left(\frac{a-tx}{a}\right)$. *Note if we can determine the function* C_1 *by imposing, say an initial condition on the solution we have found a particular solution to that problem*

$\{\{u[x,t] \to C[1]\left[\frac{at-x}{a}\right]\}\}$.

Now we define the function $f(x,t)$ *by setting*

f[x_, t_] = u[x, t]/.soln1[[1]]

$C[1]\left[\frac{at-x}{a}\right]$

f[x, 0]

$C[1]\left[-\frac{x}{a}\right]$

sol1 = DSolve[{pde1, u[0, t] == Cos[t]}, u[x, t], {x, t}]

$\{\{u[x,t] \to Cos\left[\frac{at-x}{a}\right]\}\}$

A one–dimensional, constant velocity, wave equation, defined on $(-\infty, \infty)$ that may be used to describe a solution possessing two traveling waves has the form

$$\frac{\partial^2 u}{\partial t^2} - a^2 \frac{\partial u}{\partial x^2} = 0, \quad -\infty < x < \infty. \tag{8.1}$$

Indeed, MATHEMATICA provides such a solution below. Notice that there are two independent functions $C_1(x)$ and $C_2(x)$ describing the solution. These functions may be determined in terms of the following two initial conditions, referred to as Cauchy conditions. Suppose, for example, that

$$u(x,0) = f(x), \quad \text{and} \quad \partial_t u(x,0) = g(x), \text{then}$$

$$u(x,0) = C_1(x) + C_2(x) = f(x),$$

and

$$(\partial_t u)(x,0) = -aC_1'(x) + aC21'(x).$$

These may be solved for C_1 and C_2 as

$$C_1(x) = f(x) + \frac{1}{a}C_2(x) + \int^x g(x)\,dx \quad \text{and} \quad C_2(x) = f(x) - C_1(x).$$

Now let us try this with MATHEMATICA

pde2 = *D[u[x, t], {t, 2}] − a^2 ∗ D[u[x, t], {x, 2}]* == 0

$u^{(0,2)}[x,t] - a^2 u^{(2,0)}[x,t] == 0$

soln2 = DSolve[pde2, $u[x,t]$, $\{x,t\}$]

$\left\{\left\{u[x,t] \to C[1]\left[t - \frac{x}{\sqrt{a^2}}\right] + C[2]\left[t + \frac{x}{\sqrt{a^2}}\right]\right\}\right\}$

ic2 = {$u[x,0]$ == E^(−x^2), Derivative[0, 1][u][x, 0] == 1};

DSolveValue[{pde2, ic2}, $u[x,t]$, $\{x,t\}$]

$\frac{1}{2}\left(e^{-\left(-\sqrt{a^2}t+x\right)^2} + e^{-\left(\sqrt{a^2}t+x\right)^2}\right) + t.$

Now let us repeat this process with the new initial condition

ic3 = {$u[x,0]$ == UnitStep[x] − UnitStep[x − 2], Derivative[0, 1][u][x, 0] == 1};

we = DSolveValue[{pde2, ic3}, $u[x,t]$, $\{x,t\}$]

$t + \frac{1}{2}\left(-\text{UnitStep}\left[-2 - \sqrt{a^2}t + x\right]\right.$
$\left. + \text{UnitStep}\left[-\sqrt{a^2}t + x\right] - \text{UnitStep}\left[-2 + \sqrt{a^2}t + x\right] + \text{UnitStep}\left[\sqrt{a^2}t + x\right]\right)$

We assign the velocity $a = 1$.

wequ = we/.{$a \to 1$}

$t + \frac{1}{2}(-\text{UnitStep}[-2 - t + x] + \text{UnitStep}[-t + x] - \text{UnitStep}[-2 + t + x]$
$+ \text{UnitStep}[t + x])$

Plot3D[wequ, {x, −8, 8}, {t, 0, 6}]

A plot of the solution is shown in Figure 8.1.

We try another initial condition

ic4 = {$u[x,0]$ == UnitStep[x] − UnitStep[x − 2], Derivative[0, 1][u][x, 0] == 0};

we2 = DSolveValue[{pde2, ic4}, $u[x,t]$, $\{x,t\}$]/.{$a \to 1$}

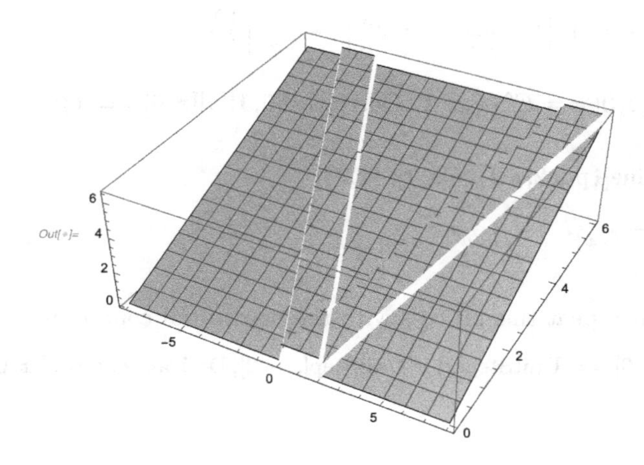

FIGURE 8.1: A plot of the solution to the boundary value problem for the one–dimensional wave equation using the initial condition 'ic3'.

$\frac{1}{2}(-\text{UnitStep}[-2-t+x]+\text{UnitStep}[-t+x]-\text{UnitStep}[-2+t+x]+\text{UnitStep}[t+x])$

Plot3D[we2, {x, −8, 8}, {t, 0, 6}]

A plot of the solution is shown in Figure 8.2.

The Cauchy problem is also natural for the first-order wave equations. As these equations are first order only, one initial, or Cauchy condition, will be required. Consequently, the initial data is given in the form

$$v(x, 0) = f(x)$$

or more generally in the form

$$v(x(s), y(s)) = \varphi(s),$$

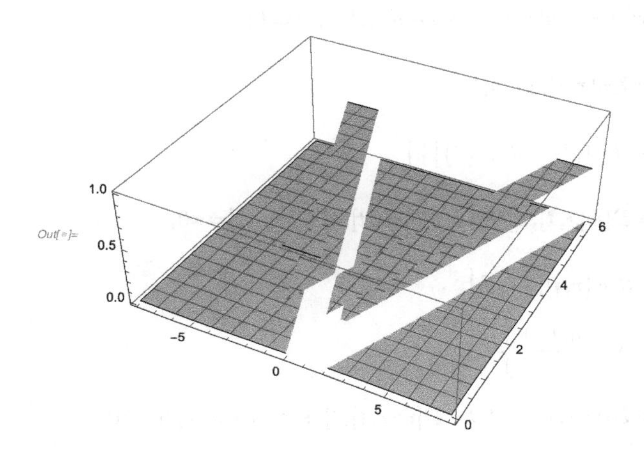

FIGURE 8.2: A plot of the solution to the boundary value problem for the one–dimensional wave equation using the initial condition 'ic4'.

where $(x(s), y(s))$ is a parametric representation for an **initial curve.**[1]

In the case of Cauchy condition of the form $v(x, 0) = f(x)$, the initial value problem can be solved by setting

$$F1(x) := f(\sqrt{2x}), \text{ where } x \geq 0.$$

Note that if $f(x)$ is not an even function of x this solution is not defined for $x < 0$. Let us now consider another first-order wave equation.

Mathematica session 8.2. *We will now use* MATHEMATICA *to solve transport equations with variable coefficients.* MATHEMATICA *does fine with it.*

```
pde6 = -x * D[u[x, t], t] + D[u[x, t], x] == 0;
```

[1]An initial curve must not be a **characteristic**, that is $\frac{dx}{ds} \neq a(x(s), y(s))$. More will be said about this shortly.

```
DSolve[pde6, u[x, t], {x, t}]
```

$\{\{u[x,t] \to C[1]\left[\frac{1}{2}\left(2t+x^2\right)\right]\}\}$

```
pde7 = -x^2 * D[u[x, t], t] + D[u[x, t], x] == 0;
```

```
DSolve[pde7, u[x, t], {x, t}]
```

$\{\{u[x,t] \to C[1]\left[\frac{1}{3}\left(3t+x^3\right)\right]\}\}$

```
pde8 = x * D[w[x, t], x] + t * D[w[x, t], t] + w[x, t]==0;
```

```
DSolve[pde8, w[x, t], {x, t}]
```

$\left\{\left\{w[x,t] \to \frac{C[1]\left[\frac{t}{x}\right]}{x}\right\}\right\}$

```
pde9 = x * D[w[x, t], x] + t * D[w[x, t], t] + x * t * w[x, t]==0;
```

```
DSolve[pde9, w[x, t], {x, t}]
```

$\left\{\left\{w[x,t] \to e^{-\frac{tx}{2}} C[1]\left[\frac{t}{x}\right]\right\}\right\}$

```
pde10 = D[v[x, y], x] + D[v[x, y], y] - v[x, y]^2 == 0;
```

```
DSolve[pde10, v[x, y], {x, y}]
```

$\left\{\left\{v[x,y] \to \frac{1}{-x-C[1][-x+y]}\right\}\right\}$

♦

The Cauchy problem for the above equation is solved by setting $F1(x) := \sqrt{-x} f(\sqrt{-x})$ which makes sense only if $f(x)$ is an odd function.

8.1.2 The method of characteristics for a first order partial differential equation

In this session we see how to use MATHEMATICA to solve first-order partial differential equations of **quasilinear** type, namely

$$a(x,y,u)\frac{\partial u}{\partial x} + b(x,y,u)\frac{\partial u}{\partial y} = c(x,y,u). \tag{8.2}$$

It is rare when we can use **Dsolve** to find an analytical expression for nonlinear problems, so we must introduce a more general approach. We do this by recognizing that a solution to (8.2) also defines a surface $u = u(x,y)$, whose normal is the vector $[u_x, u_y, -1]$. As this normal is perpendicular to the vector $[a,b,c]$, the vector $[a,b,c]$ is tangent to the surface. Because of this we define the **characteristics** of the differential equation to be the curves lying on the surface, which are defined by the ordinary differential equations

$$\frac{dx}{a(x,y,u)} = \frac{dy}{b(x,y,u)} = \frac{du}{c(x,y,u)}. \tag{8.3}$$

A natural problem to consider for the first-order partial differential equation is the Cauchy problem where the data is prescribed on a non- characteristic curve

$$x = x(t),\ y = y(t),\ u = u(t). \tag{8.4}$$

We then try to solve the Cauchy problem by solving the characteristic equations for each point which passes through the non-characteristic curve. If we use the arc-length s as the parameter to measure along the characteristic curves, we construct a surface described in terms of the parameters s and t

$$x = x(s,t),\ y = y(s,t),\ u = u(s,t). \tag{8.5}$$

To solve the characteristic equations we replace the system of ordinary equations for the characteristics by the system [2]

$$\frac{dx}{ds} = a(x,y,u),\ \frac{dy}{ds} = b(x,y,u),\ \frac{du}{ds} = c(x,y,u). \tag{8.6}$$

As this is a system of ordinary equations we may use **DSolve** to try to solve it.

In order to make this approach useful we need to specify an initial curve. The initial curve corresponds to the arc-length parameter $s = 0$. Hence, let us substitute $s = 0$ into the solution. We shall then choose the arbitrary "constants" $c1$, $c2$, $c3$ so that as functions of t these "constants" correspond to a point, indexed by t, on the initial curve. We find then that the coefficients may be replaced as a function of the parameter t.

[2]See Figure 8.3.

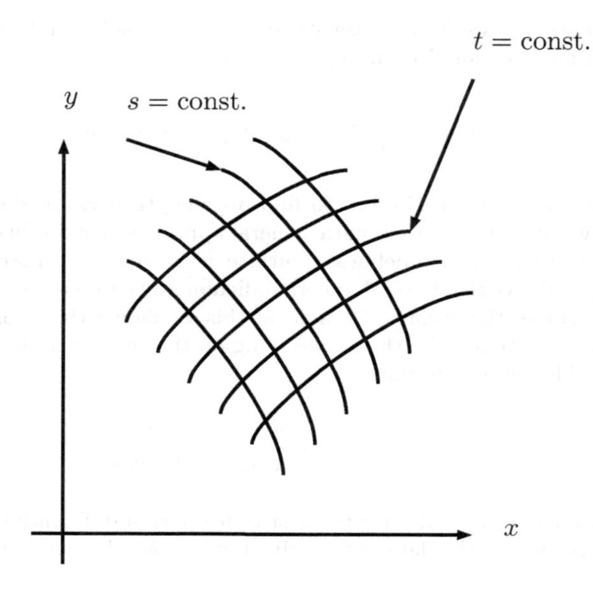

FIGURE 8.3: A family of characteristic curves (s) interlaced with curves of solutions (t).

We now want to write the solution u as a function of x and y. This is done by solving the system

$$x = x(s,t), \;\; y = y(s,t)$$

for $s = s(x,y), \;\; t = t(x,y)$ and using this to replace s and t in $u(s,t)$. See Figure 8.3This can always be done, for sufficiently small values of s, providing

$$\frac{\partial x}{\partial s}\frac{\partial y}{\partial t} - \frac{\partial x}{\partial t}\frac{\partial y}{\partial s} \neq 0$$

along the initial curve.

$$\{t = \frac{1}{2}x - \frac{1}{2}y, \, s = \frac{1}{2}y + \frac{1}{2}x\}$$

It can be seen that the solution blows up along the hyperbolic curve $x^2 - y^2 = 4$.

Mathematica session 8.3. $\boldsymbol{DSolve[\{D[x[s],s]==1, D[y[s],s]==1, D[v[s],s]}$ $\boldsymbol{==(v[s])\hat{}2\}, \{x[s],y[s],v[s]\}, s]}$

$\left\{\left\{x[s]\to s+C[1], y[s]\to s+C[2], v[s]\to \frac{1}{-s-C[3]}\right\}\right\}$

$\left\{x[s]\to s+C[1], y[s]\to s+C[2], v[s]\to \frac{1}{-s-C[3]}\right\} /.\{s\to 0\}$

$\left\{x[0]\to C[1], y[0]\to C[2], v[0]\to -\frac{1}{C[3]}\right\}$

$\{x[0]==t, y[0]==-t, v[0]==t\};$

$Solve\left[\left\{t==C[1], -t==C[2], t==-\frac{1}{C[3]}\right\}, \{C[1], C[2], C[3]\}\right]$

$\{\{C[1]\to t, C[2]\to -t, C[3]\to -\frac{1}{t}\}\}$

$\left\{\left\{x[s]\to s+C[1], y[s]\to s+C[2], v[s]\to \frac{1}{-s-C[3]}\right\}\right\} /.$
$\{\{C[1]\to t, C[2]\to -t, C[3]\to -\frac{1}{t}\}\}$

$\{\{\{x[s]\to s+t, y[s]\to s-t, v[s]\to\}\}\}$

$Solve[\{x==s+t, y==s-t\}, \{s,t\}]$

$\{\{s\to \frac{x+y}{2}, t\to \frac{x-y}{2}\}\}$

$\left\{v==\frac{1}{-s+\frac{1}{t}}\right\} /. \{s\to \frac{x+y}{2}, t\to \frac{x-y}{2}\}$

$\left\{v==\frac{1}{\frac{1}{2}(-x-y)+\frac{2}{x-y}}\right\}$

$Simplify[\%]$

$\left\{v==\frac{2(x-y)}{4-x^2+y^2}\right\}$

$Plot3D\left[\frac{2(x-y)}{4-x^2+y^2}, \{x,-2,2\}, \{y,-2,2\}\right]$

A plot is given in Figure 8.4.

$\{x[0]==t, y[0]==-t, v[0]==2/t\};$

$Solve\left[\left\{t==C[1], -t==C[2], t\hat{}2==\frac{1}{C[3]}\right\}, \{C[1], C[2], C[3]\}\right]$

$\{\{C[1]\to t, C[2]\to -t, C[3]\to \frac{1}{t^2}\}\}$

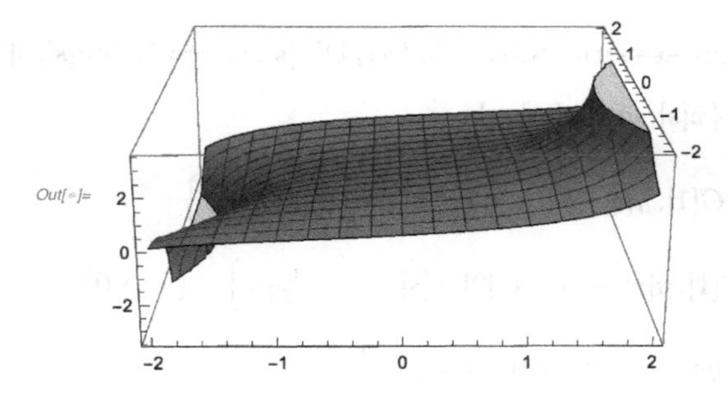

FIGURE 8.4: A plot of the solution $v(x, y) = \frac{2(x-y)}{4-x^2+y^2}$.

```
{{x[s] → s + C[1], y[s] → s + C[2], v[s] → 1/(-s - C[3])}} /.
{{C[1] → t, C[2] → -t, C[3] → -1/t^2}}
```

$\{\{\{x[s] \to s+t, y[s] \to s-t, v[s] \to \frac{1}{-s+\frac{1}{t^2}}\}\}\}$

```
Solve[{x==s + t, y==s - t}, {s, t}]
```

$\{\{s \to \frac{x+y}{2}, t \to \frac{x-y}{2}\}\}$

```
{v==1/(-s + 1/t)} /. {s → (x+y)/2, t → (x-y)/2}
Simplify[{v == 1/(1/2 (-x - y) + 2/(x - y))}]
```

$\{v == \frac{2(x-y)}{4-x^2+y^2}\}$

♦

8.2 Second order partial differential equations

We now consider several second-order equations of mathematical physics. The first will be the heat equation, otherwise known as the diffusion equation. In one space dimension it has the form

$$K\frac{\partial^2 u}{\partial x^2} = \frac{\partial u}{\partial t}.$$

Our next equation is the second-order wave equation. It differs from the first order version in that it has two traveling disturbances, each traveling in an opposite direction from the other. The second-order wave equation has the form

$$\frac{\partial^2 u(x,t)}{\partial x^2} + \frac{1}{a^2(x,t)}\frac{\partial^2 u(x,t)}{\partial t^2} = 0,$$

where $a(x,t)$ is the wave speed. Let us consider the simple case where $a = 1$[3].

The Cauchy problem in this case is to find a solution which satisfies the following initial data

$$u(x,0) = f(x) \text{ and } \frac{\partial u(x,t)}{\partial t}(x,0) = g(x),$$

where f and g are arbitrary prescribed functions. We shall also discuss the potential equation

$$\frac{\partial^2 u}{\partial x^2} + \frac{\partial^2 u}{\partial y^2} = 0.$$

8.2.1 Heat equation

In this subsection we consider the initial-boundary-value problem associated with the one-dimentional, heat equation

$$K\frac{\partial^2 u}{\partial x^2} = \frac{\partial u}{\partial t}, \quad 0 < x < L, \quad t > 0, \tag{8.7}$$

where K is the diffusivity and is measured in the units $[\text{cm}]^2[\text{sec}]^{-1}$. The solution method we use is based on the separation of variables technique, which allows the solution to be represented as a Fourier series. We take as an initial condition $u(x,0) = f(x)$, and we assume a *Dirichlet* boundary condition prescribed on the two end points of the interval $[0, L]$. For the case at hand, they are assumed to be homogeneous, namely $u(0,t) = 0$, $u(L,t) = 0$.

[3]This can always be accomplished by a stretching of the coordinates.

For the interval $[0,l] \subset \mathbf{R}^1$, the Neumann conditions require specifying the normal derivatives at the boundary, namely in the case of an interval, the homogeneous conditions will be $\frac{\partial u(x,t)}{\partial x}$ at $x = \{0, L\}$

The solution to the homogeneous Dirichlet, Initial Value Problem (DVP) may be written in the form

$$u(x,t) = \sum_{k=1}^{\infty} c_k e^{-(k^2\pi^2 Kt/L^2)} \sin\left(\frac{k\pi x}{L}\right). \tag{8.8}$$

If we have the Initial Condition $u(x,0) = f(x)$, then the Fourier coefficients are calculated using the orthogonal properties of the $\sin((\frac{k\pi x}{L})$, i.e.

$$\int_0^L \sin\left(\frac{k\pi x}{L}\right) \sin\left(\frac{j\pi x}{L}\right) dx = \frac{L}{2}\delta_{jk};$$

hence,

$$c_k = \frac{2}{L}\int_0^L f(x) \sin\left(\frac{k\pi x}{L}\right) dx. \tag{8.9}$$

As an approximate solution we consider the truncated version of the above series

$$u(x,t) = \sum_{k=1}^{N} c_k e^{\left(\frac{-k^2\pi^2 Kt}{L^2}\right)} \sin\left(\frac{k\pi x}{L}\right). \tag{8.10}$$

We write this representation as a procedure, where f is the initial data, N the number of terms in the truncated solution, and T the length of the time interval $[0,T]$ over which the solution is sought.
On the other hand, for the Neumann boundary condition we seek a solution which has vanishing derivatives at the end points of the rod. That is $\partial_x u(0,t) = \partial_x u(L,t) = 0$; hence, we seek a series solution in the form

$$u(x,t) = \sum_{1}^{\infty} b_k\, e^{-(k^2\pi^2 Kt/L^2)} \cos\left(\frac{k\pi x}{L}\right). \tag{8.11}$$

The Fourier coefficients b_k can then be computed as

$$b_k = \frac{2}{L}\int_0^L f(x) \cos\left(\frac{k\pi x}{L}\right) dx, \tag{8.12}$$

where as before $u(x,0) = f(x)$ is the initial data.

Mathematica session 8.4. *This program uses the separation of variables method to plot the solution of the heat equation in a rod, where the ends of the rod are kept at 0 degrees. The solution is truncated after N terms.*

```
heatEquationDBC[f_, α_, L_, T_, N_]:=
Module[{h}, For[k = 1, k<=N, k++, c[k] = (2/L)*
Integrate[f * Sin[x * k * π/L], {x, 0, L}]];
h = Sum[c[k] * Exp[-k^2 * π * α * t/L^2] * Sin[k * π * x/L], {k, 1, N}];
Plot3D[h, {x, 0, L}, {t, 0, T}]]
```

For expository purposes we take the the **Module** *apart. First we show how the collection of Fourier coefficients are found using the initial data* $u(x, 0) = x^3$. *Then we use this information to define a function* $h := h[x, t]$. *Then we set the coefficients* $\{L, \alpha\} = \{2, 0.5\}$. *Finally we print the solution over the prescribed* $[0 \leq x \leq 2] \times [0 \leq t \leq 6]$ *range using* **Plot3D**.

```
For[k = 1, k<=10, k++, Print[c[k] = (2/L) * Integrate[(x^3) * Sin[x * k * π/2], {x, 0, 2}]]]
```

$$\frac{32(-6+\pi^2)}{L\pi^3}$$
$$\frac{2(12-8\pi^2)}{L\pi^3}$$
$$\frac{32(-2+3\pi^2)}{9L\pi^3}$$
$$\frac{3-8\pi^2}{L\pi^3}$$
$$\frac{32(-6+25\pi^2)}{125L\pi^3}$$
$$\frac{2(4-24\pi^2)}{9L\pi^3}$$
$$\frac{32(-6+49\pi^2)}{343L\pi^3}$$
$$\frac{2\left(\frac{3}{16\pi^3}-\frac{2}{\pi}\right)}{L}$$
$$\frac{32(-2+27\pi^2)}{243L\pi^3}$$
$$-\frac{8(-3+50\pi^2)}{125L\pi^3}$$

```
h = Sum[(2/L) * c[k] * Exp[-k^2 * π * α * t/L^2] * Sin[k * π * x/L], {k, 0, 10}]
```

$$\frac{64e^{-\frac{\pi t\alpha}{L^2}}(-6+\pi^2)Sin[\frac{\pi x}{L}]}{L^2\pi^3} + \frac{4e^{-\frac{4\pi t\alpha}{L^2}}(12-8\pi^2)Sin[\frac{2\pi x}{L}]}{L^2\pi^3} + \frac{64e^{-\frac{9\pi t\alpha}{L^2}}(-2+3\pi^2)Sin[\frac{3\pi x}{L}]}{9L^2\pi^3}$$
$$+ \frac{2e^{-\frac{16\pi t\alpha}{L^2}}(3-8\pi^2)Sin[\frac{4\pi x}{L}]}{L^2\pi^3} + \frac{64e^{-\frac{25\pi t\alpha}{L^2}}(-6+25\pi^2)Sin[\frac{5\pi x}{L}]}{125L^2\pi^3} + \frac{4e^{-\frac{36\pi t\alpha}{L^2}}(4-24\pi^2)Sin[\frac{6\pi x}{L}]}{9L^2\pi^3}$$

$$+\frac{64e^{-\frac{49\pi t\alpha}{L^2}}\left(-6+49\pi^2\right)Sin\left[\frac{7\pi x}{L}\right]}{343L^2\pi^3}+\frac{4e^{-\frac{64\pi t\alpha}{L^2}}\left(\frac{3}{16\pi^3}-\frac{2}{\pi}\right)Sin\left[\frac{8\pi x}{L}\right]}{L^2}+\frac{64e^{-\frac{81\pi t\alpha}{L^2}}\left(-2+27\pi^2\right)Sin\left[\frac{9\pi x}{L}\right]}{243L^2\pi^3}$$
$$-\frac{16e^{-\frac{100\pi t\alpha}{L^2}}\left(-3+50\pi^2\right)Sin\left[\frac{10\pi x}{L}\right]}{125L^2\pi^3}$$

h/.{L → 2, α->.5}

$$\frac{16e^{-0.392699t}\left(-6+\pi^2\right)Sin\left[\frac{\pi x}{2}\right]}{\pi^3}+\frac{e^{-1.5708t}\left(12-8\pi^2\right)Sin[\pi x]}{\pi^3}+\frac{16e^{-3.53429t}\left(-2+3\pi^2\right)Sin\left[\frac{3\pi x}{2}\right]}{9\pi^3}$$
$$+\frac{e^{-6.28319t}\left(3-8\pi^2\right)Sin[2\pi x]}{2\pi^3}+\frac{16e^{-9.81748t}\left(-6+25\pi^2\right)Sin\left[\frac{5\pi x}{2}\right]}{125\pi^3}+\frac{e^{-14.1372t}\left(4-24\pi^2\right)Sin[3\pi x]}{9\pi^3}$$
$$+\frac{16e^{-19.2423t}\left(-6+49\pi^2\right)Sin\left[\frac{7\pi x}{2}\right]}{343\pi^3}+e^{-25.1327t}\left(\frac{3}{16\pi^3}-\frac{2}{\pi}\right)Sin[4\pi x]$$
$$+\frac{16e^{-31.8086t}\left(-2+27\pi^2\right)Sin\left[\frac{9\pi x}{2}\right]}{243\pi^3}-\frac{4e^{-39.2699t}\left(-3+50\pi^2\right)Sin[5\pi x]}{125\pi^3}$$

Here are the graphics for the above calculation. Please note that this uses only terms up to $n = 10$ *in the Fourier expansion of the solution.*

Plot3D[%, {x, 0, 2}, {t, 0, 6}]

A plot of the solution is given in Figure 8.5.

We now use the Module **DBC** *to produce the graphics, using the same parameter evaluations. A plot of the solution is given in Figure 8.6. Note that the graphics are identical.*

heatEquationDBC[x^3, 0.5, 2, 6, 10]

We now increase the number of Fourier terms to order 40 *and we have a more refined graphic. A plot of this solution is given in Figure 8.7.*

heatEquationDBC[x^3, 0.5, 2, 6, 40]

Finally we increase the order of Fourier coefficients to 100. *This gives very smooth results and in future calculations we fix the order at* 100. *A plot of this solution is given in Figure 8.8.*

heatEquationDBC[x^3, 0.5, 2, 6, 100]

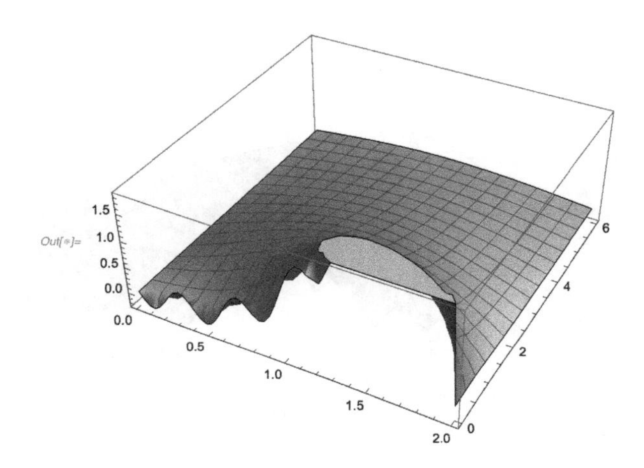

FIGURE 8.5: A plot of the truncated solution for the heat equation boundary value problem using 10 terms.

Next let us solve the heat equation using the initial data $u(x, 0) = 10x(2 - x)$ *and our command* **DCB**. *A plot of the solution is given in Figure 8.9.*

***heatEquationDBC*[10 ∗ *x* ∗ (2 − *x*), 0.5, 2, 10, 100]**

Next we try a step function, centered about $x = 1$, *as our initial data. A plot of the solution is given in Figure 8.10.*

***heatEquationDBC*[*UnitStep*[*x* − 0.9] − *UnitStep*[*x* − 1.1], 0.5, 2, 10, 100]**

Notice that the heat distribution tends to level out to zero because heat flows out of the rod on each end. We continue to take N as 100 as this seems sufficiently large to give good comparison with numerical computations.

♦

We next consider the Neumann using insulated boundary conditions, i.e. we seek the temperature distribution of a rod of length L, with insulated ends. The Fourier

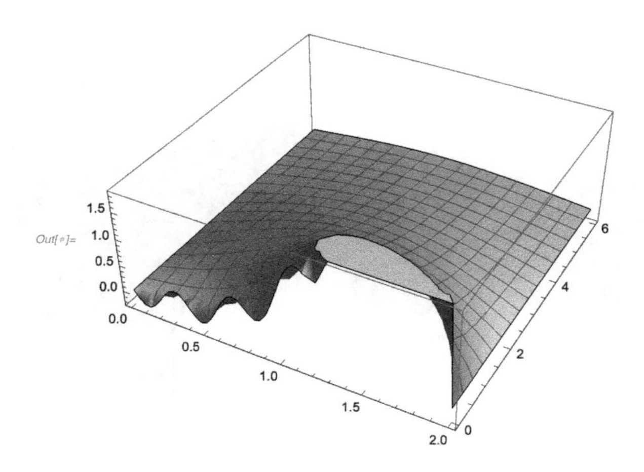

FIGURE 8.6: A plot of the solution for the heat equation boundary value problem using the Module **DBC**.

series solution for the Neumann problem is given below.

$$u(x,t) = \frac{b_0}{2} + \sum_{n=1}^{\infty} b_n exp\left(-\frac{n^2\pi^2 K\,t}{L^2}\right) \cos\left(\frac{n\pi\,x}{L}\right),$$

where the Fourier coefficients are

$$b_n = \frac{2}{L}\int_0^L u(x,0)\cos\left(\frac{n\pi\,x}{L}\right)\,dx.$$

Mathematica session 8.5. *This program uses the separation of variables method to plot the solution of the heat equation in a rod, where the ends of the rod are insulated. Notice that as $t \to \infty$ the value of the solution tends to the average temperature in the rod. We truncate the solution after N terms.*

```
heatEquationNBC[f_, α_, L_, T_, N_]:=
```

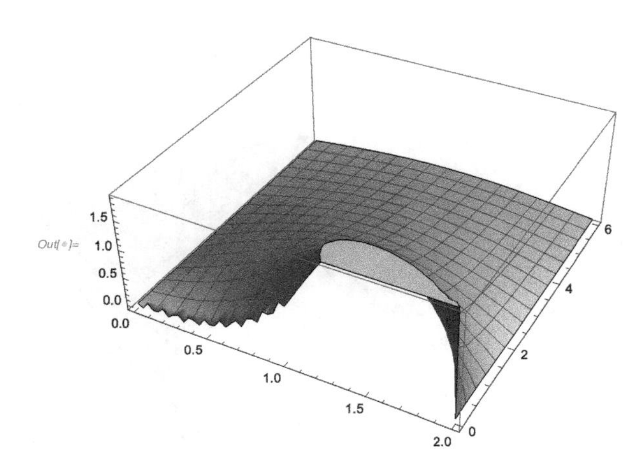

FIGURE 8.7: A plot of the truncated solution for the heat equation boundary value problem using 40 terms.

```
Module[{h}, For[k = 0, k<=N, k++, b[k] = (2/L)*
Integrate[f * Cos[x * k * π/L], {x, 0, L}]];
h = Sum[c[k] * Exp[-k^2 * π * α * t/L^2] * Cos[k * π * x/L], {k, 0, N}];
Plot3D[h, {x, 0, L}, {t, 0, T}]]
```

The Neumann boundary value problem with initial data $u(x, 0) = x^3$ *may be found using the Module* **heatEquationNBC**.

```
heatEquationNBC[x^3, 0.5, 2, 6, 100]
```

This solution, shown in Figure 8.11 seems quite smooth so we shall continue to use 100 terms for further calculations.

```
heatEquationNBC[x * (2 - x), 0.5, 2, 10, 100]
```

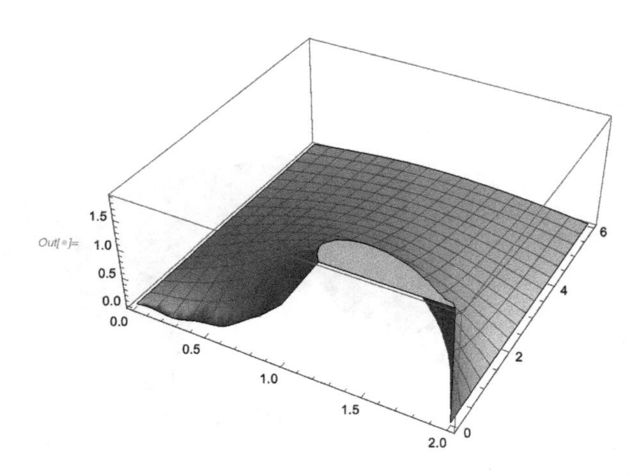

FIGURE 8.8: A plot of the truncated solution for the heat equation boundary value problem using 100 terms.

A plot of the solution is shown in Figure 8.12.

♦

8.2.2 The Laplace equation

In $\mathbb{R}^3$, using Cartesian coordinates, the Laplace equation may be written as

$$\triangle u := \frac{\partial^2 u}{\partial x^2} + \frac{\partial^2 u}{\partial y^2} + \frac{\partial^2 u}{\partial z^2} \tag{8.13}$$

It can easily be checked that Laplace's equation is invariant under translations and rotations in a Cartesian coordinate system. The reader should check this. In $\mathbb{R}^2$ the Laplace equation takes the form

$$\frac{\partial^2 u}{\partial x^2} + \frac{\partial^2 u}{\partial y^2} = 0$$

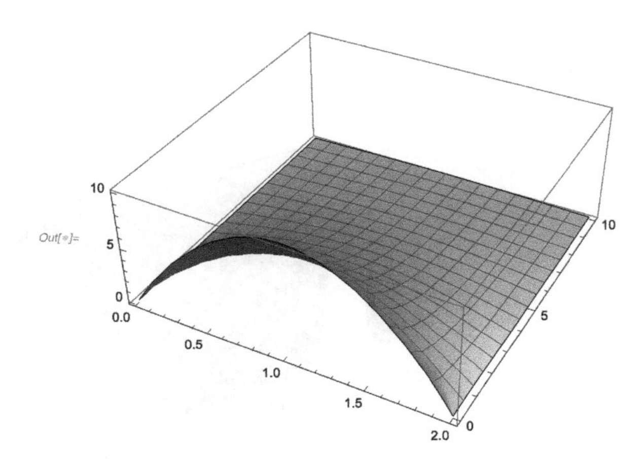

FIGURE 8.9: A plot of the solution for the heat equation boundary value problem using the initial condition $u(x,0) = 10x(2-x)$.

and in polar coordinates this becomes

$$\frac{\partial^2 u}{\partial r^2} + \frac{1}{r}\frac{\partial u}{\partial r} + \frac{1}{r^2}\frac{\partial^2 u}{\partial \theta^2} = 0. \tag{8.14}$$

This suggests, domains that are disks, we might seek solutions in the form of the series

$$u(r,\theta) = \frac{a_0}{2} + \sum_{n=0}^{\infty} r^n \left(a_n \cos(n\theta) + b_n \sin(n\theta)\right). \tag{8.15}$$

as the individual terms in the series are solutions of Laplace's equation.

Project: This suggests that the Laplace equation for a disk of radius r_0 with boundary data $u(r_0,\theta) = f(\theta)$ might be solved by computing the coefficients a_n, b_n from the formulae

$$a_n = \frac{1}{r_o^n \pi} \int_0^{2\pi} f(\theta)\cos(n\theta)\,d\theta,$$
$$b_n = \frac{1}{r_o^n \pi} \int_0^{2\pi} f(\theta)\sin(n\theta)\,d\theta.$$

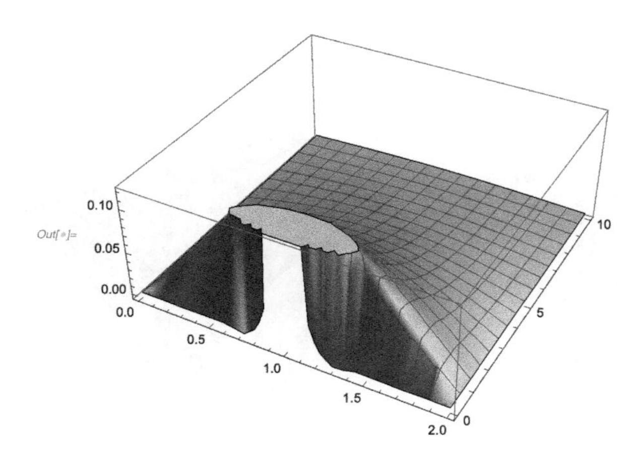

FIGURE 8.10: The initial condition is UnitStep[x-0.9] -UnitStep[x-1.1]. We have ssed 100 terms in the series solution.

Exercise 8.1. *Write a procedure which computes the Fourier coefficient a_n, b_n using s the approach suggested by the expansion (8.15).*

It was mentioned in Chapter 6, (Exercise 6.104) that in an orthogonal coordinate system the Laplacian operator takes the form

$$\triangle u := \frac{1}{h_1h_2h_3}\left[\frac{\partial}{\partial q_1}\left(\frac{h_2h_3}{h_1}\frac{\partial u}{\partial q_1}\right) + \frac{\partial}{\partial q_2}\left(\frac{h_3h_1}{h_1}\frac{\partial u}{\partial q_2}\right) + \frac{\partial}{\partial q_3}\left(\frac{h_1h_2}{h_3}\frac{\partial u}{\partial q_1}\right)\right] \tag{8.16}$$

where

$$h_i^2 = \left(\frac{\partial x}{\partial q_i}^2 + \frac{\partial y}{\partial q_i}^2 + \frac{\partial z}{\partial q_i}^2\right), \quad i = 1, 2, 3.$$

In an orthogonal system in $\mathbb{R}^2$ the Laplace equation

$$\frac{1}{h_1h_2}\left[\frac{\partial}{\partial q_1}\left(\frac{h_2}{h_1}\frac{\partial u}{\partial q_1}\right) + \frac{\partial}{\partial q_2}\left(\frac{h_1}{h_2}\frac{\partial u}{\partial q_2}\right)\right] = 0,$$

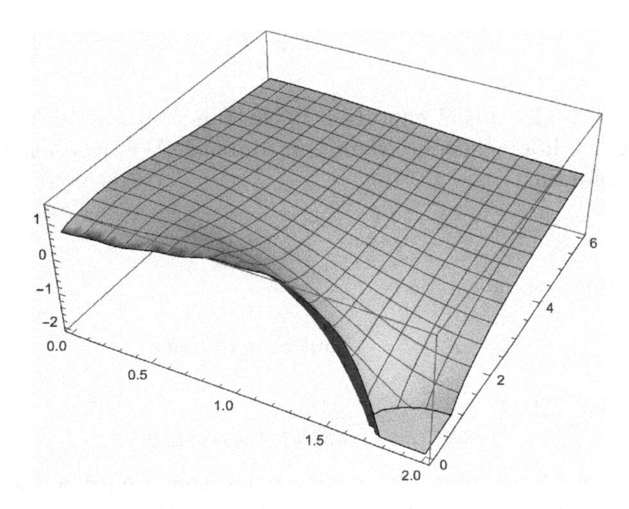

FIGURE 8.11: The initial condition for this homogeneous, Neumann problem is $u(x, 0) = x^3$. 100 terms were used for the series solution.

where

$$h_i^2 = \left(\frac{\partial x}{\partial q_i}^2 + \frac{\partial y}{\partial q_i}^2 \right), \quad i = 1, 2..$$

For example, as seen in Chapter(5), in cylindrical coordinates we have

$$q_1 = r, \quad q_2 = \theta, \quad \text{and } q_3 = z.$$

In cylindrical coordinates the Laplace equation is rewritten as

$$\triangle u = \frac{1}{r}\frac{\partial}{\partial r}\left(r\frac{\partial u}{\partial r} \right) + \frac{1}{r^2}\frac{\partial^2 u}{\partial \theta^2} + \frac{\partial^2 u}{\partial z^2}. \tag{8.17}$$

We seek separated solution of (8.17) in the form $u(r, \theta, z) = R(r)\Theta(\theta)Z(z)$. It is easy to show that we may choose separation coefficients, ℓ, μ, such that

$$\begin{aligned} r^2 R'' + rR' - \left[(\mu^2 r^2 + \ell^2) + \ell \right] R &= 0, \\ \Theta'' + \ell^2 \Theta &= 0, \\ Z'' + \mu^2 Z &= 0. \end{aligned} \tag{8.18}$$

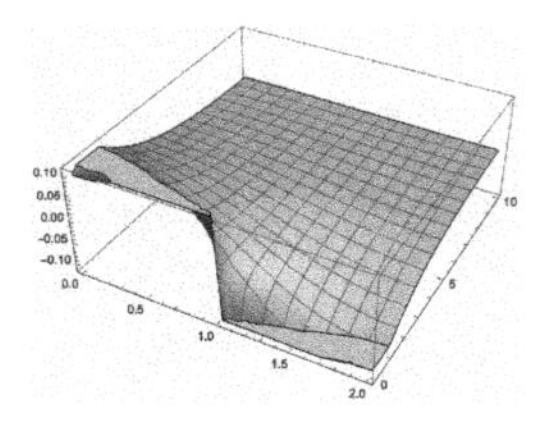

FIGURE 8.12: The initial condition is the function $u(x, 0) = x(2 - x)$. We are solving the homogeneous Neumann problem. We have used 100 terms in the series solution.

The Z-equation has the solutions

$$Z(z) = c_1 \sin(\mu z) + c_2 \cos(\mu z) \tag{8.19}$$

and the Θ-equation

$$\Theta(\theta) = b_1 \sin(\ell\theta) + b_2 \cosh(\ell\theta).$$

In order for the solution of the Θ equation to be single-valued on the interval $[0, 2\pi]$ we must require that ℓ is an integer. The radial equation of the system (8.18) upon substituting $\zeta = ikr$ becomes

$$\zeta^2 \frac{dR}{d\zeta^2} + \zeta \frac{dR}{d\zeta} + (\zeta^2 - \ell^2)R = 0. \tag{8.20}$$

The solutions to 8.20 are known as cylindrical functions and denoted by $Z_\ell(\zeta)$. For cylinders which are of bounded radius, we choose the cylindrical function $J_\ell(\zeta)$, a Bessel function of the first kind and order ℓ.[4] Hence, the radial function takes the form in the bounded cylinder

$$R(r) = J_\ell\,(ikr)\,, \tag{8.21}$$

Using the superposition principal, solutions to the Laplace equation, in a cylinder, may be written in the form

$$u(r, \theta, z) = \sum_{\ell=-\infty,\ell\neq 0}^{\infty} \sum_{k=-\infty,k\neq 0}^{\infty} (a_{k\ell})\, J_\ell\,(ikr)\, e^{\pm i(kz \pm \ell\theta)} + a_{k0} J_0(ikr) e^{ikz}. \tag{8.22}$$

If the region the equation is defined over is the cylinder

$$D := \{(r, \theta, z) : r \leq a, 0 \leq \theta < 2\pi, 0 \leq z \leq h\}\,,$$

[4]In the case where the domain is the exterior of the cylinder, the Bessel function of the second kind is used $I(_\ell\zeta)$.

a natural boundary condition would be to define U on the boundary of this cylinder, say

$$u = \begin{cases} f_T(r,\theta), & \text{on } z = L \\ f_B(r,\theta), & \text{on } z = 0 \\ f_l(z,\theta)), & \text{on } r = a \end{cases} .$$

As a special example consider the case where $u = 0$ on $z = 0, L$. In spherical coordinates the we have

$$q_1 = \rho, \quad q_2 = \theta, \quad q_3 = \varphi,$$

where the relationship between the Cartesian coordinates and the spherical coordinates is given by

$$x = \rho\cos(\theta)\sin(\varphi)\cos(\theta), \quad y = \rho\sin(\theta)\sin(\varphi)\cos(\theta), \quad z = \rho\cos(\theta)$$

and where $\rho =, \sqrt(x^2 + y^2 + z^2)$. The Laplace equation in spherical coordinates is given by

$$\triangle u = \frac{\partial}{\partial r}\left(r^2\frac{\partial u}{\partial r}\right) + \frac{1}{r^2\sin(\theta)}\frac{\partial}{\partial\theta}\left(\frac{\partial u}{\partial\theta}\right) + \frac{1}{r^2\sin^2(\theta)}\frac{\partial^2 u}{\partial\varphi^2} = 0. \tag{8.23}$$

Writing this out we have

$$\frac{1}{r^2}\frac{\partial}{\partial r}\left(r^2\frac{\partial u}{\partial r}\right) + \frac{1}{r^2\sin(\theta)}\frac{\partial}{\partial\theta}\left(\sin(\theta)\frac{\partial u}{\partial\theta}\right) + \frac{1}{r^2\sin^2(\theta)}\frac{\partial^2 u}{\partial\varphi^2} = 0$$

We now try to separate the variables by making the *Ansatz* that $u = R(r)\Theta(\theta)\Phi(\varphi)$, but if we multiply the equation by $r^2\sin^2(\theta)$ the term $\frac{\partial^2 u}{\partial\varphi^2}$ is seen to be set to a constant. We choose then

$$\frac{\partial^2 u}{\partial\varphi^2} = -m^2,$$

which permits the solutions

$$\Phi = e^{\pm m\varphi}.$$

We are left with the equation

$$\frac{\sin^2\theta}{r}\frac{d}{dr}\left(r^2\frac{dR}{dr}\right) + \frac{\sin(\theta)}{\Theta}\frac{d}{d\theta}\left(\sin(\theta\frac{du}{d\theta}\right) = m^2. \tag{8.24}$$

Hence, we obtain the system

$$\frac{1}{\sin(\theta)}\frac{d}{d\theta}\left(\sin(\theta)\frac{d\Theta}{d\theta}\right) - \left[c - \frac{m^2}{\sin^2(\theta}\right]\Theta = 0,$$

$$\frac{1}{dr}\left(r^2\frac{dR}{dr}\right) - cR = 0. \tag{8.25}$$

It will prove convenient to write the constant c as $\ell(\ell+1)$. Then the solutions of the radial equation become

$$R(r) = \left\{ \begin{array}{c} r^{\ell} \\ r^{-(l+1)} \end{array} \right\}$$

We then make the substitution $\xi = \cos(\theta)$ in the θ- Equation(8.25). This leads to

$$(1-\xi^2)\frac{d^2\Theta}{d\xi^2} - 2\xi\frac{d\Theta}{d\,xi} + \left(\ell(\ell+1)\,\frac{m^2}{1-\xi^2}\right)\Theta = 0. \tag{8.26}$$

The solutions $\Theta(\xi)$ are known as the associated Legendre function, which are denoted as

$$\Theta(\theta) = P_{\ell}^{m}(\cos(\theta), \quad m = 0, \pm 1, \pm 2, \ldots \pm n.$$

We obtain solutions of the Laplace equation in the form

$$\left\{ \begin{array}{c} r^{\ell} \\ r^{-(l+1)} \end{array} \right\} Y_{nm}(\theta,\varphi) = P_n^{(m)}(\cos(m\theta) \left\{ \begin{array}{c} \cos(m\varphi) \\ sin(m\varphi) \end{array} \right\}, \quad m = 0, \pm 1, \pm 2, \cdots \pm n.$$

If we are considering only bounded solutions in a sphere $\mathcal{B}$ then we wish to only consider non-negative powers of r; hence we seek solutions which may be expressed in terms of a series

$$u(r,\theta,\varphi) = \sum_{n=0}^{\infty}\sum_{m=-n}^{n} r^n P_n^{(m)}\,[A_{nm}cos(m\varphi) + B_{nm}\sin(m\varphi)] \tag{8.27}$$

8.2.3 MATHEMATICA package for solving Laplace's equation

In this session we want to use MATHEMATICA to solve Laplace's equation. MATHEMATICA has a built in package for solving boundary value problems for Laplace's equation. We will not give an illustration of this; however, we will develop our own software for doing this as it is more transparent for the student learning the mathematics behind the software.

Mathematica session 8.6.

```
leqn = Laplacian[u[x, y], {x, y}] == 0;

Ω = Rectangle[{0, 0}, {1, 2}];

dcond1 = DirichletCondition[u[x, y] ==
Piecewise[{{UnitTriangle[2x − 1], y == 0||y == 2}}, 0], True];

sol = DSolveValue[{leqn, dcond1}, u[x, y], {x, y} ∈ Ω]//FullSimplify
```

$$\sum_{K[1]=1}^{\infty} \frac{8 Cosh[\pi(-1+y)K[1]] Sech[\pi K[1]] Sin\left[\frac{1}{2}\pi K[1]\right] Sin[\pi x K[1]]}{\pi^2 K[1]^2}$$

asol1 = sol/.{∞ → 300}//Activate;

Plot3D[asol1//Evaluate, {x, y} ∈ Ω, PlotRange → All, PlotTheme → "Business"]

A plot of the solution is given in Figure 8.13.

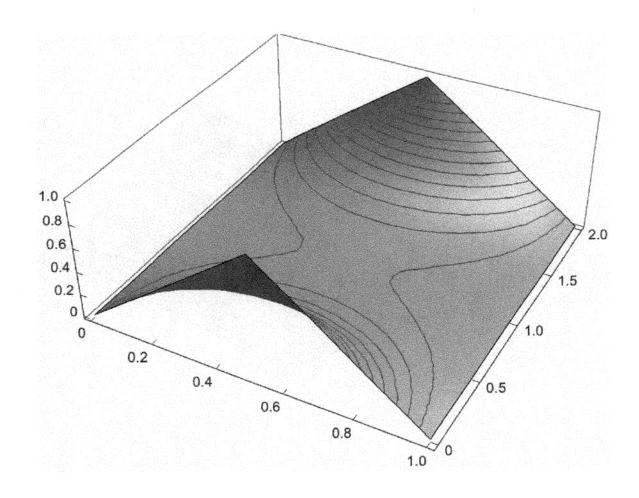

FIGURE 8.13: A plot of the solution to Laplace's equation on the rectangle $[0,0] \times [1,2]$ with **DirichletCondition[u[x, y] == Piecewise[{{UnitTriangle[2x − 1], y == 0||y == 2}}, 0], True];**.

dcond2 = DirichletCondition[u[x, y] ==

Piecewise[{{UnitTriangle[2x − 1], y == 0}}, 0], True];

sol2 = DSolveValue[{leqn, dcond2}, u[x, y], {x, y} ∈ Ω]//FullSimplify

$$\sum_{K[1]=1}^{\infty} -\frac{8\,Csch[2\pi K[1]]\,Sin\left[\frac{1}{2}\pi K[1]\right]Sin[\pi x K[1]]\,Sinh[\pi(-2+y)K[1]]}{\pi^2 K[1]^2}$$

asol2 = sol2/.{∞ → 300}//Activate;

Plot3D[asol2//Evaluate, {x, y} ∈ Ω, PlotRange → All, PlotTheme → "Business"]

A plot of the solution is given in Figure 8.14.

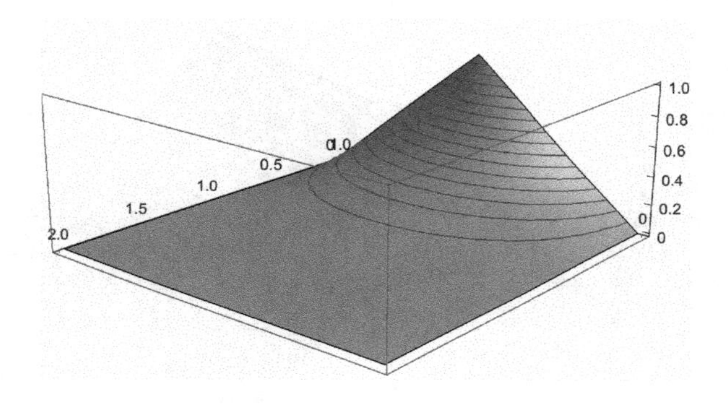

FIGURE 8.14: A plot of the solution to Laplace's equation on the rectangle $[0,0] \times [1,2]$ with **DirichletCondition[u[x, y] == Piecewise[{{UnitTriangle[2x − 1], y == 0}}, 0], True];**.

dcond3 = DirichletCondition[u[x, y] ==

Piecewise[{{UnitStep[1/2 − x], y == 0}}, 0], True];

sol3 = DSolveValue[{leqn, dcond3}, u[x, y], {x, y} ∈ Ω]//FullSimplify

$$\sum_{K[1]=1}^{\infty} -\frac{4Csch[2\pi K[1]]Sin\left[\frac{1}{4}\pi K[1]\right]^2 Sin[\pi x K[1]]Sinh[\pi(-2+y)K[1]]}{\pi K[1]}$$

asol3 = sol3/.{∞ → 300}//Activate;

The command lines appear **above** *the graphics to which they pertain. Do not confuse* **commands** *with* **captions** *here or elsewhere.*
Plot3D[asol3//Evaluate, {x, y} ∈ Ω, PlotRange → All, PlotTheme → "Business"]
A plot of the solution is given in Figure 8.15.

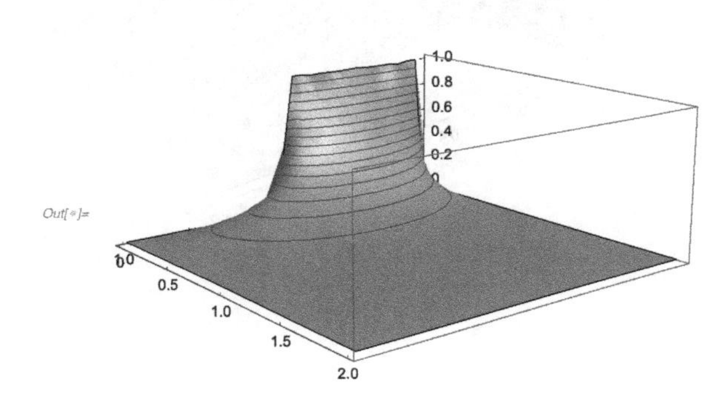

FIGURE 8.15: A plot of the solution to Laplace's equation on the rectangle $[0,0] \times [1,2]$ with **DirichletCondition[u[x, y] == Piecewise[{{UnitStep[1/2 − x], y == 0}}, 0], True];.**

dcond4 = DirichletCondition[u[x, y] ==
Piecewise[{{UnitStep[1/2 − x], y == 0||y==2}}, 0], True];
sol4 = DSolveValue[{leqn, dcond4}, u[x, y], {x, y} ∈ Ω]//FullSimplify

$$\sum_{K[1]=1}^{\infty} \frac{4Cosh[\pi(-1+y)K[1]]Sech[\pi K[1]]Sin\left[\frac{1}{4}\pi K[1]\right]^2 Sin[\pi x K[1]]}{\pi K[1]}$$

asol4 = sol4/.{∞ → 300}//Activate;

Plot3D[asol4//Evaluate, {x, y} ∈ Ω, PlotRange → All, PlotTheme → "Business"]
A plot of the solution is given in Figure 8.16.

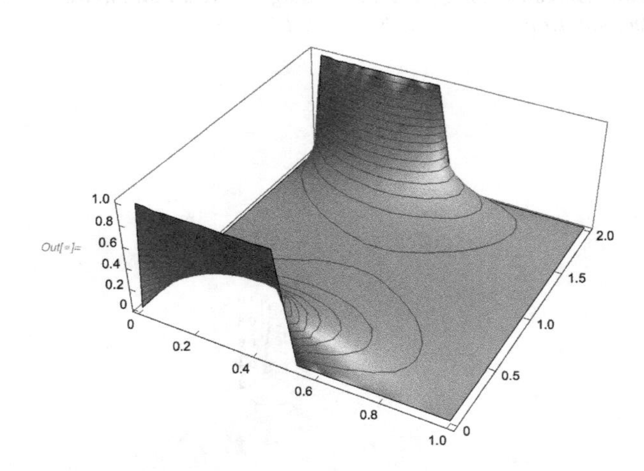

FIGURE 8.16: A plot of the solution to Laplace's equation on the rectangle $[0, 0] \times [1, 2]$ with **DirichletCondition[u[x, y] == Piecewise[{{UnitStep[1/2 − x], y == 0||y==2}}, 0], True];.**

♦

Exercises

Exercise 8.2. *Use a Taylor series expansion about $x = 0$ to find a series solution of the Legendre equation*

$$(1 - x^2)y'' - 2xy' + \ell(\ell - 1) = 0 \tag{8.28}$$

i.e. consider solutions which may be written in the form

$$y = \sum_{n=0}^{\infty} a_n x^n$$

Exercise 8.3. *Continue with the above approach to find polynomial solutions of the Legendre equation above. Hint: Try various values of the parameter ℓ by considering both positive and negative integers for ℓ.*

Exercise 8.4. *If $P_\ell(x)$ is a solution of the Legendre equation (8.28) show that*

$$y(x) = (1-x^2)^{\frac{m}{2}} \frac{d^m}{d\,x^m} P_\ell(x)$$

is a solution of the associated Legendre functions equation

$$(1-x^2)y'' - 2xy' + \left[\ell(\ell+1) - \frac{m^2}{1-x^2}\right] y = 0.$$

Suppose we wish to solve the partial differential equation with the boundary conditions $u(x,0) = 0$, $u(x,b) = 0$, $u(0,y) = 0$, and $u(a,y) = f(y)$. Then we successively substitute these boundary conditions in and solve for the coefficients. We begin with the condition at $y = 0$ We wish to compute the Fourier coefficients to account for the boundary condition $u(a,y) = f(y)$. We make use of the orthogonality properties of the trigonometric functions to do this. To this end let us declare that n and k are integers. or the sentence below? We are now in the position to write a procedure for computing the solution to the Laplace equation in the disk of radius a where the boundary data is of the form $u(a,\theta) = f(\theta)$.

8.2.4 The wave equation

We shall use d'Alembert's method to solve the string equation. The vibrating string can be modeled as a one–dimensional, second-order, wave equation, namely

$$\frac{1}{a^2}\frac{\partial^2 u(x,t)}{\partial t^2} = f(x,t), \quad -\infty < x < \infty. \tag{8.29}$$

For $f(x,y) = 0$ any function satisfying this equation may be written in the form

$$u(x,t) := p(x-at) + q(x+at) \tag{8.30}$$

and alternatively any function of this form is also a solution of the string equation. This representation is known as d'Alembert's solution. Let us now determine this solution from the Cauchy data

$$u(x,0) = f(x), \quad u_t(x,0) = g(x), \quad -\infty < x < \infty.$$

This leads to the pair of equations

$$p(x) + q(x) = f(x) \quad \text{and} \quad -ap'(x) + aq'(x) = g(x).$$

Solving these equations for p' and q' leads to

$$q'(x) = \frac{1}{2}f'(x) + \frac{1}{2a}g(x), \quad p'(x) = \frac{1}{2}f'(x) - \frac{1}{2a}g(x),$$

and, hence by integrating between 0 and $x-at$ and between 0 and $x+at$ we arrive at

$$u(x,t) = \frac{f(x-at) + f(x+at)}{2} + \frac{1}{2a}\int_{x-at}^{x+at} g(\xi)\,d\xi \tag{8.31}$$

If the string is fixed at $x = 0$, say $u(0,t) = 0$, then the solution can also be provided by (8.31), provided the functions $f(x)$ and $g(x)$ are continued as odd functions to the entire x-axis. **Why**? Likewise if at $x = 0$, $u_t(0,t) = 0$ then we may also use the representation (8.31), provided the functions $f(x)$ and $g(x)$ are continued as even, differentiable functions on the entire x-axis.

8.2.5 The Fourier method for the vibrating string

We consider the Fourier expansion method for the boundary-initial-value problem for the string equation. We seek a solution of the boundary-initial value problem for the wave equation

$$\begin{array}{ll} \frac{\partial^2 u(x,t)}{\partial x^2} - \frac{1}{a^2}\frac{\partial^2 u(x,t)}{\partial t^2} & = 0 \\ u(0,t) = 0 & u(L,t) = 0 \\ u(x,0) = f(x) & \frac{\partial u}{\partial t}(x,0) = g(x). \end{array}$$

As we found in Section 1 of this chapter the separated solutions

$$u(x,t) = X(x)T(t)$$

satisfy the equations.

$$\begin{array}{l} \frac{d^2 X(x)}{dx^2} = \lambda^2 X(x) \\ \frac{d^2 T(t)}{dt^2} = \lambda^2 a^2 T(y) = 0. \end{array}$$

This suggests that we seek a solution as an infinite series of terms

$$u(x,t) = \sum X_n(x)T_n(t),$$

where the functions $X_n(x)$, $T_n(t)$ must be of the form

$$\begin{array}{rcl} X_n(x) & = & \alpha_n \sin(x\lambda_n) + \beta_n \cos(x\lambda_n) \\ T_n(t) & = & \delta_n \sin(a\,t\lambda_n) + \gamma_n \cos(a\,t\lambda_n) \end{array}.$$

The numbers λ_n are the eigenvalues and are chosen so that $X_n(x)$ satisfy the prescribed boundary conditions. It turns out that for the present case the eigenvalues may seem to be given by $\lambda_n = \frac{n\pi}{L}, n = 1, 2, \ldots$. The solution is found in the form

$$u(x,t) = \sum_{n=1}^{\infty} \left(A_n \cos\left(\frac{n\pi\,a\,t}{L}\right) + B_n \sin\left(\frac{n\pi\,a\,t}{L}\right)\right) \sin\left(\frac{n\pi\,x}{L}\right).$$

Based on this formula we can write a program that gives an approximate solution to the string equation by truncating the series solution after N terms.

Mathematica session 8.7. *This program uses the separation of variables method to plot the displacement of a string during oscillatory motion. The displacement of the string at its end points is zero. The solution is truncated.*

```
stringEquationBIC[f_, g_, a_, L_, T_, N_]:=
Module[{h}, For[k = 1, k<=N, k++, A[k] = (2/L)
*Integrate[f * Sin[x * k * π/L], {x, 0, L}]];
For[k = 1, k<=N, k++,
B[k] = (2/k * π * a) * Integrate[g * Sin[x * k * π/L], {x, 0, L}]];
h = Sum[(A[k] * Cos[k * π * a * t/L] + B[k] * Sin[k * π * a * t/L] )*
Sin[k * π * x/L], {k, 1, N}];
Plot3D[h, {x, 0, L}, {t, 0, T}]]

Plot[x - 2 * (x - 1/2) * UnitStep[x - 1/2], {x, 0, 1}]
```

A plot of the "plucked string" function above is shown in Figure 8.17. We now

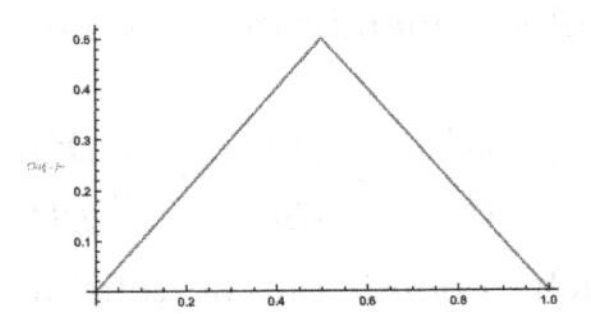

FIGURE 8.17: The plucked string in its initial position

test our program by considering the plucked string which we model by choosing the initial displacement to be given by

$$u(x,0) = \begin{cases} x & \text{when } 0 \leq x \leq \text{ where } \frac{1}{2} \\ 1-x & \text{when } 1 < x \leq 1 \end{cases}.$$

The Figure 8.17 and Figure 8.18 both show that the triangular wave repeats itself in the space-time plot.

```
stringEquationBIC[x - 2 * (x - 1/2) * UnitStep[x - 1/2], 0, 1, 1, 10, 20]


stringEquationBIC[x - 2 * (x - 1/2) * UnitStep[x - 1/2], 0, 1, 1, 10, 100]
```

♦

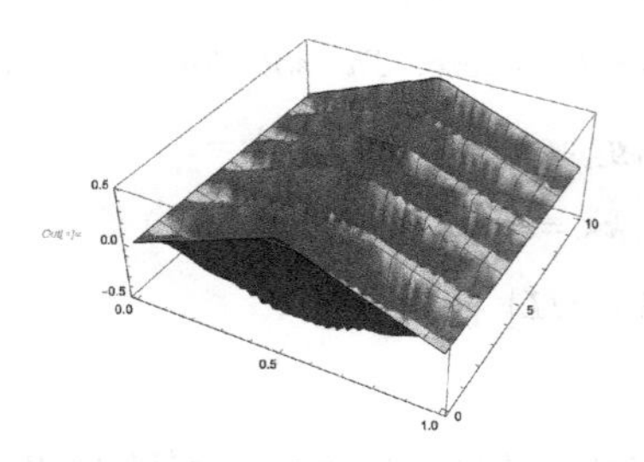

FIGURE 8.18: The periodic solution of the plucked string equation using 20 terms

8.2.6 Vibrating membrane

The vibrations of a uniform membrane may be described by a two dimensional wave equation

$$\frac{\partial^2 u}{\partial x^2} + \frac{\partial^2 u}{\partial y^2} = \frac{1}{c^2}\frac{\partial^2 u}{\partial t^2}, \tag{8.32}$$

where c is a constant, which depends on the tension and density of the membrane. Since the equation can be formally changed to the Laplace equation in $\mathbb{R}^3$ by the coordinate substitution $z = ict$, we immediately have the separated solutions by referring back to the Laplace equation. Hence, expressions of the form

$$J_m(kr)\left\{\begin{array}{c} \cos(m\theta) \times \cos(ckt) \\ \sin(m\theta) \times sin(ckt) \end{array}\right\}$$

satisfy the equations of motion of the membrane equation. If we consider the case of a circular drum, then we have the boundary condition $u(R,t) = 0$ on the rim of the drum $r = R$. In order to satisfy this condition we need to consider solutions of the above form, say for example

$$J_m(kr)\cos(m\theta)\cos(ckt),$$

where the coefficients k are chosen so that

$$J_m(kR) = 0.$$

There are an infinite number of real roots k_n which satisfy this condition. We are led therefore to consider solutions of the form

$$u(r,\varphi,t) = \sum_{m=0}^{\infty}\sum_{n=0}^{\infty} J_m(kr)\left(a_{nm}\cos(m\theta)\cos(ckt) + b_{mn}\sin(m\theta)\sin(ckt)\right) \tag{8.33}$$

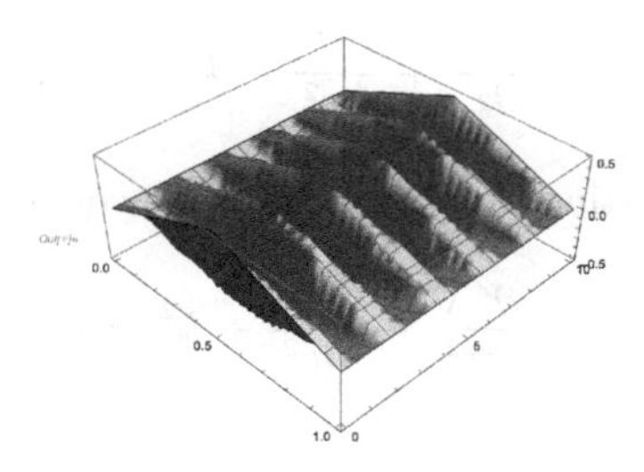

FIGURE 8.19: The periodic solution of the plucked string equation using 100 terms.

Exercises

Exercise 8.5. *Consider the two–dimensional wave equation for a cylinder of radius a. If the boundary condition is $\partial_r u$ is zero on $r = a$, find a series solution to this boundary value problem.*

Exercise 8.6. *Find a solution to the circular drum problem with an initial condition*

$$u(r, \theta, 0) = 1 - \frac{r}{R}, \quad for\, 0 < r < R.$$

Exercise 8.7. *Write a* MATHEMATICA *function which solves the above problem for arbitrary initial data. Test your function on the previously given initial data.*

Exercise 8.8. *Show that the function $u(x, y, z)$ defined by the integral*

$$u(x, y, z) := \int_{-\pi}^{\pi}\int_{-\pi}^{\pi} f\left(x\sin(\alpha)\cos(\beta) + y\sin(\alpha)\sin(\beta) + z\cos(\alpha) + ct, \alpha, \beta\right)$$

is a solution of the three–dimensional wave equation.

8.2.7 The reduced wave equation

In $\mathbb{R}^3$ the wave equation is given by

$$\frac{\partial^2 u}{\partial x} + \frac{\partial^2 u}{\partial y} + \frac{\partial^2 u}{\partial z} = c^2 \frac{\partial^2 u}{\partial t^2} \tag{8.34}$$

A special case is that of the reduced wave equation, i.e. when $u(x, y, z, t) = U(x, y, z)e^{\pm ikt}$ and the equation for U then becomes

$$\frac{\partial^2 U}{\partial x} + \frac{\partial^2 U}{\partial y^2} + \frac{\partial^2 U}{\partial z} + k^2 U = 0. \tag{8.35}$$

These solutions are harmonic in the time variable t. We consider below variations of the reduced wave equation. In two dimensions, the resulting equation

$$\frac{\partial^2 U}{\partial x} + \frac{\partial^2 U}{\partial y^2} + k^2 U = 0.$$

may be separated by setting $k^2 = k_1^2 + k_2^2$ in the equation below

$$\frac{\partial^2 X}{\partial x^2} + \frac{\partial^2 Y}{\partial y^2} + k^2 = 0,$$

thereby achieving

$$X(x)'' + k_1^2 X(x) = 0, \quad Y(y)'' + k_2^2 Y(y) = 0.$$

Separated solutions for (8.35) are then available in the form

$$U_{k_1,k_2}(x,y) = e^{\pm i(k_1 x \pm k_2 y)}.$$

In the case of polar coordinates the separation method leads to

$$r^2 \frac{R''}{R} = r\frac{R'}{R} + \frac{|phi''}{\Phi} + r^2 k^2 = 0.$$

If we set $\frac{\varphi''}{\Phi} = -m^2$ (Why reader?), the radial equation becomes

$$r^2 R'' + rR' + \left(k^2 r^2 - m^2\right) R = 0. \tag{8.36}$$

The equation (8.36) has the solution $J_m(kr)$ in the interval $[0, a]$; hence the functions defined by the series

$$U(x,y) = \sum_{m=0}^{\infty} J_m(kr)\left(a_m \cos(m\theta) + b_m \sin(m\theta)\right) \tag{8.37}$$

providing the series converges, represents a solution of equation (8.35). Suppose we wish to find the solutions to the reduced wave equations where the boundary conditions are $U = f(\theta)$ on $r = a$. We may use a combination of the functions

$$\left\{ \begin{array}{c} \cos(m\varphi) \\ sin(m\varphi) \end{array} \right\} J_m\left(m\pi\alpha_{mn} r/a\right); \quad J_m\left(m\pi\alpha_{mn}\right) = 0.$$

[5] It is known that the functions $\{J_m\left(m\pi\alpha_{mn} r/a\right)\}$ for different n are orthogonal. Moreover

$$\int_0^a \left[J_m\left(m\pi\alpha_{mn} r/a\right)\right]^2 r\, dr = \frac{1}{2} a^2 \left[J_{m-1}\left(\pi\alpha_{mn}\right)\right]^2.$$

[5]A large number of the α_{mn} roots are known to MATHEMATICA, hence, permitting computations to be made with good accuracy.

Verify that the d'Alembert formula (8.31) holds for the situation where either the boundary condition $u(0,t) = 0$ or, on the other hand, the boundary condition $\partial_t u(0,t) = 0$ holds, providing in the first case odd extensions of the data are necessary, whereas in the second case even extensions of the data are necessary.

Exercise 8.9. *Show that*

$$u(x,y) = \int_0^\infty f(\xi)e^{ik(x\cos(\xi)+y\sin(\xi))}\,d\xi$$

is a solution of the reduced wave equation

$$\frac{\partial^2 u}{\partial x} + \frac{\partial^2 u}{\partial y^2} + k^2 u = 0.$$

Exercise 8.10. *The gamma function is defined by the infinite integral* $\Gamma(\nu+1) := \int_0^\infty x^\nu e^{-t}\,dt$*. Using this definition show that* $\Gamma(\nu+1) = \mu\Gamma(\nu)$ *and, hence, deduce that* $\Gamma(\nu+1)$ *is a generalization of* $n!$*.*

Exercise 8.11. *The method of Frobenius is discussed in the Appendix. Seeking a series solution in the form*

$$y = \sum_{k=0}^\infty a_k x^{k+\alpha},$$

show that Bessel's equation

$$x^2y'' + xy' + (x^2 - \nu^2)y = 0$$

has a series solution

$$y = J_\nu(x) := \sum_{k=0}^\infty \frac{(-1)^k}{\Gamma(k+1)\Gamma(\nu+k+1)}\left(\frac{x}{2}\right)^{\nu+2k}.$$

Exercise 8.12. *Show if* ν *is not an integer a second solution to Bessel's equation is given by*

$$J_{-\nu}(x) := \sum_{k=0}^\infty \frac{(-1)^k}{\Gamma(k+1)\Gamma(-\nu+k+1)}\left(\frac{x}{2}\right)^{-\nu+2k}.$$

Exercise 8.13. *Use* MATHEMATICA *to find the solution to (8.37) in a disk* $r = 1$*, when*

$$U = \begin{cases} U = 1 & \text{when} \quad 0 \le \theta \le \pi, \\ U = 0 & \text{when} \quad \pi < \theta < 2\pi \end{cases}$$

8.3 Series methods for ordinary differential equations

For solving partial differential equations using the method of separation of variables we sometimes arrive at ordinary differential equations which are not immediately solvable to the eye. Many of these equations, however, may be solved using series expansions. In this section we show how to use MATHEMATICA to obtain series solutions to differential equations [6] with analytic coefficients.For the first example we consider a second order differential equation with **analytic** coefficients, and obtain an expansion about a **regular** point.[7] Let us consider the case where $x = 0$ is a regular point and seek an approximate solution in the form of a truncated series

$$y(x) = \sum_{k=0}^{N} a_k x^k.$$

In order to collect terms effectively and truncate terms we take the Taylor series of the above output. We shall seek two solutions, one with the initial conditions

$$y(0) = 1 \quad \text{and} \quad \frac{dy}{dx}(0) = 0$$

and the other with the initial conditions

$$y(0) = 0 \quad \text{and} \quad \frac{dy}{dx}(0) = 1.$$

It is well-known that these two solutions will be linearly independent as their Wronskian is not zero at $x = 0$. To solve for the **Taylor coefficients** for the first series solution, we set $a_0 = 1$ and $a_1 = 0$. Having done this we collect the coefficients of the various powers of x, by using the command **coeff**. To collect these as a list this is done using a `do` loop.We then solve the first six equations for the coefficients $a_k, \ k = 2 \ldots 7$. These coefficients are substituted into the power series representation. Next we plot the Taylor solution and compare this with a numerically solved solution. To solve numerically we use `dsolve` with the flag `numeric`. The numerical solution is plotted and the two compared with `display`. The Taylor solution is seen to be quite accurate up to $x = 2$ and then the polynomial starts to blow up. To compute the second solution by the Taylor method we set the coefficient to satisfy the second initial condition and then repeat the previous steps. Again the Taylor

[6] We consider a coefficient to be analytic in x about a point x_0 if the coefficient has a Taylor series expansion in terms of positive powers of $(x - x_0)$. For example all polynomials are analytic.

[7] A regular point is a point x_0 where the function has a power series in terms of non negative powers of $(x - x_0)$.

series solution compares well with the numerical solution until $x = 2$ and then it starts to blow up. **y = Sum[a[i] * x^i, {i, 9}] + O[x]^(11)**

$a[1]x + a[2]x^2 + a[3]x^3 + a[4]x^4 + a[5]x^5 + a[6]x^6 + a[7]x^7 + a[8]x^8 + a[9]x^9 + O[x]^{11}$

D[y, {x, 2}] + Cos[x] * y == 0

$$2a[2] + (a[1] + 6a[3])x + (a[2] + 12a[4])x^2 + \left(-\frac{a[1]}{2} + a[3] + 20a[5]\right)x^3 + \left(-\frac{a[2]}{2} + a[4] + 30a[6]\right)x^4 + \left(\frac{a[1]}{24} - \frac{a[3]}{2} + a[5] + 42a[7]\right)x^5 + \left(\frac{a[2]}{24} - \frac{a[4]}{2} + a[6] + 56a[8]\right)x^6 + \left(-\frac{a[1]}{720} + \frac{a[3]}{24} - \frac{a[5]}{2} + a[7] + 72a[9]\right)x^7 + \left(-\frac{a[2]}{720} + \frac{a[4]}{24} - \frac{a[6]}{2} + a[8]\right)x^8 + O[x]^9 == 0$$

LogicalExpand[%]

$$2a[2] == 0 \&\& a[1] + 6a[3] == 0 \&\& a[2] + 12a[4] == 0 \&\& -\frac{a[1]}{2} + a[3] + 20a[5] == 0 \&\& -\frac{a[2]}{2} + a[4] + 30a[6] == 0 \&\& \frac{a[1]}{24} - \frac{a[3]}{2} + a[5] + 42a[7] == 0 \&\& -\frac{a[2]}{720} + \frac{a[4]}{24} - \frac{a[6]}{2} + a[8] == 0 \&\& \frac{a[2]}{24} - \frac{a[4]}{2} + a[6] + 56a[8] == 0 \&\& -\frac{a[1]}{720} + \frac{a[3]}{24} - \frac{a[5]}{2} + a[7] + 72a[9] == 0$$

Solve[%]

$$\left\{\left\{a[2] \to 0, a[3] \to -\frac{a[1]}{6}, a[4] \to 0, a[5] \to \frac{a[1]}{30}, a[6] \to 0, a[7] \to -\frac{19a[1]}{5040}, a[8] \to 0, a[9] \to \frac{29a[1]}{72576}\right\}\right\}$$

sol1 = 1 − 1/6 * x^3 + 1/(30) * x^5 − (19)/(5040) * x^7 + (29)/(72576) * x^9

$1 - \frac{x^3}{6} + \frac{x^5}{30} - \frac{19x^7}{5040} + \frac{29x^9}{72576}$

ode1 = {Y"[x] + Cos[x] * Y[x] == 0, Y[0] == 0, Y'[0] == 1}

$\{\text{Cos}[x]Y[x] + Y''[x] == 0, Y[0] == 0, Y'[0] == 1\}$

sol2 = NDSolve[ode1, Y, {x, −10, 10}]

{{Y → InterpolatingFunction[]}}

p1 = Plot[Y[x]/.sol2, {x, −10, 10}]

A plot is shown in Figure 8.20.

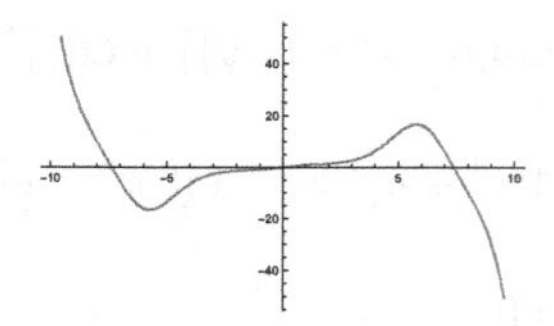

FIGURE 8.20: A plot of $p1$.

$$\mathbf{p2 = Plot}\left[1 - \frac{x^3}{6} + \frac{x^5}{30} - \frac{19x^7}{5040} + \frac{29x^9}{72576}, \{x, -10, 10\}, \mathbf{PlotStyle \to Red}\right]$$

A plot is shown in Figure 8.21.

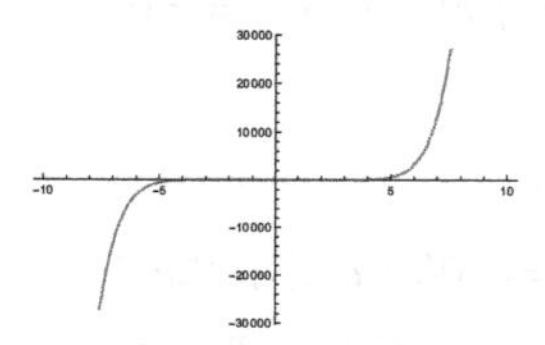

FIGURE 8.21: A plot of p2.

Show[p1, p2]

A plot is shown in Figure 8.22.

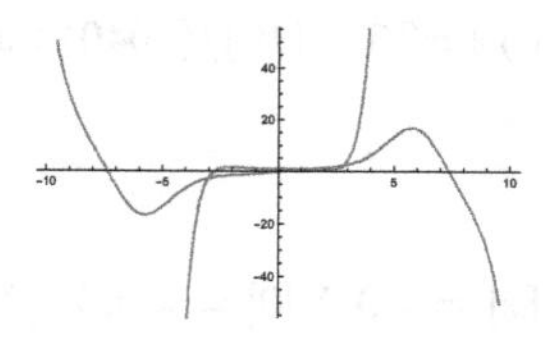

FIGURE 8.22: A plot of both p1 and p2.

8.4 Regular-singular points

Our next example of how to solve an equation with analytical coefficients involves finding a solution about a **regular-singular** point. For the second order differential equation

$$\frac{d^2y(x)}{dx^2} + p(x)\frac{dy(x)}{dx} + q(x)y(x) = 0,$$

whose coefficients are singular at the point x_0, x_0 is considered a regular-singular point if both

$$p(x)(x - x_0) \quad \text{and} \quad q(x)(x - x_0)^2$$

have Taylor series at x_0 . As an example, let us consider the differential equation

$$x^2\frac{d^2y(x)}{dx^2} + x\frac{dy(x)}{dx} + 4y(x) = 0.$$

We notice if we divide this differential equation by x^2, the coefficient of the leading derivative is 1, and the coefficients of the other terms are singular at $x = 0$. Moreover, the orders are such that the coefficients of the first-order derivative and also the unknown have first– and second–order singularities respectively. For expansions about regular singular points we seek power series of the form

$$y(x) = \sum_{k=0}^{\infty} a_k\, x^{k+m};$$

about the regular singular point at $x = 0$. Hence, we substitute the truncated series into the differential equation.

Mathematica session 8.8. *We begin by considering the Bessel equation with parameter* α, *namely*

$$\frac{d^2y}{dx^2} + \frac{1}{x}\frac{dy}{dx} + \left(1 - \frac{\alpha^2}{x^2}\right)y = 0. \tag{8.38}$$

We seek a power series solution which contains an arbitrary power m of x. This power m will be determined below. [8]

y[x] = Sum[a[n] * x^(n + m), {n, 0, 10}]

This series is then substituted into the Bessel equation of index α leading to

$x^m a[0] + x^{1+m} a[1] + x^{2+m} a[2] + x^{3+m} a[3] + x^{4+m} a[4] + x^{5+m} a[5] + x^{6+m} a[6] + x^{7+m} a[7] + x^{8+m} a[8] + x^{9+m} a[9] + x^{10+m} a[10]$

p = Expand[(D[y[x], {x, 2}] + 1/x * D[y[x], x] + (1 − α^2/x^2) * y[x])/x^(m − 2)]

$m^2 a[0] + x^2 a[0] - \alpha^2 a[0] + xa[1] + 2mxa[1] + m^2 xa[1] + x^3 a[1] - x\alpha^2 a[1] + 4x^2 a[2] + 4mx^2 a[2] + m^2 x^2 a[2] + x^4 a[2] - x^2\alpha^2 a[2] + 9x^3 a[3] + 6mx^3 a[3] + m^2 x^3 a[3] + x^5 a[3] - x^3\alpha^2 a[3] + 16x^4 a[4] + 8mx^4 a[4] + m^2 x^4 a[4] + x^6 a[4] - x^4\alpha^2 a[4] + 25x^5 a[5] + 10mx^5 a[5] + m^2 x^5 a[5] + x^7 a[5] - x^5\alpha^2 a[5] + 36x^6 a[6] + 12mx^6 a[6] + m^2 x^6 a[6] + x^8 a[6] - x^6\alpha^2 a[6] + 49x^7 a[7] + 14mx^7 a[7] + m^2 x^7 a[7] + x^9 a[7] - x^7\alpha^2 a[7] + 64x^8 a[8] + 16mx^8 a[8] + m^2 x^8 a[8] + x^{10} a[8] - x^8\alpha^2 a[8] + 81x^9 a[9] + 18mx^9 a[9] + m^2 x^9 a[9] + x^{11} a[9] - x^9\alpha^2 a[9] + 100x^{10} a[10] + 20mx^{10} a[10] + m^2 x^{10} a[10] + x^{12} a[10] - x^{10}\alpha^2 a[10]$

The coefficient of the term of lowest power in x contains a_0, which we assume is non-zero. Then in order for the power series to vanish identically the terms multiplying the a_0 in this lowest order term must vanish. These terms lead to an equation in the **index** *m, namely*[9]

Solve [m² − α² == 0, m]

$\{\{m \to -\alpha\}, \{m \to \alpha\}\}$

indicial equation $m^2 = 0$*; hence, $m = 0$ is a double root. We substitute this value for the index m into each of the equations and solve for the coefficients of*

[8]Ferdinand Georg Frobenius (26 October 1849 -August 1917) was a German mathematician known for his work on differential equations and elliptic functions.

[9]This quadratic equation in m is called the indicial equation.

the power series.. As we will see later, if the two roots, in this case α and $-\alpha$ do not differ by an integer each index will lead to an independent solution. We usually choose the larger index (or index with the larger real part) when finding the first solution. We substitute therefore $m = \alpha$ to find $y_1(x)$.

tp = Simplify[p/.{m → α}]

$$x\left((1+2\alpha)a[1]+x(a[0]+4(1+\alpha)a[2])+x^2(a[1]+3(3+2\alpha)a[3])+\right.$$
$$x^3(a[2]+8(2+\alpha)a[4])+\ (x^4(a[3]+5(5+2\alpha)a[5])+$$
$$x^5(a[4]+12(3+\alpha)a[6])+x^6(a[5]+7(7+2\alpha)a[7])+$$
$$x^7(a[6]+16(4+\alpha)a[8])+x^{10}a[9]+x^8(a[7]+9(9+2\alpha)a[9])+$$
$$\left.x^{11}a[10]+x^9(a[8]+20(5+\alpha)a[10])\right)$$

MATHEMATICA *wants to work with terms created by the command* **Series**; *however, one may add an order of magnitude onto the end of the polynomial and turn it into a* MATHEMATICA *Series , in our case a term $O[x]^{11}$.*

tpp = Series[tp, {x, 0, 10}]

$$(a[1]+2\alpha a[1])x+(a[0]+4a[2]+4\alpha a[2])x^2+(a[1]+9a[3]+6\alpha a[3])x^3+(a[2]+16a[4]+8\alpha a[4])x^4+(a[3]+25a[5]+10\alpha a[5])x^5+(a[4]+36a[6]+12\alpha a[6])x^6+(a[5]+49a[7]+14\alpha a[7])x^7+(a[6]+64a[8]+16\alpha a[8])x^8+(a[7]+81a[9]+18\alpha a[9])x^9+(a[8]+100a[10]+20\alpha a[10])x^{10}+O[x]^{11}$$

Since we have a series we may use the operation **SeriesCoefficient**. *Try this with a polynomial.*

SeriesCoefficient[tpp, {x, 0, 5}]

We check that it will produce a coefficient by inputting

$a[3]+25\ a[5]+10\alpha\, a[5]$

Using a **For** *command we generate the coefficient of x^i*

For[i = 1, i < 10, i++, Print[SeriesCoefficient[tpp, {x, 0, i}] == 0]]

$a[1] + 2\alpha a[1] == 0$ $a[0] + 4a[2] + 4\alpha a[2] == 0$ $a[1] + 9a[3] + 6\alpha a[3] == 0$ $a[2] + 16a[4] + 8\alpha a[4] == 0$ $a[3] + 25a[5] + 10\alpha a[5] == 0$ $a[4] + 36a[6] + 12\alpha a[6] == 0$ $a[5]+49a[7]+14\alpha a[7] == 0$ $a[6]+64a[8]+16\alpha a[8] == 0$ $a[7]+81a[9]+18\alpha a[9] == 0$

We may now solve for the coefficients a_i using the **Solve** *command*

```
Solve[{a[1] + 2αa[1] == 0, a[0] + 4a[2] + 4αa[2] == 0, a[1] + 9a[3] + 6αa[3] == 0,
a[2] + 16a[4] + 8αa[4] == 0, a[3] + 25a[5] + 10αa[5] == 0, a[4] + 36a[6] + 12αa[6] == 0,
a[5] + 49a[7] + 14αa[7] == 0, a[6] + 64a[8] + 16αa[8] == 0,
a[7] + 81a[9] + 18αa[9] == 0},
{a[1], a[2], a[3], a[4], a[5], a[6], a[7], a[8], a[9], a[10]}]
```

$$\Big\{\Big\{a[1] \to 0, a[2] \to -\frac{a[0]}{4(1+\alpha)}, a[3] \to 0,$$
$$a[4] \to \frac{a[0]}{32(1+\alpha)(2+\alpha)}, a[5] \to 0, a[6] \to -\frac{a[0]}{384(1+\alpha)(2+\alpha)(3+\alpha)}$$
$$, a[7] \to 0, a[8] \to \frac{a[0]}{6144(1+\alpha)(2+\alpha)(3+\alpha)(4+\alpha)}, a[9] \to 0\Big\}\Big\}$$

In order to obtain the Bessel function of order 0 *now input the coefficients a_i into the series and set $\alpha = 0$. We obtain an approximate solution:*

```
x^α * Sum[a[n] * x^n, {n, 0, 9}]/.
```

$$\Big\{a[0] \to 1, a[1] \to 0, a[2] \to -\frac{a[0]}{4(1+\alpha)}, a[3] \to 0, a[4] \to \frac{a[0]}{32(1+\alpha)(2+\alpha)}, a[5] \to 0,$$
$$a[6] \to -\frac{a[0]}{384(1+\alpha)(2+\alpha)(3+\alpha)}, a[7] \to 0, a[8] \to \frac{a[0]}{6144(1+\alpha)(2+\alpha)(3+\alpha)(4+\alpha)}, a[9] \to 0\Big\}$$

$$x^\alpha \left(1 - \frac{x^2 a[0]}{4(1+\alpha)} + \frac{x^4 a[0]}{32(1+\alpha)(2+\alpha)} - \frac{x^6 a[0]}{384(1+\alpha)(2+\alpha)(3+\alpha)} + \frac{x^8 a[0]}{6144(1+\alpha)(2+\alpha)(3+\alpha)(4+\alpha)}\right)$$

♦

Mathematica session 8.9. *We could obtain the zero order Bessel function $J_0(x)$ directly from the above by setting $\alpha = 0$. Do this! However, for didactic purposes it might not hurt to see this once more with $\alpha = 0$ in the Bessel equation of zeroth order. To this end, we substitute into*

$$\frac{d^2y}{dx^2} + \frac{1}{x}\frac{dy}{dx} + \left(1 - \frac{\alpha^2}{x^2}\right) y = 0. \tag{8.39}$$

y[x] = Sum[a_n * x^(n + m), {n, 0, 10}]

$x^m a_0 + x^{1+m} a_1 + x^{2+m} a_2 + x^{3+m} a_3 + x^{4+m} a_4 + x^{5+m} a_5 + x^{6+m} a_6 + x^{7+m} a_7 + x^{8+m} a_8 + x^{9+m} a_9 + x^{10+m} a_{10}$

p = Expand[(D[y[x], {x, 2}] + 1/x * D[y[x], x] + y[x])/x^(m − 2)]

$m^2 a_0 + x^2 a_0 + x a_1 + 2mx a_1 + m^2 x a_1 + x^3 a_1 + 4x^2 a_2 + 4mx^2 a_2 + m^2 x^2 a_2 + x^4 a_2 + 9x^3 a_3 + 6mx^3 a_3 + m^2 x^3 a_3 + x^5 a_3 + 16x^4 a_4 + 8mx^4 a_4 + m^2 x^4 a_4 + x^6 a_4 + 25x^5 a_5 + 10mx^5 a_5 + m^2 x^5 a_5 + x^7 a_5 + 36x^6 a_6 + 12mx^6 a_6 + m^2 x^6 a_6 + x^8 a_6 + 49x^7 a_7 + 14mx^7 a_7 + m^2 x^7 a_7 + x^9 a_7 + 64x^8 a_8 + 16mx^8 a_8 + m^2 x^8 a_8 + x^{10} a_8 + 81x^9 a_9 + 18mx^9 a_9 + m^2 x^9 a_9 + x^{11} a_9 + 100x^{10} a_{10} + 20mx^{10} a_{10} + m^2 x^{10} a_{10} + x^{12} a_{10}$

tp = Simplify[p/.{m → 0}]

$x\left(x a_0 + \left(1 + x^2\right) a_1 + x\left(\left(4 + x^2\right) a_2 + x\left(\left(9 + x^2\right) a_3 + x\left(\left(16 + x^2\right) a_4 + x\left(\left(25 + x^2\right) a_5 + x\left(\left(36 + x^2\right) a_6 + x\left(\left(49 + x^2\right) a_7 + x\left(\left(64 + x^2\right) a_8 + x\left(\left(81 + x^2\right) a_9 + x\left(100 + x^2\right) a_{10}\right)\right)\right)\right)\right)\right)\right)\right)\right)$

tpp = Series[tp, {x, 0, 10}]

$a_1 x + (a_0 + 4a_2) x^2 + (a_1 + 9a_3) x^3 + (a_2 + 16a_4) x^4 + (a_3 + 25a_5) x^5 + (a_4 + 36a_6) x^6 + (a_5 + 49a_7) x^7 + (a_6 + 64a_8) x^8 + (a_7 + 81a_9) x^9 + (a_8 + 100a_{10}) x^{10} + O[x]^{11}$

For[i = 1, i < 10, i++, Print[SeriesCoefficient[tpp, {x, 0, i}] == 0]]

$a_1 == 0\ a_0 + 4a_2 == 0\ a_1 + 9a_3 == 0\ a_2 + 16a_4 == 0\ a_3 + 25a_5 == 0\ a_4 + 36a_6 == 0\ a_5 + 49a_7 == 0\ a_6 + 64a_8 == 0\ a_7 + 81a_9 == 0$

j_0= Solve[{a_0==1, $a_1 == 0, a_0 + 4a_2 == 0, a_1 + 9a_3 == 0, a_2 + 16a_4 == 0, a_3 + 25a_5 == 0, a_4 + 36a_6 == 0, a_5 + 49a_7 == 0, a_6 + 64a_8 == 0, a_7 + 81a_9 == 0$}, {$a_1, a_2, a_3, a_4, a_5, a_6, a_7, a_8, a_9$}]

g1 = Plot[BesselJ[0, x], x, 0, 2, PlotStyle -> Red]

A plot of g1 is shown in Figure 8.23.

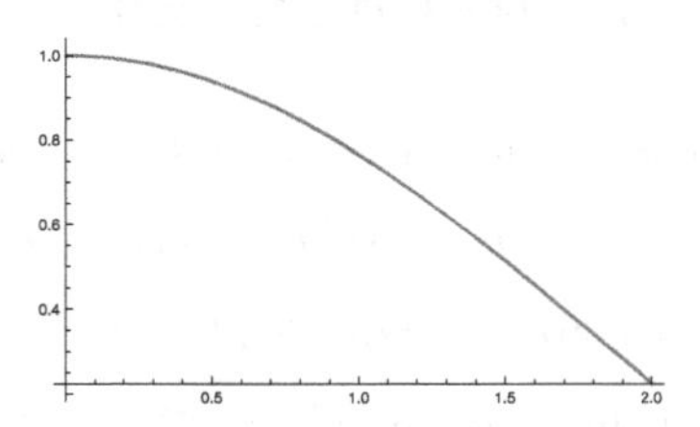

FIGURE 8.23: A plot of g1.

We now obtain the truncated solution of the zeroth order Bessel equation

g2 = Plot[1 − 1/4 * x^2 − 1/16 * x^4 − 1/36 * x^6 − 1/64 * x^8, {x, 0, 2}]

A plot of g1 is shown in Figure 8.24.

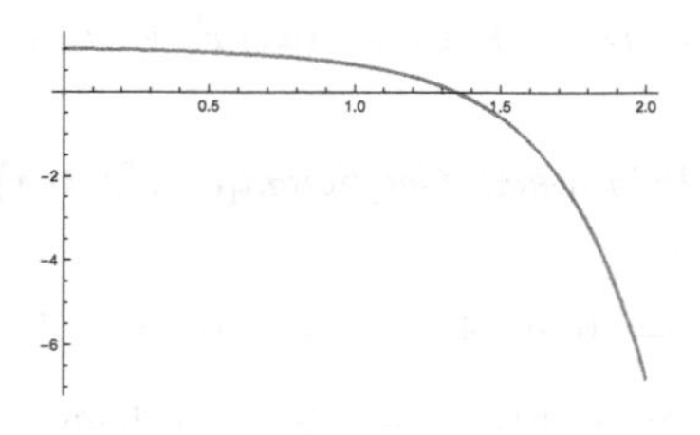

FIGURE 8.24: A plot of g2.

Show[g1, g2]

A plot of both g1 and g2 is shown in Figure 8.25.

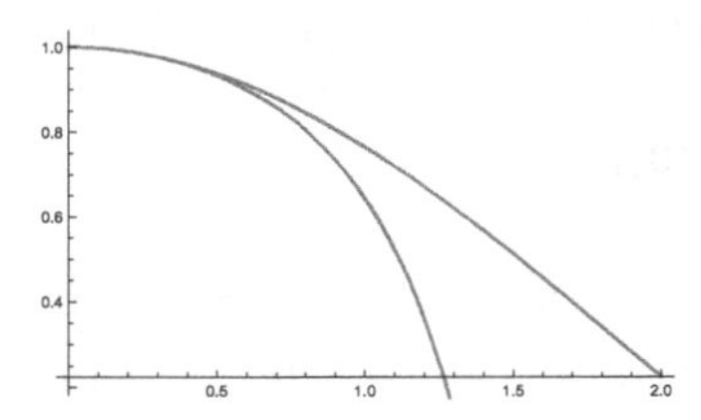

FIGURE 8.25: A plot of both g1 and g2.

p = Expand[(D[y[x], {x, 2}] + 1/x * D[y[x], x] + y(1 − 1/x^2) * [x])/x^(m − 2)]

♦

Mathematica session 8.10. **y[x] = Sum[a_n * x^(n + m), {n, 0, 10}];**

p2 = Expand[(D[y[x], {x, 2}] + 1/x * D[y[x], x] + (1 − 4/x^2) * y[x])/x^(m − 2)]

$$-4a_0 + m^2a_0 + x^2a_0 - 3xa_1 + 2mxa_1 + m^2xa_1 + x^3a_1 + 4mx^2a_2 + m^2x^2a_2 + x^4a_2 + 5x^3a_3 + 6mx^3a_3 + m^2x^3a_3 + x^5a_3 + 12x^4a_4 + 8mx^4a_4 + m^2x^4a_4 + x^6a_4 + 21x^5a_5 + 10mx^5a_5 + m^2x^5a_5 + x^7a_5 + 32x^6a_6 + 12mx^6a_6 + m^2x^6a_6 + x^8a_6 + 45x^7a_7 + 14mx^7a_7 + m^2x^7a_7 + x^9a_7 + 60x^8a_8 + 16mx^8a_8 + m^2x^8a_8 + x^{10}a_8 + 77x^9a_9 + 18mx^9a_9 + m^2x^9a_9 + x^{11}a_9 + 96x^{10}a_{10} + 20mx^{10}a_{10} + m^2x^{10}a_{10} + x^{12}a_{10}$$

Solve[−4a_0 + m^2a_0 == 0, m]

$\{\{m \to -2\}, \{m \to 2\}\}$

tp2 = Simplify[p2/.{m → 2}]

$x\left(xa_0+\left(5+x^2\right)a_1+x\left(\left(12+x^2\right)a_2+x\left(\left(21+x^2\right)a_3+\right.\right.\right.$
$x\left(\left(32+x^2\right)a_4+x\left(\left(45+x^2\right)a_5+x\left(\left(60+x^2\right)a_6+x\left(\left(77+x^2\right)a_7+\right.\right.\right.\right.$
$\left.x\left(\left(96+x^2\right)a_8+x\left(\left(117+x^2\right)a_9+x\left(140+x^2\right)a_{10}\right)\right)\right)\right)\right)\right)\right)\right)$

ttp2 = Series[tp2, {x, 0, 10}]

$5a_1x+(a_0+12a_2)x^2+(a_1+21a_3)x^3+(a_2+32a_4)x^4+(a_3+45a_5)x^5+(a_4+60a_6)x^6+$
$(a_5+77a_7)x^7+(a_6+96a_8)x^8+(a_7+117a_9)x^9+(a_8+140a_{10})x^{10}+O[x]^{11}$

For[i = 1, i < 10, i++, Print[SeriesCoefficient[ttp2, {x, 0, i}] == 0]]

$5a_1 == 0$ $a_0+12a_2 == 0$ $a_1+21a_3 == 0$ $a_2+32a_4 == 0$ $a_3+45a_5 == 0$
$a_4+60a_6 == 0$ $a_5+77a_7 == 0$ $a_6+96a_8 == 0$ $a_7+117a_9 == 0$

j_2 = Solve[{$5a_1 == 0, a_0+12a_2 == 0, a_1+21a_3 == 0,$
$a_2+32a_4 == 0, a_3+45a_5 == 0,$
$a_4+60a_6 == 0, a_5+77a_7 == 0, a_6+96a_8 == 0, a_7+117a_9 == 0$},
{$a_1, a_2, a_3, a_4, a_5, a_6, a_7, a_8, a_9$}]

$\left\{\left\{a_1\to 0, a_2\to -\frac{a_0}{12}, a_3\to 0, a_4\to \frac{a_0}{384}, a_5\to 0,\right.\right.$
$\left.\left.a_6\to -\frac{a_0}{23040}, a_7\to 0, a_8\to \frac{a_0}{2211840}, a_9\to 0\right\}\right\}$

$j = (1 - \frac{a_0}{12} * x$^2 $+ \frac{a_0}{384} * x$^4 $- \frac{a_0}{23040} * x$^6 $+ \frac{a_0}{2211840} * x^8) * x$^2/. {$a_0$ → 1}

$x^2\left(1-\frac{x^2}{12}+\frac{x^4}{384}-\frac{x^6}{23040}+\frac{x^8}{2211840}\right)$

g3 = Plot[j, {x, 0, 1}]

A plot of g3 is shown in Figure 8.26.

g4 = Plot[BesselJ[2, x], {x, 0, 1}, PlotStyle → Red]

A plot of g4 is shown in Figure 8.27.

Show[g4, g3]

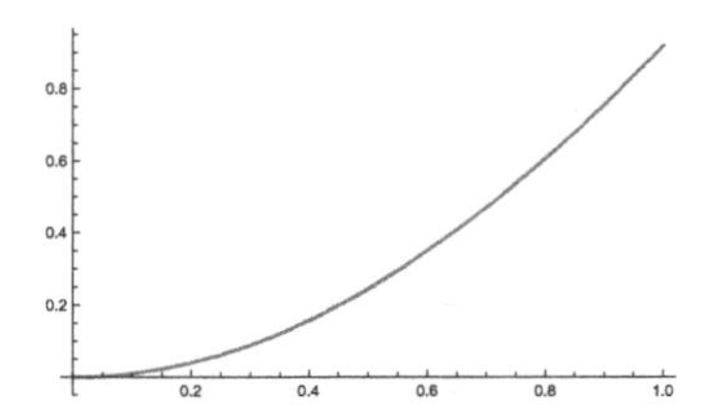

FIGURE 8.26: A plot of g3.

A plot of both g3 and g4 is shown in Figure 8.28.

♦

8.4.1 Project

Having seen how to construct a solution line by line let us attempt to write a program which does this. As this will not be so easy, we begin by partitioning the problem into sections which will each perform part of the task. The first part is a mini-program to construct the Taylor coefficients of the coefficients $p(x)$ and $q(x)$. If the two roots of the indicial equation do not differ by an integer then we may find solutions in the form

$$y_1(x) = x^{m_1}\left(1 + \sum_{n=1}^{\infty} a_n\, x^n\right)$$
$$y_2(x) = x^{m_2}\left(1 + \sum_{n=1}^{\infty} b_n\, x^n\right)$$

However, if the two indices differ by an integer we may not get a second solution in this way. If the two indices are the same, i.e. a double root, then we do **NOT** get a second solution in this form! Hence, the choice of the index we use to construct the first solution must be carefully chosen. In the case where roots differ by an integer the root with the larger real part always gives rise to a solution. Hence, we wish to first

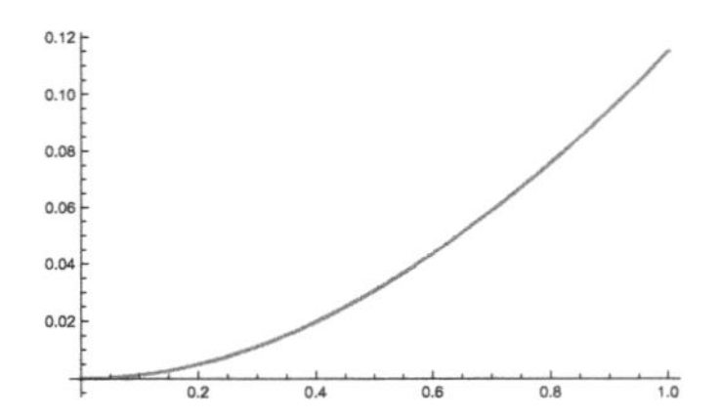

FIGURE 8.27: A plot of g4.

construct this solution. We make use of the command **Head**. The command **Head**[x] tests for different types of numbers. A particular case which must be distinguished is when the difference between the two roots of the indicial equation differ by an integer. (See [4] page 148.) A subcategory which is of further interest occurs when the roots are identical, i.e. they differ by zero. Let us experiment a bit. Let us write a mini-program to choose one of the roots. For the moment we consider only the possibility of *real* roots. **Max**$[x_1, x^2, \ldots]$ picks out the largest real number in a list. Once we understand how to incorporate the choosing mechanism into `frobenius` we will consider the case of complex roots. If the root is a double root, i.e. $m_1 = m_2$ then the second solution may be found in the form

$$y_2(x) = y_1(x) \ln(|x|) + |x|^{m_1} \sum_{n=1}^{\infty} a_n x^n$$

whereas, if $m_1 - m_2$ is a positive integer then the second solution is found in the form

$$y_2(x) = a\, y_1(x) \ln(|x|) + |x|^{m_2} \sum_{n=1}^{\infty} a_n x^n.$$

We now consider the case of complex roots and how to choose the root with the larger real part to start with. If the two roots differ by an integer, we need to find our first solution using that root with largest real-part. MATHEMATICA has a command **Re**$[z]$ which chooses the real part of the complex number z. It also allows one to decide whether an index is complex, real, or an integer, thus making it possible to augment previous programs to decide whether the two indices are equal, differ by an integer and whether they are real or complex.

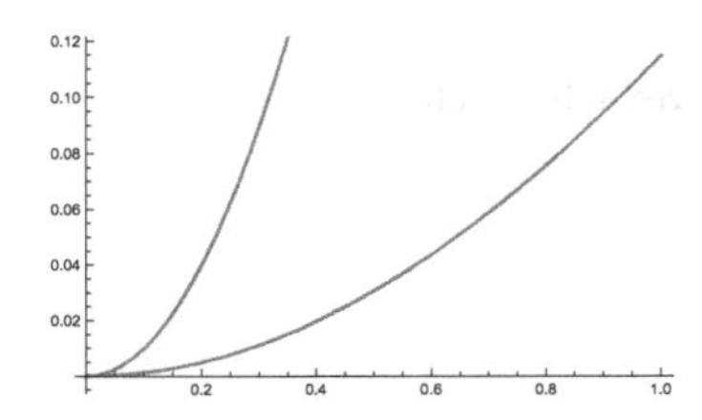

FIGURE 8.28: A plot of both g3 and g4.

We exhibit some MATHEMATICA commands which will be useful in determining the type of number we are dealing with. The reader should, as an exercise, use the commands below to augment one of the previous programs using this information to decide in what form to seek a solution.
Hint: First check to see if $m_1 - m_2 = 0$. If this is not the case, then search for the index with larger real part, using the command

$$\text{Positive}[Re[m_1] - Re[m_2]]$$

If the output to this line is **False** then m_2 is the index with the largest real part, or they have equal real parts. Providing the $m_1 \neq m_2$ we may use each root of the indicial equation to generate a solution. If they are equal to each other we have a method, mentioned above, to find the second solution.

Mathematica session 8.11. *Head*[0]

Integer

***Head*[5.6]**

Real

***Head*[2 + 3 * *I*]**

Complex

***Re*[2 + 3 * *I*]**

2

***Positive*[*Re*[3 + 5 * *I*] − *Re*[4 + 7 * *I*]]**

False

♦

Using these ideas write a program which inputs a differential equation of the regular singular type and computes the first and second solution.

Bibliography

[1] A.Ben-Israel & R. P. Gilbert: *Computer-Supported Calculus*,Texts ,Monographs in Symbolic Computation, Springer, Wien, (2002).

[2] R. Courant & D. Hilbert: *Methods of Mathematical Physics*, Vol. 1, Interscience, New York,(1953).

[3] P.R.Garabedian: *Partial Differential Equations* Wiley, New York, (1964).

[4] R.P. Gilbert & G. C. Hsiao: *Maple Projects for Differential Equations*, Prentice Hall, Upper Saddle River, (2003).

[5] W. Koepf, Adi Ben-Israel & R. Gilbert: *Mathematik mit DERIVE*, in Computeralgebra, Vieweg, Braunzweig, Germany (1993).

[6] A. Lichnerowicz: *Elements of Tensor Calculus*, Trans J. W. Leech & D. J. Newman, Dover, Mineola, (2016).

[7] H. Margenau & G. M. Murphy: em The Mathematics of Physics and Chemistry, Van Nostrand, New York, (1943).

[8] J.L. Synge & A. Schild: *Tensor Calculus*, University of Toronto Press, Toronto, (1949).

[9] B. Spain: *Tensor Calculus: A concise Course*, Dover, Mineola, (2003)

Index